Lecture Notes in Civil Engineering

Volume 636

Lecture Notes in Civil Engineering (LNCE) publishes the latest developments in Civil Engineering—quickly, informally and in top quality. Though original research reported in proceedings and post-proceedings represents the core of LNCE, edited volumes of exceptionally high quality and interest may also be considered for publication. Volumes published in LNCE embrace all aspects and subfields of, as well as new challenges in, Civil Engineering. Topics in the series include:

- Construction and Structural Mechanics
- Building Materials
- Concrete, Steel and Timber Structures
- Geotechnical Engineering
- Earthquake Engineering
- Coastal Engineering
- Ocean and Offshore Engineering; Ships and Floating Structures
- Hydraulics, Hydrology and Water Resources Engineering
- Environmental Engineering and Sustainability
- Structural Health and Monitoring
- Surveying and Geographical Information Systems
- Indoor Environments
- Transportation and Traffic
- Risk Analysis
- Safety and Security

To submit a proposal or request further information, please contact the appropriate Springer Editor:

- Pierpaolo Riva at pierpaolo.riva@springer.com (Europe and Americas);
- Swati Meherishi at swati.meherishi@springer.com (Asia—except China, Australia, and New Zealand);
- Wayne Hu at wayne.hu@springer.com (China).

All books in the series now indexed by Scopus and EI Compendex database!

Brydon T. Wang · Chien Ming Wang ·
Kim Weinert · Rutger de Graaf-van Dinther
Editors

Large Floating Solutions

Design, Construction, Legality of Offshore Structures and Buoyant Urbanism

Springer

Editors
Brydon T. Wang
Deal Advisory and Infrastructure
KPMG Australia
Brisbane, QLD, Australia

Centre for Policy Futures
The University of Queensland
St. Lucia, QLD, Australia

Kim Weinert
T. C. Beirne School of Law
The University of Queensland
St. Lucia, QLD, Australia

Chien Ming Wang
School of Civil Engineering
The University of Queensland
St. Lucia, QLD, Australia

Blue Economy Cooperative Research
Centre
Launceston, Australia

Rutger de Graaf-van Dinther
Blue21
Delft, The Netherlands

ISSN 2366-2557 ISSN 2366-2565 (electronic)
Lecture Notes in Civil Engineering
ISBN 978-981-96-5434-5 ISBN 978-981-96-5435-2 (eBook)
https://doi.org/10.1007/978-981-96-5435-2

This Springer imprint is published by the registered company Springer Nature Singapore Pte Ltd.
The registered company address is: 152 Beach Road, #21-01/04 Gateway East, Singapore 189721, Singapore

Preface: Surface

In 2008, the authors contributed to the first edited collection on *Very Large Floating Structures* (VLFS), led by Chien Ming Wang, Eiichi Watanabe and Tomoaki Utsunomiya. The book comprised nine chapters that focused on the technological advances in engineering and construction that would enable careful use of our marine-based resources. The chapters explored wave phenomenon and properties, the hydroelastic analysis of VLFS, structural analysis and design of VLFS, mooring systems, breakwaters, maintenance and anti-corrosion systems of large steel floating structures.

Our chapter set out three core arguments for expanding sensitively over adjacent water bodies using floating structures. First, the scarcity of available coastal land necessitates alternative solutions. Second, with growing urban populations, there is an increased appetite for better integration of our cities with the nearshore and offshore environments. Third, floating structures present opportunities to harness renewable energy, develop innovative aquaculture installations and capitalise on the *Blue Revolution* (Wang and Wang 2006).

We also reflected on humanity's persistent fascination with the sea as a site for settlement and exploration. The idea of building cities on or beneath the sea has a long history, but it was reignited between the 1960s and 1970s when the technological optimism of the preceding decade met the anxieties of overpopulation and environmental degradation. The oil shocks of 1973 and 1979 and the discovery of the ozone layer depletion intensified a desire to build settlements in extreme environments: deserts, the poles, the sea and even outer space. Today, as we confront similar anxieties over climate change and resource scarcity, these ideas are resurfacing, as discussed in this book's introduction.

Our research into floating cities emerged from the convergence of two projects: CM's work at Kyoto University supporting the *Megafloat* project, which examined floating alternatives to mitigate subsidence risks as observed in the artificial island constructed for the Kansai International Airport; and Brydon's 2004 research at the University of Melbourne that explored marine utopian proposals. This collaboration marked the beginning of our work on buoyant urbanism, where the buoyancy force of water is used to support urban structures, whether as permanently floating cities or amphibious architecture that adapts to the rise and fall of temporary incursions of water. By integrating modular floating components into coastal cities, we can create hybrid floating cities that ensure critical infrastructure remains resilient to flooding and sea level rise.

In the intervening two decades, there has been increasing activity in this research space on large floating structures. A basic search on Science Direct indicates a growing number of publications related to 'large floating structures', rising from approximately 2,400 in 2002 to around 16,000 in 2024, with a cumulative total of 200,000 since 2002. We have argued that floating cities can help us meet our *Sustainable Development Goals* (Wang and Wang 2020), that these schemes are not washed-up ideas (Wang 2019), and that floating structures can help enhance our aquatic ecosystems and protect our vulnerable coastlines (Wang 2023; Luo et al. 2022). It has been our privilege seeing this research field grow from strength to strength and we set out below some of the highlights.

To date, we have put together two edited volumes, with Springer, pertinent to this field of research. The first, *Large Floating Structures: Technological Advances* (2015) undertook a comprehensive examination of the engineering, design, and implementation of large-scale floating structures, featuring contributions from researchers across architecture, civil engineering, offshore and marine engineering, and urban planning. The book documented key floating infrastructure projects worldwide, including floating bridges, spar drilling and production platforms, floating breakwaters, offshore wind turbines, oil storage bases and Vincent Callebaut's seminal floating city proposal: *Lilypad*—a floating ecopolis for climate-displaced populations.

The second, *Automating Cities: Design, Construction, Operation and Future Impact* (Wang and Wang 2021) presented a multidisciplinary exploration of urban automation, integrating contributions from researchers across architecture, engineering, law, urban studies and digital technology. The book examined the automation of design, construction and operational processes, focusing on the increasing datafication of cities and its socio-political consequences. Notably, it expanded discussion on climate-resilient floating cities by contextualising these urban solutions within the gamut of data-driven urban technologies: digital twins, automated structural monitoring, blockchain governance; and how these technologies are reshaping infrastructure delivery (Wang 2021). As offshore and coastal developments become increasingly automated, the book underscores the critical role of digital infrastructure in facilitating the expansion of cities onto adjacent water bodies, ensuring sustainability while navigating emerging regulatory and ethical challenges. This second book arose out of Brydon's research on the datafication and automation of construction processes and legal frameworks around such innovations, as well as evolving regulatory challenges and the global increase in AI regulation. He returns to this theme with his co-authors Michael Guihot and Bernadette Hyland-Woods in the Chapter "Navigating the Legal and Regulatory Frontiers of Urban AI, Autonomous Maritime Robots and Floating Cities", where they explore the integration of Urban AI and Agentic AI in autonomous maritime robots.

This edited volume provides an update to the field and follows these two aforementioned edited collections, continuing our strong relationship with Springer. Once again, it is a multidisciplinary exploration of buoyant urbanism as a response to climate change, integrating perspectives from engineering, law, architecture, and urban planning. The book positions buoyant urbanism as a critical climate adaptation strategy, covering floating cities, aquaculture systems, and offshore energy infrastructure, but also covers the granular aspects of a floating city: floating homes and suburbs, exploring how these should be designed at a modular level. We note that *amphibious architecture* would also be included in our definition of 'buoyant urbanism'. These are structures that would typically sit on the foundations or the ground and rise with the incursion of water. However, given our focus on VLFS and floating cities, we have excluded this innovative urban solution from a detailed study in this book.

We also want to draw attention to Joseph Lim's excellent work *Oil Rig and Superbarge Floating Settlements* (2020) that set out to consider how we repurpose decommissioned semi-submersible oil rigs and large floating platforms as habitable offshore settlements. The book is bold and represents the best in speculative architecture. Drawing on a spread of architecture, engineering and urban planning explorations, the chapters examined how these innovative repurposing of decommissioned maritime structures can produce high-density floating cities, offshore aquaculture and facilities for disaster relief.

Likewise, Koen Olthuis and David Keuning produced a deeply compelling narrative of urban development on water in their 2010 book *Float! Building on Water to Combat Urban Congestion and Climate Change*. In it, Olthuis and Keuning (2010) make the case for *Hydrocity*, 'where both land and water are the carriers of urban functions, infrastructure and identity … where the boundary between wet and dry is no longer relevant'. We have approached a similar concept that we term 'hybrid floating cities'—where cities progressively adopt buoyant urbanism in stages, allowing incremental adaptation to the growing demand to respond to floods and rising sea levels. Olthuis' chapter in this book, co-written with his colleagues at Waterstudio, Subramani and Montezuma, explores the *Maldives Floating City*, a project flagged in his original book, and the ten design principles that shaped the development of the project.

Kim Weinert joins us as an editor and expert in applying property law concepts to novel scenarios. This is a critical piece in the jigsaw puzzle of how we align the law and policy to enable the design and deployment of VLFS and other buoyant structures to improve the climate resilience of cities. Kim ran a research project, involving law students from the TC Beirne School of Law at the University of Queensland that explored how the concept of home is critical to legal conceptions of floating homes and its aggregation into floating suburbs and floating cities (Chapter "From Land to Water: Policy Considerations Supporting Why Floating Cities Can Help Combat Climate Change"). Her own Chapter "Home is Where the Heart Floats: Anchoring Legal Property Concepts for Environmental Migrants" takes an expansive view of different legal classifications of floating structures and artificial islands. Examining the different treatment of property from common law and civil jurisdiction perspectives, she concludes that we will need to re-conceptualise property theory to address how property law classifies ownership and values these objects.

In terms of continuing this sharing of knowledge with the next generation of designers, engineers, urban planners, legislators and policymakers, Brydon also ran a number of architectural design studios for the Master of Architecture students at the University of Queensland exploring: *Surveillance, Smart cities and the Sensorised Building* (2018), *The Automated City* (2019) and *Floating Cities: Buoyant Urbanism, Ecological Engineering and Marine Utopian Proposals* (2022). The content of discussions and resultant participant output of the last architectural design studio have been captured in Chapter "Designing for Buoyant Urbanism: Pedagogy, Architectural Paradigms and Social Acceptance of Floating Cities". The 2022 studio paralleled the architecture design studio held at the School of Architecture, Royal Melbourne Institute of Technology, by Nic Bao exploring artificial reefs and computational design (Chapter "Designing for Artificial Reefs: The Role of Architecture in Marine Ecosystem Restoration").

Rutger also joins us as an editor, bringing extensive experience in floating structures across industry and academia. His 2021 edited volume, *Climate Resilient Urban Areas: Governance, Design and Development in Coastal Delta Cities*, advanced discussions on integrating water management and the circular economy with climate-resilient cities. His work highlights five key capacities that climate-resilient cities must develop: *threshold capacity*, how a city prevents damage through proactive infrastructure intervention; *coping capacity*, its management and mitigation strategies to resist extreme weather impacts; *recovery capacity*, the ability to return to or improve upon pre-disaster conditions; *adaptive capacity*, how a city anticipates and responds to future uncertainties; and *transformative capacity*, how a city embraces systemic change and resilience-building. Floating cities primarily sit within the transformative capacity of a city but intersects with the other capacities. That is, these innovative urban solutions will allow a city to prevent damage (threshold capacity), allow cities to absorb the impact of climate change, floods, storms, incursion of water (coping capacity), minimise disruptions to return to business-as-usual (recovery capacity), and rise to the challenge of any further developments with water-related disruptions (adaptive capacity).

Finally, these past two decades have also seen the development of international conferences focused on exploring buoyant urbanism. To date, there have been four World Conferences on Floating Solutions (WCFSs). These have been held in Singapore, Rotterdam, Tokyo and Hong Kong. As members of the international scientific committee for these conferences, we look forward to the Fifth World Conference on Floating Solutions in Espoo, Finland. The proceedings of these WCFSs have been published by Springer Nature under the *Lecture Notes for Civil Engineering* book series. They contain a vast amount of information and knowledge on floating solutions for solving global challenges facing mankind. This growing global engagement underscores the urgency of advancing floating solutions as a response to climate change and urban resilience. We invite readers to explore this edited volume, which examines buoyant urbanism as a critical pathway for the future of our cities and underscores the need for rapid legal and policy adaptation to implement these innovative urban solutions effectively.

Brisbane, Australia

Brydon T. Wang
Brydon.Wang@uq.edu.au

Chien Ming Wang
cm.wang@uq.edu.au

References

De Graaf-van Dinther R (ed) (2020) Climate resilient urban areas: governance, design and development in coastal delta cities. Springer Nature

Lim J (2020) Oil rig and superbarge floating settlements, vol 82. Springer Nature

Luo D, Wang B, Bao DW, Ward S (2022) Opportunities for further development of 3D-printed floating artificial reefs. J Aquac Mar Biol 11(2):58–63

Olthuis K, Kuening D (2010) Float! Building on water to combat urban congestion and climate change. Frame Publishers

Wang B (2019) Floating cities: the future or a washed-up idea? The Conversation

Wang BT (2021) The machine metropolis: Introduction to the automated city. In: Automating cities: design, construction, operation and future impact, pp 1–21

Wang B (2023) Regulatory reform and the potential use of floating artificial reefs under blue carbon and ocean carbon methodologies. In: International conference on ocean, offshore and arctic engineering (No. 42nd)

Wang BT, Wang CM (eds) (2021) Automating cities: design, construction, operation and future impact. Springer Nature

Wang CM, Wang BT (2006) Colonization of the ocean and VLFS technology. In: Very large floating structures. CRC Press, pp 15–34

Wang CM, Wang BT (2015) Large floating structures. Ocean Eng Oceanogr 3

Wang CM, Wang BT (2020) Floating solutions for challenges facing humanity. In: ICSCEA 2019: proceedings of the international conference on sustainable civil engineering and architecture. Springer Singapore, pp 3–29

Contents

Large Floating Solutions: Buoyant Urbanism Is Our Key Solution to the Emergent Global Water Crisis ... 1
Brydon T. Wang, Chien Ming Wang, Kim D. Weinert, and Rutger de Graaf-van Dinther

Technological and Engineering Foundations of Floating Cities

Exploring Design of Floating Urban Developments: Integrating Density, Engineering and Environmental Considerations ... 29
Barbara Dal Bo Zanon, Rui P. F. Gomes, and Rutger de Graaf-van Dinther

Brain Coral: Floating City in the Maldives ... 53
Koen Olthuis, Sridhar Subramani, and Mila Avellar Montezuma

Mega Floating City: Green Float ... 73
Donghee Ko

Analysis and Design of Floating Bridges: Recent Advances and Challenges ... 105
Torgeir Moan, Ole David Økland, and Ole Andre Øiseth

Modular Floating Structures for Offshore Urban Development in Real Sea Conditions ... 135
Gil Wang

Environmental and Sustainability Considerations

Industrialization Challenges in Floating Wind ... 171
Carlos A. Garrido-Mendoza

Hydrodynamic Responses of a Modular Floating Solar Platform ... 195
Xuanlie Zhao, Lidong Zhang, Jia Wang, and Yao Zhou

Innovative Aquaculture "Penghu": the First Semi-submersible Wave-Energy-Powered Aquaculture Platform in China 211
Wanzhen Yue, Songwei Sheng, and Zhenxin Huang

Modelling and Testing of Hinged Connected VLFS with Applicability to Offshore Renewables 231
Abel Arredondo-Galeana, Shen Li, Saishuai Dai, and Maurizio Collu

Large Offshore Floating Fish Pens: Ocean Farm 1 and Havfarm 1 249
Per Arild Åland, Anders Fjellheim, and Torgeir Torgersen

Floating Leisure Infrastructure: Integrating the World's Largest Floating Infinity Pool in Lake Como 275
Kimmo Kristian Saharinen and Tytti Sirola

Submersible Platforms for Offshore Marine Permaculture 303
Brian von Herzen, Huu Phu Nguyen, Chien Ming Wang, Chenxuan Huang, Perfecto Tubal, Eric Smith, Joseph Rauch, Sam Donohue, Lucy Schlueter, Sergey Ingram, and Theresa Theuretzbacher

Design and Fabrication of Floating Artificial Reefs: Enhancing Biodiversity and Food Production 335
Dan Luo and Valeriya Komyakova

Socio-legal Considerations to Address Social Acceptance and Legalities of Floating Structures

Designing for Buoyant Urbanism: Pedagogy, Architectural Paradigms and Social Acceptance of Floating Cities 353
Brydon T. Wang

Designing for Artificial Reefs: The Role of Architecture in Marine Ecosystem Restoration 395
Ding Wen Bao

City Parts: Urban Form Design Guidelines for Floating Cities 417
Kirsty Volz

Navigating the Legal and Regulatory Frontiers of Urban AI, Autonomous Maritime Robots and Floating Cities 435
Brydon T. Wang, Bernadette Hyland-Wood, and Michael Guihot

From Land to Water: Policy Considerations Supporting Why Floating Cities Can Help Combat Climate Change 453
Lucy M. Cradduck

Home is Where the Heart Floats: Anchoring Legal Property Concepts for Environmental Migrants 467
Kim D. Weinert, Apoorva Abeysundera, Gauravi Baral, Liluz Maria Biscocho, Rose Foster, Zoe Hennessy, Kit Leach, Shiina Oiwa, Sophie Oudyn, Amber Sheridan, and Leanna Zhang

The 'Bundle of Rights' of Property: A Close Examination of the Jural Relationship in Artificial Landscapes and Floating Structures as Land Replacements 487
Kim D. Weinert

Buoyant Futures: Our Global Duty to Recognise Floating Cities as 'Territory' in International Law 507
Brydon T. Wang

Editors and Contributors

About the Editors

Dr. Brydon T. Wang is an expert on the trustworthy development and regulation of AI and Construction technology. He researches on trustworthy decision-making in infrastructure delivery and data governance, with a focus on how technology can help increase climate-resilience in cities. He is dually qualified in law and architecture and advises on multi-billion dollar infrastructure projects with KPMG. Brydon practised as a construction and technology lawyer with global top-tier law firm Allens Linklaters and has a previous career in architecture and contract administration on award-winning construction projects. He has over 15 years in the construction industry in both client-side (government) and contractor-side roles.

Brydon was recently the lead editor of the book, *Automating Cities: Design, Construction, Operation and Future Impact* (Springer, 2021), and co-edited *Large Floating Structures* (Springer, 2015), a book exploring how we can embrace floating structures in our cities to respond to the increased incidences of floods and sea-level rises. He was featured on ABC Radio National's *Future Tense* discussing offshore architecture and marine urban sprawl and on Seeker's popular documentary, 'How close are we to Living in the Ocean?' Brydon's expertise in technology and cities also extends to data-focused technologies such as digital twins, BIMs, smart construction contracts, and the blockchain. He developed and taught the contract and technology law subjects at the Queensland University of Technology (data privacy, regulating the internet, EU AI Act) and taught architecture and law at the University of Queensland. His Ph.D. thesis on the 'Role of Trustworthiness in Automated Decision-making Systems and the Law' received the *QUT Faculty of Business and Law Executive Dean's Commendation for Outstanding Doctoral Thesis Award*.

Chien Ming Wang is a professor in Structural Engineering, The University of Queensland. He is a fellow of the Australian Academy of Technology and Engineering, a member of the European Academy of Sciences and Arts, a fellow of Academy of Engineering Singapore, a fellow of the Institution of Structural Engineers, a fellow of the Institution of Engineers Singapore and a fellow of the Society of Floating Solutions Singapore. His research interests are in the areas of structural mechanics and floating structures, and he has published over 520 journal papers and 10 books on these subjects. He is the leader of the Offshore Engineering Program of the Blue Economy Cooperative Research Centre that conducts research projects that combine seafood, renewable energy and offshore engineering, underpinned by a $329 million grant from the Australian Government and industry partners over a 10-year period. He is a founding member and engineering science leader of the International Engineering Science Consortium, the vice president of the Society of Floating Solutions Singapore and the Emeritus chairman of East Asia Pacific Conference on Structural Engineering and Construction steering committee. He has won many awards that include the 2019 Nishino Medal, 2019 JN Reddy medal, Singapore's Minister of National Development's R&D Award, IStructE Singapore Structural Awards, Keith Eaton Award, Lewis Kent Award, IES Prestigious Engineering Achievement Award, IES/IStructE Best Paper Award and the Grand Prize of the Next Generation Container Port Challenge. He was the consultant and advisor on many structural and floating projects that include the world's largest floating performance stage at Marina Bay, multi-purpose floating structures research project funded by the Land and Liveability National Innovation Committee and JTC Corporation, Maritime and Port Authority's project on uses of the underground space underneath container ports in Tuas, JTC project on floating hydrocarbon storage facility and HDB project on floating wetlands. He is currently conducting research on the next-generation offshore fish pens and seaweed cultivation platforms.

Dr. Kim Weinert is a legal academic and lawyer. Kim researches the intersection of law, society, and culture. She has a particular focus on charity law, not-for-profit governance, and property law. Kim's not-for-profit research has been cited by the Queensland Supreme Court and the Western Australian parliament. Kim was the recipient of the Australian Postgraduate Award (APA) Scholarship at Griffith University, where she completed her doctoral thesis on speech legality in Australian culture.

Kim's research interest in law and culture in Australia and the South Pacific has been published in international journals. The exploration of cultural legalities in Kim's research provides a unique perspective on the impact doctrinal law has on minority groups. Through this lens, Kim can identify and address numerous shortcomings in the law which paves the way for legal and governance reform. Exploring the traditional boundaries of doctrinal law and its theoretical foundations in relation to emerging and new societal challenges, her research unites several strands of legal inquiry linked to social justice and equality. Her charity law research has been cited by the Queensland Supreme Court, the Western Australian Parliament, and eminent legal charity scholars. The impact of Kim's research has produced an impact on how charity law is discussed and applied in Australia.

Rutger de Graaf-van Dinther is a visionary entrepreneur and researcher specializing in water management and floating urbanization. With a background in civil engineering and a passion for sustainable urban development, Rutger has been at the forefront of innovative solutions for building on water, addressing the challenges of climate resilience and urban growth. As a co-founder and director of Blue21 (2015–present), Rutger spearheads global advancements in floating infrastructure, following his earlier entrepreneurial success with DeltaSync (2007) and Indymo (2015–present). His projects include the Floating Pavilion Rotterdam (2010), cited by the IPCC as a transformative climate adaptation model, and the Bluelands Floating Platform prototype (2023) at TU Delft's Green Village.

Rutger's research credentials are equally impressive, with an M.Sc. (2005) and Ph.D. (2009) in Civil Engineering from Delft University of Technology, both completed cum laude. He served as an applied research professor on adaptive urban development at Rotterdam University of Applied Sciences (2011–2021) and co-initiated the Floating Future Research Project, securing €5.3 M from the Dutch Research Council (2023).

Beyond academia and entrepreneurship, Rutger co-founded the Blue Revolution Foundation (2013) to promote interdisciplinary research on floating solutions and the Thinktank on Floating Urban Development (2017) to address governance challenges. A prolific author and speaker, Rutger has delivered keynotes on five continents, contributed to over 30 publications, and featured in global media outlets like The New York Times, BBC, and Financial Times. His work continues to shape the future of sustainable urbanization worldwide.

Contributors

Apoorva Abeysundera T.C Beirne School of Law, University of Queensland, Brisbane, Australia

Per Arild Åland DNV, Høvik, Norway

Abel Arredondo-Galeana Department of Naval Architecture, Ocean and Marine Engineering, University of Strathclyde, Glasgow, UK

Ding Wen Bao School of Architecture and Urban Design, RMIT University, Melbourne, Australia

Gauravi Baral T.C Beirne School of Law, University of Queensland, Brisbane, Australia

Liluz Maria Biscocho T.C Beirne School of Law, University of Queensland, Brisbane, Australia

Maurizio Collu Department of Naval Architecture, Ocean and Marine Engineering, University of Strathclyde, Glasgow, UK

Lucy M. Cradduck QUT Law School, Queensland University of Technology, Gardens Point Campus, Brisbane, QLD, Australia

Saishuai Dai Department of Naval Architecture, Ocean and Marine Engineering, University of Strathclyde, Glasgow, UK

Barbara Dal Bo Zanon Blue21, Delft, The Netherlands;
Department of Urbanism, Delft University of Technology (TU Delft), Delft, The Netherlands

Rutger de Graaf-van Dinther Blue21, Delft, The Netherlands

Sam Donohue The Climate Foundation, Compostela, Cebu, Philippines

Anders Fjellheim Salmar Aker Ocean AS, Trondheim, Norway

Rose Foster T.C Beirne School of Law, University of Queensland, Brisbane, Australia

Carlos A. Garrido-Mendoza Saitec Offshore Technologies S.L., Leioa, Spain

Rui P. F. Gomes Blue21, Delft, The Netherlands

Michael Guihot School of Law, Queensland University of Technology, Brisbane, QLD, Australia

Zoe Hennessy T.C Beirne School of Law, University of Queensland, Brisbane, Australia

Chenxuan Huang School of Civil Engineering, The University of Queensland, St. Lucia, QLD, Australia;
Blue Economy Cooperative Research Centre, Launceston, TAS, Australia

Zhenxin Huang Guangzhou Institute of Energy Conversion, Chinese Academy of Sciences, Guangzhou, China

Bernadette Hyland-Wood Centre for Data Science, Queensland University of Technology, Brisbane, QLD, Australia

Sergey Ingram The Climate Foundation, Boulder, CO, USA

Donghee Ko Shimizu Corporation, Tokyo, Japan

Valeriya Komyakova Centre for Marine Socioecology, University of Tasmania, Hobart, TAS, Australia;
Oceans Institute, The University of Western Australia, Crawley, WA, Australia

Kit Leach T.C Beirne School of Law, University of Queensland, Brisbane, Australia

Shen Li Department of Naval Architecture, Ocean and Marine Engineering, University of Strathclyde, Glasgow, UK

Dan Luo School of Architecture, Design and Planning, University of Queensland, Brisbane, Australia

Torgeir Moan Department of Marine Technology, NTNU, Trondheim, Norway

Mila Avellar Montezuma ABE, TU Delft, Delft, The Netherlands;
IHE Delft Institute for Water Education, Delft, The Netherlands

Huu Phu Nguyen School of Civil Engineering, The University of Queensland, St. Lucia, QLD, Australia;
Blue Economy Cooperative Research Centre, Launceston, TAS, Australia

Ole Andre Øiseth Department of Structural Engineering, NTNU, Trondheim, Norway

Shiina Oiwa T.C Beirne School of Law, University of Queensland, Brisbane, Australia

Ole David Økland SINTEF Ocean, Trondheim, Norway

Koen Olthuis Waterstudio.NL, Rijswijk, The Netherlands;
ABE, TU Delft, Delft, The Netherlands

Sophie Oudyn T.C Beirne School of Law, University of Queensland, Brisbane, Australia

Joseph Rauch The Climate Foundation, Compostela, Cebu, Philippines

Kimmo Kristian Saharinen Bluet, Vantaa, Finland

Lucy Schlueter The Climate Foundation, Cincinnati, OH, USA

Songwei Sheng Guangzhou Institute of Energy Conversion, Chinese Academy of Sciences, Guangzhou, China

Amber Sheridan T.C Beirne School of Law, University of Queensland, Brisbane, Australia

Tytti Sirola Bluet, Vantaa, Finland

Eric Smith The Climate Foundation, Compostela, Cebu, Philippines

Sridhar Subramani Waterstudio.NL, Rijswijk, The Netherlands;
ABE, TU Delft, Delft, The Netherlands

Theresa Theuretzbacher The Climate Foundation, Bergen, Norway

Torgeir Torgersen NSK Ship Design AS, Harstad, Norway

Perfecto Tubal The Climate Foundation, Compostela, Cebu, Philippines

Kirsty Volz School of Architecture and Built Environment, Queensland University of Technology, Brisbane, Australia

Brian von Herzen The Climate Foundation, Woodford, QLD, Australia

Brydon T. Wang Deal Advisory and Infrastructure, KPMG Australia, Brisbane, QLD, Australia;
Centre for Policy Futures, The University of Queensland, St. Lucia, QLD, Australia

Chien Ming Wang School of Civil Engineering, The University of Queensland, St. Lucia, QLD, Australia;
Blue Economy Cooperative Research Centre, Launceston, Australia

Gil Wang Coastal and Marine Engineering Research Institute (CAMERI), Technion – Israel Institute of Technology, Haifa, Israel

Jia Wang College of Shipbuilding Engineering, Harbin Engineering University, Harbin, China

Kim D. Weinert TC Beirne School of Law, The University of Queensland, St. Lucia, Brisbane, QLD, Australia;
Centre for Public, International and Comparative Law, Brisbane, Australia;
Visiting Fellow 2025, State Library of New South Wales, Sydney, Australia

Wanzhen Yue Guangzhou Institute of Energy Conversion, Chinese Academy of Sciences, Guangzhou, China

Leanna Zhang T.C Beirne School of Law, University of Queensland, Brisbane, Australia

Lidong Zhang College of Shipbuilding Engineering, Harbin Engineering University, Harbin, China

Xuanlie Zhao College of Shipbuilding Engineering, Harbin Engineering University, Harbin, China

Yao Zhou College of Shipbuilding Engineering, Harbin Engineering University, Harbin, China

Large Floating Solutions: Buoyant Urbanism Is Our Key Solution to the Emergent Global Water Crisis

Brydon T. Wang, Chien Ming Wang, Kim D. Weinert, and Rutger de Graaf-van Dinther

Abstract This introductory chapter sets the foundation for *Large Floating Solutions: Design, Construction, Legality of Offshore Structures and Buoyant Urbanism* by presenting buoyant urbanism as a timely and necessary response to the escalating global water crisis driven by climate change. As urban centres increasingly face recurring floods and sea-level rises, floating urban developments offer a transformative approach to climate adaptation, resilience and sustainable urbanisation. The chapter draws on historical and contemporary precedents, showing how communities have long adapted to life on water through floating and amphibious settlements. It then examines the technological, environmental and legal frameworks that underpin the viability of modern floating cities. These include advances in Very Large Floating Structures (VLFS), modular construction systems and the integration of marine ecosystems, alongside emerging legal and policy mechanisms to regulate these developments. Positioned within the framework of the United Nations Sustainable Development Goals, buoyant urbanism is framed not only as a strategy for flood resilience but as a new frontier for reimagining urban futures. As an entry point to the book's interdisciplinary exploration, the chapter outlines the structure and themes of the volume and calls for a paradigm shift in urban planning that embraces the adaptive and generative potential of living on water.

B. T. Wang (✉)
Centre for Policy Futures, The University of Queensland, St. Lucia, QLD, Australia
e-mail: Brydon.Wang@uq.edu.au

Deal Advisory & Infrastructure, KPMG Australia, Brisbane, QLD, Australia

C. M. Wang
School of Civil Engineering, The University of Queensland, St. Lucia, QLD, Australia
e-mail: cm.wang@uq.edu.au

Blue Economy Cooperative Research Centre, Launceston, Australia

K. D. Weinert
TC Beirne School of Law, The University of Queensland, St. Lucia, QLD, Australia
e-mail: k.weinert@uq.edu.au

R. de Graaf-van Dinther
Blue21, Delft, The Netherlands

B. T. Wang et al. (eds.), *Large Floating Solutions*, Lecture Notes in Civil Engineering 636, https://doi.org/10.1007/978-981-96-5435-2_1

Keywords Buoyant urbanism · Floating cities · Climate adaptation · Flood resilience · Sustainable development · VLFS

1 Introduction

The critical necessity for innovative floating urban solutions is increasingly evident as climate change reshapes our urban landscapes, particularly in cities vulnerable to rising seas and frequent floods. The urgency is apparent with the increased incursion of water into our cities despite our efforts to dam, bund and engineer new levees. We are even considering design standards to accept water flowing into our homes [4]. But a key solution to addressing our emergent global problem with water and its unguided incursion into our cities is to consider how we can use the buoyancy of water to float above these challenges, to utilise water to support our buildings, to help us connect the city in different ways, and to open new possibilities to engage sustainably with the resources that our adjacent waters hold within the framework of the blue economy.

This chapter serves as a comprehensive introduction to the book, which explores the design, engineering, policy and legal frameworks necessary to realise buoyant urbanism. This volume addresses pressing climate-related challenges and demonstrates how floating structures—from floating homes and urban developments, to large-scale aquaculture systems, offshore floating windfarms and leisure facilities—can help cities worldwide achieve resilience, sustainability and adaptability in a changing environment. At its core, the book makes the case that these technologies and policies are no longer speculative; they are urgent and practical responses to a global crisis, positioning buoyant urbanism and floating structures as a central pillar of climate adaptation strategies.

The book is divided into chapters that systematically cover the design and construction of floating structures, environmental integration, economic feasibility, and governance requirements. The chapters trace the evolution of buoyant urbanism from conceptual designs to real-world applications. They underscore how traditional land-based planning must adapt to support urbanism on water. Each chapter contributes to a cohesive narrative that positions floating urban solutions not only as a means to adapt to climate impacts but as transformative urban spaces that offer alternative pathways toward sustainability.

Following this introduction, the book is organised thematically into three major sections, each addressing essential facets of buoyant urbanism. These sections provide a systematic exploration of the engineering, environmental, socio-economic, and legal considerations necessary to realise and sustain floating cities. The first section focuses on the *technological and engineering foundations* of floating cities. Chapters here engage with the design and construction of Very Large Floating Structures (VLFS) and modular floating systems, which form the physical basis of buoyant urban environments. Through case studies and technical analyses, this section covers structural resilience, hydrodynamic stability, and the innovative materials and energy solutions required to sustain floating cities in dynamic water conditions.

The second section transitions into *environmental and sustainability considerations*. Here, chapters explore the integration of natural ecosystems within floating structures, emphasising methods that minimise environmental impact and support biodiversity. The role of floating cities in climate resilience and sustainable development is central, as these urban spaces are positioned as adaptive solutions that align with the United Nations Sustainable Development Goals (SDGs), especially those concerning climate action, life below water, and sustainable urban communities. This section highlights how floating cities can contribute to food security, renewable energy production and job creation, serving as integrated systems that meet essential human needs while promoting economic resilience. By reviewing floating aquaculture and other offshore farming projects, as well as renewable energy platforms, this section demonstrates how floating solutions can foster sustainable livelihoods in vulnerable coastal regions.

The third section examines the *socio-legal dimensions* of buoyant urbanism includes *legal and policy frameworks* necessary to support floating urban development. It explores how floating structures intersect with legal frameworks governing property, land use, environmental regulations, and urban planning, highlighting key ambiguities that must be resolved to facilitate large-scale adoption. The chapters analyse zoning challenges, the evolution of property rights for floating developments, and the implications of new technologies, such as AI-driven urban management and autonomous maritime systems.

Beyond legal considerations, this section also examines the social and cultural dimensions of floating cities, including public perceptions of home, community identity and urban liveability in water-based settlements. The role of architectural design and governance mechanisms in fostering acceptance and sustainability is critically assessed, alongside discussions on environmental migration and climate adaptation strategies. By integrating legal, regulatory and societal perspectives, this section positions buoyant urbanism as an urgent policy and governance challenge, requiring interdisciplinary solutions to ensure its legitimacy and long-term success.

Through these themes, the book provides a comprehensive roadmap for buoyant urbanism, presenting it not only as an innovative floating solution to climate change but as a transformative approach to urban planning and sustainability. This thematic organisation guides the reader through each facet of this proposed floating solution, from technical foundations to social implications and policy requirements, laying out the multidimensional pathway to flood-resilient cities.

2 A Key Urban Solution with a Long History

We have a sustained history of ingeniously adapting our settlements to take advantage of the myriad opportunities presented by being located adjacent to or on the water [20]. For millennia, communities have developed floating and amphibious architectures to survive and thrive in flood-prone and water-abundant areas. Floating fishing villages, for instance, have been a staple of life in Southeast Asia and other regions, where communities live on stilted or buoyant structures, following the ebb and flow of tides and floods. Notable examples include the floating villages of Cambodia's *Tonle Sap* lake, where houses are built to rise and fall with water levels, enabling communities to sustain livelihoods despite frequent floods. The floating villages of *Halong Bay* (*Ba Hang*, *Vung Vieng* and *Cong Dam*) are comprised of floating structures that sit on boats, similar to those in *Aberdeen,* Hong Kong [15]. This lifestyle of living on the water, including conducting aquacultural, transport and other economic activities in close proximity to the water, underscores a persistent form of urban resilience, where adaptability to water is interwoven with cultural practices and economic activity. Beyond these examples of urban adaptation to water, the 'water nomads' or *Bajau* people in Southeast Asia represent a unique lifestyle where entire communities live semi-nomadically on water, moving seasonally and relying on the sea for sustenance [6]. These communities have developed a profound relationship with water, adapting their shelters, fishing methods, and social organisation to exist in harmony with their aquatic environment.

In other parts of the world, water bodies have been used as a defensive measure against invasions. The *Uros* in Peru fled from the incoming incursion of the Inca Empire by building islands on *Lake Titicaca* [12]. The community built floating islands of reeds that could be launched from the banks deep into the water body away from land-based threats. This modular approach to exploiting the inherent mobile nature of floating structures underscores an ingenuity that continues to be used by engineers in thinking about how cities can be responsive to programmatic requirements. Likewise, the *Tofinu* tribe in Benin, West Africa, chose to settle on *Lake Nokoue* to escape from the Fon tribe, creating *Ganvié*—what is now the largest water-based community in Africa [10]. While many of these structures are based on stilts, the surrounding infrastructure that supports urban life within these settlements rely on the buoyancy force of water. These villages and settlements surrounded themselves with water as a defensive mechanism against external threats, creating natural moats that protected inhabitants from invasions. This strategic use of water showcases early examples of buoyant and amphibious communities that turned potential vulnerabilities into protective assets.

Similarly, the founding of Venice was one that sought to use the natural defences of the Venetian lagoon around the various islets, later establishing the more permanent settlement on the Rialto islands. However, what is significant about Venice is in the way the city-state conceptualised its urban form and territories. In its ascent, the Venetian state had three subdivisions, '*Dogado*' (Venice proper), holdings on the Italian peninsula that it termed '*Domini di Terraferma*' and it had a separate grouping of assets under '*Domini da Mar*' (Domains of the Sea) or '*Stato da Mar*' (State of the Sea) [2]. It is this third subdivision of a state spread over the maritime or 'over'-seas holdings of the Venetian state, that has particular bearing on how we conceptualise buoyant urbanism in this book. While we explore the notion of satellite floating cities and villages sited in international waters and various water bodies, the concept of buoyant urbanism also requires us to examine hybridisation of the city:

to take its key assets, its climate-vulnerable infrastructure: social (schools, hospitals), transport networks (bridges, ports, ferry and boat terminals), energy infrastructure (wind farms, power stations), food infrastructure (aquaculture, mariculture, fish pens), leisure facilities; and begin to think about how we can float these infrastructure assets to ensure they remain flood-resilient. This way of thinking about urban development is what is advanced in this book as a key solution to the emergent global problem of water and its incursion in cities through floods, unpredictable weather patterns brought on by the climate emergency, and sea-level rises.

Such urban adaptation strategies to deal with the incursion of water are, as with the technology of floating structures, not novel. For example, in Amsterdam, a city deeply connected to water management, the Dutch pioneered the use of polder systems and extensive canal networks to control and repurpose floodwaters to protect the city. The city's canals have been utilised as essential infrastructure to mitigate floods and optimise urban water flow, laying the groundwork for modern water management practices. These historical precedents reveal the human ingenuity that has long been directed toward water adaptation strategies, setting the stage for contemporary buoyant urbanism. However, in contrast to these examples where water was largely predictable, today's climate change–induced water challenges have introduced unprecedented variability and risk.

2.1 Escalating Flooding Challenges Worldwide

The accelerating impacts of climate change have intensified the problem of water management for cities worldwide. Rising global temperatures contribute to more frequent and intense flooding events, which disrupt urban centres, coastal communities and rural areas alike [25]. The current prevalence of extreme weather patterns, including hurricanes, monsoons and unpredictable rainfall, has created a water management crisis that is fundamentally reshaping urban infrastructure needs. Floods are no longer isolated incidents but have become recurrent threats in cities across diverse geographic regions. This problem is pressing in cities with long histories of flooding but is also increasingly relevant for those previously unaccustomed to severe flood events.

In Asia, for instance, countries like Bangladesh experience annual monsoon floods that displace millions and destroy vast areas of farmland [3]. Rapid urbanisation, combined with inadequate flood control infrastructure, exacerbates these disasters, turning natural events into humanitarian crises. Similarly, Tokyo and Singapore, two of Asia's most developed cities, have found themselves vulnerable to extreme weather events, including typhoons and heavy rains that have overwhelmed their sophisticated drainage and flood management systems. Singapore has invested heavily in a comprehensive network of drains, canals, and retention ponds to combat urban flooding, yet the increasing intensity of rainstorms challenges even these advanced defences [5]. Tokyo, on the other hand, has built extensive underground water storage facilities to mitigate flood risks but faces persistent threats from typhoons and seasonal storms [14].

Across Europe, cities such as Venice and parts of Spain have long grappled with high water levels and seasonal flooding. Venice, in particular, contends with 'acqua alta' or high tide flooding, which is now more frequent and severe due to 'changes in atmospheric circulation' [9]. The city's historical canals, initially designed for transport and drainage, are no longer sufficient to cope with the surging Adriatic Sea, necessitating complex engineering solutions such as the Experimental Electromechanical Module (MoSE) project: a system of mobile gates aimed at temporarily isolating Venice's lagoon from the rising tides [9]. In Spain, irregular yet intense storms that result in flash floods and damaged property and infrastructure are exacerbated by an aging population. This lethal mix of flood risk and an elderly population with mobility challenges resulted in close to half of Valencia's flood victims in 2024 being older than 70 years of age [11].

The Americas also confront significant flood risks. Coastal cities in the United States, including Miami and New York, face recurrent flooding due to rising sea levels, storm surges and hurricanes [7, 24]. Despite advances in levee systems and storm surge barriers, these urban areas are vulnerable to extreme weather events that lead to catastrophic floods. Meanwhile, in Latin America, cities such as Rio de Janeiro contend with flash floods triggered by torrential rains, often resulting in deadly landslides in densely populated hillside neighbourhoods [1].

Australia is not exempt from these threats, with Brisbane serving as a critical example of a city facing severe flood risk. Brisbane has experienced numerous major floods in recent history, notably in 2011 and 2022, when significant portions of the city were submerged, displacing residents and damaging critical infrastructure. As a river city, Brisbane's flood challenges are exacerbated by climate variability, necessitating both short-term flood defences and long-term adaptive urban planning to reduce vulnerability.

2.2 *The Climate Crisis as a Water Crisis: United Nations Observations*

Given the global escalating challenge of floods, it is no surprise that the United Nations (UN) has highlighted the critical link between climate change and water, describing the climate crisis as fundamentally a water crisis [17]. Climate change impacts manifest most visibly through water: from melting ice caps and glaciers to extreme floods and droughts (the lack of water). Rising global temperatures disrupt precipitation patterns to intensify storm systems and cause unprecedented fluctuations in sea levels, directly affecting freshwater resources, agriculture and urban areas. According to the UN, the water cycle has been destabilised, resulting in more intense and frequent weather-related disasters [17]. In this context, floods not only threaten human lives and property but also pollute water resources, degrade soil and disrupt ecosystems.

The effects of climate change on water resources are particularly severe for vulnerable populations and low-lying coastal areas. Flooding and saltwater intrusion endanger freshwater supplies, while damaged infrastructure can result in unsafe drinking water and sanitation systems. For cities, the stakes are high: infrastructure disruptions, increased energy consumption for water management and public health risks are but a few of the cascading effects associated with the climate-induced water crisis. The UN's call for sustainable water management thus extends beyond simple flood control and advocates for innovative solutions, including floating cities, as part of a broader strategy to mitigate and adapt to climate change [16].

'*Buoyant urbanism*', as an umbrella term to describe both floating cities, floating buildings and amphibious architecture, presents a viable alternative that aligns with several Sustainable Development Goals (SDGs), such as Goal 11 (Sustainable Cities and Communities) and Goal 13 (Climate Action). These floating structures can lessen the environmental pressures on terrestrial resources [22], offer new habitats for marine life [13], and potentially reduce carbon emissions associated with land-based urbanisation [18]. Floating urban development can serve as a means to safeguard communities while providing essential services, jobs, and housing, advancing both environmental resilience and social stability.

In light of the UN's emphasis on sustainable water solutions, this book advocates for floating structures as essential components of future urban resilience. As cities worldwide confront the increasing risks of floods and rising sea levels, buoyant urbanism represents a forward-thinking, adaptive strategy that aligns human needs with ecological imperatives. The subsequent chapters explore how these floating solutions—through innovative engineering, policy reforms and environmental integration—can reshape urbanism for an uncertain future, responding to the water crisis we face today with practical and sustainable interventions.

3 Key Technological and Engineering Considerations

The development of floating cities requires cutting-edge technological and engineering innovations to address the unique challenges posed by life on water. These considerations include the structural integrity of floating platforms, environmental adaptability, scalability, and the integration of sustainable materials and energy sources. Each technological decision has implications not only for the resilience and habitability of floating cities but also for their environmental impact and long-term viability. This section explores key engineering aspects, including the design of *Very Large Floating Structures* (VLFS), modularity, hydrodynamic stability, environmental resilience, and the role of material science in constructing sustainable floating urban environments.

3.1 Very Large Floating Structures

Very Large Floating Structures (VLFS) represent the foundational technology for building sustainable floating urban spaces [21]. VLFS are massive floating platforms that can support extensive infrastructure and large populations, from housing and commercial zones to energy and water treatment facilities. A critical aspect of VLFS is their ability to remain stable under various weather conditions, including waves, storms, and tides. Hydrodynamic simulations and multi-platform configurations are essential to ensuring that VLFS remain resilient and comfortable for residents.

Designing VLFS involves extensive modelling to account for water movement, wave interactions and environmental load effects, which influence stability and comfort. These platforms use advanced materials and structural design techniques to distribute weight and withstand marine forces. Additionally, VLFS configurations can be adapted based on urban density requirements and environmental conditions. By optimising platform design for various urban functions, engineers and architects can ensure that VLFS meet both technical and urban planning needs.

3.2 Modularity and Flexibility

Modular construction is a crucial engineering strategy for buoyant urbanism, allowing for scalable, adaptable, and resilient structures [23]. Modular Floating Structures (MFS) are composed of individual modules that can be assembled to form larger complexes. Each module can serve different urban functions, from residential units to green spaces, enabling cities to grow and adapt as needed. Modularity offers flexibility in configuration, which is essential for responding to changing environmental and social conditions. For example, individual modules can be detached or repositioned to accommodate seasonal weather patterns or to allow for maintenance.

Modularity also enhances resilience by isolating potential damage. In the case of storms or extreme waves, modular configurations enable damaged sections to be repaired or replaced without affecting the entire structure. The modular approach, widely adopted in floating projects like those in the Netherlands, provides a sustainable model for urban growth on water, as it allows cities to expand incrementally and adjust to diverse site conditions.

3.3 Hydrodynamic Stability and Comfort

The hydrodynamic stability of floating structures is critical for ensuring safe and comfortable living conditions. Floating cities face unique challenges in maintaining stability, as they must withstand wave actions, water currents and varying tides. Advanced engineering techniques, such as hydrodynamic modelling and simulations, allow engineers to predict and manage the forces acting on floating structures. These models account for factors like wave direction, frequency and amplitude, enabling precise adjustments to enhance stability.

To improve stability, floating cities often rely on designs with low centres of gravity and distributed weight. Additionally, some floating structures use stabilisation technologies, such as dampening systems or ballast tanks, which adjust the structure's weight distribution in response to wave movements. This technology reduces the sensation of motion, creating a stable environment that mimics land-based conditions, which is crucial for resident comfort and for ensuring the viability of infrastructure, including transport systems, power generation and waste treatment plants.

3.4 *Integration with Natural and Environmental Systems*

Floating cities must be designed not only to coexist with, but to actively support marine ecosystems. The integration of environmental systems into the design of floating structures can help mitigate ecological impacts and enhance the sustainability of these developments [13]. For instance, many floating projects incorporate artificial coral reefs or marine habitats beneath their platforms to support biodiversity and stabilise the structure. These artificial reefs provide habitat for marine life, promote coral growth and contribute to ecosystem resilience.

Further, floating cities can utilise natural elements, such as mangroves or floating vegetation, to serve as buffers against waves and storms. These natural systems not only reduce wave energy but also sequester carbon, enhance water quality and protect coastal areas from erosion [18]. Such environmental integration aligns with the United Nations Sustainable Development Goals (SDGs), particularly SDG 14 (Life Below Water) and SDG 13 (Climate Action), as it fosters biodiversity and contributes to climate mitigation efforts.

3.5 *Sustainable Materials and Energy Solutions*

Material science plays a significant role in the durability, sustainability and environmental impact of floating cities. Given the exposure to harsh marine environments, materials used in floating structures must be resilient to corrosion, be lightweight and designed to be environmentally sustainable. Innovative materials, such as marine-grade aluminium, reinforced concrete and advanced polymers, are commonly used for their durability and resistance to saltwater corrosion. Additionally, materials with low environmental footprints, such as recycled or locally sourced materials, are prioritised to reduce the ecological impact of constructing these floating structures.

Sustainable energy generation is another key consideration, as floating cities aim to minimise reliance on land-based energy resources. Many floating cities incorporate renewable energy sources, such as solar panels, wind turbines and wave energy converters, to meet their power needs. In projects like the *Penghu* wave-energy-powered platform in China (Chapter "Innovative Aquaculture "Penghu": The First Semi-submersible Wave-Energy-Powered Platform in China"), renewable energy solutions are integrated with floating aquaculture, illustrating how floating cities can harness natural resources sustainably. This self-sufficiency in energy aligns with SDG 7 (Affordable and Clean Energy), fostering a circular, low-carbon urban ecosystem.

3.6 Water and Waste Management

Water management and waste treatment systems are essential components of floating cities, as these urban areas lack traditional connections to terrestrial water and sewage infrastructure. Floating cities often employ decentralised systems for water purification and wastewater treatment. Technologies like desalination plants, rainwater harvesting systems and greywater recycling ensure a steady supply of clean water, which is critical for urban resilience. Some floating cities also use constructed wetlands and other natural filtration systems to treat wastewater, reducing the environmental impact and enhancing sustainability.

By addressing water and waste management as core aspects of floating city design, engineers and urban planners can ensure that these communities function independently and sustainably, meeting the daily needs of residents while protecting surrounding marine ecosystems. These systems embody sustainable urban practices that contribute to key public health objectives, resource efficiency and environmental protection.

4 Legal and Policy Implications

As the concept of floating cities gains momentum as a solution to climate-induced challenges, legal and policy frameworks must evolve to address the unique implications of buoyant urbanism. The development of floating structures introduces complex legal questions around property rights, jurisdiction, environmental regulations, and citizenship [8]. Policy considerations, including zoning, infrastructure governance and resource management must also be adapted to support sustainable and equitable floating communities. This section examines the legal and policy challenges posed by buoyant urbanism, focusing on the need for clear property rights, regulatory frameworks and international cooperation to support the establishment of these innovative urban spaces.

4.1 Property Rights and Ownership Frameworks

One of the most fundamental legal challenges associated with floating cities is the question of property rights. Unlike traditional land-based developments, floating cities occupy water space, which raises issues about ownership, jurisdiction and the classification of property. In many legal systems, property laws are based on terrestrial land, creating ambiguities when it comes to structures in and on water. Floating cities blur the lines between real and personal property, requiring new frameworks to govern ownership, leasing and usage rights [19].

Three primary models can be considered for property rights in floating cities: first, private ownership, where individuals or corporations hold rights over or in the floating structures and their use; second, communal ownership, inspired by an Aristotelian view of shared resources, where water space and floating structures are governed collectively; and third, a hybrid model, where private ownership is granted for specific structures, but the water space remains state-controlled. The hybrid model aligns with property laws, where water bodies are often considered public resources, while private entities may lease or use portions for specified licenced purposes. Establishing clear property rights is essential for promoting investment in floating cities and ensuring legal certainty for residents, investors and stakeholders.

At the same time, legal theory strives to understand the notion of property in other ways: looking beyond an individual's substantive legal rights in the property of a thing speaks to the outward social and relational aspects of property. However, these bundle of rights in the property of a 'thing' is complex and connected to norms and relationships an individual property owner has with the rest of the world (and for floating cities, this could be a global community). Proprietary rights are central to an owner's well-being and autonomy and in recent times, there is growing scholarly attention paid to how the concept of 'home' is a dimension that broadens the understanding of an individual's autonomy and well-being, one that is intrinsically connected to property. For many, the loss of land also means a painful loss of their 'home'—a point both Weinert and Wang explore separately in their chapters in this book.

4.2 Zoning and Land Use Policies

Zoning and land use policies are integral to successfully integrating floating urban developments within existing urban frameworks. Traditional zoning laws are often ill-suited for water-based developments, as they are primarily designed for terrestrial spaces. Floating urban spaces require specialised zoning that accounts for their mobility, multi-use capabilities and potential far-reaching environmental impact. Innovative zoning frameworks can help manage the potential conflicts between residential, commercial, recreational and industrial uses of water, ensuring that floating urban developments align with local environmental goals and social needs.

For instance, zones within floating cities can be designated for multi-use purposes, much like traditional urban planning. However, these zones must also consider the unique risks associated with water-based living, including environmental sensitivity, wave exposure and storm vulnerability. Given the limited footprint of these floating structures, it is likely that zoning laws must be calibrated vertically across the structure, dividing various floors into different functional zones, with zones on external and higher floors needing more robust planning principles to address climatic considerations. In Brisbane, flood-prone areas are subject to specific development restrictions; and similar policies could be adapted to regulate floating structures in terms of site selection, resilience and other engineering requirements and environmental impact. Establishing zoning guidelines for floating developments will provide clarity and ensure that these new urban spaces are developed in a manner that protects both human communities and marine ecosystems.

4.3 *Environmental Regulations and Sustainability Standards*

Buoyant urban development must comply with environmental regulations to minimise their impact on marine ecosystems and contribute to sustainable urbanisation goals. While these structures are less environmentally damaging than land reclamation and do not alter water flow patterns, they can still impact local marine life particularly where their footprint does not permit sufficient light penetration into the underwater environment [21, see Chap. 2]. Thus, robust regulatory frameworks are required to manage the ecological footprint of floating cities, ensuring that they support rather than disrupt surrounding ecosystems.

Environmental impact assessments should be mandatory for floating developments, especially in ecologically sensitive regions. These assessments can guide developers in incorporating sustainability measures, such as artificial reefs, to support biodiversity and reduce erosion. Additionally, floating cities could be required to meet sustainability standards aligned with international environmental agreements, such as the Paris Agreement and the United Nations' Sustainable Development Goals (SDGs). Floating cities offer opportunities for carbon-neutral development through renewable energy sources, green infrastructure and waste management practices, with environmental policy and regulations that incentivise and compel developers to incorporate these features.

4.4 *Governance and Infrastructure Management*

Floating urban developments require new governance models to address their infrastructure and service needs, from water and waste management to energy supply and transportation. Unlike traditional cities, which often rely on centralised governance structures, these floating settlements may necessitate hybrid models that combine public and private sector management. These cities can function as semi-autonomous zones where governance structures are tailored to the specific requirements of water-based living.

One governance approach is to establish floating cities as 'special administrative zones', where flexible regulations allow for experimentation with urban planning, environmental standards, and social policies in a floating regulatory sandbox. This approach has parallels with existing governance models for special economic zones (SEZs) or regulatory sandboxes used to permit startups pilot emerging technologies to identify potential risks and allow regulation to be more responsive to the technology. In this way, governments can support innovative, adaptable systems for managing floating infrastructure. Alternatively, some floating settlements may operate under international governance frameworks, particularly if they are sited in international waters or span multiple jurisdictions. This model would require multilateral cooperation to regulate infrastructure, environmental impact, and the rights of inhabitants.

Infrastructure management in floating cities also demands specific regulatory oversight. Critical services, including energy, sanitation and transportation, may need to be decentralised and delivered through self-sustaining systems. For example, energy policies could encourage the use of renewable sources like solar, wind and wave energy to meet sustainability goals. Water and waste management systems could prioritise closed-loop designs, ensuring minimal impact on marine environments. Establishing governance and infrastructure standards will be crucial for maintaining resilient and self-sustaining floating cities.

4.5 International Law and Jurisdictional Challenges

The establishment of floating cities in international waters introduces complex jurisdictional challenges that require coordination under international law. Traditional concepts of territory, state sovereignty and jurisdiction do not neatly apply to floating cities, especially those located beyond a country's territorial sea. This raises questions about legal authority, citizenship and the rights of residents, particularly if floating cities are positioned as semi-autonomous or self-governing entities.

The United Nations Convention on the Law of the Sea (UNCLOS) provides a framework for defining territorial waters, exclusive economic zones, and the rights of states over maritime resources. However, UNCLOS does not fully address the status of artificial islands or floating structures as potential sovereign territories. The development of floating cities calls for an expanded legal interpretation of 'territory' that recognises these structures as habitable zones with specific governance rights. For instance, Brydon explores the issue of Small Island Developing States (SIDS) facing sea-level rises and advocates in his Chapter "Buoyant Futures: Our Global Duty to Recognise Floating Cities as 'Territory' in International Law" for floating cities to be considered extensions of their sovereign territory, helping them retain statehood and maritime rights even as their traditional landmasses are submerged.

In addition, international cooperation will be essential for establishing floating cities as recognised entities within the global legal framework. Recognising floating cities under international law could help ensure that their residents have defined rights, protections and obligations. This may also involve establishing floating cities as 'safe havens' for climate-displaced populations, supporting the UN's vision of addressing climate-induced migration within an organised, rights-based framework.

The legal and policy considerations associated with buoyant urbanism are as complex and multi-dimensional as the structures themselves. From property rights and zoning to environmental standards and international law, floating cities challenge conventional legal frameworks and demand innovative approaches to governance and regulation. As the chapters ahead will demonstrate, addressing these legal and policy implications is crucial for ensuring that floating cities are not only technically and environmentally feasible, but are also socially just, legally recognised and aligned with global climate adaptation and sustainability goals. Through adaptive policy frameworks and international cooperation, floating cities can serve as resilient, inclusive spaces that embody the future of urban development in a water-stressed world.

5 Structure of the Book

The structure of the book proceeds in three sections.

Part I: Technological and Engineering Foundations of Floating Cities

This section introduces readers to the core technological and engineering principles underpinning buoyant urbanism, highlighting the foundational elements essential to floating city design. Through in-depth analyses of *Very Large Floating Structures* (VLFS), modular systems and case studies from pioneering projects, these chapters provide a comprehensive look at how floating cities can achieve structural resilience, hydrodynamic stability and be deployed in ocean environments.

Chapter 2: Exploring the Design of Floating Urban Developments: Integrating Density and Environmental Considerations

This chapter outlines a multidisciplinary framework for designing large-scale floating urban developments, addressing rising sea levels and land scarcity. It integrates urban density, structural engineering and environmental considerations to balance human needs with ecological impacts. Three floating neighbourhood configurations are compared, highlighting the trade-offs between stability, structural safety, comfort and shading effects on aquatic ecosystems. The study emphasises modularity and scalability, using hydrodynamic simulations and environmental impact assessment to guide design decisions. Findings suggest that spacing and configuration of platforms influence light penetration and wave dynamics, which are critical for sustainability. The chapter underscores the importance of integrating engineering and environmental tools, supported by site-specific analyses, to ensure that floating developments are both resilient and harmonious with their surroundings. By providing a replicable methodology, it advances buoyant urbanism as a viable response to climate change and urban expansion.

Chapter 3: Brain Coral: Floating City in the Maldives

Focusing on the Maldives' ambitious Maldives Floating City (MFC), this chapter presents a case study on rapid, modular construction designed to address rising sea levels and urban land scarcity through large-scale buoyant urbanism. Developed collaboratively by the Maldivian government, Waterstudio, and Dutch Docklands, the MFC integrates sustainability, modularity and resilience across its design and operations. Located within an atoll near Male, the city is guided by ten foundational parameters, including legal compliance, modular construction, renewable energy systems, waste management, and ecological harmony. These principles ensure adaptability to dynamic environmental and social conditions. The chapter highlights key features of the MFC, such as modular floating platforms anchored with telescopic piles, decentralised utility systems and wave-buffering coral reefs that enhance marine ecosystems. The city intends to support approximately 20,000 residents, fostering socio-economic growth through tourism and community engagement while providing climate-adaptive housing solutions. Environmental strategies include rooftop gardens, seaweed cultivation and advanced waste management systems, while phased development allows for incremental expansion, maintaining flexibility in response to environmental and infrastructural challenges. The MFC exemplifies how modular floating solutions can create resilient, liveable spaces that integrate with the local marine environment, demonstrating a scalable approach to water-based urbanisation in vulnerable coastal regions.

Chapter 4: Mega Floating City: Green Float

The Green Float project, developed by Shimizu Corporation in Japan, represents a visionary approach to creating self-sustaining zero-carbon floating cities powered by renewable energy. Located in equatorial waters, these mega floating cities provide a bold concept of how advances in floating structures technology and green technology can support the development of a 1000-m-tall megastructure atop a 2000-m-diameter floating base to house between 30,000 and 50,000 residents. These technological advancements include honeycomb base units and automated processes to reduce high-altitude work, while at the same time addressing wave resilience and disaster preparedness. By incorporating sustainable systems for energy, water, and food production, including vertical farms, renewable energy technologies and advanced recycling, the Green Float seeks to achieve a closed-loop ecosystem that is scalable and relocatable. Extensive modelling and numerical analyses set out in the chapter demonstrate how structural safety can be achieved where such a mega floating city is deployed under equatorial and shallow bay conditions. In doing so, this chapter presents a vision of a sustainable and resilient floating city at an unprecedented scale.

Chapter 5: Analysis and Design of Floating Bridges: Recent Advances and Challenges

This chapter provides an in-depth look at floating bridge technology, which is integral to the connectivity and accessibility of floating cities. Using Norway's Bjørnafjord floating bridge as a case study, the chapter covers recent advancements in floating bridge engineering, including techniques for handling environmental load effects and risk-based design against ship impacts as well as detailing methods for estimating damage, repair scenarios and traffic disruption. Key engineering challenges are examined, such as wave-current interactions, structural damping, and long-term variability in environmental conditions. By addressing these issues through both numerical and experimental methods, this chapter highlights the structural innovations necessary to maintain stability and safety in floating bridges, such as curved end-anchored floating bridges, as well as advanced damping techniques. These innovations allow floating bridges to meet serviceability, ultimate limit state (ULS), fatigue limit state (FLS), and accidental limit state (ALS) criteria and demonstrate how critical infrastructure like transportation can be adapted to meet the unique demands of floating urban environments. The findings underscore the importance of interdisciplinary collaboration and technology transfer from offshore and maritime sectors to achieve sustainable and resilient designs for floating bridges.

Chapter 6: Modular Floating Structures for Offshore Urban Development in Real-Sea Conditions

This chapter explores the design and deployment of Modular Floating Structures (MFS), which allow for flexible, sustainable urban development in open water conditions. MFS are proposed as a scalable and adaptable alternative to traditional land reclamation, focusing on modularity, environmental integration, and compliance with existing maritime and civil regulations. The chapter discusses the structural and legal considerations of MFS, including combining maritime standards with civil engineering codes, ensuring stability, and accommodating zoning and urban planning needs. It highlights Module9000, an MFS module designed for residential use, examining its structural integrity, hydrostatic properties and occupant comfort under real-sea conditions. Advanced analyses assess its motion response and compliance with residential comfort standards, revealing the importance of wave direction and module configuration in mitigating motion-induced discomfort. The chapter also introduces the concept of MFS urban blocks that are spatially optimised and arranged in various formations to meet urban and environmental requirements. At the same time, this chapter presents a novel floating seawall concept using retired ships to attenuate waves and enable MFS deployment in open water. By integrating smart city principles and providing a sustainable, adaptable approach to coastal and offshore urbanisation, this chapter offers MFS as a practical and sustainable method for urban expansion and climate-resilient coastal development.

Part II: Environmental and Sustainability Considerations

This section examines the environmental and ecological dimensions of buoyant urbanism, exploring how floating cities can be designed to harmonise with natural ecosystems, support biodiversity and promote sustainable resource management. This section also examines the socio-economic aspects of buoyant urbanism, highlighting how floating cities can support sustainable livelihoods, enhance food security, provide renewable energy and create resilient communities. Each chapter provides insights into sustainable development practices that integrate floating cities with their marine environments, aligning these innovations with global climate action and conservation goals.

Chapter 7: Industrialization Challenges in Floating Wind

Floating wind technology holds significant potential as a renewable energy source for floating cities. This chapter examines the industrialisation challenges of floating wind energy, focusing on technological, economic, regulatory, and infrastructural barriers across different regions. Technological hurdles include the development of cost-effective floating platforms, durable mooring systems, and robust subsea cables, alongside innovations to enhance stability and maintenance efficiency. Economic challenges centre on high capital expenditure and achieving cost-competitiveness through economies of scale and advanced manufacturing. Regulatory issues involve fragmented policies, lengthy permitting processes, and the need for international collaboration to streamline frameworks that foreshadow the later chapters of this book Infrastructural requirements include upgrading ports, supply chains, and grid integration systems to support large scale deployment. Regional variations are analysed: Europe's challenges with supply chain and grid integration; Asia's focus on deepwater deployment and typhoon resilience; the Americas' regulatory and financing complexities; and Australia's emerging interest, hindered by competition from established renewables. This chapter underscores the need for technological innovation, regulatory streamlining, and stakeholder engagement to unlock floating wind's potential in the global energy transition and the creation of sustainable, energy-independent floating cities.

Chapter 8: Hydrodynamic Responses of a Modular Floating Solar Platform

This chapter examines the hydrodynamic behaviour of modular floating solar platforms, assessing their viability for offshore renewable energy applications. It presents a numerical analysis of wave-structure interactions using the boundary element method, focusing on the motion responses and structural loads of modular photovoltaic platforms and their arrays. The study reveals that modular platforms exhibit multiple resonance peaks due to hydrodynamic interactions between modules, in contrast to the single peak observed in isolated modules. Notably, the vertical forces on connectors in weather-facing modules are significantly higher, highlighting key structural stress points. The findings contribute to the optimisation of floating solar platform designs, supporting the development of resilient and efficient offshore photovoltaic systems that align with global sustainability objectives.

Chapter 9: Innovative Aquaculture "Penghu": The First Semi-submersible Wave Energy-Powered Platform in China

This chapter presents China's 'Penghu' project, a floating platform that integrates wave energy and solar power with offshore fish farming, in operation since 2019. The chapter details the engineering of wave-energy converters and solar photovoltaic systems that supply renewable energy to the platform's aquaculture facilities. The platform's hybrid energy system reduces reliance on traditional energy sources and enhances the sustainability and autonomy of offshore aquaculture by creating a self-sufficient system that co-locates food production and renewable energy generation. Penghu incorporates features for typhoon defence, including a significant draft capacity for submersion and locking mechanisms for wave energy converters during extreme conditions. The project exemplifies the feasibility of co-locating wave energy and aquaculture systems, offering valuable insights for future offshore renewable energy applications and sustainable aquaculture development.

Chapter 10: Modelling and Testing Techniques of Rigid and Hinged Connected VLFS for Offshore Renewable Energy Applications

This chapter explores the use of hinged VLFS platforms to support offshore renewable energy applications, particularly wind energy. Hinged VLFS designs address the challenges of high wave loads by incorporating flexible joints that reduce internal stresses and improve platform stability, allowing for the deployment of shared platforms for multiple wind turbines. The chapter examines both the environmental benefits of this approach, such as reduced energy costs and emissions, and the ecological implications of deploying large-scale hinged structures. Numerical modelling approaches, such as discrete beam element methods, and experimental testing of rigid and hinged VLFS prototypes are discussed and reveal how hinged VLFS can significantly reduce structural strain while maintaining stability, allowing the structure to be deployed in diverse marine conditions crucial for offshore renewable energy production. The technological solution supports broader applicability of VLFS as multipurpose platforms supporting industries such as aquaculture, tourism and buoyant urban development to achieve sustainable, innovating ocean resources utilisation.

Chapter 11: Large Offshore Floating Fish Pens: Ocean Farm 1 and Havfarm 1

This chapter examines the shift of aquaculture from traditional coastal settings to offshore environments. As demand for sustainable food sources grows, innovation in fish farming opens the possibility for us to meet this growing demand, address global food security while mitigating environmental impacts. This chapter examines two pioneering installations, Ocean Farm 1 and Havfarm 1, detailing key operational considerations, including site selection, resourcing, and the integration of advanced technologies like remote operations, hybrid power systems, and digital monitoring for efficiency and sustainability. The chapter details the technological and environmental strategies behind these projects, including adaptations for wave stability, fish welfare, and ecological sustainability. It also addresses the challenges of operating in harsh marine conditions, such as biofouling and waste management. By examining these floating aquaculture systems, the chapter highlights their role in advancing sustainable food production in a way that aligns with marine conservation goals.

Chapter 12: Floating Leisure Infrastructure: Integrating the World's Largest Floating Infinity Pool at Lake Como

This chapter examines the design of floating leisure infrastructure, focusing on the world's largest floating infinity pool at Lake Como, Italy. Designed in collaboration with architects Herzog & de Meuron, Bluet's innovative floating infinity pool transformed the hotel's waterfront into an iconic feature. The floating structure required bespoke technical solutions to enhance stability and mitigate motion-induced discomfort. Advanced hydrodynamic assessments, including wave tank testing and in-situ data collection, informed the pool's design, which incorporates a catamaran-based platform, damping plates and a precision mooring system to reduce movement. The chapter explores the challenges of achieving the infinity effect on a floating platform, requiring increased pumping capacity and tailored gutter design to maintain visual and functional integrity. Additionally, it discusses regulatory and environmental considerations, including water quality management and compliance with maritime and building standards. The project underscores the growing viability of floating leisure infrastructure as a sustainable and adaptive model for waterfront development, offering insights into the future of buoyant urbanism.

Chapter 13: Submersible Platforms for Offshore Marine Permaculture

This chapter explores the innovative use of submersible platforms for offshore marine permaculture, focusing on their role in addressing climate change, restoring marine ecosystems, and enhancing seaweed cultivation. It introduces marine permaculture as a sustainable method to regenerate nutrient supply through deep cycling and regenerative upwelling, fostering biodiversity and improving carbon sequestration. The chapter details the Climate Foundation's development of submersible platforms, highlighting successful trials in the Philippines from 2020 to 2024, where significant seaweed growth and ecosystem restoration were achieved. It compares past and present marine infrastructure designs, emphasising the evolution towards energy-efficient and storm-resilient systems. Advanced features, such as modular Aqua-Tractors for depth adjustment and renewable energy integration, are examined for their operational and environmental benefits. The chapter also addresses regulatory challenges and outlines future directions, including optimising designs for different environments, exploring new seaweed applications, and integrating marine permaculture with offshore wind and aquaculture. This multidisciplinary approach underlines marine permaculture's potential to contribute to food security, climate mitigation and sustainable ocean use.

Chapter 14: Design and Fabrication of Floating Artificial Reefs: Enhancing Biodiversity and Ecosystem Health

Floating artificial reefs (FARs) represent an innovative approach to marine habitat restoration, supporting both aquaculture and biodiversity. This chapter examines the ecological and engineering advancements in artificial reef design, including modular construction, 3D printing, and material innovations that enable FARs to function as standalone habitats or integrated elements within offshore energy projects. The chapter highlights the regulatory challenges associated with FAR implementation, underscoring the need for interdisciplinary collaboration among engineers, ecologists, policymakers, and industry stakeholders to establish appropriate governance frameworks. Case studies from the Blue Economy CRC illustrate the potential of floating artificial reefs to contribute to sustainable aquaculture and offshore development. By integrating ecological restoration with technological innovation, floating artificial reefs offer a transformative approach to balancing economic development with marine conservation, supporting the resilience of marine ecosystems while advancing the objectives of the blue economy.

Part III: Socio-legal Considerations to address Social Acceptance and Legalities of Floating Structures

This final section explores the socio-legal dimensions of floating cities, addressing regulatory, governance and societal challenges that influence their viability as climate adaptation solutions. It examines how floating structures intersect with legal frameworks on property, sovereignty and international law, while also considering social perceptions of home, community and urban governance in buoyant urbanism. These chapters assess the role of architecture in shaping public engagement with floating cities, the implications of autonomous maritime robots and AI-driven urban planning, and the potential for floating cities to mitigate climate change impacts. Additionally, the section analyses land-use planning, environmental migration and the evolution of property rights to accommodate floating structures. By integrating legal and social perspectives, these chapters contribute to a holistic understanding of how floating cities can be legally recognised and socially accepted as a response to rising sea levels and urban resilience.

Chapter 15: Designing for Buoyant Urbanism: Pedagogy, Architectural Paradigms and Social Acceptance of Floating Cities

This chapter examines the role of architectural education in advancing buoyant urbanism, arguing that integrating floating city design into architectural curricula is essential for fostering innovation and public acceptance. It presents the University of Queensland's *Floating Cities: Buoyant Urbanism, Ecological Engineering and Marine Utopias* design studio as a case study, highlighting how students engaged with historical and contemporary paradigms of floating urbanism to develop speculative and practical solutions. The chapter outlines six key architectural paradigms that have shaped floating city proposals, from traditional floating settlements to climate-resilient urbanism, emphasising how these approaches inform contemporary practice. By exposing students to interdisciplinary collaboration and ecological design principles, the studio contributed to the development of adaptive, sustainable and culturally responsive floating structures. This chapter underscores the importance of architectural pedagogy in legitimising buoyant urbanism as a viable response to climate change, urban expansion and socio-political challenges.

Chapter 16: Designing for Artificial Reefs: The Role of Architecture in Marine Ecosystem Restoration

This chapter explores the role of architecture in marine ecosystem restoration, examining how design innovation can contribute to ecological resilience and biodiversity conservation. It presents the outcomes of the *Martine Utopia* Master of Architecture Design Studio at RMIT University, which challenges students to integrate computational design, biomimicry, and environmental data-driven methods to develop floating artificial reefs. Through case studies of student projects, including *Laminarosity*, *Undamaging*, *Heronia*, and *Mantaverse*, the chapter illustrates how architecture can function as an agent of ecological intervention rather than merely a spatial discipline. It highlights the potential for artificial reefs to serve as both environmental and cultural artefacts, fostering engagement with marine ecosystems while supporting sustainable urban development. By positioning architecture as an active contributor to environmental stewardship, this chapter reinforces the interdisciplinary potential of buoyant urbanism in addressing climate change-induced degradation of marine habitats.

Chapter 17: City Parts: Urban Form Design Guidelines for Floating Cities

This chapter examines the application of urban form design guidelines to floating cities, arguing that governance structures are necessary to ensure sustainability, affordability and social equity in buoyant urbanism. Drawing on case studies from Vietnam, the Philippines and the Netherlands, it contrasts the idyllic floating developments in wealthy nations with the informal, resource-dependent floating settlements in Southeast Asia, highlighting the socio-economic disparities between these models. The chapter applies Oswald Ungers and Rem Koolhaas's *City Parts* concept to the *Oceanix* floating city prototype in South Korea, exploring how clustering, public space provision and density regulation impact urban liveability and ecological sustainability. It finds that while *Oceanix City* integrates strong mobility strategies and open space planning, its high level of shared amenities risks inflating housing costs, limiting accessibility. By advocating for urban design frameworks that balance affordability, environmental impact, and liveability, this chapter positions urban form governance as a critical enabler of sustainable floating cities.

Chapter 18: Navigating the Legal and Regulatory Frontiers of Urban AI, Autonomous Maritime Robots and Floating Cities

This chapter examines the role of artificial intelligence (AI) and autonomous maritime robots in the development and governance of floating cities, assessing the legal, regulatory and ethical challenges posed by their deployment. It explores the use of *Urban AI* in planning and managing floating cities, optimising resource distribution and enabling dynamic adaptation to environmental changes. Additionally, the chapter analyses *Agentic AI* and autonomous maritime robotics, particularly in offshore wind energy, fish farming and infrastructure maintenance, highlighting their potential to enhance efficiency and sustainability. However, the integration of AI-driven systems into offshore environments raises complex legal questions concerning liability, jurisdiction, data governance and cybersecurity risks. The chapter critically evaluates existing international legal frameworks, including the United Nations Convention on the Law of the Sea (UNCLOS), and recent AI regulatory developments in the European Union and the United States, arguing that urgent regulatory adaptations are required to address emerging governance gaps. By advocating for new international standards and ethical guidelines, this chapter underscores the need for a coordinated, proactive approach to integrating AI and automation into the future of buoyant urbanism.

Chapter 19: From Land to Water: Policy Considerations Supporting Why Floating Cities Can Help Combat Climate Change

As climate change accelerates, traditional land-based property models face increasing challenges. This chapter explores the role of floating cities as a proactive policy solution to mitigate the adverse impacts of climate change, particularly in coastal and riverine areas. It emphasises the critical challenges posed by rising sea levels and extreme weather events, which threaten existing land-based communities and infrastructure. The chapter critiques historical and current land planning policies for their inability to address these risks adequately and highlights the unsustainable reliance on reactive measures like barricades and forced relocations. Floating cities are presented as a dynamic, adaptable alternative, capable of preserving community connections and enabling orderly transitions for affected populations. The concept aligns with international goals for sustainable urban development and integrates innovating uses of water bodies for habitation, energy, and food production. While acknowledging the legal and practical challenges, the chapter underscores the necessity for new approaches in land use planning to ensure resilience against climate change impacts.

Chapter 20: Home Is Where the Heart Floats: Anchoring Legal Property Concepts for Environmental Migrants

This chapter examines the socio-legal implications of floating structures as a response to climate change-induced displacement, focusing on how they can preserve and redefine the concept of 'home' for environmental migrants. It argues that traditional property frameworks, which prioritise land-based ownership, fail to accommodate the lived realities of displaced communities, necessitating a broader understanding of home that includes cultural, psychological, and social dimensions. Through case studies of the Solomon Islands and Florida, the chapter contrasts communal, customary approaches to land tenure with individualistic, market-driven property models, highlighting the tensions between legal definitions of property and the lived experience of displacement. It further explores how international law, particularly human rights and refugee frameworks, could evolve to recognise floating structures as legitimate, stable housing solutions for climate migrants. By integrating legal, architectural, and social perspectives, this chapter contributes to the broader discourse on buoyant urbanism as a means of fostering resilience, continuity, and belonging in the face of environmental change.

Chapter 21: The 'Bundle of Rights' of Property: A Close Examination of the Jural Relationship in Artificial Landscapes and Floating Structures as Land Replacements

This chapter critically evaluates the viability of floating structures and artificial islands as residential options; a conclusive property framework regarding these constructs still needs to be developed. While there might be a natural tendency to eliminate any perceived inconsistencies to categorise these engineered spaces as real property, it overlooks context and the strains of theoretical tensions in a system of private property. Unlike land, the utility of artificial islands and floating, which are attached to or located within a coastal zone of a territorial interior, manifests a dialectical relationship between natural resources, like water, air and land, and manufactured objects. The cultivation and exploration of natural resources must be controlled (as best it can be) for the manufactured object to have value as the purported solution to the issues of land deviation.

Chapter 22: Buoyant Futures: Our Global Duty to Recognise Floating Cities as 'Territory' in International Law

This final chapter explores the international legal framework required to support floating cities as adaptive solutions to rising sea levels and as alternatives to the displacement of coastal populations. It examines the existential threat faced by Small Island Developing States due to rising sea levels, which challenge established principles of statehood, sovereignty and maritime rights under international law. The chapter argues that recognising floating cities as sovereign extensions of submerged territories should be considered an *erga omnes* obligation, reflecting the collective responsibility of states to address climate-induced displacement and uphold fundamental human rights. It critically analyses the legal ambiguities surrounding the classification of floating cities under the United Nations Convention on the Law of the Sea (UNCLOS) and explores emerging legal norms that could provide a framework for their governance. By advocating for international cooperation in recognising floating cities as territory, this chapter underscores the necessity of developing a unified legal framework to ensure the preservation of maritime zones, support the rights of vulnerable populations and foster global resilience to environmental change.

References

1. Avelar AS, Netto ALC, Lacerda WA, Becker LB, Mendonça MB (2013) Mechanisms of the recent catastrophic landslides in the mountainous range of Rio de Janeiro, Brazil. In: Landslide science and practice: Volume 4: Global environmental change, pp 265–270
2. Benton L (2009) A search for sovereignty: law and geography in European empires, 1400–1900. Cambridge University Press
3. Blasco F, Bellan MF, Chaudhury MU (1992) Estimating the extent of floods in Bangladesh using SPOT data. Remote Sens Environ 39(3):167–178

4. Brisbane Sustainability Agency (2023) Floor-resilience strategies. Flood Resilient Homes Program. https://www.sustainablebrisbane.com.au/wp-content/uploads/2023/05/FRSP-Resilience-Strategies_2023.pdf
5. Chan FKS, Chuah CJ, Ziegler AD, Dąbrowski M, Varis O (2018) Towards resilient flood risk management for Asian coastal cities: lessons learned from Hong Kong and Singapore. J Clean Prod 187:576–589
6. Clifton J, Majors C (2012) Culture, conservation, and conflict: perspectives on marine protection among the Bajau of Southeast Asia. Soc Nat Resour 25(7):716–725
7. Colle BA, Rojowsky K, Buonaito F (2010) New York City storm surges: climatology and an analysis of the wind and cyclone evolution. J Appl Meteorol Climatol 49(1):85–100
8. Cradduck LM, Wang BT, Weinert KD (2024) Underpinning the value of buoyant urbanism: returning torrens to the seas. In: International conference on offshore mechanics and Arctic engineering, vol 87875. American Society of Mechanical Engineers, p V009T13A006
9. Faranda D, Ginesta M, Alberti T, Coppola E, Anzidei M (2023) Attributing Venice Acqua Alta events to a changing climate and evaluating the efficacy of MoSE adaptation strategy. NPJ Climate Atmos Sci 6(1):181
10. Glandon DM (2015) Measuring resilience is not enough; we must apply the research. Researchers and practitioners need a common language to make this happen. Ecol Soc 20(2)
11. Jones S (2024) Almost half of Valencia's flood victims were aged over 70, figures show. The Guardian, 15 Nov
12. Kent M (2013) The importance of being Uros: indigenous identity politics in the genomic age. Soc Stud Sci 43(4):534–556
13. Luo D, Wang B, Bao DW, Ward S (2022) Opportunities for further development of 3D-printed floating artificial reefs. J Aquaculture Mar Biol 11(2):58–63
14. Nakamura H, Oosawa M (2021) Effects of the underground discharge channel/reservoir for small urban rivers in the Tokyo area. IOP Conf Ser Earth Environ Sci 703(1):012029
15. Trang NTT (2016) Architectural approaches to a sustainable community with floating housing units adapting to climate change and sea level rise in Vietnam. Int J Civil Environ Struct Constr Archit Eng 10(2):168–179
16. UN Habitat (2022) The second UN roundtable on sustainable floating cities: meeting the rising seas with floating infrastructure
17. UN Water (2019) Climate change and water. Policy Brief by the UN-Water Expert Group on Water and Climate Change
18. Wang B (2023) Regulatory reform and the potential use of floating artificial reefs under blue carbon and ocean carbon methodologies. In: International conference on ocean, offshore and arctic engineering (No. 42nd)
19. Wang B (2024) Development of the legal definition of the floating city: judicial interpretation of structural characteristics of floating homes and developments. In: Proceedings of the third world conference on floating solutions: WCFS2023; 28–29 Aug, Tokyo, Japan. Springer, pp 139–157
20. Wang CM, Wang BT (2006) Colonization of the ocean and VLFS technology. In: Wang CM, Watanabe E, Utsunomiya T (eds) Very large floating structures. CRC Press, pp 15–34
21. Wang CM, Wang BT (2015) Large floating structures. In: Ocean engineering & oceanography, vol 3. Springer
22. Wang CM, Wang BT (2020) Floating solutions for challenges facing humanity. In: ICSCEA 2019: proceedings of the international conference on sustainable civil engineering and architecture. Springer, Singapore, pp 3–29
23. Wang G, Goldfeld Y, Drimer N (2019) Expanding coastal cities—proof of feasibility for modular floating structures (MFS). J Clean Prod 222:520–538
24. Wdowinski S, Bray R, Kirtman BP, Wu Z (2016) Increasing flooding hazard in coastal communities due to rising sea level: Case study of Miami Beach, Florida. Ocean Coast Manag 126:1–8
25. Whitfield PH (2012) Floods in future climates: a review. J Flood Risk Manage 5(4):336–365

Dr. Brydon Wang is an expert on the trustworthy development and regulation of AI and Construction technology. He researches on trustworthy decision-making in infrastructure delivery and data governance, with a focus on how technology can help increase climate-resilience in cities. He is dually qualified in law and architecture and advises on multi-billion dollar infrastructure projects with KPMG. Brydon practised as a construction and technology lawyer with global top-tier law firm Allens Linklaters and has a previous career in architecture and contract administration on award-winning construction projects. He has over 15 years in the construction industry in both client-side (government) and contractor-side roles.

Brydon was recently lead editor of the book, *Automating Cities: Design, Construction, Operation and Future Impact* (Springer, 2021) and co-edited *Large Floating Structures* (Springer, 2015), a book exploring how we can embrace floating structures in our cities to respond to the increased incidences of floods and sea-level rises. He was featured on ABC Radio National's *Future Tense* discussing offshore architecture and marine urban sprawl, and on Seeker's popular documentary, 'How close are we to Living in the Ocean?' Brydon's expertise in technology and cities also extends to data-focused technologies such as digital twins, BIMs, smart construction contracts, and the blockchain. He developed and taught the contract and technology law subjects at the Queensland University of Technology (data privacy, regulating the internet, EU AI Act), and taught architecture and law at the University of Queensland. His Ph.D. thesis on the 'Role of Trustworthiness in Automated Decision-making Systems and the Law' received the *QUT Faculty of Business and Law Executive Dean's Commendation for Outstanding Doctoral Thesis Award*.

Chien Ming Wang is a Professor in Structural Engineering, The University of Queensland. He is a Fellow of the Australian Academy of Technology and Engineering, a Member of the European Academy of Sciences and Arts, a Fellow of Academy of Engineering Singapore, a Fellow of the Institution of Structural Engineers, a Fellow of the Institution of Engineers Singapore and a Fellow of the Society of Floating Solutions Singapore. His research interests are in the areas of structural mechanics and floating structures and has published over 520 journal papers and 10 books on these subjects. He is the Leader of the Offshore Engineering Program of the Blue Economy Cooperative Research Centre that conducts research projects that combine seafood, renewable energy and offshore engineering, underpinned by a $329 million grant from the Australian Government and industry partners over a 10-year period. He is a founding member and Engineering Science Leader of the International Engineering Science Consortium, Vice President of the Society of Floating Solutions Singapore and the Emeritus Chairman of East Asia Pacific Conference on Structural Engineering and Construction steering committee. He has won many awards that include the 2019 Nishino Medal, 2019 JN Reddy medal, Singapore's Minister of National Development's R&D Award, IStructE Singapore Structural Awards, Keith Eaton Award, Lewis Kent Award, IES Prestigious Engineering Achievement Award, IES/IStructE Best Paper Award and the Grand Prize of the Next Generation Container Port Challenge. He was the consultant and advisor on many structural and floating projects that include the world's largest floating performance stage at Marina Bay, multi-purpose floating structures research project funded by the Land and Liveability National Innovation Committee and JTC Corporation, Maritime and Port Authority's project on uses of the underground space underneath container ports in Tuas, JTC project on floating hydrocarbon storage facility and HDB project on floating wetlands. He is currently conducting research on the next generation offshore fish pens and seaweed cultivation platforms.

Dr. Kim Weinert is a legal academic and lawyer. Kim researches the intersection of law, society, and culture. She has a particular focus on charity law, not-for-profit governance, and property law. Kim's not-for-profit research has been cited by the Queensland Supreme Court and the Western Australian parliament. Kim was the recipient of the Australian Postgraduate Award (APA) Scholarship at Griffith University, where she completed her doctoral thesis on speech legality in Australian culture.

Kim's research interest in law and culture in Australia and the South Pacific has been published in international journals. The exploration of cultural legalities in Kim's research provides a unique perspective on the impact doctrinal law has on minority groups. Through this lens, Kim can identify and address numerous shortcomings in the law which paves the way for legal and governance reform. Exploring the traditional boundaries of doctrinal law and its theoretical foundations in relation to emerging and new societal challenges, her research unites several strands of legal inquiry linked to social justice and equality. Her charity law research has been cited by the Queensland Supreme Court, the Western Australian Parliament, and eminent legal charity scholars. The impact of Kim's research has produced an impact on how charity law is discussed and applied in Australia.

Rutger de Graaf-van Dinther is a visionary entrepreneur and researcher specializing in water management and floating urbanization. With a background in civil engineering and a passion for sustainable urban development, Rutger has been at the forefront of innovative solutions for building on water, addressing the challenges of climate resilience and urban growth. As co-founder and director of Blue21 (2015–present), Rutger spearheads global advancements in floating infrastructure, following his earlier entrepreneurial success with DeltaSync (2007) and Indymo (2015–present). His projects include the Floating Pavilion Rotterdam (2010), cited by the IPCC as a transformative climate adaptation model, and the Bluelands Floating Platform prototype (2023) at TU Delft's Green Village.

Rutger's research credentials are equally impressive, with an MSc (2005) and Ph.D. (2009) in Civil Engineering from Delft University of Technology, both completed cum laude. He served as Applied Research Professor on adaptive urban development at Rotterdam University of Applied Sciences (2011–2021) and co-initiated the Floating Future Research Project, securing €5.3M from the Dutch Research Council (2023).

Beyond academia and entrepreneurship, Rutger co-founded the Blue Revolution Foundation (2013) to promote interdisciplinary research on floating solutions and the Thinktank on Floating Urban Development (2017) to address governance challenges. A prolific author and speaker, Rutger has delivered keynotes on five continents, contributed to over 30 publications, and featured in global media outlets like The New York Times, BBC, and Financial Times. His work continues to shape the future of sustainable urbanization worldwide.

Technological and Engineering Foundations of Floating Cities

Exploring Design of Floating Urban Developments: Integrating Density, Engineering and Environmental Considerations

Barbara Dal Bo Zanon, Rui P. F. Gomes, and Rutger de Graaf-van Dinther

Abstract Rising sea levels are placing increasing strain on densely populated coastal areas, challenging the sustainability of land reclamation as a long-term strategy. To address these pressures, a combination of innovative approaches is needed to create climate-resilient spaces for human activities. Among these, floating technologies show significant potential as part of a broader suite of adaptive solutions. While promising, the large-scale implementation of floating structures in coastal environments is still in its early stages. Scaling these technologies involves overcoming significant hurdles, including engineering and design complexities, governance and legislative constraints, ecological considerations, and social acceptance. To tackle these challenges, we propose a comprehensive methodology for designing floating developments. This methodology considers urban density, wave interactions and environmental conditions, acknowledging the complex interplay among these factors. Its broader scope reflects a multidisciplinary perspective, combining urbanism, environmental science, and hydrodynamics. By synthesizing multidisciplinary considerations, this methodology provides a structured framework to support the design, development, and governance of large-scale floating structures.

Keywords Climate adaptation · Floating urban development · Density · Environmental design · Comfort · Stability

B. Dal Bo Zanon (✉) · R. P. F. Gomes · R. de Graaf-van Dinther
Blue21, Delft, The Netherlands
e-mail: barbara@blue21.nl

R. P. F. Gomes
e-mail: rui@blue21.nl

R. de Graaf-van Dinther
e-mail: rutger@blue21.nl

B. Dal Bo Zanon
Department of Urbanism, Delft University of Technology (TU Delft), Delft, The Netherlands

B. T. Wang et al. (eds.), *Large Floating Solutions*, Lecture Notes in Civil Engineering 636, https://doi.org/10.1007/978-981-96-5435-2_2

1 Introduction

Coastal and delta regions face mounting challenges from both human activities and environmental changes. These low-lying areas are increasingly attracting a significant share of the global population, a trend expected to continue in the coming decades. According to the IPCC, over one billion people could be living in these vulnerable regions by 2050 [8]. This population growth amplifies the risks posed by rising sea levels and extreme weather events, threatening both communities and critical infrastructures.

The challenges, however, extend beyond climate change. Extensive human modification of coastal landscapes—often aimed at supporting urban and industrial development—has dramatically altered local ecosystems. Recent studies estimate that approximately 2530 km^2 of land has been reclaimed from coastal areas since the start of the century [21]. While land reclamation has supported economic growth and urban expansion, it has often come at the expense of critical habitats and natural barriers that protect against storm surges [1, 5]. The resulting loss of wetlands and other coastal features has left these regions increasingly vulnerable to erosion and storm damage.

Considering the ongoing urban and industrial expansion along with the uncertainties posed by climate change, there is a pressing need for climate-adaptive solutions in urban development. Two promising approaches are amphibious and floating technologies. Amphibious buildings have foundations that can rise with floodwaters, while floating buildings are permanently buoyant. Depending on the local context, these technologies can be used to mitigate flood risks while avoiding the displacement of communities or ecosystems.

1.1 Scaling up of Floating Development

In coastal regions that are struggling to balance urban and industrial expansion with the encroachment of the sea, floating developments offer a viable solution. They provide additional space on the water without the need for land reclamation, which is increasingly unsustainable due to the scarcity of sand and other resources [22]. Floating developments present a climate-resilient alternative that can adapt to rising sea levels.

Despite being perceived as innovative, floating developments are not a novel concept. Many communities have lived on water for centuries, and numerous floating projects have been realized [14, 18]. However, most floating technologies are still applied on a small scale. To effectively respond to the challenges faced by coastal and delta areas worldwide, it is imperative to scale up floating technologies in the coming years.

Scaling up requires addressing technical aspects closely related to the design and spatial organization of floating developments. The interaction between waves and floating structures, and the interrelation among structures, necessitate a deep understanding of the dynamic qualities of the water environment. At the same time, a better understanding of the qualities of urban space on the water and the effects of floating structures on the environment is needed.

1.2 Objective of the Study

This study aims to develop a comprehensive and interdisciplinary design framework for floating developments by integrating key parameters from urban design, engineering, and environmental considerations. This methodology is intended to support feasibility studies and policy making by offering a structured approach that balances technical, environmental, and urban planning requirements, serving as a valuable tool for the conceptualization and planning of floating developments.

2 Methodology

This study integrates key aspects related to urban design, engineering and environmental analysis, focusing on density (urban design), structural safety, stability, comfort (engineering), and shading (environmental analysis) (Fig. 1). While this chapter does not provide an exhaustive examination of each domain, the chosen parameters are considered important for guiding design decisions.

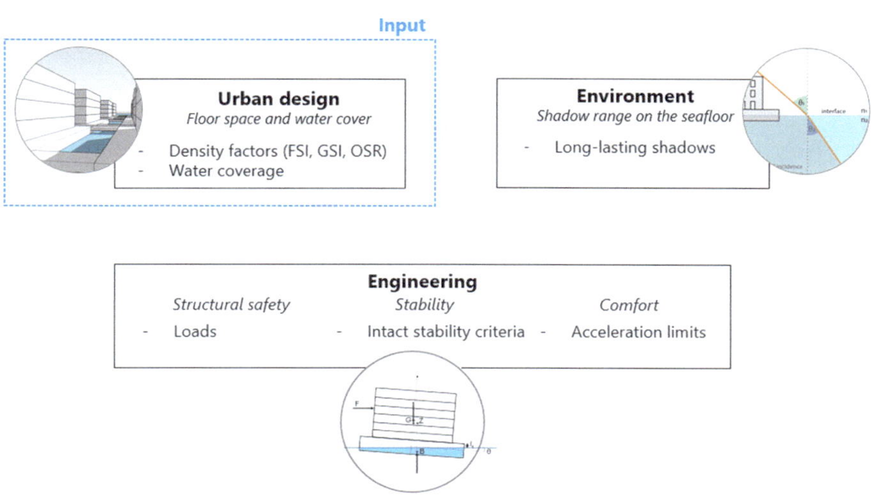

Fig. 1 Schematic overview of the methodology used in this study

Density, a key concept in urbanism, measures the concentration of entities like people or buildings within a specific area. Factors influencing density include Floor Space Index (FSI), Ground Space Index (GSI), Open Space Ratio (OSR), and Layers (L). Network Density (N) further describes the length of networks per square meter [19]. In floating developments, compactness relative to the water area is considered a relevant additional variable in describing density.

Engineering requirements aim to ensure structural safety and comfort, evaluating stability (ability to return to the upright position after disturbance) and comfort (being within acceleration limits). Wang et al. [23] detail these requirements, with stability and comfort being crucial in small neighbourhood-scale designs and platform interactions. In addition to comfort, loads on the main structural components should be evaluated to guarantee structural safety. To carry out these analyses, the software HydroMEC+ has been used to solve hydrostatic and hydrodynamic problems.

The environmental impacts of floating structures on aquatic ecosystems, particularly at larger scales, are not yet fully understood. Research indicates that these structures can affect water quality and ecology, primarily through their surface-covering effects. Unlike temporary shading caused by natural features like tall land structures or trees, permanent shading from floating structures may induce localized environmental impacts that require careful site-specific evaluation. The shading caused by floating platforms can lead to a localized reduction in light intensity, which in turn impacts primary production and oxygen concentration in the water [3, 11, 16]. This shading effect is of relevance since light is essential for photosynthesis, the process by which aquatic plants produce food and oxygen. Czapiewska et al. [2] proposed metrics that consider environmental and ecological aspects, with shading selected as the main parameter for analysis due to its influence on multiple water quality variables. As the understanding of these impacts evolves, it becomes increasingly important to integrate environmental considerations into the design and placement of floating developments to mitigate potential negative effects on aquatic ecosystems.

2.1 Case Study Definition

Three multi-platform floating neighbourhood configurations have been defined and will be compared. The three configurations are illustrated in Fig. 2. As shown in Table 1, the three block typologies are characterized by the same Floor Space Index (FSI) but different Ground Space Indices (GSI). The water surface covered by floating structures varies considerably among the three configurations. The overall development footprint is the same (270 m by 270 m). Table 1 includes the main dimensions of each block type.

The complete configuration of the studied floating urban development is presented in Fig. 3. The development considers a floating breakwater (BW) to attenuate the effect of waves on the floating neighbourhood (C_X, where X is A, B or C), and a bottom-fixed structure representing a vertical quay wall (QW). The presence of the

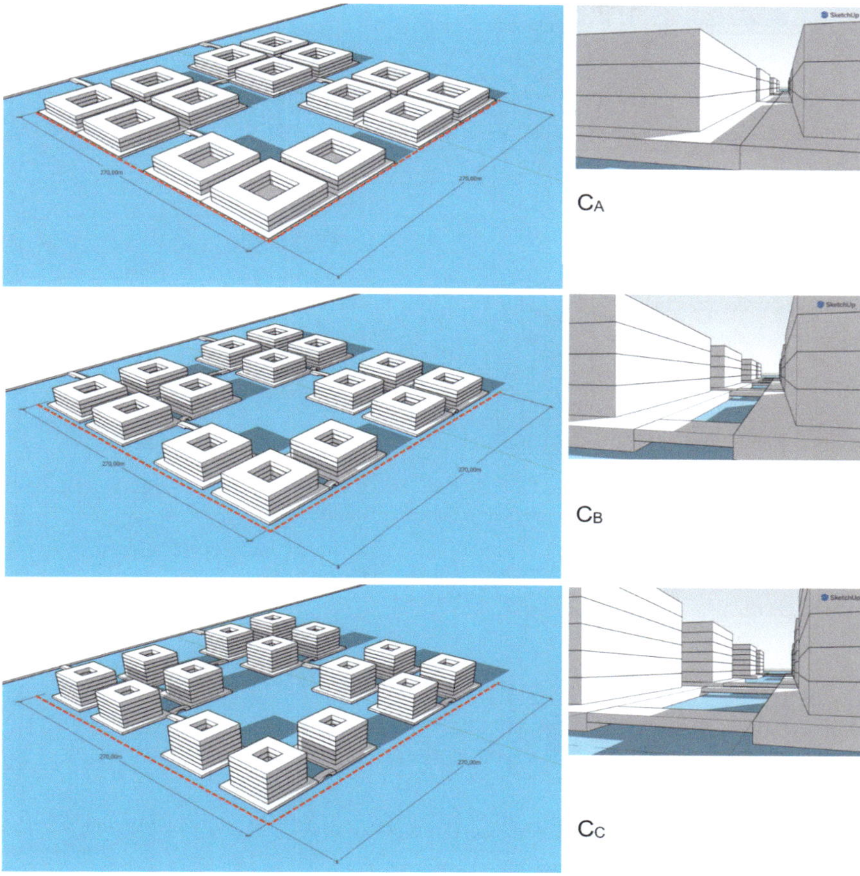

Fig. 2 Left: Overview of block configurations with blocks C_A, C_B and C_C. Right: street view with shading (LAT 45°, 2:00 P.M., March 21st). Configurations have the same FSI

bottom-fixed structure is intended to simulate the reflection of waves on a coastal structure. In this study, four different compositions are analysed:

- Isolated floating neighbourhood (C_X);
- Floating neighbourhood with a vertical quay wall (C_X + QW);
- Floating neighbourhood with a floating breakwater (C_X + BW);
- Floating neighbourhood with a vertical quay wall and floating breakwater (C_X + BW + QW).

All wave calculations consider head waves.

Table 1 Block typologies and dimensions

	3-floor (C_A)	4-floor (C_B)	6-floor (C_C)
Platform dimensions (L × W × H)	60 × 60 × 6 m	51.5 × 51.5 × 6 m	43 × 43 × 6 m
Building dimensions (L × W × H)	50 × 50 × 10.5 m	41.5 × 41.5 × 14 m	33 × 33 × 21 m
Building depth (m)	12	11.3	9.8
Floor Space Index (FSI)[a]	1.05	1.05	1.05
Ground Space Index (GSI)[a]	0.4	0.3	0.2
Open Space Ratio (OSR)[a]	0.6	0.7	0.8
Uncovered water area ratio (%)	31	49	64
Covered water area ratio (%)	69	51	36
Floor size to platform area ratio	1.5	2.1	3.0
Platform weight (ton)	4105	3026	2111
Building weight (ton)	3879	3819	3779
Platform freeboard (m)	3.8	3.5	2.9

[a] Platform (basement) space is not included in the calculations

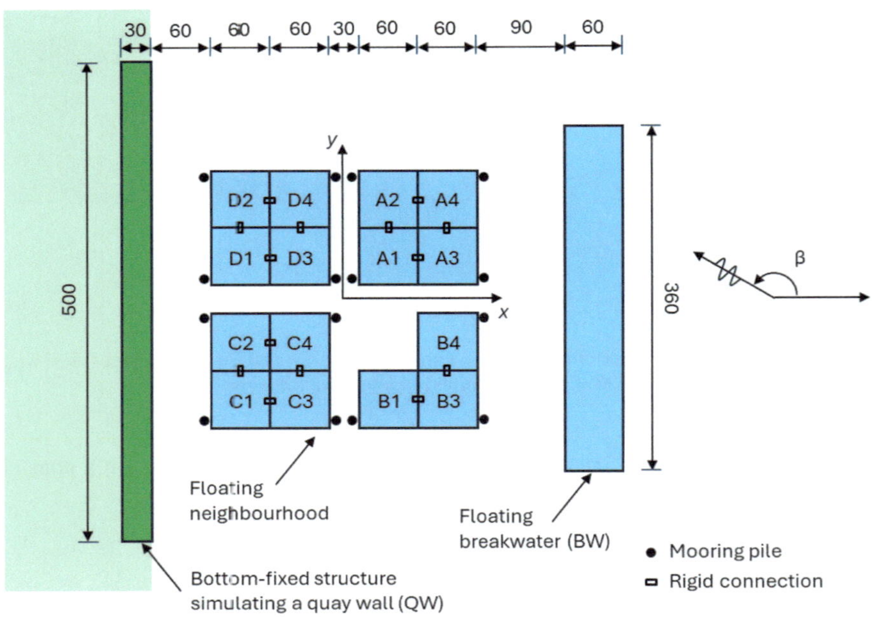

Fig. 3 Top view of platform configuration scheme with Module IDs, bottom fixed structure simulating a quay wall (QW) and a floating breakwater (BW). Dimensions are in meters

2.2 Assumptions

In the analysis, the following assumptions are made:

- All configurations have one N-S and one E-W canal. Platforms with three-floor buildings are assumed to have no gaps in-between them. 4-Floor and 6-Floor block typologies are assumed to have gaps around all platforms.
- The water depth is 15 m.
- The wave-structure interaction is based on potential flow theory, considering linear wave theory and small floating body motions.
- Platforms within the 4- or 3-unit clusters are connected to neighbouring platforms via a connector with very high stiffness that restricts relative motions, thus acting as a rigid connector.
- The horizontal displacement of each platform is limited by mooring piles simulated by horizontal springs.
- The floating breakwater (BW) has a draft of 9 m.
- The quay wall (QW), which has a width of 500 m, is assumed to be long enough to represent the wave reflection effect of a long vertical wall on the floating neighbourhood, even though the wall end effects are present for large wave periods.
- The shading analysis does not consider effects such as light scattering/ absorption, refraction angle variations due to waves or turbidity.

3 Results

This section details the steps within the proposed approach and the results of the Stability, Motion, Comfort and Structural Safety analyses, as well as the Shading analysis.

3.1 Stability

The stability analysis evaluates the ability of a floating structure to maintain equilibrium and prevent overturning when subjected to environmental loading, such as those caused by waves, currents or wind. To analyse the stability of the platform, the case of a single module with the building installed is considered. This is the least favourable situation due to the reduced waterplane area, as rigidly connecting the modules enhances the assembly's stability. This case typically occurs during transport of modules to their final location. To evaluate stability, we follow the stability criteria suggested by Wang et al. [23], which are based on IMO's intact stability criteria for pontoon platforms [9]. The following three criteria are considered:

- The area under the righting lever curve to the angle of maximum righting lever should not be less than 4.6 m deg.
- The static angle of heel due to a uniformly distributed wind load of 540 Pa (corresponding to a wind speed of 30 m/s) should not exceed an angle corresponding to half the freeboard.
- The minimum range of stability should be 20 degrees.

The effect of a horizontal wind load acting on a module, and its effect on reducing freeboard is shown in Fig. 4. From the figure, it is also possible to deduce that imposing a heel angle produces a righting moment proportional to the righting lever GZ. The righting lever curve as a function of the heel angle is presented in Fig. 5, for the single module of the three configurations (C_A, C_B, C_C). For positive heel angle values, a positive righting lever indicates that the module is stable and able to return to still water equilibrium. In the case of a negative righting lever, the module will overturn. The curves show the positive impact of wider platforms on stability, as the righting lever increases with the platform width. The summary of stability results is presented in Table 2, where the three stability criteria are analysed. All three criteria are satisfied for all three platform-building arrangements.

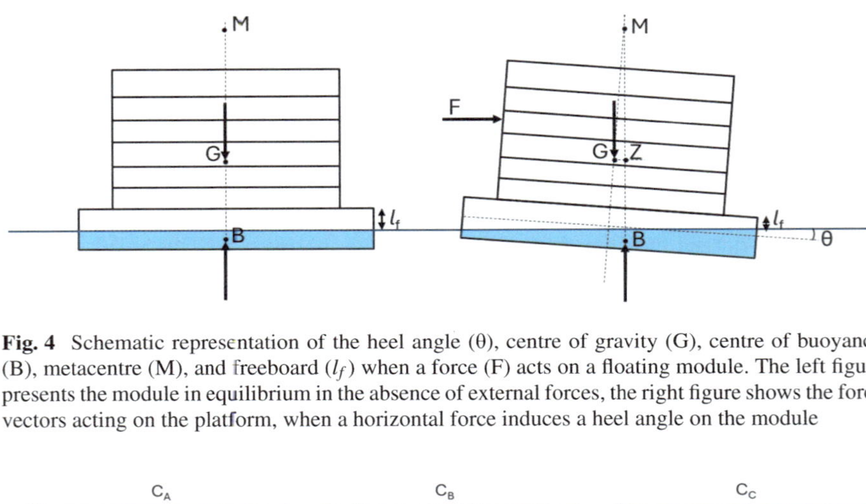

Fig. 4 Schematic representation of the heel angle (θ), centre of gravity (G), centre of buoyancy (B), metacentre (M), and freeboard (l_f) when a force (F) acts on a floating module. The left figure presents the module in equilibrium in the absence of external forces, the right figure shows the force vectors acting on the platform, when a horizontal force induces a heel angle on the module

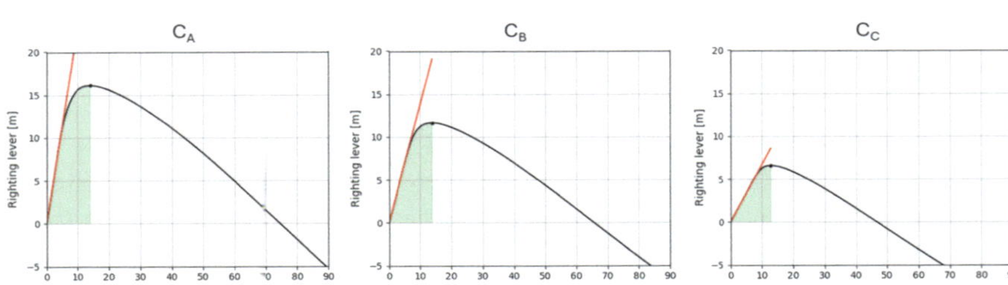

Fig. 5 Righting lever (GZ) as a function of the heel angle, for the 3 different block typologies. The red curve represents the slope of the curve at the origin

Table 2 Intact stability results for individual modules of the 3 configurations

Config	Platform length (m)	Platform height (m)	Freeboard (m)	z-coord. CoG (m)	Wind-induced angle (m)	Freeboard reduction with wind (%)	Max. range of stability (deg)	Righting lever area (m deg)
C_A	60	6	3.84	6.1	0.12	2.0	74.5	161.0
C_B	51.5	6	3.48	7.17	0.35	5.9	65.7	108.3
C_C	43	6	2.89	9.49	1.41	23.5	46.7	52.1
Intact stability criteria						< 50	> 20	> 4.6

3.2 Motion

The analysis of comfort is focused on the interaction of incoming waves with the floating structure. The software HydroMEC+, based on potential flow theory, linear wave radiation and diffraction forces, is used to compute the motion of floating platforms in waves. The motion of the modules is obtained by solving a system of linear equations in the frequency domain. To obtain statistically representative values for the dynamics of floating structures in irregular-wave sea states, a stochastic (spectral) approach is used, which is based on the superposition of elementary regular wave components. This assumption implies that the instantaneous free-surface elevation at a given position can be described as a stationary, ergodic and Gaussian process [6, 20]. The irregular-wave sea state is characterized by a JONSWAP spectrum (with a shape factor of 3.3) defined by a significant wave height H_S and a peak period T_P [6].

Figure 6 presents the normalized root-mean-square (RMS) values of surge (displacement in the x-axis direction), heave (displacement in the vertical direction) and pitch (rotation around an axis parallel to the y-axis) as a function of the peak period. The figure shows the maximum value found for all 15 modules. The normalization of these values considers the RMS value of the incident wave free-surface elevation, which in this case is $\sigma_\eta = H_S/4$. Results shows the dependency of the response on the peak period. In general, the response magnitude increases with the peak period. The results show a relatively small surge motion due to the horizontal motion constraint imposed by the mooring piles. Heave and pitch seems to be slightly amplified by the presence of the quay wall (QW) and significantly reduced by the floating breakwater (BW).

To understand the behaviour of floating platforms and the wave fields around them, wave amplitudes were computed for a regular-wave scenario considering radiated, diffracted and incident waves. Figure 7 presents the wave amplitude ratio for the 4 different configurations and 3 platform-building arrangements, for head waves ($\beta = 180°$) and a regular-wave period of 6.67 s (wavelength equal to 62.9 m). The red areas represent wave amplification, and the blue areas represent wave attenuation. The presence of the quay wall (QW), located leeward of the floating neighbourhood generates reflections that amplify waves around the modules, when compared to

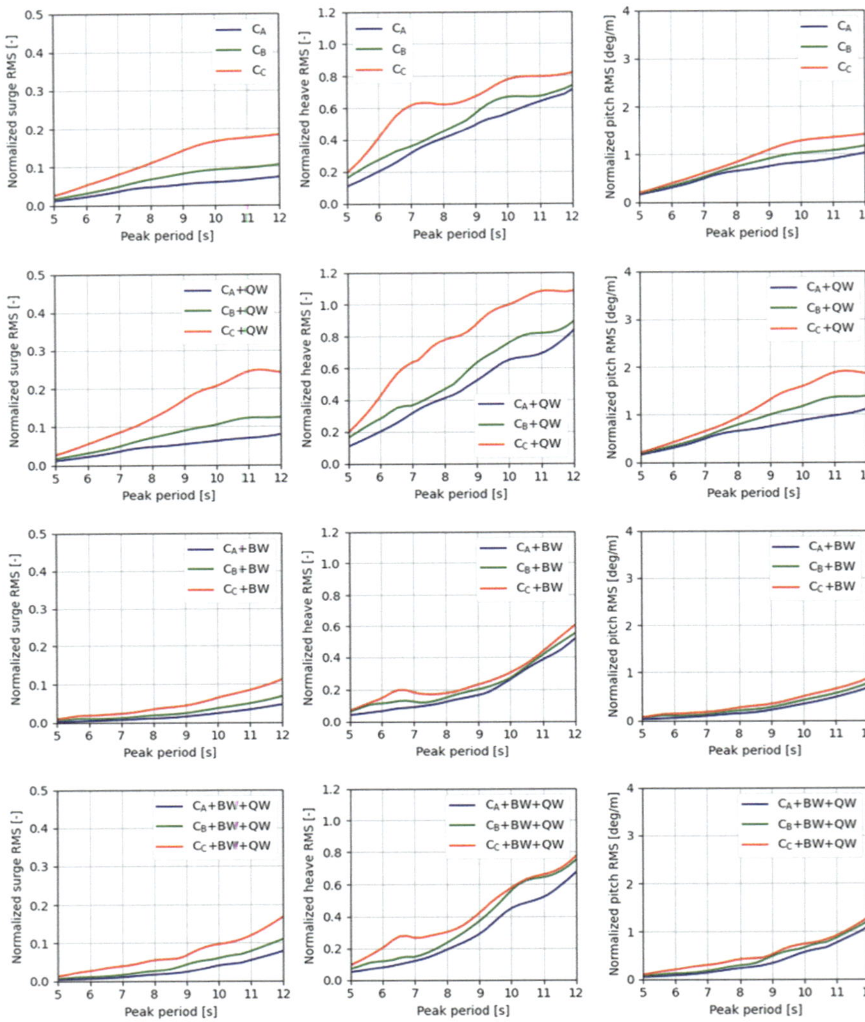

Fig. 6 Normalized maximum surge, heave and pitch root-mean-square (RMS) values as a function of the peak period, for all configurations

the base scenario of the isolated neighbourhood. This effect is also displayed in the motion analysis in Fig. 6, where the displacements are amplified by the presence of the quay wall (QW). Theoretically, wave reflection from an infinitely long vertical wall generates a standing wave in front of the wall with maximum amplitudes twice the incident wave amplitude, with a distance between maxima equal to half wavelength [4]. In this case, the end effect of the quay wall and the presence of other structures in the domain result in a more complex wave pattern even though the distance between amplitude maxima in the front of the wall is the expected half wavelength. The presence of a large floating breakwater (BW) with a length of 60 m and a depth of 9 m produces a significant wave attenuation effect on its leeward side. Such attenuation effect is also visible in the responses shown in Fig. 6, where the motions are clearly attenuated by the presence of the breakwater around the peak period of 6.67 s.

The wave fields presented in Fig. 7 show the phenomenon of gap resonance, which occurs when trapped waves in a gap between structures have their amplitude amplified at specific frequencies corresponding to the natural frequencies of the gap [13]. Such frequencies depend on the gap geometry and wave conditions [7]. The current scenarios have gaps between neighbouring modules within the same island (17 m for C_B and 34 m for C_C) and gaps between islands (30 m), which can contribute to the appearance of gap resonance. In Fig. 7, the arrangement C_B shows the highest amplification of wave amplitudes in the vicinity of the modules caused by the presence of the gap, for a regular-wave period of 6.67 s. This amplification is not observed in the motion results of Fig. 6 because gap resonance effects are very concentrated around the gap natural frequencies, showing a typically spiky variation over frequency, and the irregular-wave responses from Fig. 6 combine the contributions from a range of frequencies, thereby smoothing the effect. It is worth mentioning that the approach considered here, based on linear potential flow theory, has a tendency to overestimate the free-surface amplification near the resonant frequencies, as it neglects viscous and nonlinear effects [15].

3.3 *Comfort*

The evaluation of comfort requires the dynamic analysis of the whole urban development when subjected to waves, which was presented in the previous section. In this case, we consider unidirectional irregular waves characterized by a JONSWAP spectrum, defined by a significant wave height H_S and a peak period T_P. The irregular-wave response is computed through a spectral analysis based on the regular-wave response of the entire 15-module system. Comfort is evaluated here by measuring the accelerations in the usable areas within the platform and buildings. Since no guidelines exist to evaluate comfort in urban floating structures, two criteria are considered based on suggested satisfactory horizontal accelerations of building used for general purposes from the international standard ISO 6897 [10] and on seakeeping criteria for cruise liners according to NORDFORSK [17]. Overviews of existing guidelines that can be applied to assess the comfort in floating structures can be found in the literature (see [24] and [12]).

Figure 8 presents the variation of the maximum root-mean-square acceleration found within the usable areas of the floating neighbourhood with the peak period of a sea state defined by a JONSWAP spectrum, for a significant wave height of 1.5 m. The results vary significantly with the peak period. Vertical accelerations are significantly reduced with the floating breakwater. In general, the platform-building

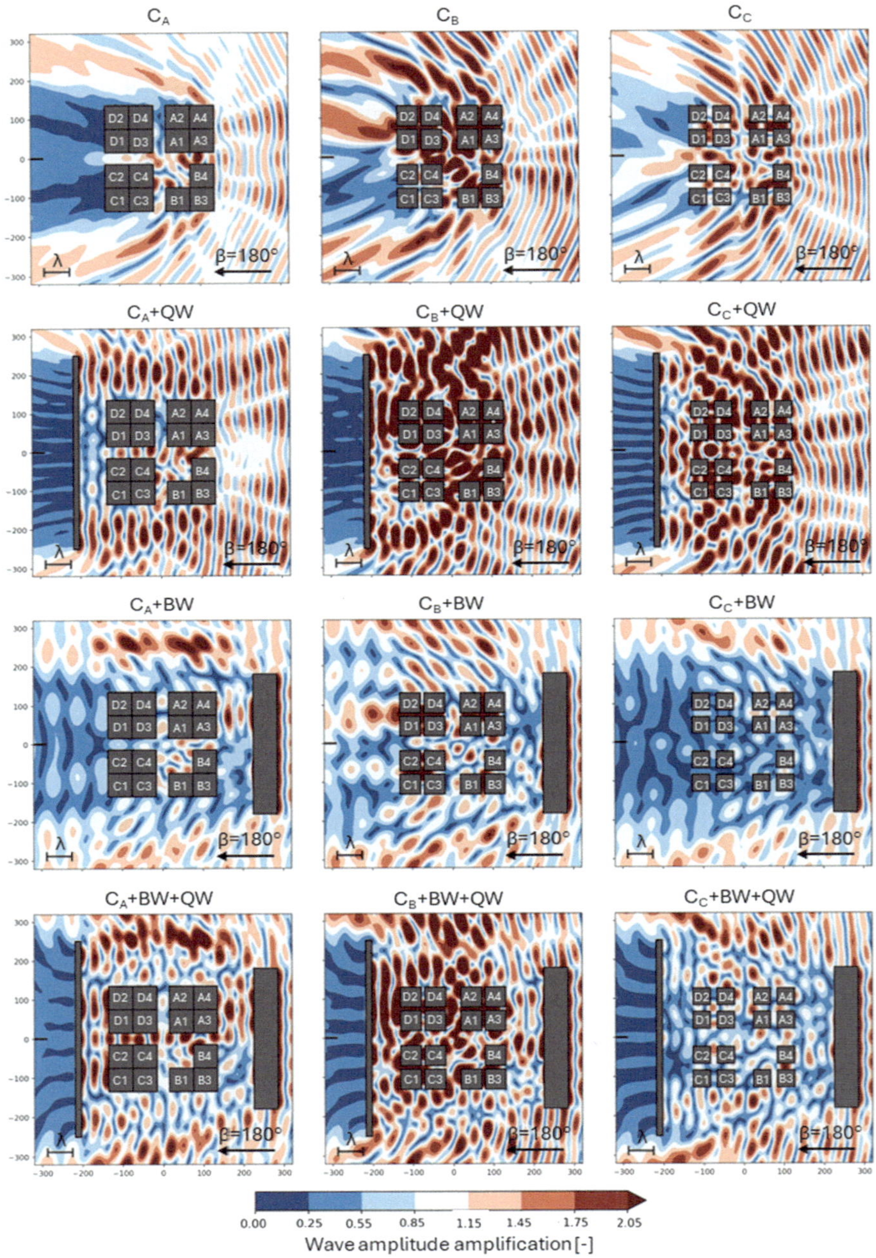

Fig. 7 Wave amplitude ratio for all configurations and compositions, under head waves (β = 180°), with a regular-wave period of 6.67 s. The columns represent a different platform-building arrangement of the floating neighbourhood and each row shows a different configuration, with and without a floating breakwater and quay wall. Blue areas represent wave attenuation and red areas display wave amplification

arrangement C_C presents the highest accelerations. Horizontal accelerations are relatively low due to the high horizontal stiffness of the mooring piles and connectors. All accelerations displayed in the graphs are below the horizontal acceleration limit recommended by NORDFORSK for cruise liners and by ISO 6897 for offshore fixed structures. For the platform-building arrangement C_A, the vertical acceleration suggested for cruise liners (NORDFORSK) is verified for $T_P < 7$ s for the configurations without floating breakwater and for all T_P for the cases with the floating breakwater. The other arrangements (C_B and C_C) present a slightly more restrictive range of T_P for the configurations without breakwater.

Table 3 presents the maximum root-mean-square acceleration on the platforms and its location for each platform configuration (see Fig. 3 for a schematic overview of Module IDs). These results consider a sea state defined by $H_S = 1.5$ m and $T_P = 6.67$ s. Using the most restrictive recommendations presented in Fig. 8 for horizontal and vertical accelerations, we observe that the comfort criteria are satisfied for C_A.

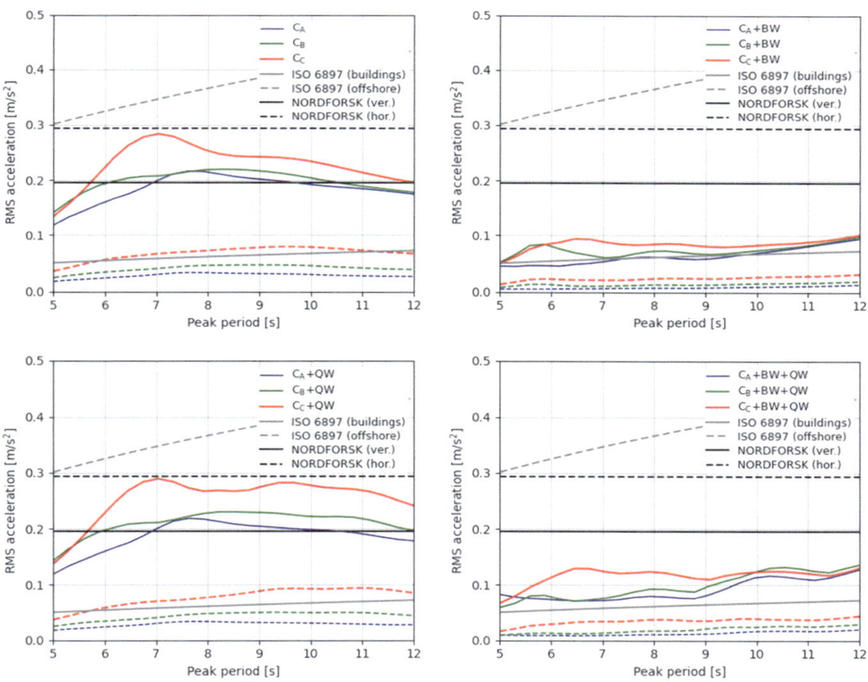

Fig. 8 Maximum root-mean-square acceleration found in the usable areas of the floating neighbourhood for each configuration, shown as a function of the sea state peak period. Blue, green and red solid lines represent vertical acceleration, while dashed lines indicatet horizontal acceleration. The graphs include reference curves with recommendations from ISO 6897 [10] for maximum horizontal accelerations in buildings and fixed offshore strictures, and from NORDFORSK for cruise liners [17]

Table 3 Evaluation of comfort based on the maximum root-mean-square acceleration measured in the three configurations, for 2 different compositions

Direction	Max. value	Configurations					
		C_A	C_A + QW + BW	C_B	C_B + QW + BW	C_C	C_C + QW + BW
x	RMS acc. (m/s^2)	0.028	0.008	0.038	0.012	*0.063*	0.034
	Module ID	B1	B1	B1	B1	B1	B1
y	RMS acc. (m/s^2)	0.008	0.009	0.014	0.011	0.038	0.026
	Module ID	B4	B3	B1	B4	B1	B4
z	RMS acc. (m/s^2)	0.184	0.071	*0.206*	0.073	*0.274*	0.129
	Module ID	B3	B4	B3	B3	B3	B3

Values in italic exceed acceleration values recommended by ISO 6897 [10] and NORDFORSK [17]

For platform-building arrangement C_B and C_C, comfort criteria are only satisfied in the floating breakwater configuration.

3.4 *Structural Safety*

The analysis of the structural safety of the platform presented here is focused on the evaluation of the bending moment acting on the floating structures. Here we consider still water bending moments and moments induced by wave action on the rigid connectors linking neighbouring modules. Figure 9 presents the still water bending moment as a function of the longitudinal coordinate. A single module is analysed, as these bending moment curves are equal for all modules of the same configuration. The bending moment depends on the location of columns, beams and load-bearing walls in the building. The curves show the benefit of having the building more evenly distributed along the platform, providing a better balance between gravity and buoyancy forces.

The still water bending moment is based on static loads, however the motion of the floating structures in waves is significant, and dynamic bending moments can become considerable. For this reason, the bending moment caused by wave-structure interaction on the connectors between neighbouring modules was computed. Figure 10 shows the maximum vertical bending moment RMS value measured in the rigid connectors as a function of the peak period. The results show that the presence of the quay wall slightly increases the bending moment, and the presence of the floating breakwater reduces it substantially.

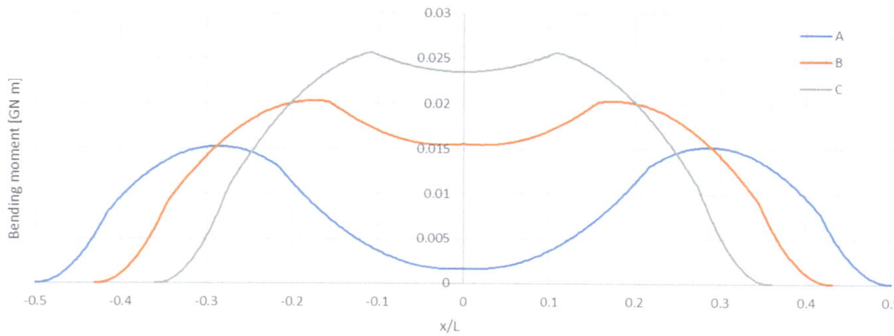

Fig. 9 Still water bending moment along the *x*-axis coordinate for the 3 different block typologies

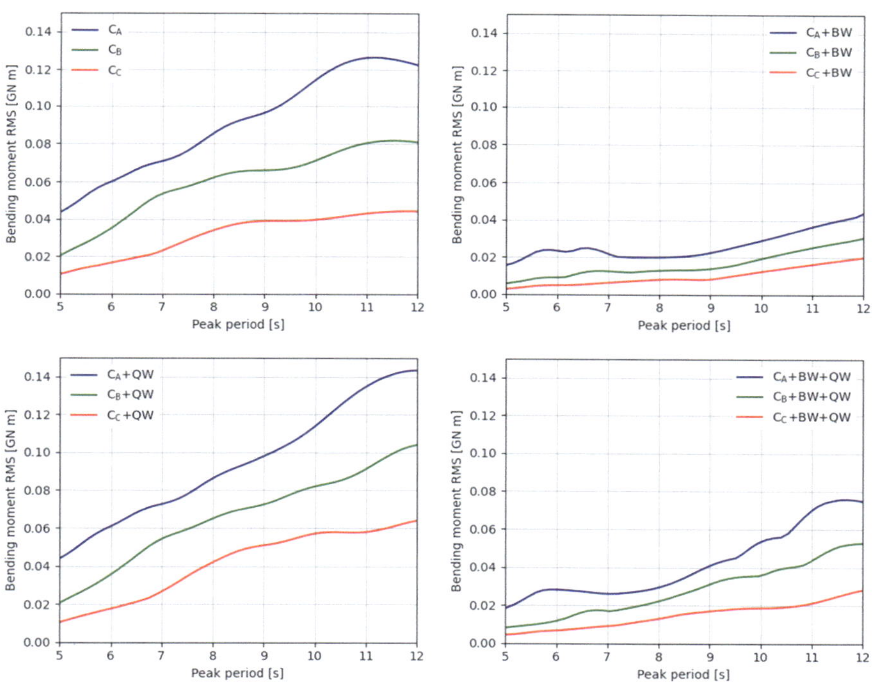

Fig. 10 Maximum vertical bending moment root-mean-square (RMS) value acting on the rigid connectors between neighbouring modules as a function of the irregular-wave sea state peak period, for all four configurations, for a significant wave height of 1.5 m

Among the platform-building arrangements, C_A has the highest dynamic bending moment for the range of peak periods presented. For the configurations without a breakwater, the RMS value of the dynamic bending moment is higher than the still water bending moment. The comparison between still water bending moment and

RMS bending moment should be carefully considered as the RMS value does not represent a peak value of the bending moment.

3.5 Underwater Shadow Range Analysis

For buildings, shading analyses are often conducted to understand how light access in the surroundings is impacted. A floating structure will cast shadows over the water surface and its shading will influence the underwater light intensity. For shading above water, existing shadow analysis tools can be used. For underwater shadows, however, the behavior of light passing through the water medium must be considered.

According to Snell's Law, light (or any other wave) changes direction when it passes from one medium into another medium with a different refractive index. This principle is fundamental in optics and can be stated as follows:

$$n_1 \sin \theta_1 = n_2 \sin \theta_2$$

where

- n_1 is the refractive index of the first medium,
- θ_1 is the angle of incidence (the angle between the incident ray and the normal to the surface at the point of incidence),
- n_2 is the refractive index of the second medium,
- θ_2 is the angle of refraction (the angle between the refracted ray and the normal to the surface at the point of refraction).

A schematic overview of the Law of Refraction is illustrated in Fig. 11. To account for the alteration of light ray paths upon entering the water column, a novel solar model has been developed by the authors. This model uses NOAA Solar Calculations as input to determine the sun position based on the time of year. With this input along with the platform dimensions and the water depth, a simplified 2D shadow contour is projected on the bottom of the water body. For a given day, multiple shadows are cast at regular intervals of time to generate a 'shadow range' diagram.

Figure 12 illustrates the shadow range diagram for the three block types A, B and C and the respective configurations. Shadows are cast on the spring equinox (March 21st) at intervals of 2 h. Colors from light blue to dark purple show the duration of the shade, which can vary from less than 2 h to more than 8 h. The diagrams show that platforms with smaller footprints allow a larger water area to be accessed by direct sunlight. Placing platforms spaced from each other also seem to have benefits in increasing light access. In the N-S direction, shadows are shorter but tend to persist for longer.

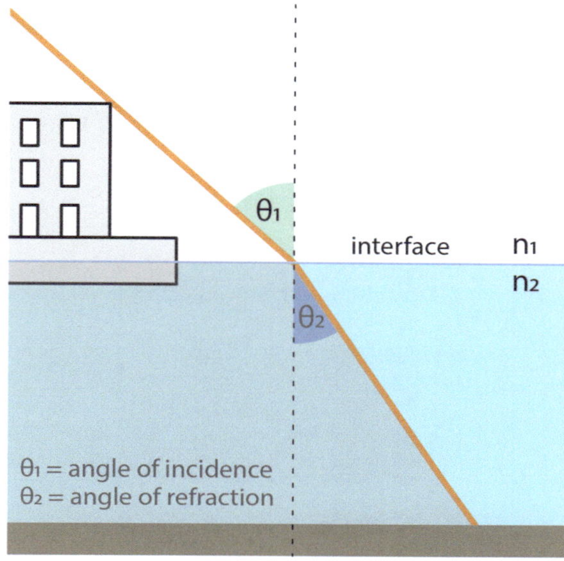

Fig. 11 Representation of Snell's Law of refraction

4 Overview and Discussion of Results

A comparison of the results for the three configurations (C_A, C_B and C_C) for the case including a floating breakwater (BW) and a quay wall (QW) is presented in Table 4 considering a sea state with $H_S = 1.5$ m and $T_P = 6.67$ s. All block types meet the stability criteria for the single modules. Regarding comfort, the maximum vertical accelerations within each configuration meet the standard for cruise ships. According to Wang et al. [24], vertical acceleration is the primary cause of discomfort at sea. Contrary to horizontal accelerations, there are currently no standards for onshore buildings that can serve as reference for evaluating vertical accelerations. Vertical accelerations in Table 4 are below the standard for cruise ships. As the design develops and comfort is to be addressed in further detail, Motion Sickness Dose Value (MSDVz) could also be considered as proposed by Wang et al. [24].

When considering the structural requirements, it can be concluded that less compact structures present higher bending moments and therefore require heavier structural elements.

Regarding shading effects, three-floor blocks have the largest footprint, resulting in a greater seafloor area impacted by prolonged shading. On the spring equinox, shadow range analysis at a latitude of 45° shows that approximately 77% of the seafloor within the platform footprint remains shaded for over 8 h. In contrast, this area decreases to 49% for configurations with six-floor blocks, which feature larger gaps between structures. While shading by floating structures is known to influence water temperature and primary production [3] drawing definitive conclusions about these impacts requires site-specific data. Kitazawa et al. [11], investigated shading

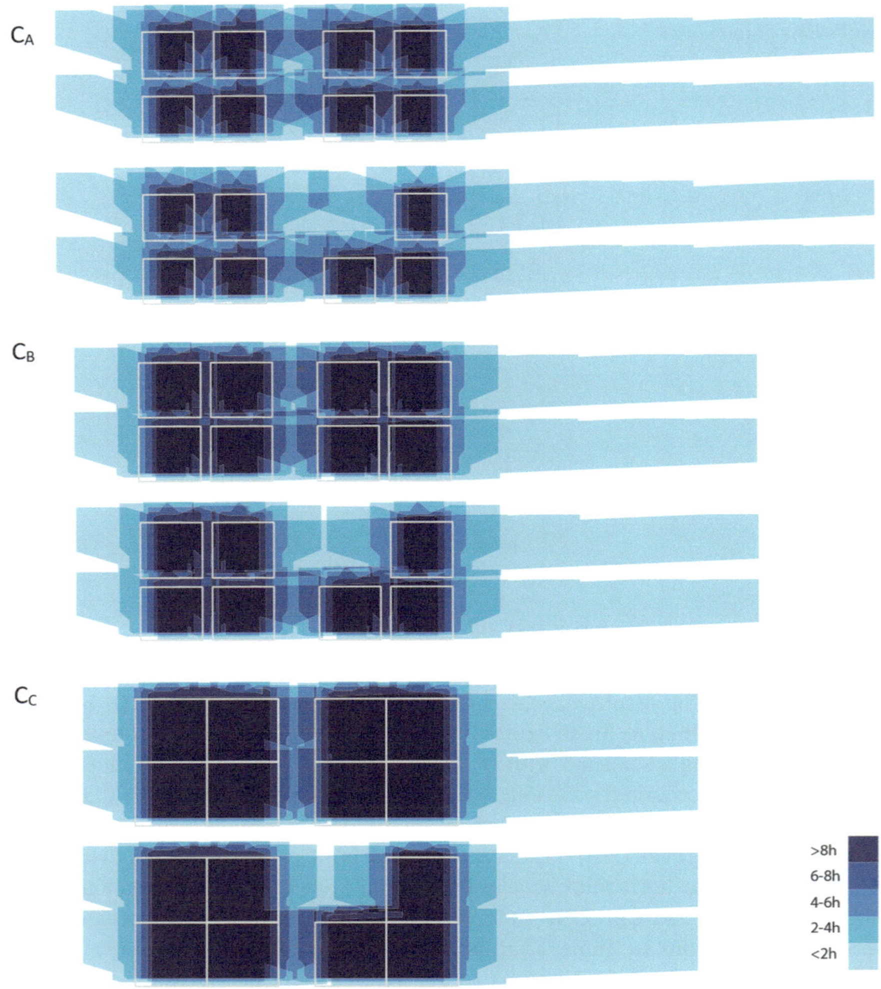

Fig. 12 Representation of the shadow patterns for the analysed 3 block typologies and configurations

effects around the Mega-Float structure in Tokyo Bay and found that the impact of the 120 m wide, 1 km long platform was primarily confined to the upper 5 m of the water column. Key variables, such as water exchange rates and organic matter productivity, are crucial for predicting the effects of platforms covering part of the water surface. Further research on large floating structures is essential to deepen our understanding of these impacts and inform evidence-based design guidelines.

This study defines three blocks with equal FSI, arranged in three similar configurations. The results indicate there is a trade-off to be made between meeting the engineering requirements for stability, structural safety and comfort, the water coverage

Table 4 Overview of results per block type

	3-floor (C_a + BW + QW)	4-floor (C_b + BW + QW)	6-floor (C_c + BW + QW)
Stability	Verified	Verified	Verified
Vertical RMS acceleration (comfort) (m/s^2)	0.071	0.073	0.129
Relative dynamic bending moment (%)	100	64	31
Area with > 8 h of shading versus platform footprint[a] (%)	77	62	49
Area with > 8 h of shading versus development area[a] (%)	57	34	19

[a] Calculated throughout the spring equinox, assumed on March 21st

by the presence of platforms and the light accessibility under the platforms (which influences primary production). While the more compact configuration with larger gaps (Configuration C_C with the 6-Floor block) allows light access to reach a much larger part of the sea bottom and present less demanding dynamic loads between modules, it also presents slightly higher vertical accelerations. Configuration C_B may be a compromise, as it still allows light to reach the seafloor in-between platforms and presents a better level of comfort. While spacing platforms may be beneficial, closer spacing may lead to wave reflections and interference between platforms, potentially amplifying wave heights and impacting platform stability. This study also briefly addressed the wave behaviour in the gaps between platforms. The phenomenon of gap resonance should be carefully considered in the design of these structures.

5 Limitations, General Applicability and Recommendations for Future Research

The presented integrated approach to the design of floating urban developments can be applied to both coastal regions and inland water bodies. A limitation of the selected case study is its reliance on hypothetical conditions, which limits the specificity and applicability of its findings. Location-specific results could be achieved by selecting an existing site and allowing for a thorough investigation of site-specific environmental conditions, water quality, and ecological characteristics. Such analyses would provide crucial data for better predicting the impacts of shading by floating structures (such as turbidity, water exchange rates, etc.).

The tool HydroMEC+, specifically developed to conduct engineering feasibility studies for floating urban developments, proved instrumental in the design and structural analysis of these projects. However, alongside engineering-focused tools, there is a critical need for the development of complementary tools that can predict the environmental effects of design choices. The solar analysis tool presented here offers an initial assessment of the spatial extent of long-lasting shadows, but further development is necessary. Future advancements could include modeling light intensity under and around floating structures, as well as flow modeling.

For these tools to provide meaningful and actionable insights, it is essential to establish design guidelines supported by comprehensive water quality and ecological monitoring data from existing projects. Detailed modeling simulations are also needed to predict how larger-scale, yet-to-be-built floating structures will interact with their surrounding environment, both above and below the water surface. By integrating engineering and environmental tools, urban planners and developers can create floating developments that not only meet human needs but also safeguard and enhance the surrounding natural environment.

While the current discourse on floating urban development has predominantly focused on addressing technical challenges—such as structural stability, hydrodynamics, connectors, etc.—it is equally important to extend research into the spatial design aspects of these developments. Key areas of focus for spatial design research should include the connection of these new urban areas with pre-existing settlements (on land), the design and functionality of public spaces, and the overall arrangement of settlements. This should consider not only environmental conditions but also the intended uses, mobility patterns, and the broader social and cultural context.

6 Conclusions

The presented approach marks an important step forward in considering floating urban development from a multidisciplinary perspective. Historically, research in this area has often been siloed, with a focus on either engineering challenges, architectural design, or the impacts of floating structures on water quality and ecosystems. However, the complexity and interconnectedness of these factors necessitate an integrated approach that bridges these disciplines.

As floating developments are planned to be scaled up, it is essential to integrate environmental considerations from the concept phase. By testing different configurations, designers can examine how spatial arrangements within the built environment interact with waterfront areas. At the same time, hydrostatic and hydrodynamic analyses provides insights into the stability and the wave-structure interaction, respectively. The shadow range analysis provides insight into how floating structures may affect the underwater environment below and around them. Impacts can vary depending on local environmental conditions (e.g.: currents, turbidity, etc.).

As floating developments scale up, monitoring of underwater conditions is essential to build better models for the environmental analyses and to inform the design process.

References

1. Costanza R, Pérez-Maqueo O, Martinez ML, Sutton P, Anderson SJ, Mulder K (2008) The value of coastal wetlands for hurricane protection. Ambio 37(4):241–248. https://doi.org/10.1579/0044-7447(2008)37[241:tvocwf]2.0.co;2
2. Czapiewska K, Roeffen B, Dal Bo Zanon B, Lin F-Y, Härtwich H (2017) Environmental assessment framework for floating development
3. De Lima RLP, De Graaf-van Dinther RE, Boogaard FC (2022) Impacts of floating urbanization on water quality and aquatic ecosystems: a study based on in situ data and observations. J Water Climate Change 13(3):1185–1203. https://doi.org/10.2166/wcc.2022.325
4. Dean RG, Dalrymple RA (1991) Water wave mechanics for engineers and scientists. World Scientific, Singapore ; Teaneck, NJ
5. Ferrario F, Beck MW, Storlazzi CD, Micheli F, Shepard CC, Airoldi L (2014) The effectiveness of coral reefs for coastal hazard risk reduction and adaptation. Nat Commun 5:3794. https://doi.org/10.1038/ncomms4794
6. Goda Y (2010) Random seas and design of maritime structures, 3rd edn. World Scientific, Singapore
7. He G, Jing P, Jin R, Zhang W, Zhang J, Liu T (2021) Two-dimensional numerical study on fluid resonance in the narrow gap between two rigid-connected heave boxes in waves. Appl Ocean Res 110:102628. https://doi.org/10.1016/j.apor.2021.102628
8. Intergovernmental Panel On Climate Change (2022) The ocean and cryosphere in a changing climate: special report of the intergovernmental panel on climate change, 1st edn. Cambridge University Press. https://doi.org/10.1017/9781009157964
9. International Maritime Organization (1993) Code on intact stability for all types of ships covered by IMO instruments. Resolution A.749(18) adopted on 4 Nov 1993
10. International Organization for Standardization (1984) ISO 6897:1984—Guidelines for the evaluation of the response of occupants of fixed structures, especially buildings and off-shore structures, to low-frequency horizontal motion (0.063 to 1 Hz), Geneva, Switzerland
11. Kitazawa D, Tabeta S, Fujino M, Kato T (2010) Assessment of environmental variations caused by a very large floating structure in a semi-closed bay. Environ Monit Assess 165(1–4):461–474. https://doi.org/10.1007/s10661-009-0959-9
12. Lin F-Y (2019) A catalogue of technical requirements and best practices for the design. D7.2, Space@Sea project No. D7.2
13. Molin B (2001) On the piston and sloshing modes in moonpools. J Fluid Mech 430:27–50. https://doi.org/10.1017/S0022112000002871
14. Moon CH (2015) 플로팅 건축, 새로운 건축 패러다임 [Floating architecture, a new architectural paradigm]. 이음스토리
15. Ning D, Su X, Zhao M, Teng B (2015) Hydrodynamic difference of rectangular-box systems with and without narrow gaps. J Eng Mech 141(8):04015023. https://doi.org/10.1061/(ASCE)EM.1943-7889.0000935
16. Nobre R, Boulêtreau S, Colas F, Azemar F, Tudesque L, Parthuisot N, Favriou P et al (2023) Potential ecological impacts of floating photovoltaics on lake biodiversity and ecosystem functioning. Renew Sustain Energy Rev 188:113852. https://doi.org/10.1016/j.rser.2023.113852
17. NORDFORSK (1987) Assessment of Ship performance in a seaway: the nordic co-operative project: 'seakeeping performance of ships'. Nordforsk, Copenhagen, Denmark

18. Okamura H, Kikutake T, Daiki K (2000) "メガフロート空港の研究開発 [Megafloat airport research and development]
19. Pont MB, Haupt P (2023) Spacematrix: space, density and urban form—revised edition. TU Delft OPEN Books, TU Delft OPEN Books, https://doi.org/10.59490/mg.38
20. Sarpkaya T, Isaacson M (1981) Mechanics of wave forces on offshore structures. Van Nostrand Reinhold, New York, NY
21. Sengupta D, Choi YR, Tian B, Brown S, Meadows M, Hackney CR, Banerjee A et al (2023), Mapping 21st century global coastal land reclamation. Earth's Future 11(2):e2022EF002927. https://doi.org/10.1029/2022EF002927
22. UNEP (2022) Sand and sustainability: 10 strategic recommendations to avert a crisis
23. Wang G, Goldfeld Y, Drimer N (2019) Expanding coastal cities—proof of feasibility for modular floating structures (MFS). J Clean Prod 222:520–538. https://doi.org/10.1016/j.jclepro.2019.03.007
24. Wang G, Rosenfeld Y, Drimer N, Goldfeld Y (2021) Occupant comfort analysis for rigid floating structures—methodology and design assessment for offshore dwelling module. Ships Offshore Struct 16(2):184–199. https://doi.org/10.1080/17445302.2020.1718267

Barbara Dal Bo Zanon is a senior architect at Blue21, a Dutch multidisciplinary team of experts in water-based urban development. She graduated in 2010 from the IUAV University of Venice (Italy) with a master's degree in Sustainable Architecture. Since 2011, Barbara has been actively involved in research, concept design, and project management for (inter)national projects on floating development. Her work focuses on environmental design, urban metabolism, and resource-efficient systems in water-based urban development. Currently pursuing her Ph.D. within the Floating Future project at the Department of Urbanism of TU Delft, Barbara's research explores the upscaling of floating urban developments as a forward-thinking adaptation to the challenges posed by climate change.

Rui P. F. Gomes is a lead engineer at Blue21 (2021–present), specializing in the design of innovative floating structures and sustainable solutions for water-based living. His work focuses on hydrodynamics, structural design, hydroelasticity in very large floating structures, and modular floating systems, addressing pressing global challenges such as climate change and rising sea levels. Rui earned an M.Sc. in Mechanical Engineering (2007) and a Ph.D. in Sustainable Energy Systems (2013) from Instituto Superior Técnico (IST), University of Lisbon. His academic and professional expertise spans floating structures, marine renewable energy, and fluid mechanics. Rui has made significant contributions to the development of wave energy converters, advancing key areas such as control, turbomachinery, optimization, numerical modelling, and experimental testing. These contributions have established him as a respected authority in the fields of ocean engineering and renewable energy. Before joining Blue21, Rui worked as a researcher at IST, University of Lisbon, focusing on marine renewable energy and participating in several European-funded projects. He also served as an Invited Assistant Professor in the Department of Mechanical Engineering at IST, where he mentored and educated future engineers.

Rutger de Graaf-van Dinther is a visionary entrepreneur and researcher specializing in water management and floating urbanization. With a background in civil engineering and a passion for sustainable urban development, Rutger has been at the forefront of innovative solutions for building on water, addressing the challenges of climate resilience and urban growth. As co-founder and director of Blue21 (2015–present), Rutger spearheads global advancements in floating infrastructure, following his earlier entrepreneurial success with DeltaSync (2007) and Indymo (2015–present). His projects include the Floating Pavilion Rotterdam (2010), cited by the IPCC as a transformative climate adaptation model, and the Bluelands Floating Platform prototype (2023) at TU Delft's Green Village.

Rutger's research credentials are equally impressive, with an MSc (2005) and Ph.D. (2009) in Civil Engineering from Delft University of Technology, both completed cum laude. He served as Applied Research Professor on adaptive urban development at Rotterdam University of Applied Sciences (2011–2021) and co-initiated the Floating Future Research Project, securing €5.3 M from the Dutch Research Council (2023).

Beyond academia and entrepreneurship, Rutger co-founded the Blue Revolution Foundation (2013) to promote interdisciplinary research on floating solutions and the Thinktank on Floating Urban Development (2017) to address governance challenges. A prolific author and speaker, Rutger has delivered keynotes on five continents, contributed to over 30 publications, and featured in global media outlets like The New York Times, BBC, and Financial Times. His work continues to shape the future of sustainable urbanization worldwide.

Brain Coral: Floating City in the Maldives

Koen Olthuis, Sridhar Subramani, and Mila Avellar Montezuma

Abstract The Maldives Floating City (MFC), an innovative collaboration between Maldivian government, Waterstudio and Dutch Docklands, offers a groundbreaking response to the pressing need for climate-resilient urban solutions in the Maldives. Guided by a structured framework of ten foundational parameters, the MFC seamlessly integrates environmental preservation with economic growth, adapting sustainable urban practices to an offshore setting uniquely suited to the Maldives' geographic and climatic context. This chapter examines the MFC's development through these ten core principles, which shaped its design, operational strategies, and adaptive approaches. The framework addresses critical aspects of floating urbanism, including resilience to sea-level rise, ecological harmony, community inclusivity, and renewable energy systems. Key features—such as modular spatial design, wave-buffering barrier island reefs, and robust socioeconomic integration—highlight the project's potential to achieve long-term urban sustainability. By applying these guiding principles, the MFC establishes a replicable model for future floating developments, adaptable to various environmental and cultural contexts. This holistic approach not only consolidates the project's critical components but also paves the way for the widespread adoption of floating and amphibious technologies, marking a pivotal moment in the evolution of climate-adaptive urban design.

Keywords Climate resiliency · Floating urbanism · Adaptive design · Maldives Floating City

K. Olthuis (✉) · S. Subramani
Waterstudio.NL, Rijswijk, The Netherlands
e-mail: j.olthuis@waterstudio.nl

S. Subramani
e-mail: s.subramani@waterstudio.nl

K. Olthuis · S. Subramani · M. A. Montezuma
ABE, TU Delft, Delft, The Netherlands

M. A. Montezuma
IHE Delft Institute for Water Education, Delft, The Netherlands

B. T. Wang et al. (eds.), *Large Floating Solutions*, Lecture Notes in Civil Engineering 636, https://doi.org/10.1007/978-981-96-5435-2_3

1 Introduction

As coastal areas face increasing flooding and rising sea levels, the demand for resilient infrastructure is expected to grow significantly. This trend could drive the development and adoption of floating and amphibious housing, particularly in vulnerable regions such as low-lying territories [4, 7]. Floating and amphibious structures—among humanity's oldest methods of habitation and mobility—are poised to become widespread and mainstream. A tipping point for their adoption [3] could occur if current environmental, economic, and social demand trends persist, including sea-level rise, more frequent extreme weather events, and rapid advancements in material technologies [2, 8].

The acceleration of urban population growth and intensifying climate change could hasten this shift, either as a proactive preparedness measure or, in the event of a sudden disaster, as a reactive game-changer. Conversely, if environmental threats recede, the urgency for widespread adoption might diminish, delaying the tipping point [5]. In either scenario, floating and amphibious infrastructure represents a critical strategy for addressing the challenges of an uncertain future.

In the Maldives, Sea Level Rise poses serious risks to infrastructure, the economy, livelihoods, and public health. Coastal communities face heightened threats of displacement, income loss, and reduced access to essential services. Nevertheless, the study highlights various adaptation strategies to bolster resilience, such as coastal protection efforts, sustainable land-use planning, livelihood diversification, and community-driven initiatives [1]. Tourism is the main driver for the country's economic growth, it accounts for about 10% of global GDP and employs one in ten people worldwide [6]. In 2023, the economy expanded by 4.1%, with a notable 9.8% growth in the first quarter of 2024. Tourist arrivals reach 1.3 million by August, with projections suggesting a record-breaking 2 million arrivals in 2024, marking an 8.6% increase over 2023 [9]. To sustain this momentum and foster economic resilience, it should incentivize innovative city development approaches—especially by aligning tourism with strategic urban zoning dedicated to tourism growth and sustainable infrastructure.

The Maldives Floating City (MFC) serves as a testing ground for integrating cutting-edge floating and amphibious technologies with practical, climate-adaptive urban solutions. Positioned as an extension of Male, the MFC aims to accommodate the Maldives' unique environmental challenges and support the country's growth by attracting residents to live and work in a new type of water-based urban environment. By combining innovative wave-breaking structures, sustainable energy systems, and environmentally sensitive design, MFC addresses both climate resilience and the increasing demand for urban space near coastal cities, as well as their key requirements.

Moving from a narrative of "climate refugees" to "climate innovators," MFC exemplifies a proactive approach to transforming existential threats into opportunities for sustainable urban growth. As the Maldives' population growth begins to stabilize, the focus shifts towards improving living conditions while creating an appealing destination for global citizens. The MFC, designed for 5000 floating units across a 200-ha lagoon and set for completion by 2027, represents a concrete step toward scalable, large-scale floating urbanism. This project not only reimagines urban development in the Maldives but also lays a foundation for widespread floating urbanism as a viable, resilient solution to rising seas and limited land availability worldwide.

1.1 Problem Statement

How can floating urbanism provide scalable, resilient solutions to address land scarcity, rising sea levels, and sustainable city development in climate-vulnerable regions like the Maldives, given that existing urban planning frameworks are primarily land-based and lack provisions for floating environments? This calls for a new framework—guided by the 10 key requirements of the Maldives Floating City (MFC)—to set a precedent for adapting urban planning to the unique challenges of climate change and sustainable economic growth. The 10 key requirements and key focuses are given in Table 1.

1.2 Assignment

This assignment explores the development project of the MFC, a pioneering vision conceived through a collaborative effort between the Maldivian government, Dutch Docklands, and Waterstudio. The project's vision and conceptualization are rooted in leveraging floating structures to meet spatial and socio-economic demands, while fostering urban and maritime integration. MFC's design embraces modularity and flexibility, allowing for adaptable city systems that aim to house around 20,000 residents, generate employment, boost tourism, and provide economic opportunities. Situated within a naturally formed atoll, reinforced by surrounding barrier islands, the project incorporates a unique brain coral design, symbolizing environmental harmony and coral conservation.

The development encompasses several key milestones: from initial design phases and obtaining essential regulatory approvals to launching prototypes and advancing scaled development. This assignment will delve into each aspect of the MFC, examining how it accommodates a diverse range of stakeholders—including government officials, architects, environmental specialists, and developers—toward realizing a sustainable floating urban environment.

Table 1 10 key requirements and key focuses of the Maldives Floating City (MFC)

Domain	Requirement	Key focus	Priority level
1	Legality	Legal status, real estate classification, law enforcement	
2	Modularity	Prefabricated modular units, scalable and flexible design	
3	Accessibility	Multi-modal transportation, canals for internal mobility	
4	Mooring, stability, and safety	Mooring systems, structural stability, safety protocols	
5	Environmental impact management	Minimizing environmental footprint, enhancing marine life	
6	Power supply	Smart grids, renewable energy (solar, blue energy)	
7	Supply and waste management	Integrated waste and resource management systems	
8	Defense, law, and order	Establishing safety and governance structures	
9	Community-based city	Cultural integration, community engagement	
10	Flexibility and response time	Adaptable design for future challenges	

1.3 Background and Existing Conditions

The Netherlands has established multiple maritime zones in the North Sea and the Caribbean Sea, following the United Nations Convention on the Law of the Sea (UNCLOS). These zones—territorial sea, contiguous zone, exclusive economic zone, and continental shelf—define the country's rights, including natural resource exploitation and maritime traffic inspection.

Building up on the definition from the Dutch Ministry of Defence, a classification is set up for the Floating Constructions based on its legal regions—assuming that legality is the most important requirement among the 10 requirements. The Floating Legal Regions, based on the maritime zone are the following (see Fig. 1):

1. the territorial sea (up to 12 miles off the coast); Maldives Case (see Fig. 1);
2. the contiguous zone (between 12 and 24 miles off the coast);
3. the exclusive economic zone (up to 200 miles off the coast);
4. Internal waters;
5. Archipelagic waters (AW).

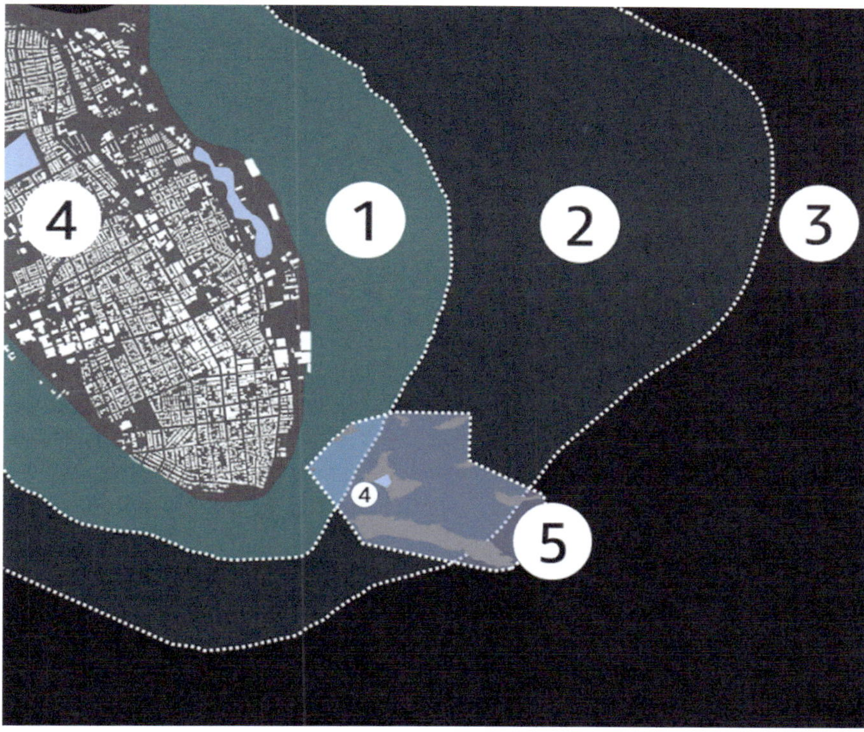

Fig. 1 Maritime Legal Regions which provide the legality scope for floating projects, in the following order: 1—Exclusive Economic Zones (EEZ); 2—Territorial Seas (TS); 3—Contiguous Zones (CZ); 4—Internal Waters (IW); 5—Archipelagic Waters (AW). *Source* Authors, based on the Dutch Ministry of Defense, 2024

Floating legal regions provide a classification framework that informs urban design by establishing location-based legal parameters for floating cities. In the case of the MFC, the project is situated within a leased atoll near Hulhumale, extending Male's urban footprint onto water. This lease arrangement and the protected status of nearby reefs establish guidelines that shape both spatial planning and environmental management, ensuring alignment with marine ecosystems. Site-specific analyses by Royal HaskoningDHV (RHDHV) further detail factors like water depth, wind patterns, and wave dynamics, which influence structural resilience and adaptability. This regulatory and environmental framework lays the foundation for balancing ecological sensitivity with innovative urban growth.

2 Methodology

This research examines the design process of the MFC through a research-by-design approach, applying a structured framework of ten key requirements that establish a foundation for a scalable and adaptable floating city model. These requirements address the critical challenges of climate resilience, environmental adaptability, and urban viability in response to evolving environmental, social, and economic conditions. As a pioneering example of large-scale floating infrastructure, the MFC demonstrates how these requirements can be synthesized into a cohesive model that meets the unique challenges posed by climate change and rising sea levels and provides a replicable pathway for similar developments worldwide.

The study focuses on three core domains within the Floating City Development Framework: Regulatory and Legal Integration, Design and Engineering, and Sustainability and Social Integration. In the regulatory domain, the framework emphasizes legal legitimacy for floating infrastructure, including real estate classification, legal protections, and governance. The design and engineering domain prioritizes modular construction, advanced power systems, and adaptable structures to ensure resilience and scalability. Finally, the sustainability and social integration domain promotes environmental stewardship, social cohesion, cultural relevance, and economic viability through integrated waste management, community development, and accessibility.

This methodology allows for the iterative testing and refining of each requirement (see Fig. 2) within the specific context of the Maldives, where unique climate and social challenges are present. The research ultimately offers a comprehensive, replicable model for future floating city developments, combining legal, technical, and social elements to support sustainable urbanism in regions at risk from climate change.

3 Key Requirements of Design Process

The Maldives Floating City's design is guided by 10 key-requirements that establish boundaries and ensure its feasibility within a floating urban framework. With a foundation in existing legal structures, the design incorporates modularity to allow scalability and easy adaptation, enhancing accessibility and integration with Male. Elements like mooring systems, stability measures, and safety protocols provide structural resilience in dynamic marine conditions, while environmental impact management ensures harmony with surrounding ecosystems. A reliable power supply, efficient supply and waste systems, and provisions for defense and law enforcement uphold functionality and security. Built as a community-centric city, the design fosters engagement, while flexibility and rapid response capabilities address evolving needs, making the city both adaptable and resilient.

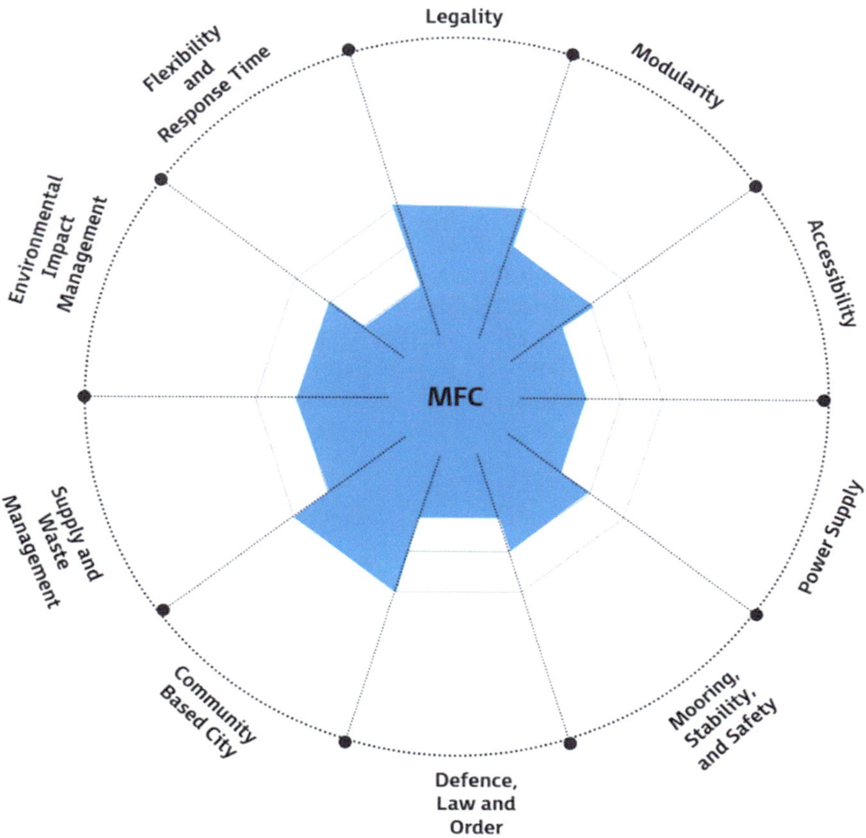

Fig. 2 Matrix of 10 key requirements for floating architecture and urbanism, analyzing the importance level of each of the parameters. *Source* Authors, 2024

3.1 Legality

A floating city must be part of the legal framework of a city or country. This gives the inhabitants a legal status. The floating homes and other structures must be labeled as real estate, like in Maldives, in order to make finance and insurance possible. To establish the MFC within a robust legal framework, site-specific factors and regulatory compliance are paramount. The location's bathymetry, existing coral habitats, wave exposure, and potential coastal erosion necessitate a design approach sensitive to both marine biodiversity and physical stability. MFC's primary function is to support tourism and provide sustainable housing options, aiming to attract diverse socio-economic groups by offering an innovative urban experience in a climate-adaptive city. Under the Developer Domain, MFC incentivizes foreign ownership, granting residence permits for property purchases, thus reinforcing its appeal as both an investment and a home base.

Public institutions, such as police and emergency services, are integrated to ensure safety and maintain order, reinforcing MFC's alignment with national standards. Moreover, the project adheres to relevant national and environmental laws. For instance, the Climate Emergency Act (Act no. 9/2021) and Maldives Climate Change Policy Framework establish foundational climate-resilience targets, aligning MFC with national climate commitments. Similarly, compliance with the Regulation on Protected Areas (2018/R-78) and Protected Species Regulation (2021/R-25) safeguards MFC's impact on protected marine and terrestrial species, while the Decentralization Law supports coordination with Atoll, City, and Island Councils, ensuring the city's sustainable integration with surrounding communities and ecosystems. Together, these policies underscore MFC's alignment with Maldives' legal standards, being legality the most determinant requirement among the 10 requirements (Fig. 3).

3.2 *Modularity*

The floating city's design is underpinned by modularity, ensuring that each element of the project is standardized and adaptable. The city's form is composed of prefabricated floating foundations, primarily in the form of tripod and rectangular platforms. These standardized platforms host both infrastructural services and modular superstructure housing units, enabling efficient construction and flexible expansion. The decentralized infrastructure systems—including water, sewage, energy, and communication—are organized by sectors, with each sector connected by bridges, allowing for seamless integration across the city while supporting efficient maintenance and reduced environmental impact.

This modular approach facilitates easy transportation and assembly on-site; thereby significantly minimizing ecological disruption. Construction is carried out in phases, allowing for gradual expansion while addressing the evolving needs of the city and its inhabitants. Mooring piles, installed on-site, anchor the fully constructed and equipped platforms that are towed into position. Structural spines—future roadways embedded with piping—connect the platforms, establishing the city's foundational grid. Housing units, assembled in modules of five homes, are similarly towed and connected to these spines (see Figs. 4a and b). Quays for water bus stops and bridges linking the platforms are then towed into place and integrated with the structural spines. This modular, phased construction process ensures scalability, adaptability, and efficient infrastructure management, aligning with the city's long-term vision for sustainable growth.

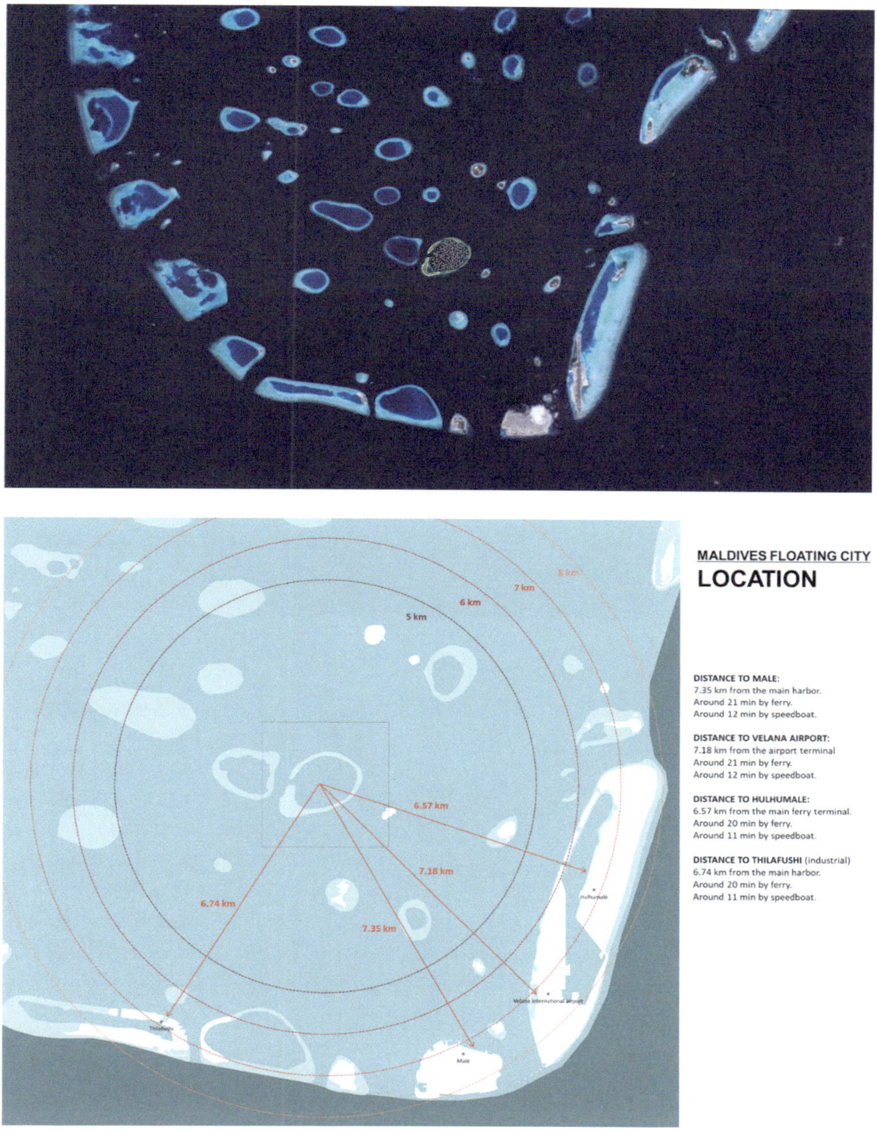

Fig. 3 Location of MFC within the broader archipelago (top) and location (bottom). *Source* Authors, 2024

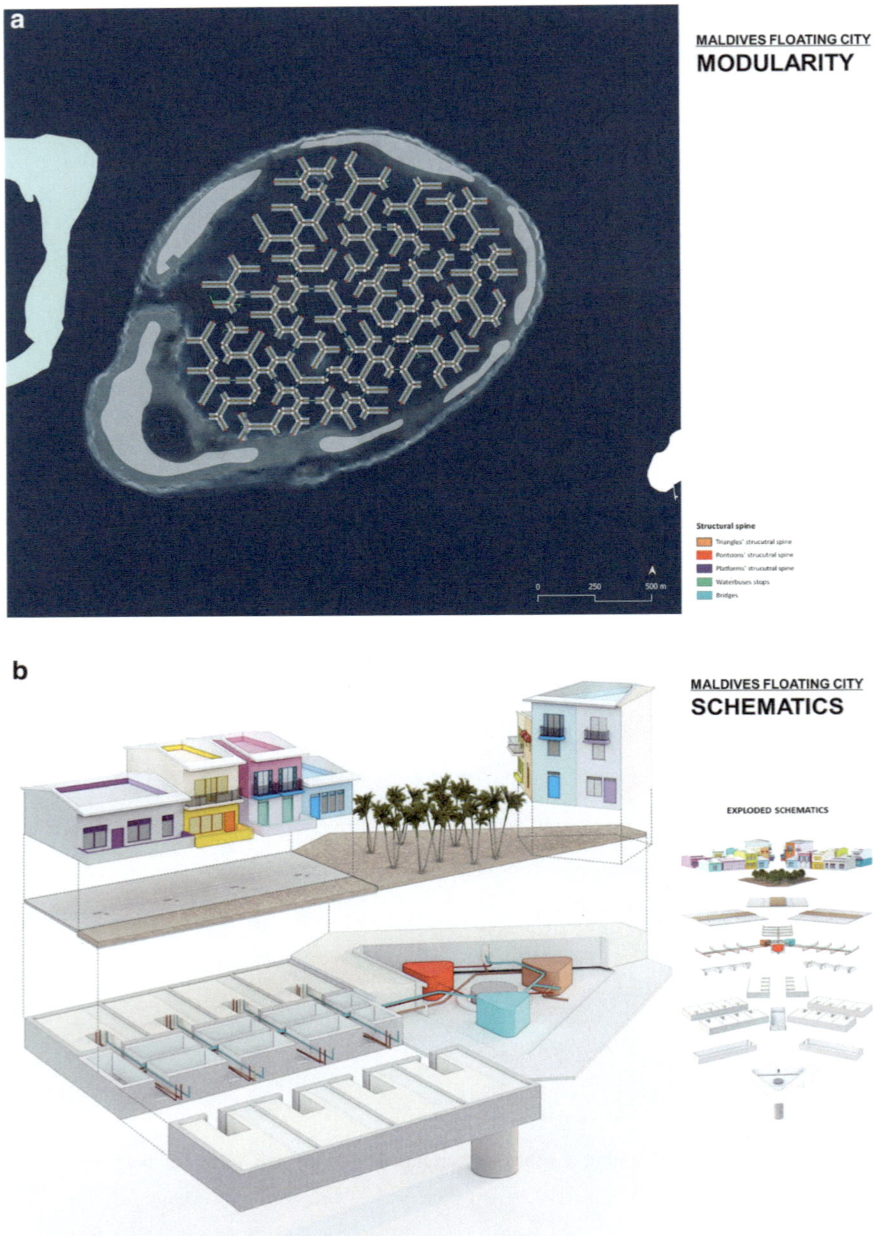

Fig. 4 **a** Modularity of MFC's floating foundation systems. **b** Modularity schematics of MFC's constructive systems

3.3 *Accessibility*

The MFC employs a structured, pedestrian-focused mobility infrastructure, creating a car-free, walkable environment where essential services are accessible. Approximately 32% of the island's area is allocated to navigating channels. Connectivity to Malé, Hulhumale, and its surrounding lagoons is achieved through three ferry terminals positioned at natural access points, where passenger and cargo flows are efficiently managed. The West and south terminals serve as the primary logistical hub, handling goods distribution to waterbus stations throughout the city as secondary hubs.

The city's waterway network consists of primary, secondary, and tertiary channels (see Fig. 5). Primary channels connect ferry terminals to Malé and surrounding islands, secondary channels link to waterbus stations within MFC, and tertiary channels provide local access within residential areas. Each sector includes one or two waterbus stations for comprehensive coverage.

The 18.5-km street network comprises three levels: boulevards and avenues (35%), main streets (20%), and residential streets (38%). Boulevards, 8 m wide, accommodate high pedestrian traffic with shops and cafes; main streets (also 8 m wide) distribute traffic toward residential zones, featuring spaces for street furniture; and 5-m residential streets serve as community spaces and playgrounds. Designated

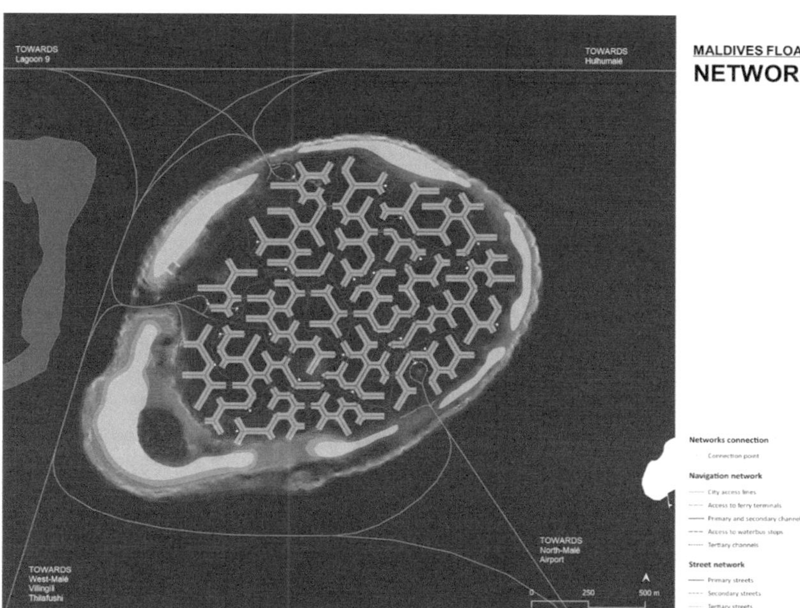

Fig. 5 Proposed accessibility network connections via navigation and within the streets

water access points ensure seamless integration with the water-based transit system, supporting sustainable mobility across the floating city.

3.4 Mooring, Stability, and Safety

The MFC's location near the equator minimizes exposure to cyclones, while natural atoll formations provide wave protection and enhancing stability. Designed to expand horizontally, the city's stability is further supported by tripods with telescopic piles that anchor the platforms and adjust with water levels (see Fig. 6). Rigidly connected tripods and rectangular platforms feature cushioned joints to dampen horizontal forces. Initial structural analysis confirms that the architectural design of pontoons and pile moorings is feasible. Most pontoons maintain stability without damping, but those near lagoon entrances may require additional damping, necessitating further assessment. Freeboard requirements are generally 1.0 m, with 1.25–1.5 m needed near entrances, achieved through wave walls. Pontoons exhibit acceptable movement levels for residential comfort, though additional analysis is needed at the entrances. Single pile moorings are feasible with diameters ranging from 2.5 to 3.5 m depending on water depth, requiring boring in deeper areas. Joint movements of 3 cm to 6 cm are anticipated, dependent on comfort standards and damping. Overall, these measures ensure that the city's mooring, stability, and safety systems meet the demands of its marine environment, supporting a secure and resilient floating infrastructure.

3.5 Environmental Impact Management

The floating city incorporates several strategies to minimize environmental impact and enhance local ecosystems. Coral reefs surrounding the city are widened to act as natural wave barriers, while the edges of floating platforms mimic reef structures to promote coral growth beneath (see Fig. 7). Each unit includes around 36 m2 of rooftop gardens designed to support local vegetation and foster biodiversity. The foundation's undersides are optimized to host seaweed and coral structures, enriching underwater habitats and reducing shadowing on the seabed.

Water management includes rainwater harvesting systems, with collection tanks integrated across the city. Cooling needs are met by a deep-water cooling system (see Fig. 7), utilizing water from depths of 200 m to maintain a consistent energy exchange. The city prioritizes native flora to support local biodiversity, addressing Maldives' limited plant diversity.

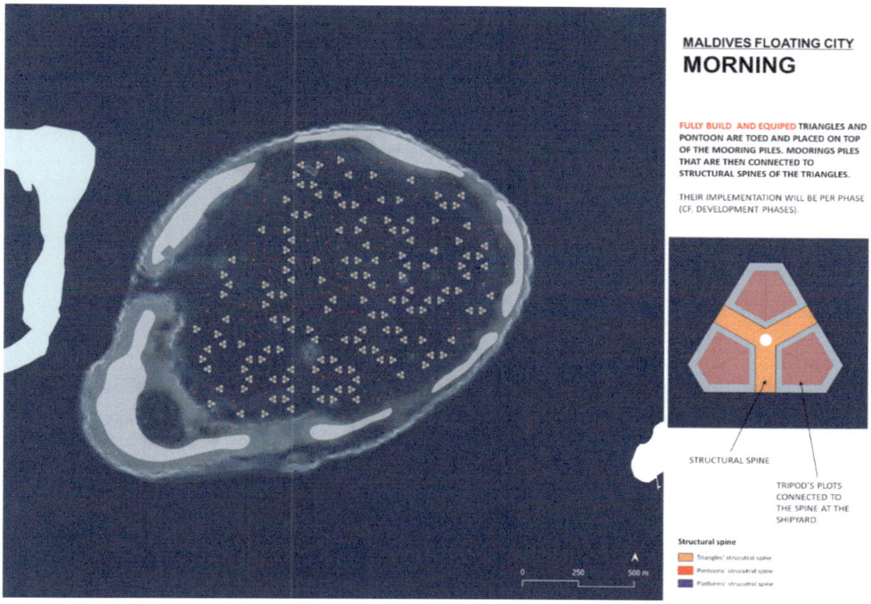

Fig. 6 Proposed mooring types, triangular and pontoons

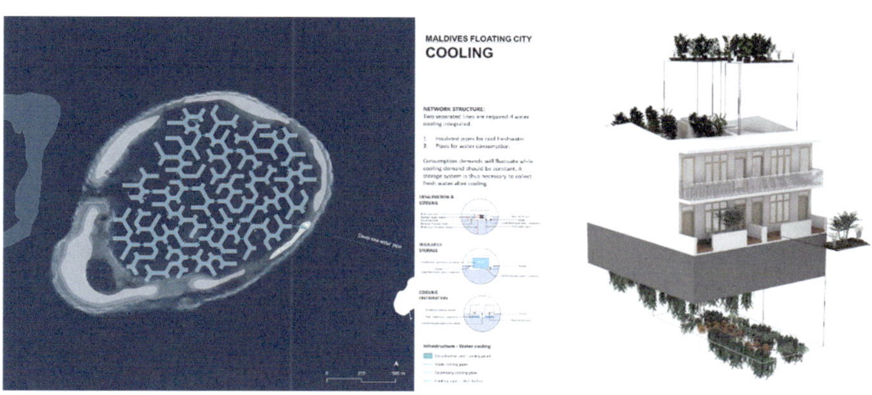

Fig. 7 Deep sea cooling system (left) and terrace garden and ecological growth underneath the platform (right)

3.6 Power Supply

Power and utilities are foundational to the viability of a floating city, with energy management focused on reducing consumption and generating green energy through a smart, interconnected grid. Estimated daily energy requirements for the city's

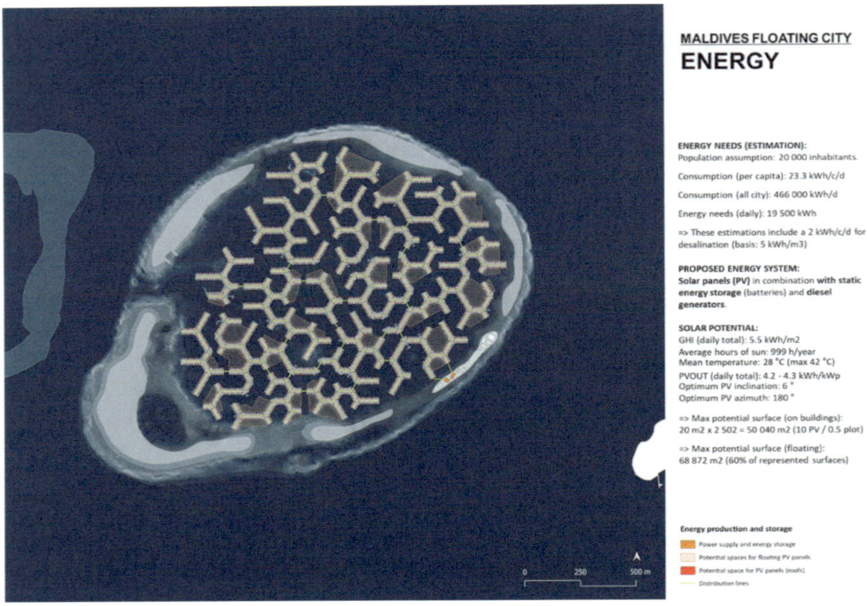

Fig. 8 Proposed power supply network, estimated energy needs, proposed energy system, solar potential

20,000 residents are around 466,000 kWh. To sustainably meet this demand, the city leverages solar panels on building's rooftop (see Fig. 8) generating 45% of the requirement. A combined surface area of 118,912 m^2 with open water area in the atoll and rooftop generates sufficient energy to meet the city's daily demands, ensuring reliable power for all essential functions. Integrating solar energy with static storage batteries and backup diesel generators, the smart grid efficiently shares and redistributes surplus energy between units, fostering a balanced, self-sustaining energy ecosystem critical for the city's long-term success.

3.7 *Supply and Waste Management*

Supply and waste management in the floating city incorporates innovative systems designed for sustainability, efficiency, and minimal environmental impact. Cooling is achieved through a deep-sea cooling system, with cool water arriving from desalination plants. Initially, this water is used to cool buildings via heat exchangers, thereby reducing the need for conventional air conditioning. Once it serves its cooling function, the water is made available for resident use and stored in strategically distributed tanks. These interconnected tanks not only provide a stable water supply but also

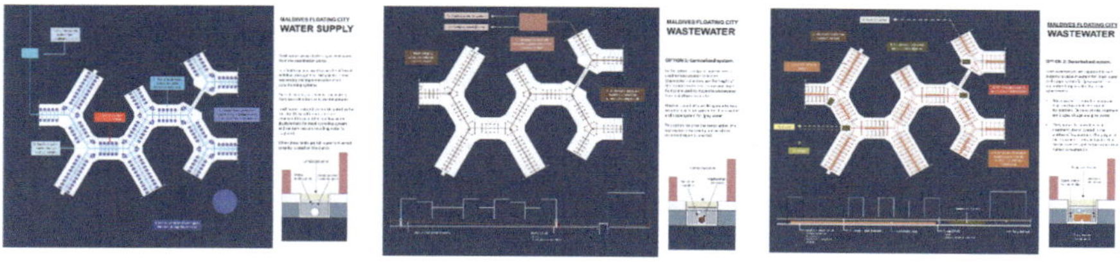

Fig. 9 Water Supply needs and network structure (left) and wastewater centralized (middle) and decentralized systems (right)

support city stability through a ballast system, preventing stagnation and maintaining sanitation. Excess water is redirected to tanks on nearby barrier islands.

The city's sewage system is designed with a black and gray water separation, with two treatment approaches under consideration: a centralized or decentralized system (see Fig. 9). In the centralized approach, all wastewater is collected via a singular pipe system, which relies on gravity and pumps to move wastewater between platforms. This collected wastewater is directed to a central treatment plant located on a nearby reclaimed island. The centralized system may also utilize separate vacuum and pipe networks to manage black and grey water flows independently. Alternatively, the decentralized system separates black and grey water directly at the source. Black water is treated in anaerobic digesters located within platform foundations, generating biogas and sludge as by-products. Meanwhile, grey water undergoes treatment in localized facilities, producing reclaimed water suitable for non-potable uses. Both treatment approaches are designed to produce clean water and usable by-products—such as sludge for potential fertilizer and reclaimed water for various applications—contributing to a closed-loop, sustainable approach to waste and resource management.

3.8 Defense, Law and Order

Safety and security in the floating city operate under the jurisdiction of Maldivian authorities, integrating with existing systems of law enforcement and emergency response already established in the nation's water territories. Police stations are strategically located to facilitate efficient boat patrols, ensuring that all city areas can be reached within a maximum travel time of 10 min. This floating city functions as an extension of existing neighborhoods in Malé, increasing the patrol area rather than requiring entirely new governance systems.

A centrally located fire station and police station in the middle sector further strengthen emergency response capabilities, ensuring rapid coordination for incidents across the city. Fire ambulances, adapted for water-based operations, provide specialized support for fire-related emergencies. This integration of safety infrastructure demonstrates how the floating city expands existing urban networks while maintaining effective and reliable security measures, ensuring that it functions as a natural and secure extension of the Maldivian urban system.

3.9 Community-Based City

The floating city is designed with a community-centric approach, reflecting the environment of the Maldivian islands while enhancing residents' quality of life. By incorporating elements of the existing inland landscape, the city creates vibrant communal spaces (see Fig. 10), such as squares and pedestrian-friendly pathways, that encourage social interaction and active lifestyles. Mixed-use development within the tripod structures combines residential, commercial, and public services, with essential facilities strategically located at street intersections to enhance accessibility. Economic opportunities are woven into the city's framework, with job creation in sectors like tourism, hospitality, and retail. Affordable housing options are integrated to accommodate a diverse resident base, including local fishing families, who are invited to maintain their cultural heritage within this new urban landscape.

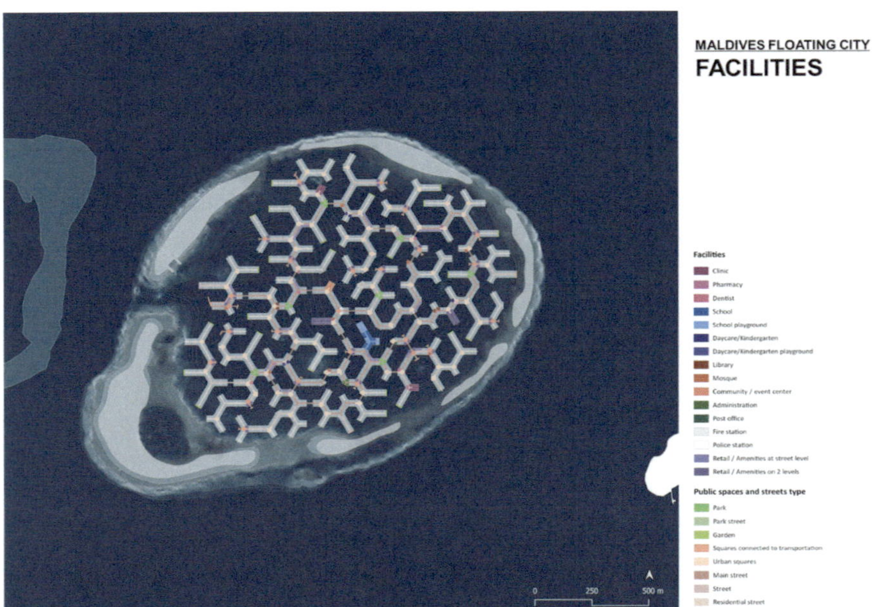

Fig. 10 Facilities, public spaces, and street type

3.10 *Flexibility and Response Time*

The city's modular layout enhances its adaptability, enabling incremental development through distinct phases (see Fig. 11). Comprising nine development phases, this modular approach allows for ongoing maintenance, expansion, and adjustments to the city layout as needs evolve. The modular design also supports rapid response to environmental or infrastructural challenges, ensuring that the city can adapt efficiently to both planned expansions and unforeseen events. This flexibility, combined with phased growth, fosters a resilient urban environment that can evolve with the needs of its residents and the ecological conditions of its unique setting.

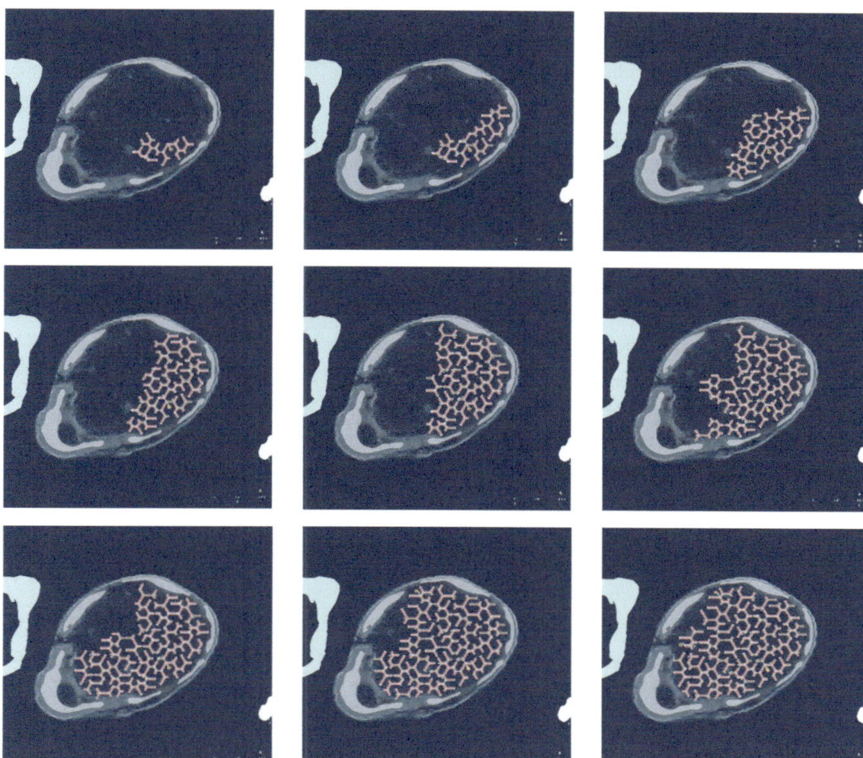

Fig. 11 Phased development of the city

4 Discussion and Conclusion

The Maldives Floating City (MFC) is not only a conceptual vision but a pioneering model in execution, with its first floating module already established. This project highlights the practicality of floating cities as viable urban typologies in the face of climate emergencies, particularly for populations most vulnerable to rising sea levels. By moving beyond traditional land-based urban solutions, MFC redefines urban development in climate-vulnerable regions, emphasizing adaptability, sustainability, and resilience.

The MFC's design is guided by 10 foundational requirements, tailored specifically to the Maldivian context. Anchored by a strong legal framework, these requirements address the country's unique socio-environmental and economic challenges, ensuring feasibility and alignment with national regulations. For example, the classification of floating homes as real estate facilitates financing and ownership, while compliance with environmental laws safeguards protected marine and terrestrial ecosystems. This process underscores the importance of site-specific frameworks that balance ecological sensitivity with urban functionality. While these requirements were designed for the Maldives, the framework provides a replicable methodology that can be adapted to other contexts, allowing future floating cities to address diverse challenges with context-specific solutions.

The development and testing of the MFC prototype provided critical insights, enabling the refinement of design decisions and adaptation to local conditions. For instance, the optimal layout solution reflected the Maldivian family structure of 4–6 persons, while the modular superstructure allowed housing units to be combined, mirroring traditional settlements. These design decisions were validated through community testing in Malé, demonstrating the importance of engaging stakeholders in the development process.

Environmental learnings from the prototype informed strategies for vegetation and cooling systems. The limited soil depth for planting larger vegetation highlighted the need to optimize planting depths while mitigating the "dead load" impact on foundations. Similarly, the potential for enhancing passive cooling through cross-ventilation became evident, reinforcing the project's commitment to reducing energy demand and operational costs. These insights emphasize the value of iterative testing during the early stages of development.

The MFC's phased construction and bankability framework are critical to its success. By integrating economic resilience with phased development, the project ensures adaptability and financial sustainability over time. This structured approach is supported by a self-reflecting matrix that ranks critical system parameters, allowing adjustments in response to evolving conditions. This adaptability positions MFC as a flexible solution for similar environments worldwide.

Significantly, the MFC demonstrates how climate threats can inspire innovation. By adopting a proactive approach, the project shifts the narrative from viewing vulnerable populations as "climate refugees" to "climate innovators." The strategic integration of the 10 key requirements ensures the city is both adaptable to its environment and a replicable blueprint for large-scale floating urbanism globally. This framework can be recalibrated to align with the specific needs and contexts of other regions, supporting the establishment of floating cities tailored to future challenges.

In conclusion, the MFC is a landmark initiative that transforms existential threats into opportunities for sustainable development. By combining technical innovation, socio-economic inclusivity, and environmental stewardship, the project exemplifies how floating urbanism can address pressing global challenges. More than a singular project, the MFC serves as a critical stepping stone, demonstrating the feasibility of floating cities as a mainstream urban solution. Its success not only validates the concept but also has the potential to accelerate the tipping point for widespread floating urbanism. The scalable and replicable framework developed for the MFC provides a pathway for integrating adaptability, resilience, and sustainability into urban planning, setting a precedent for future developments worldwide. The MFC offers not only a solution to immediate climate challenges but also a transformative model that redefines urban living, paving the way for cities worldwide to adapt and thrive in the face of rising seas and changing environmental conditions. By shortening the timeline for the mainstream adoption of floating cities, the MFC positions itself as a catalyst for global innovation in sustainable urban development.

References

1. Ali MH, Jaleel M (2024) Socioeconomic ramifications of sea level rise in the Maldives: a holistic assessment of impacts and adaptation strategies. J Environ Geogr Stud 3(1):1–16
2. Abou Chakra M, Bumann S, Schenk H, Oschlies A, Traulsen A (2018) Immediate action is the best strategy when facing uncertain climate change. Nat Commun 9, Article 2566. https://doi.org/10.1038/s41467-018-04968-1
3. Fuchs S, Thaler T (2017) Tipping points in natural hazard risk management: how societal transformation can provoke policy strategies in mitigation. J Extreme Events 4(1):1750006. https://doi.org/10.1142/S2345737617500063
4. Kappenthuler S, Seeger S (2019) Addressing global environmental megatrends by decoupling the causal chain through floating infrastructure. Futures 113:102420. https://doi.org/10.1016/j.futures.2019.04.007
5. Lenton TM (2013) Environmental tipping points. Annu Rev Environ Resour 38:1–29. https://doi.org/10.1146/annurev-environ-102511-084654
6. Matzana V, Oikonomou A, Polemis M (2022) Tourism activity as an engine of growth: lessons learned from the European Union. J Risk Financ Manage 15(4):177. https://doi.org/10.3390/jrfm15040177
7. United Nations (2019) Sustainable floating cities can offer solutions to climate change threats facing urban areas. In: Deputy Secretary-General tells first high-level meeting. UN Press Release. https://press.un.org/en/2019/dsgsm1269.doc.htm
8. Winkelmann R, Donges JF, Smith EK, Milkoreit M, Eder C, Heitzig J, Katsanidou A, Wiedermann M, Wunderling N, Lenton TM (2020) Social tipping processes for sustainability: an analytical framework. SSRN. https://doi.org/10.2139/ssrn.3708161
9. World Bank (2024) Maldives development update, October 2024: seeking stability in turbulent times. World Bank, Washington, DC. http://hdl.handle.net/10986/42238

Koen Olthuis With his family roots in shipbuilding, architecture, and engineering, Dr. Koen Olthuis founded Waterstudio in 2003 as the first architectural firm fully dedicated to floating architecture. As a senior researcher in Floating Cities and Delta Urbanism at Delft University of Technology, he and his team explore strategies and solutions for The Netherlands to expand onto water as part of the country's adaptation to its below-sea-level geography. Dr. Olthuis advises governments and cities worldwide on combating the effects of climate change and urbanization. His signature projects include the Floating City in the Maldives, as well as floating parks, houses, and theaters for canals, lakes, and open oceans. He is the author of FLOAT! and City Apps, which explore innovative water-based urban solutions for communities across economic spectrums, aiming to enhance the resilience and performance of cities near water.

Sridhar Subramani Currently pursuing his PhD in Delta Urbanism at Delft University of Technology, his research focuses on floating dynamics, aiming to enhance urban performance through floating solutions and multi-agent modeling. For the past seven years, he has worked with Waterstudio.nl, contributing to research and computational design. With expertise in urban planning, supported by data analytics and advanced tools, he now focuses on data-driven urban and generative design concepts. His work enables developers and clients to fine-tune custom parameters, optimizing, analyzing, and making informed decisions based on hundreds of data points and simulation results.

Mila Avellar Montezuma is an Architect, Urbanist, and Landscape Designer (UFPE) with an M.Sc. in Water Science and Engineering, specializing in Sustainable Urban Water Management and Climate-Resilient Cities, from UNESCO-IHE, TU Delft, and IHS Erasmus. She also holds a postgraduate degree in Urban Heritage Strategies for Water Challenges from IHS, TU Delft, and RCE. Passionate about advancing climate change adaptation, Mila employs a research-by-design approach focusing on nature-driven technological solutions. Her diverse experience spans public, private, and nonprofit sectors, leading and contributing to transdisciplinary projects involving prototyping, laboratories, and international workshops. She has developed water-adaptive projects in Brazil, China, Bangladesh, and the Netherlands. As a practitioner and researcher, Mila emphasizes resilience and adaptation, collaborating with institutions such as TU Delft, UNESCO-IHE, WaterStudio, UFPE, and Rede Clima MCTI.

Mega Floating City: Green Float

Donghee Ko

Abstract The world is facing a range of critical challenges linked to population growth, including increasing demands for energy, water, and food, rising CO_2 emissions, and the depletion of natural resources. Covering 70% of the Earth's surface, the oceans present significant potential for solutions to these global challenges. The "Green Float" project, developed by Shimizu Corporation in Japan, introduces an innovative concept for sustainable, floating urban environments. This project aims to create self-sufficient, zero-carbon, and zero-emission cities on artificial islands. The envisioned structure comprises a 1000-m-tall mega-tower and a 2000-m-diameter floating base, strategically located near the Equator to reduce risks from waves and strong winds while optimizing renewable energy potential. Construction planning and technical development have been carried out to realize the floating structure offshore. Extensive research, including hydraulic model experiments and numerical simulations, has demonstrated the feasibility and safety of a scaled-down version, Green Float II, designed for shallow coastal waters. These studies confirm the structure's resilience against extreme weather conditions and tsunamis, supporting the project's vision of sustainable and innovative urban development.

Keywords Mega floating city · Green Float · Construction planning · Hydraulic model experiment · Numerical computation

1 Introduction

Green Float is a new conceptual model for an environmentally friendly island city and is under development by the Shimizu Corporation, Japan (Fig. 1) [1]. The idea is to construct artificial islands on the ocean with the aim of achieving a self-sufficient, zero-carbon, zero-emission city. This project is a mega floating circular structure with a diameter of 2000 m, and it carries a high-rise building that is approximately 1000 m tall.

D. Ko (✉)
Shimizu Corporation, Tokyo, Japan
e-mail: dh.ko@shimz.co.jp

B. T. Wang et al. (eds.), *Large Floating Solutions*, Lecture Notes in Civil Engineering 636, https://doi.org/10.1007/978-981-96-5435-2_4

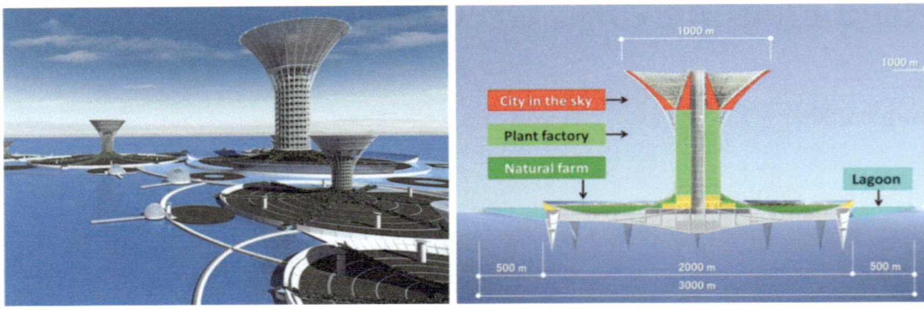

Fig. 1 Mega floating city Green Float

To improve our technological capabilities, the Shimizu Corporation developed Green Float II which is roughly one-tenth the scale of the original planned Green Float. Several experiments and numerical computations have been conducted to assess feasibility with respect to safety against winds, waves, and tsunamis.

1.1 Why Create Floating Cities?

The world is currently facing dual population explosions: the overall growth of the Earth's population and the excessive concentration of people in urban areas. This has led to rapid overcrowding and significant disruption in cities. Environmental challenges related to energy, water, food, CO_2, and natural resources are overwhelming the planet's capacity to manage them.

At the Shimizu Corporation, we have turned our focus to the oceans that cover 70% of the Earth's surface and embarked on the design of a mega floating city. Many unresolved problems on land can be more effectively addressed by developing these floating cities at sea; considered the last frontier on Earth. With the population explosion in developing countries, over 500 new cities are being planned worldwide, primarily on land. Hence, it is crucial to explore the potential of the Green Float mega floating city, which aims to understand, coexist with, and utilize the ocean's potential.

1.2 Green Float: A Mega Floating City

This artificial island, with a diameter of 3000 m, floats on the ocean. It features a 500-m-wide lagoon around its periphery, resulting in a net land area diameter of 2000 m. The skyscraper city rises to 1000 m tall, with the city in the sky (Fig. 2) starting at 700 m above sea level. The middle levels house a plant factory, making the island self-sufficient in food. The ground level supports facilities such as a natural farm, playground, and beach resort (Fig. 3).

The peripheral area of the city in the sky serves as a residential zone for 30,000 to 50,000 inhabitants, while the central area functions as a business zone with offices and research centres. This zone is designed mainly for research, development, and branch offices of companies engaged in biotechnology, renewable energy, and marine enterprises. Residents will have the opportunity to enjoy the lagoon and green forests at the ground level.

Many large-scale port cities worldwide have expanded through land reclamation, significantly impacting local environments and ecosystems. A mega floating city can mitigate these impacts on the marine environment and ecosystem when compared to land reclamation. Additionally, while reclaimed land remains permanent, a mega floating city can be easily returned to the sea if downsizing becomes necessary. The city itself can also be relocated if needed, enabling sustainable urban development that is more environmentally and ecologically friendly.

Fig. 2 City in the sky

Fig. 3 Ground level of floating islands

Conventional cities rely on fossil fuel-based energy systems that, although efficient, are geographically limited. In contrast, cities in the new environmental era, which are self-sufficient with renewable energy, may have lower efficiency but can be established anywhere globally. The mega floating city Green Float aims to be self-sufficient in energy and water by harnessing the ocean's potential, serving as a pioneering model for sustainable marine cities in the new environmental era.

2 Concept and Technological Innovations

The Green Float concept embodies two areas of innovation, namely, green innovation and float innovation (Fig. 4).

2.1 *Float Innovation*

Locating the city at a low latitude near the Equator would minimize the float risk posed mainly by wave heights and strong winds. Because typhoons, hurricanes, and cyclones rarely occur there (Fig. 5), the equatorial region is considered the most advantageous for a super-high-rise mega floating city.

This Green Float innovation allows expansion of existing large harbor cities that face difficulties in securing vacant land onto the adjacent water bodies. For an island country that is short of land area, new areas for development could be secured.

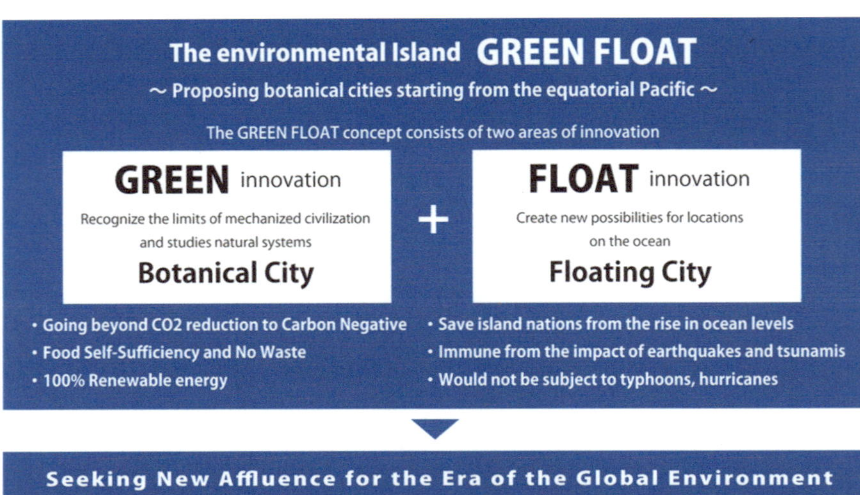

Fig. 4 Green and float innovations

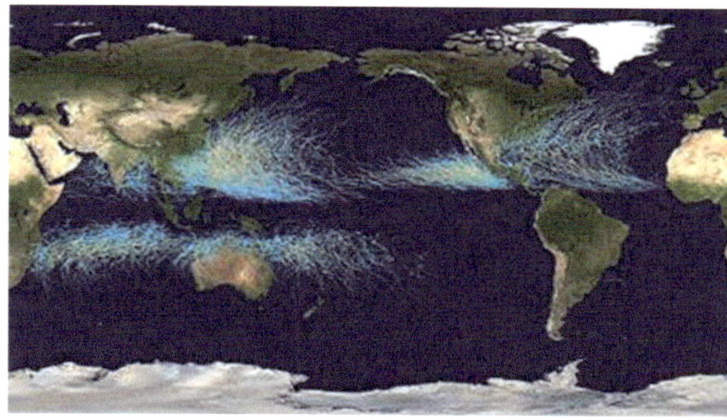

Fig. 5 Paths and location

Compared with landfills, development would be possible that is environmentally and ecologically friendly.

The float innovation is unaffected by rising sea levels due to climate change. Many island nations are at risk of being submerged as rising sea levels, driven by climate change, threaten their very existence. These countries are on the frontlines of a global crisis, where the effects of melting ice caps and warming oceans could erase entire communities and cultures from the map. A mega floating city is one way to save such countries. It also avoids the influence of earthquakes and tsunamis. When a tsunami reaches a shallow shore, the wave height increases. However, a megacity floating offshore is entirely immune from the impact of earthquakes and tsunamis.

The Green Float is a transferable and flexible city. Because of durability issues, many cities and buildings on land cease to be functional within their lifetimes. By contrast, a mega floating city is easily moved once its role ends, allowing the location to be redeveloped. The life cycle cost can be reduced by relocating and continuing to use the lifetime of matter. Also, to meet the needs of the expansion of urban area, it is possible for the city to grow naturally as cells, modules, and units, much like a water lily grows on water (Fig. 6).

2.2 *Green Innovation*

In the future global environmental age, most energy that will be used will be renewable energy with no CO_2 emissions. Because the equatorial regions are rich in sunlight, the green potential, centred on renewable energy, would be maximized.

The Pacific equatorial islands experience a constant temperature of around 28 °C. Because the temperature drops by 0.6 °C for every 100 m of height, it will be 6 °C lower at 1000 m. With the floating sky city rising 700–1000 m above the Equator, the temperature there would be a refreshing 26 °C all year round (see Figs. 7 and 8).

Solar power can provide electricity efficiently in regions with favorable sunshine hours and solar radiation angle. In addition, because the sea surface is at a much different temperature than the seabed, the efficiency of power generation based on

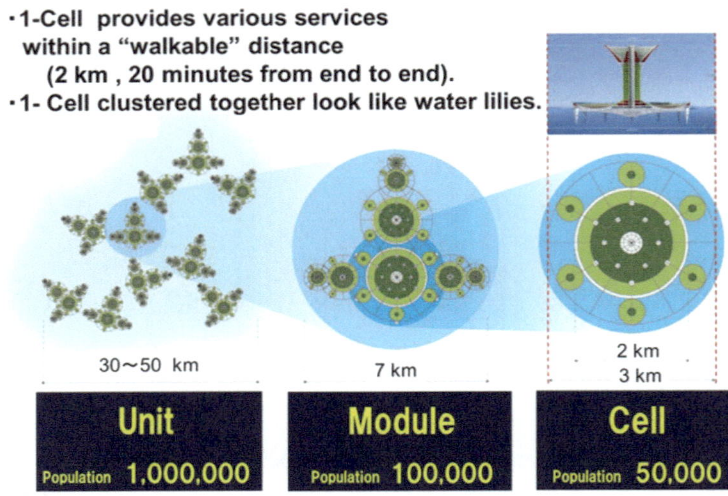

Fig. 6 Growth of mega floating city

Fig. 7 Temperature above the Equator

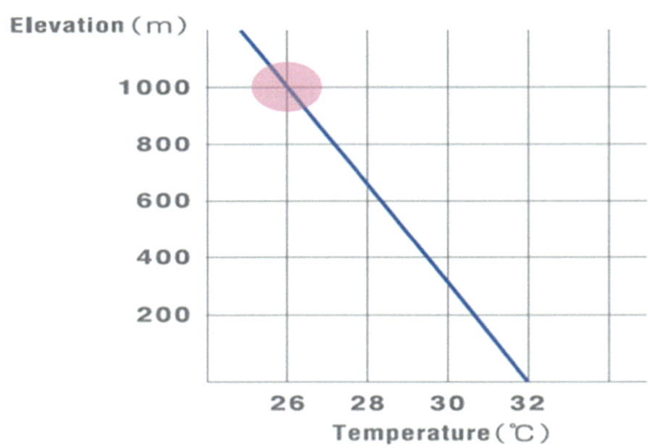

the ocean temperature difference also improves. Taking the CO_2 emissions data for Japan as a benchmark, Fig. 9 shows how Green Float could reduce such emissions.

Regarding fresh water supply, the annual precipitation is high in the equatorial Pacific. However, precipitation levels differ considerably between the rainy and dry seasons. By storing water in the high-rise mega floating city and controlling water use throughout the city, it would be possible to supply almost the same amount of water required for living.

With respect to food self-sufficiency, one can harness the power of the sun to create a vertical vegetable factory that contributes to food self-sufficiency, food mileage, and food traceability (Fig. 10).

Fig. 8 Pleasant and comfortable temperature

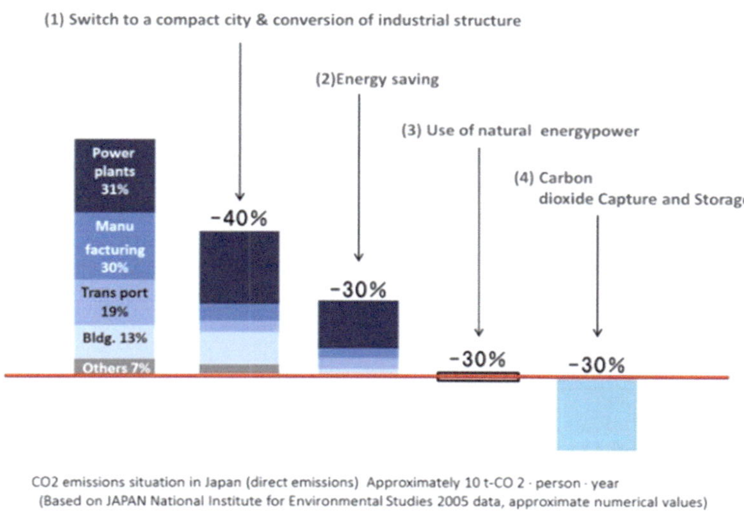

Fig. 9 CO_2 reduction of Green Float

The floating city is also a green city with zero emissions and make use of recycling technologies. Human waste in the form of water, garbage, and CO_2 can be nutritious for plants. By exploiting the height of the sky city to redistribute human waste to lower levels, plants, and other living things would be supplied with nutrition. This raises the possibility of complete resource recycling from waste to food (Fig. 11).

Fig. 10 Vertical vegetable factory

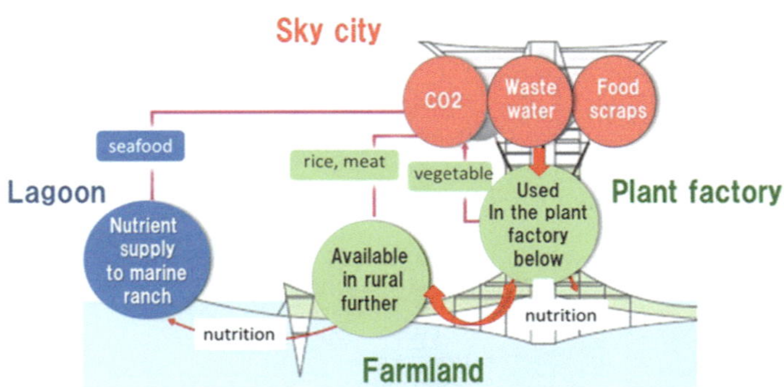

Fig. 11 Waste supplies food sources with nutrition

3 Structure of Green Float

A mega-floating structure is composed of a mega-tower with a height of 1000 m and a floating base with a diameter of 2000 m. In addition, the mega-ring with a diameter of 1000 m is attached to the upper part of the mega-tower (Fig. 12). The mega-tower and mega-ring are made of steel trusses. The floating circular base is an assembly of the concrete honeycomb cells (Fig. 13).

The main materials used for construction are given in Table 1 in view of future material development.

The thickness of the floating base varies from 200 m near the centre and 50 m near the edge. These thicknesses were determined from hydrostatic and structural analyses. In structural analysis, we considered the self-weight of the floating structure, wind load (westerlies) and water pressure by buoyancy. An example of the structural analysis result is shown in Figs. 14 and 15.

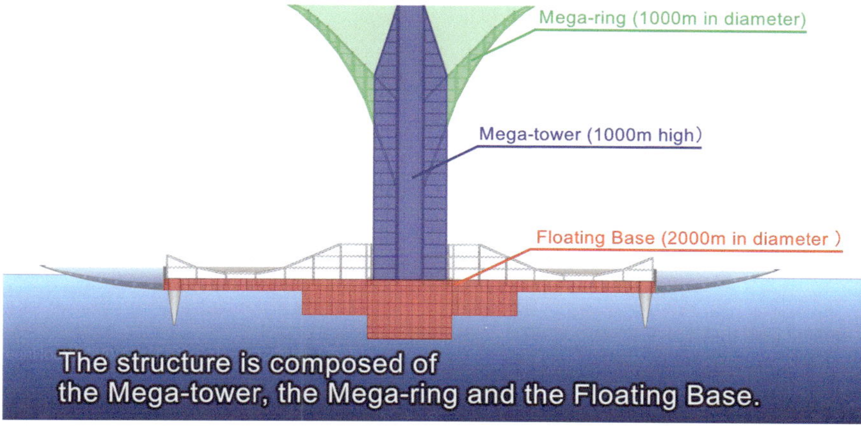

Fig. 12 Components of Green Float

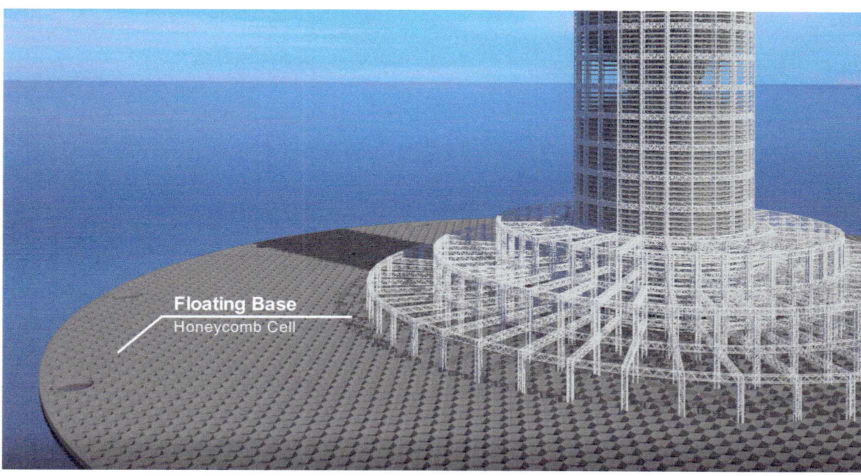

Fig. 13 Floating base (honeycomb cells)

Table 1 Main materials used for construction of Green Float

Material	Specification	Strength (MPa)	Remarks
Steel	SN490	325	
	HT3000	3000	Under research HT1000
Rebar	SD490	490	
	SD1000	1000	Equivalent to PC steel
Concrete	Fc150	150	
	Fc1000	1000	Under research

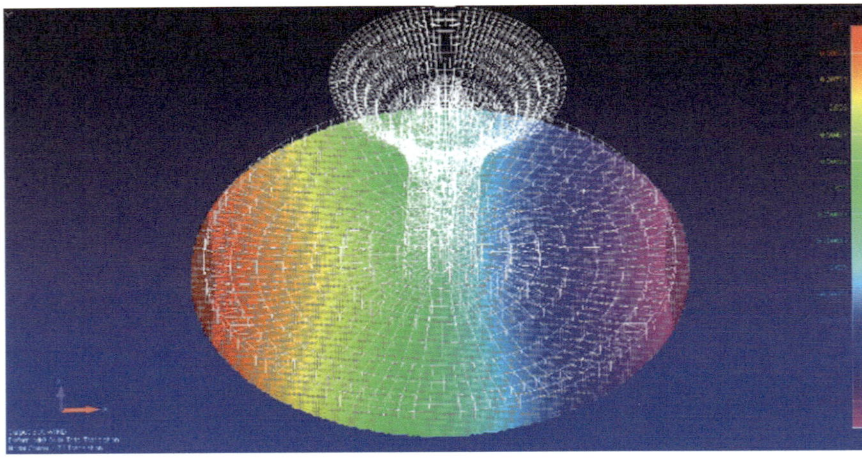

Fig. 14 Analysis of whole model

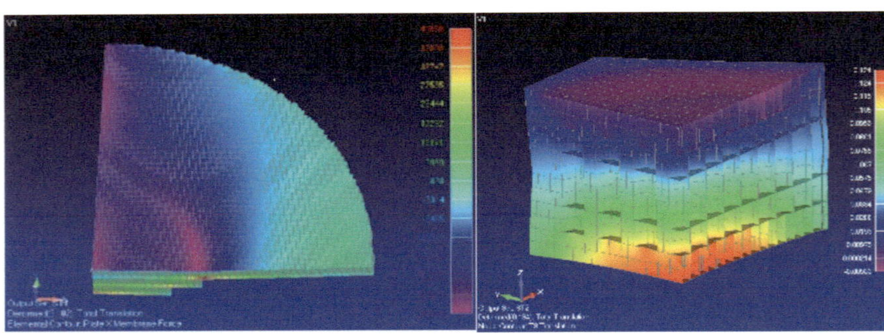

Fig. 15 Analysis of partial models

4 Construction Planning

Various studies are needed to realize mega floating cities. The development of construction methods and materials is also required and so construction plans are essential. Therefore, the following points were noted in the construction planning.

First is a rationalization of construction, such as labour-saving. In general, transportation of equipment and materials is difficult, resulting in higher construction costs on the sea than on the land. Next, in the construction of floating cities having high-rise buildings, it was essential to develop a construction method that drastically reduces work at high places. Even in this construction planning, automated construction by machines assuming future technology development was considered.

4.1 Construction Procedure and Process

In the construction planning of mega floating structures, it is important to keep relatively small ones under construction safety and stability. It is desirable to make as large units as possible in dry docks and to join them in a calm ocean space. The construction site of the dry dock was set as a calm bay in an area close to the equator.

The construction steps are shown below. Step 1 involves construction at the coastal dry dock, whereas Steps 2 to 4 involve construction offshore. Details of these will be explained in the following sections.

- Step 1. Production of honeycomb units in dry-docks

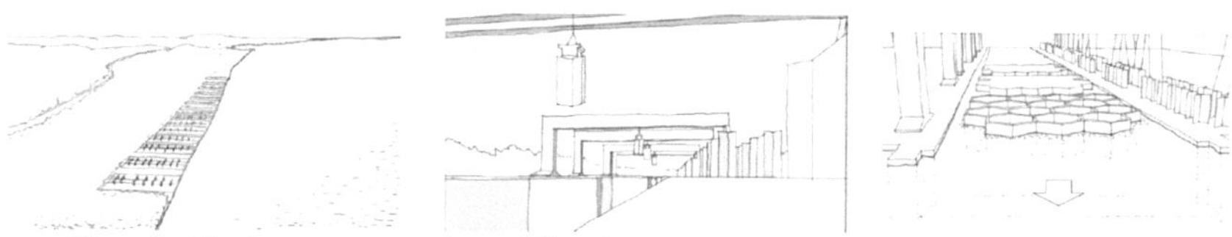

- Step 2. Marine construction of floating base

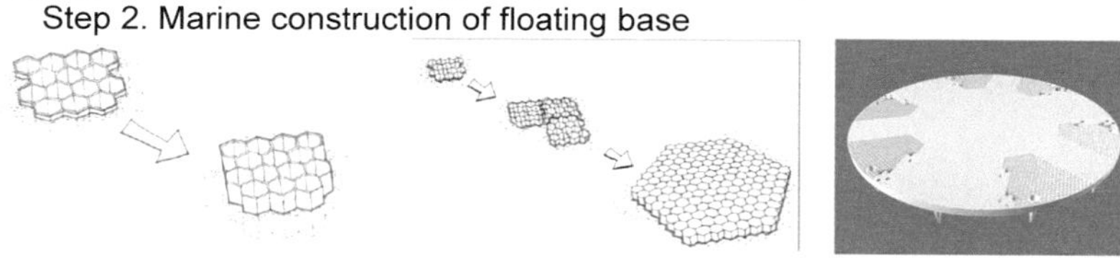

- Step 3. Automated construction of Mega-tower by Ocean Smart System

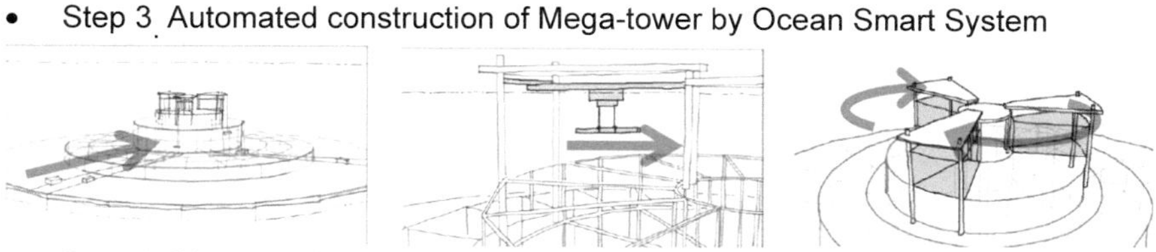

- Step 4. Dive and rise-up construction method of Mega-tower

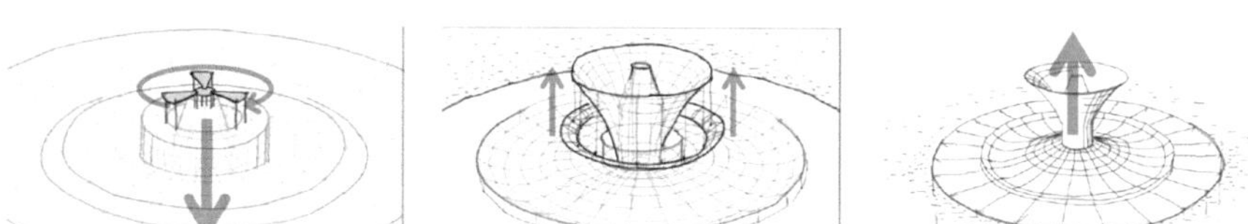

The construction process was assumed to be 10 years from the production of honeycomb units to the completion of the mega-tower. The assumption process is shown in Fig. 16. However, construction of dry-docks and urban facilities were not included. The construction plans shown in the following sections were based on the quantity and scale necessary to achieve this process.

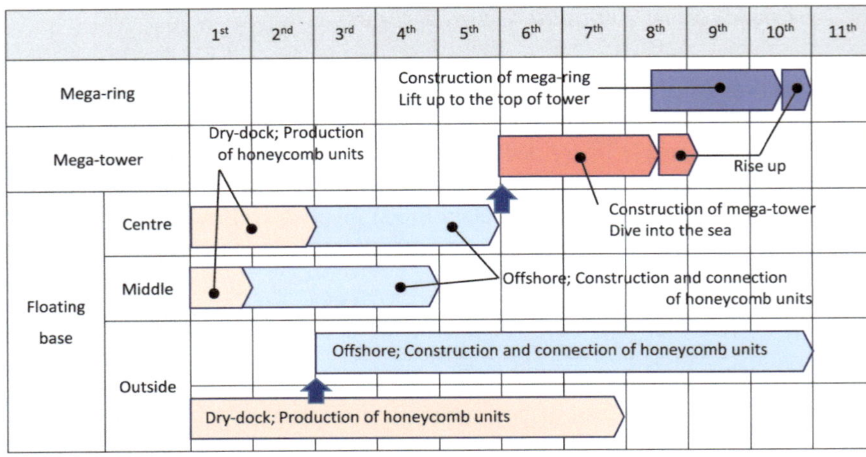

Fig. 16 Overall construction process chart of Green Float

4.2 Floating Structure

The floating base is constructed by connecting honeycomb units offshore to form large floating platform. Honeycomb units are built at dry-docks. This section shows dry-dock and offshore construction of floating structure.

4.2.1 Production of Honeycomb Units in Dry-Docks

Dry docks will be built on the coastal part of the bay in the equatorial region. Since the honeycomb unit to be manufactured has a height of 15 m and a diameter of 100 m, the dry-dock has a depth of 15 m and a width of 110 m. At first, the precast method was planned, but it is assumed that concrete floating will be made with large 3D printers in the future. Therefore, the equipment plan of the dock is as shown in Fig. 17. Honeycomb units will be constructed by the 3D printer up to 15 m high. Completed Honeycomb units will be launched to the sea (Fig. 18).

4.2.2 Marine Construction of Floating Base

The construction of the honeycomb units will be automated by using the Honeycomb printer on a deeper water area. The honeycomb printer consists of two sets of cranes and a 3D printer, and it is characterized by being able to change its position for construction. While re-installing the honeycomb printer, the honeycomb units will be layered to a height of 50 m (Fig. 19). The core base of 300 m in diameter provides the foundation of the tower and this base is constructed by connecting units (Fig. 20).

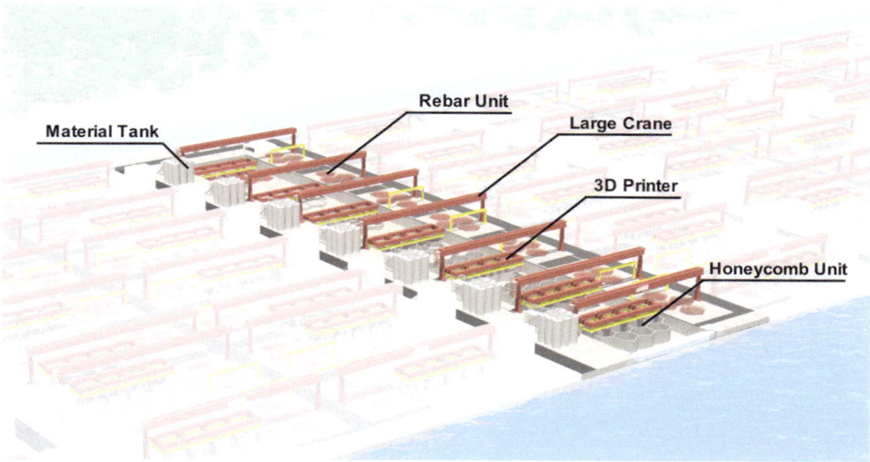

Fig. 17 Equipment dry-dock

Fig. 18 Docking out of honeycomb units

4.2.3 Underwater Connection

Honeycomb units will be connected in the ocean. The connecting technology of submerged tunnels will be applied for this. The concept of the connecting part is shown in Fig. 21. There is a gap with rubber gaskets in the periphery on the connecting surface. First, by using a strong-back hydraulic robot arm installed at the top of the connecting part, units are pulled from each other until the rubber gaskets come into contact (Fig. 22). Also, since the connection surface has positioning cotters, it can

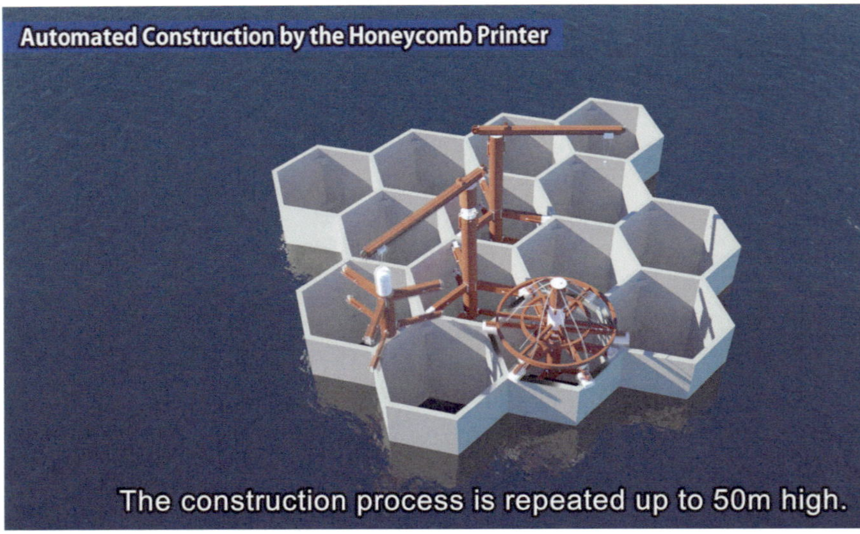

Fig. 19 Built-up honeycomb units offshore

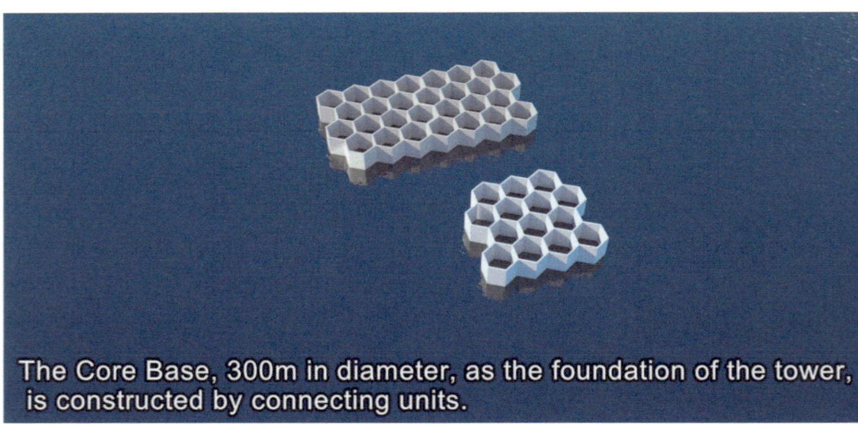

Fig. 20 Connecting honeycomb units offshore

connect to the position as designed. A gap is brought to a near-vacuum state by draining the water. After the connecting parts are in close contact, the connecting surface is applied to compression force by tensioning PC steel materials. Finally, the gap was filled with concrete.

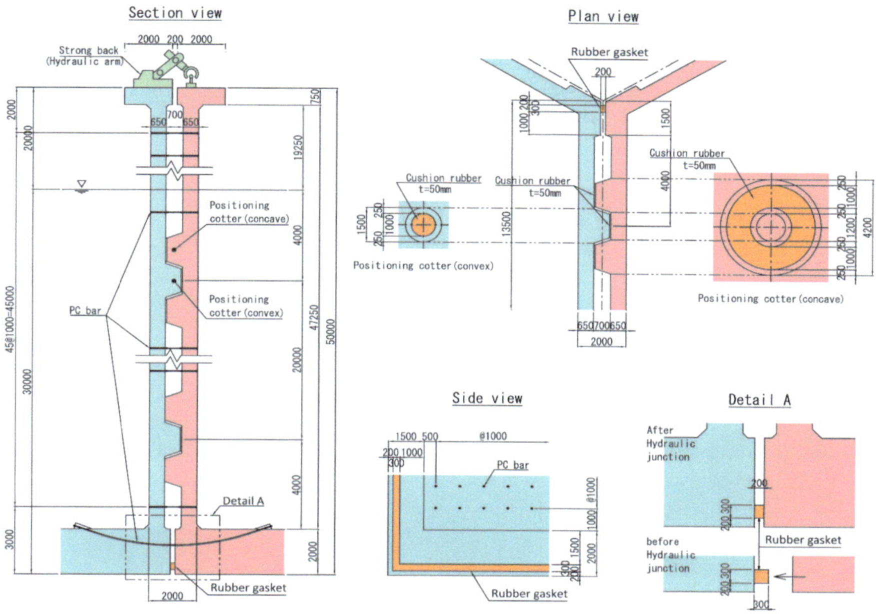

Fig. 21 Details of connection part

Fig. 22 Connection of honeycomb units

4.3 Superstructure

Construction of the superstructure is carried out automatically and continuously on the floating base. At the same time, the construction method that a steel frame can be assembled on the sea surface was devised.

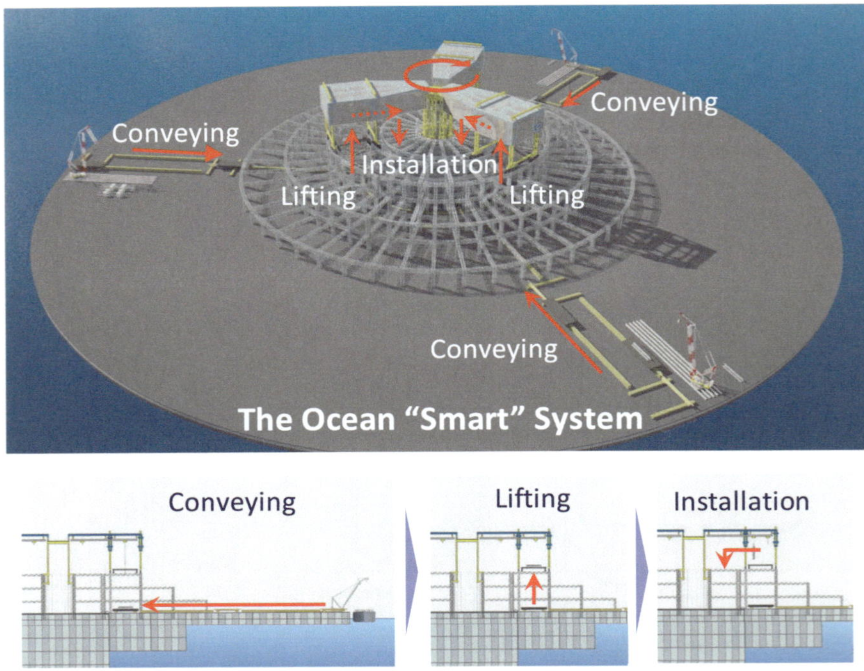

Fig. 23 The Ocean "smart" system

4.3.1 Marine Construction of Floating Base

Steel frame erection of the mega-tower starts when the Floating Base becomes 700 m in diameter. The Ocean "Smart" System is installed after the completion of the three-layer mega-tower. Structural components unloaded are carried and lifted, and then fixed at designated positions automatically by this system. The configuration and workflow of this system is shown in Fig. 23. The Ocean "Smart" System rotate and moves to the next blocks when every fifteen-degree part is completed. Detail of the Ocean smart system is shown in Figs. 24 and 25. In addition, this system combined with the dive and rise-up construction method reduces the work at the high places.

4.3.2 Dive and Rise-Up Construction Method of Mega-Tower

The procedure of the dive and rise-up construction method is shown in Figs. 26 and 27. The mega-tower goes down under the sea when every one layer of the structure is completed by the ocean smart system. It is made possible by buoyancy adjustment. The base of the tower is equipped with a ballast function for it. While controlling the state with a guide jack, the tower with a height of 1000 m is submerged once. Meanwhile, by repeating assembly and diving, it is possible to construct the tower

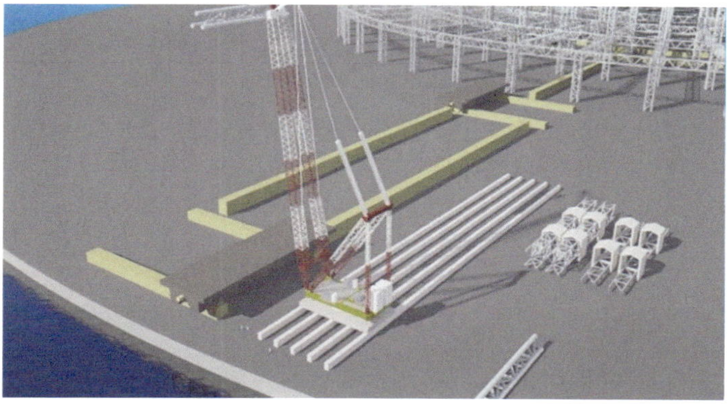

Fig. 24 Conveying system

Fig. 25 Lifting and installation system

at the same height all the time. Next, the mega-tower is raised up to 350 m so as not to interfere with the assembling mega ring (Fig. 28). This levitation is from the buoyancy obtained by discharging the ballast water. Preassembled components of the mega-ring are lifted and installed in the mega tower (Fig. 29). As the mega-ring is lifted at a low position, the materials for lifting, such as cables and jacks, can be reduced. After completing the mega-ring, the mega-tower is raised to the remaining 650 m by using buoyancy.

In the construction of high-rise structures as high as 1000 m, the construction scale will be large due to the weight of lifting equipment itself. It is very rational to control construction level of height as diving and raising by utilizing buoyancy forces generated in the floating structure.

Fig. 26 Rise up using buoyancy

Fig. 27 Lifting mega ring

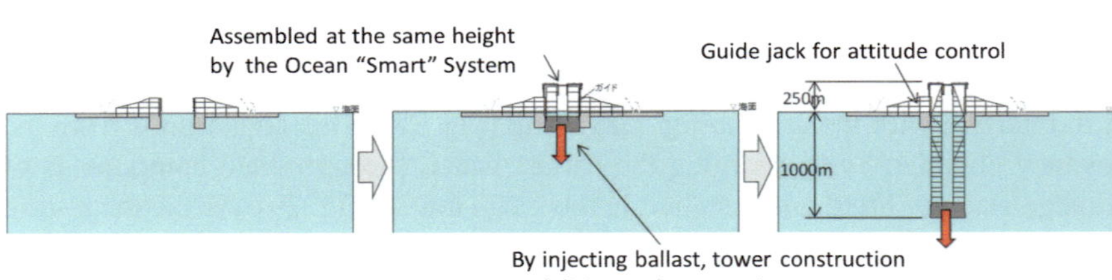

Fig. 28 Dive construction method

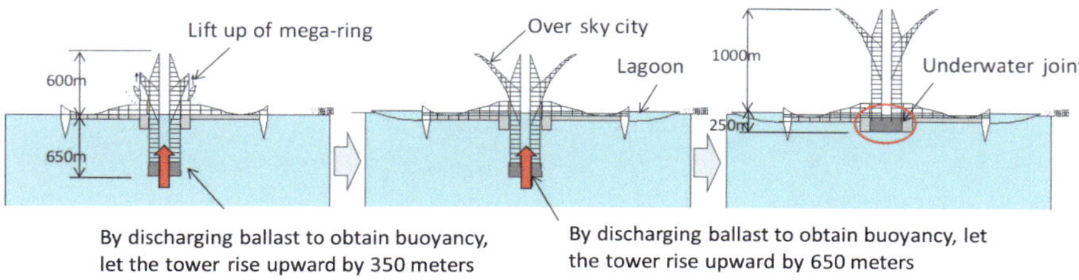

Fig. 29 Rise-up construction method

5 Verification by Experiments and Analysis

To perform basic verification of the Green Float technology, Shimizu has also proposed Green Float II on a scale that is one-tenth that of the original concept (Fig. 30). Green Float II is intended to be built on shallow water at mid-latitudes, such as in a bay near a large city. As such, the effects of tsunamis and the strong winds and high waves caused by typhoons must be considered in its structural design.

In this study, hydraulic model experiments and numerical computations were conducted to assess whether the structure of Green Float II is sufficiently safe against winds and waves that have an expected return period of 200 years and also the tsunami that would result from the largest earthquake scenario envisioned for the area.

Fig. 30 Green Float II

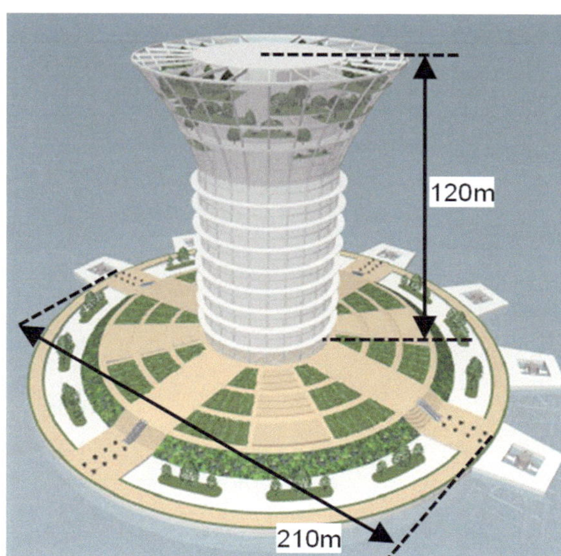

5.1 *Hydraulic Experiments: Winds and Waves*

5.1.1 Experimental Apparatus and Conditions

Shimizu assumes Green Float II to be a 200-m-diameter floating structure carrying a 120-m-tall high-rise building and assume the water depth of the construction area to be 20 m as in the Tokyo Bay. The experimental model was made to one-hundredth scale, and its specifications as given in Table 2 were set according to Froude's similarity law.

To reproduce the elasticity of the full-scale high-rise building, the model building was divided into three parts that were each fixed to a stainless steel rod (Fig. 31). The natural frequency of the building's elastic vibrations is 0.28 Hz at model size (2.8 Hz at full size).

In the actual plan, seven mooring devices were used that comprise fenders and dolphins. In this experiment, four linear springs with equivalent spring coefficients were used to moor the floating structural model.

Green Float II is supposed to operate for 100 years, from which a return period of 200 years was determined for the expected winds and waves. For instance, the wind speed is 46 m/s and the significant wave height and the significant wave period are 3.7 m and 6.1 s, respectively, at full size. The experimental situation is shown in Fig. 32.

Table 2 Specifications of model

Item		Full Size	1/100 Model
Building	Height	120 m	1200 mm
	Diameter (top)	100 m	1000 mm
	Diameter (base)	56 m	560 mm
	Mass	57,192 ton	55.8 kg
	Stiffness	2.46×10^{10} kNm^2	2.4 kNm^2
	Natural frequency	0.28 Hz	2.8 Hz
Floating body	Height	10 m	100 mm
	Diameter	210 m	2100 mm
	Draft	9 m	90 mm
	Mass	262,808 ton	256.2 kg
Whole structure	Displacement	320,000 ton	312 kg
	Center height	19.6 m	196 mm
	Radius of gyration	48.7 m	487 mm

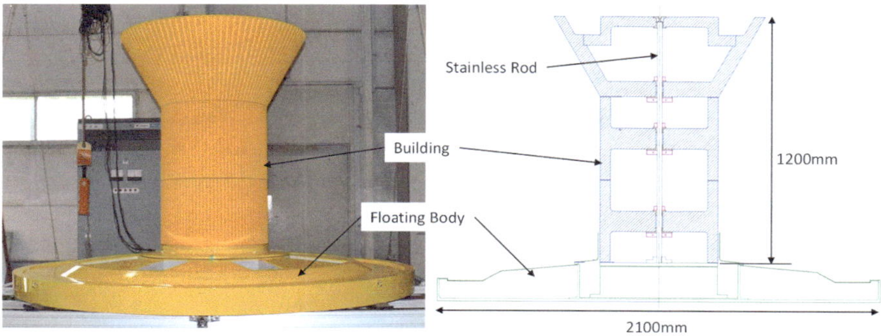

Fig. 31 Experimental model for winds and waves

Fig. 32 Experimental apparatus for winds and waves

5.1.2 Method of Numerical Computations

The motions of the floating body in irregular waves were determined by solving the following equation of motion:

$$(\boldsymbol{M} + \boldsymbol{A}_W + \boldsymbol{A}_E) \cdot \ddot{\boldsymbol{X}}(t) + \int_0^t \boldsymbol{R}(t-\tau) \cdot \dot{\boldsymbol{X}}(t-\tau) d\tau + (\boldsymbol{C}_W + \boldsymbol{K}) \cdot \boldsymbol{X}(t)$$
$$= \boldsymbol{F}_{WE}(t) + \boldsymbol{F}_{WD}(t) + \boldsymbol{F}_{wind}(t), \quad (1)$$

where X is the vector displacement of the centroid of the floating body, namely,

$$X = (x_1, x_2, x_3, x_4, x_5, x_6) \quad (2)$$

Here, suffixes 1–6 denote the modes of surge, sway, heave, roll, pitch, and yaw, respectively, with the surge motion being in the same direction as the wave direction. Furthermore, M is the mass matrix of the floating body, A_W is the matrix of added mass coefficients associated with the dynamic fluid forces, A_E is the vibrational force of the building, R is the matrix of the retarded function, C_W is the matrix of static buoyancy coefficients, K is the matrix of spring coefficients of the mooring system, F_{WE} is the wave diffraction force, F_{WD} is the wave drifting force, and F_{wind} is the wind force. The terms A_W, R, and F_{WE} are computed using a three-dimensional Green's function method developed by Kobayashi [2], and F_{WD} is estimated using a method proposed by Pinkster [3].

In order to focus on evaluating the external forces on the whole structure and its mooring system, no account is taken of any resonance between the building and the floating body. However, the forces acting on the floating body due to the motion of the building (e.g., shear stress and bending moment at the base of the building) are considered as substitution by added mass coefficient A_E. The only components of A_E that are considered are those in directions 1 (surge), 3 (heave), and 5 (pitch):

$$A_{E11} = \frac{M_{SS}^2 H_{SS}^3 \omega^2}{20EI}, \tag{3}$$

$$A_{E15} = \frac{M_{SS}^2 H_{SS}^3 \omega^2}{20EI}\left(\frac{13H_{SS}}{18} - BG\right), \tag{4}$$

$$A_{E51} = \frac{91 M_{SS}^2 H_{SS}^4 \omega^2}{2520EI}, \tag{5}$$

$$A_{E55} = \frac{M_{SS}^2 H_{SS}^4 \omega^2}{2520EI}(66H_{SS} - 91BG), \tag{6}$$

Here, M_{ss} is the mass of the building, H_{ss} is its height, ω is the angular frequency, EI is the effective stiffness of the building, and BG is the distance between the base of the building and the centroid of the floating body.

5.1.3 Results and Discussion

The irregular wave experiments were conducted in two runs with different phases for 90 s each. The maximum measured values are summarized in Table 3 (in full-size values). The maximum response displacement of the centroid of the floating structural model in the horizontal direction (same as the wave direction) was 0.6 m. However, at the top of the building, the response displacement was amplified to 1.6 m due to the elastic motion of the building.

The maximum mooring force was approximately 23,000 kN. The measured force was below 27,300 kN, which is the design condition of the fenders. Therefore, it was confirmed that the mooring system of Green Float II could withstand those winds and waves whose return period is 200 years.

Table 3 Maximum measured values

Items	Run 1	Run 2
Horizontal displacement (at the center) [m]	0.593	0.601
Horizontal displacement (at the top) [m]	1.647	1.643
Vertical displacement (at the center) [m]	– 0.889	– 0.813
Vertical displacement (at the top) [m]	– 0.583	– 0.573
Rotation angle (pitching) [degree]	1.06	1.06
Horizontal acceleration (at the top) [m/s^2]	0.933	1.004
Vertical acceleration (at the top) [m/s^2]	0.176	0.189
Horizontal mooring forces [kN]	22,671	23,336
Shear stress (at the base) [MN]	27.7	29.6
Bending moment (at the base) [MN m]	2509.6	2676.2

Figure 33a–f show the frequency spectra of the various measured values. Regarding the horizontal displacement (Fig. 33b) and mooring force (Fig. 33d), below 0.1 Hz. there are remarkably high low-frequency components close to the natural frequency of the mooring system. This is attributed to the long-period components of varying wave drifting forces. Regarding the horizontal displacement of the top of the tower (Fig. 33c), the shear stress and moment at the base of the building (Fig. 33e and f, respectively) show relatively high components around 0.3 Hz, which is almost the natural frequency of elastic vibrations of the building. In view of these results, it is believed that resonance occurred between the floating body and the building.

5.2 *Hydraulic Experiments: Tsunami*

5.2.1 Experimental Apparatus and Conditions

The experimental apparatus is shown in Fig. 34. The wave flume is 2 m wide, 25 m long, and 80 mm deep, with a tsunami generator at the upstream end and a slope at the downstream end. The tsunami generator comprises a water tank and a turbulence-reduction device. The reservoir tank is primed by reducing the pressure therein, and then a tsunami is generated by opening one ball valve and seven solenoid valves sequentially. The experimental model was made to 1/250 scale, and its specifications, as given in Table 4, were set according to Froude's similarity law. The floating model is moored 10 m downstream from the tsunami generator using four springs (see Figs. 35 and 36).

The experimental cases are summarized in Table 5. The maximum water level and the rise time were set with reference to observations (Tsunami observation result by MPAT [4] & Tsunami survey results by Tohoku Electronic Power [5] and the greatest earthquake scenario envisioned by the Japanese government [6]. Figure 37 compares (i) the water level observed at Onagawa Nuclear Power Plant for the Tohoku

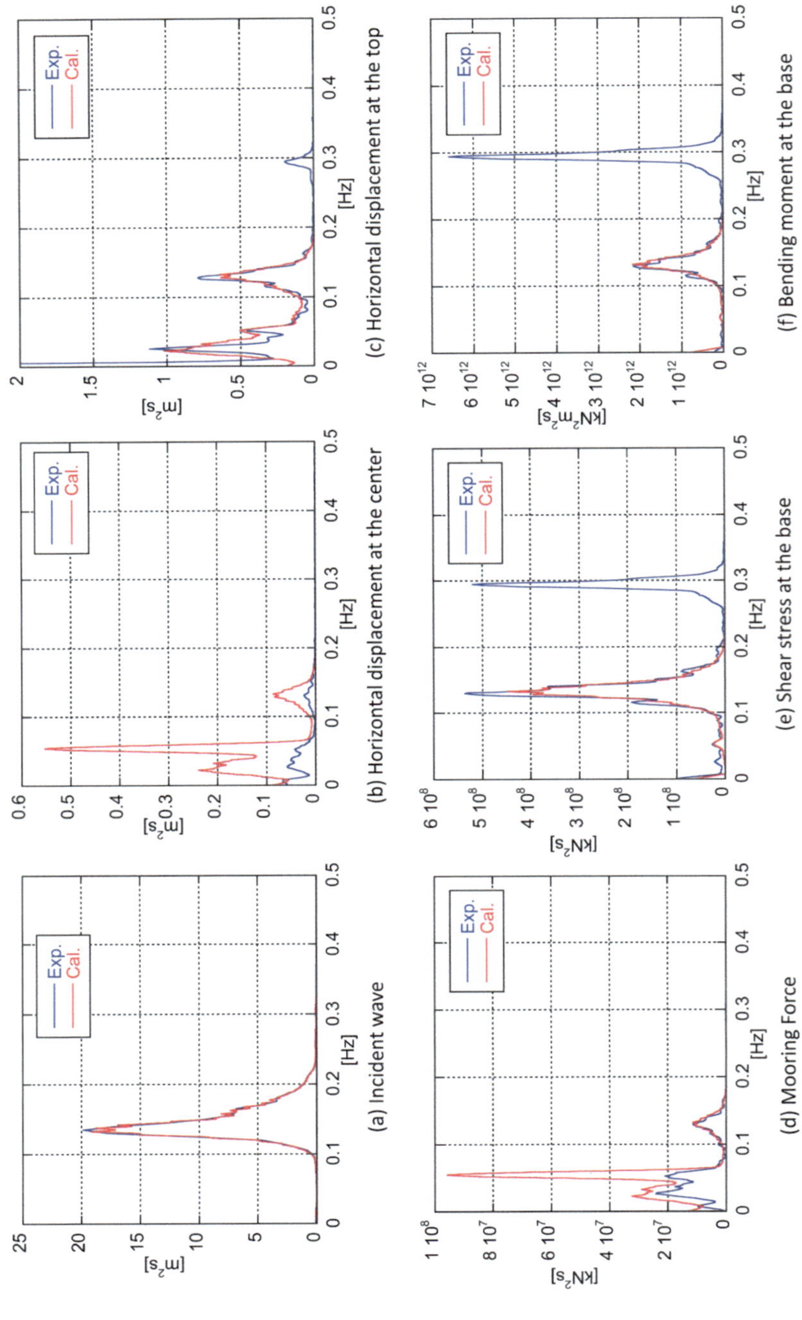

Fig. 33 Frequency spectra of measured values

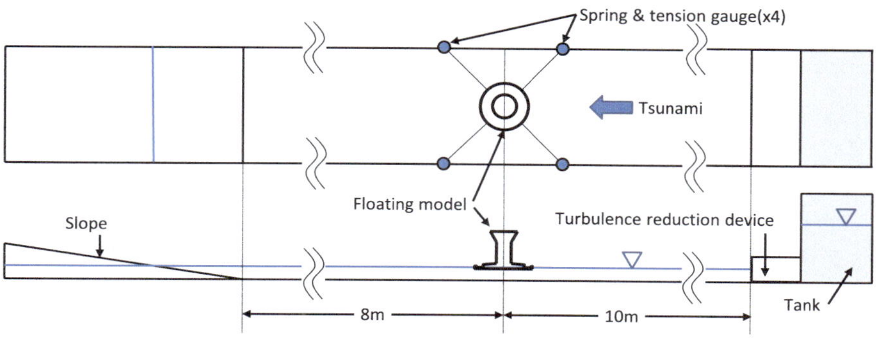

Fig. 34 Experimental apparatus for tsunami

Table 4 Model specifications

Items	Full size	1/250 model
Diameter of floating body	210 m	840 mm
Height of floating body	10 m	40 mm
Draft	9 m	36 mm
Displacement	320,000 t	19.98 kg
Height of gravity center	19.41 m	77.7 mm
Radius of gyration	60.0 m	240 mm
Height of building	120 m	480 mm

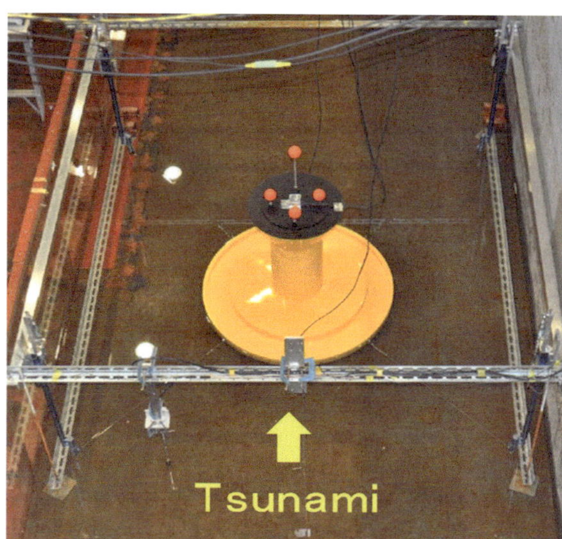

Fig. 35 Experimental model for tsunami

Fig. 36 Tsunami experiment in situ

earthquake and tsunami on 3 November 2011, (ii) the target waveform, and (iii) the wave reproduced by the tsunami generator. The reproduced wave resembles a series of small bores because the speed at which each valve was opened could not be controlled. Nevertheless, the maximum water elevation and the rise time were reproduced well. In addition to these observed or estimated waves, we also conducted experiments with minimized rise time (referred to hereinafter as the bore case). In the case of Tokyo Bay, the tsunami generator could not produce long-period estimated waves and so only the bore case was investigated.

Table 5 Experimental cases

Cases (wave type)	Site	Full size		1/250 model	
		Max. height (m)	Rise time (min)	Max. height (mm)	Rise time (s)
Observed/estimated wave	Port of Shimizu	3.6	15	14	57
	Port of Sendai	7.7	15	31	57
	Onagawa	13.0	15	52	57
Bore	Tokyo Bay	1.1	–	4.4	–
	Port of Shimizu	3.6	–	14	–
	Port of Sendai	7.7	–	31	–
	Onagawa	13.0	–	52	–

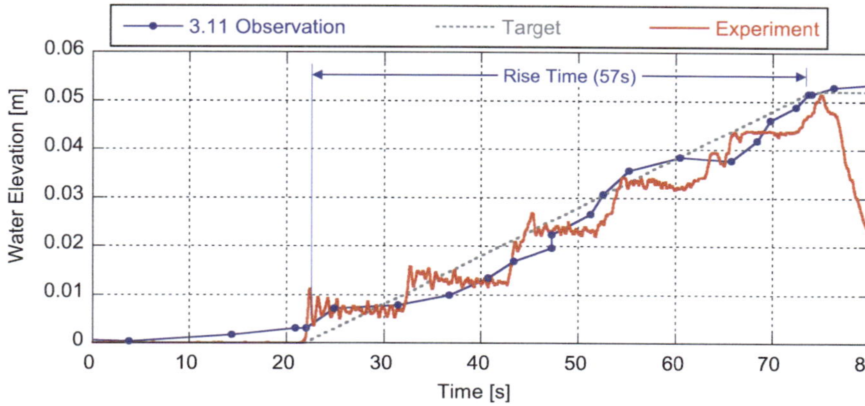

Fig. 37 Waveform of tsunami (at Onagawa)

5.2.2 Results and Discussion

Table 6 summarizes the maximum measured mooring forces (at full size). For Tokyo Bay, the maximum measured mooring force was 19,078 kN. This value is below 27,300 kN, which is the design condition of the fenders. This confirms that the mooring system of Green Float II could withstand the tsunami that would result from the greatest earthquake scenario envisioned for Tokyo Bay.

Figure 38a–f show examples of time histories measured in the case of observed waves at Onagawa. The horizontal mooring force (Fig. 38d) clearly rises sharply several times, forming peaks that correspond to those of the speed of water elevation (Fig. 38c), not the water elevation itself (Fig. 38a) or the flow velocity (Fig. 38b). Figure 39 shows the correlations between the peak values of the horizontal mooring force and (a) the water elevation, (b) the flow velocity, and (c) the speed of water elevation. The horizontal mooring force shows the highest correlation with the speed of water elevation ($R = 0.994$).

Table 6 Maximum measured mooring forces

Cases (wave type)	Site	Max. height (m)	Rise time (min)	Maximum majored mooring force (kN)
Observed/estimated wave	Port of Shimizu	3.6	15	67,172
	Port of Sendai	7.7	15	67,445
	Onagawa	13.0	15	67,728
Bore	Tokyo Bay	1.1	–	19,078
	Port of Shimizu	3.6	–	65,995
	Port of Sendai	7.7	–	180,524
	Onagawa	13.0	–	302,024

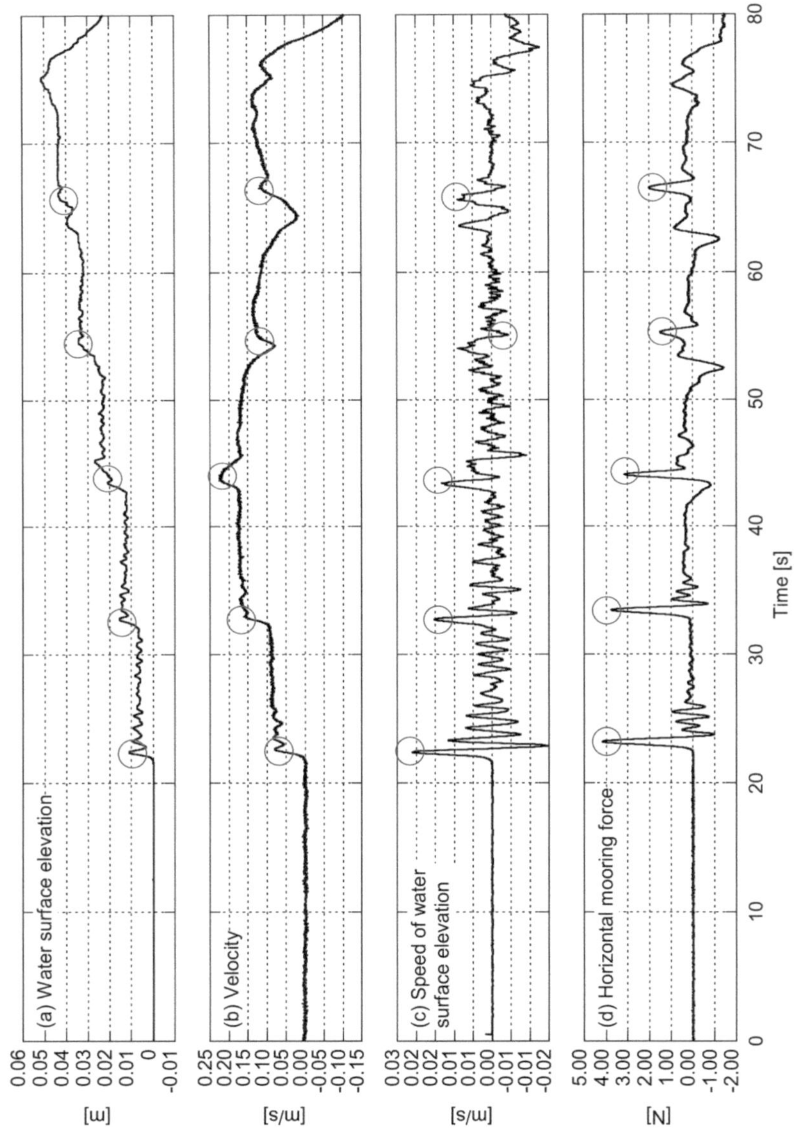

Fig. 38 Time histories measured in case of site waves at Onagawa

Fig. 39 Correlations with peak values of horizontal mooring force

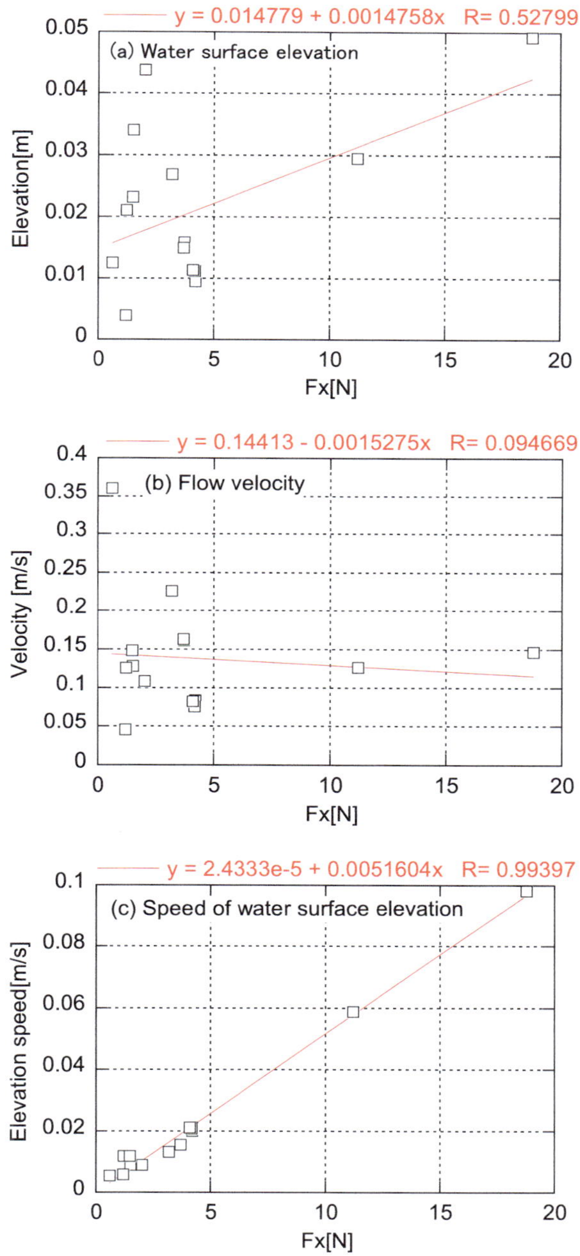

The experiments and computations confirmed that the mooring system of Green Float II could withstand winds and waves with a 200-year return period in the Tokyo Bay. However, it was also found that resonance might occur at around 3.0 Hz between the building and the floating body. This resonance could amplify the horizontal oscillation at the top of the building. In addition, around this resonance frequency, there could be large spectral components of shear stress and bending moment at the base of the building. It is important to consider these phenomena carefully in the structural design. There is also the problem of establishing a way to evaluate the resonance effect for the structural design.

Furthermore, it was confirmed that Green Float II is sufficiently safe against the tsunami that would result from the greatest earthquake scenario envisioned for the Tokyo Bay. It was found that the horizontal mooring force depends on the speed of water elevation rather than on the water elevation itself or the flow velocity. A theoretical study of the tsunami forces acting on the floating body is planned for the near future.

6 Concluding Remarks

In this chapter, the mega floating city "Green Float," with the capacity to accommodate 30,000–50,000 people, was reported. This mega floating city is cutting-edge not only in floating technology but also in green technology. Shimizu proposed the concept of a completely innovative new city for the future of the global environment.

Regarding Float technology, structural safety was evaluated by experiments and analyses modeling the calm ocean conditions in the equatorial area and the bay of a port city, and sufficient structural safety was ensured even in natural disasters such as typhoons and tsunamis, making Green Float worthy of a structural safety certificate by a third-party organization.

Under circumstances where global population growth and the population city concentration will be accelerated in the future, Green Float is a useful solution against rising sea levels due to climate change and for the big cities expansion to offshore and is expected to improve networking between islands. The Green Float proposed by the Shimizu Corporation is expected to contribute to human sustainability.

References

1. Shimizu Corporation, Shimizu Dream, Green Float. No date [cited on 31 July 2024]. Available from: https://www.shimz.co.jp/en/topics/dream/content03/
2. Kobayashi M, Shimada K, Fujihira T (1985) A simulation program of dynamic response of a moored structure subjected to compound external force. Technol Bull Mitsui Eng Shipping Co., Ltd. 125:10–16

3. Pinkster JA (1974) Low-frequency phenomena associated with vessels moored at sea. Paper SPE 4837. Paper presented at the Spring Meeting of the Society of Petroleum Engineering of AIME, Amsterdam, The Netherlands
4. National Institute of Maritime (2011) Port and Aviation Technology, Outline of tsunami observation result (In Japanese)
5. Tohoku Electric Power Co., Inc. (2011) Survey results of the tsunami caused by the 2011 Tohoku earthquake at Onagawa nuclear power plant (In Japanese)
6. Cabinet Office (2012) Japan disaster management in Japan (In Japanese). Available from: http://www.bousai.go.jp/jishin/nankai/model/

Dr. Donghee Ko is a seasoned civil engineer with over 25 years of expertise spanning project management, design leadership, and technical consulting across diverse civil engineering domains. Specializing in marine and offshore engineering, Dr. Ko has contributed significantly to the advancement of floating and offshore energy projects, drawing from a wealth of experience in consulting engineering, construction management, and innovative design solutions. Currently a key member of the frontier division of Shimizu Corporation, Dr. Ko has previously held pivotal roles at Samsung C&T and Samsung Heavy Industries, showcasing a career defined by excellence in large-scale infrastructure and pioneering offshore developments. Academically, Dr. Ko holds a Ph.D. in Civil Engineering from the University of Tokyo, an MBA from the Business School of KAIST, and both B.Sc. and M.Sc. degrees in Civil Engineering from KAIST. This extensive academic foundation, coupled with hands-on industry experience, positions Dr. Ko as a leading authority in the development and implementation of large floating structures and cutting-edge offshore solutions.

Analysis and Design of Floating Bridges: Recent Advances and Challenges

Torgeir Moan, Ole David Økland, and Ole Andre Øiseth

Abstract This chapter presents recent research and developments on floating bridges, with a particular focus on the analysis of environmental load effects for ULS design checks and analyses for risk-based design against ship impacts of the Bjørnafjord floating bridge in Norway. Both numerical and experimental methods are used in addressing challenging features associated with handling inhomogeneity in environmental conditions, wave interaction between pontoons, wave-current interaction and viscous damping, parametric excitation generating horizontal motion of the bridge as well as proper and efficient account of the long-term variability in environmental conditions. A risk analysis procedure to deal with the major accidental load for floating bridges, namely ship impacts, is presented. It involves an estimate of the frequency of impact scenarios, the immediate structural damage and possible flooding, the expected time for repair and ultimate consequences in terms of fatalities and road traffic downtime. To reduce the relatively high road traffic disruption rate estimated, it is proposed that the impact frequency for relatively frequent low energy events (in addition to the initial focus on 10^{-4} events relevant for the ALS criterion) be scrutinized. Additionally, consider introducing extended ship traffic control measures, as well as to possibly introduce a limit state in addition to the ALS to address the risk associated with frequent impacts.

Keywords Floating bridge · Serviceability · Safety · Sea loads · Ship impacts · Analysis for design

T. Moan (✉)
Department of Marine Technology, NTNU, Trondheim, Norway
e-mail: tormo@ntnu.no

O. D. Økland
SINTEF Ocean, Trondheim, Norway
e-mail: Ole.D.Okland@sintef.no

O. A. Øiseth
Department of Structural Engineering, NTNU, Trondheim, Norway
e-mail: ole.oiseth@ntnu.no

B. T. Wang et al. (eds.), *Large Floating Solutions*, Lecture Notes in Civil Engineering 636, https://doi.org/10.1007/978-981-96-5435-2_5

1 Introduction

Floating bridges have been employed for centuries. By utilizing the buoyancy force offered by the water, floating bridges can carry dead loads and traffic loads over a very long span; thereby making them an effective means of transportation across deep channels, lakes, fjords and straits. They might be a continuous floating "beam" or supported by discrete pontoons. Modern floating bridges can be traced back to the pontoon bridge design that was implemented in the Hobart Bridge, Australia in 1943. Examples of modern floating bridges are described by Wang and Wang [58]. A main reason for choosing floating bridges was the long span and the depth of water or a thick layer of soft soil. Currently, the Norwegian Public Road Administration (NPRA) is undertaking technology development to cut the travelling time by 50% on the Coastal Highway Route E-39 from Trondheim to Kristiansand in Norway (e.g. [34]). This development especially involves projects to replace ferry transport across 8 fjords by providing bridges, submerged tunnels or subsea rock tunnels. As shown by Moan and Eidem [34] the main bridge concepts for strait crossing in the 8 fjords in the E39 project in Norway, include:

A: Curved, end-anchored floating bridge on discrete pontoons (see Fig. 1).
B: Straight, floating bridge on discrete pontoons anchored by mooring lines.
C: Suspension bridge with pylons supported on land, seabed or supported on floating bodies.

Bridge types A and B would typically include a cable stayed high bridge to provide a navigation channel. Recently, the focus has been on research and engineering analysis and design of the bridge across the nearly 5 km wide Bjørnafjord, which has a maximum water depth of about 550–600 m.

Bridges make up a part of the road infrastructure to improve the efficiency of the transport system. At the same time, bridges should in general fulfill sociopolitical criteria that address the aesthetics, environmental sustainability, budgetary and legal constraints. Moreover, they should not represent a hindrance to ship traffic or other activities in the fjord or strait. The design of floating bridges must satisfy serviceability and safety criteria to achieve a safe function of the bridge. Moreover, floating

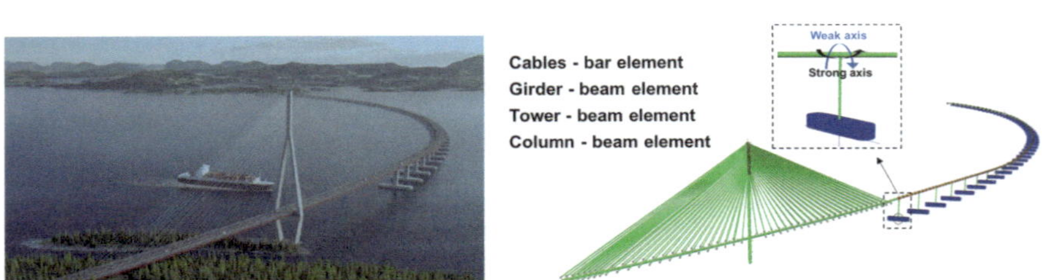

Fig. 1 Artist impression and structural model of a proposed floating bridge concept for Bjørnafjord crossing in Norway. At the South End (in the bottom of the figure) a high cable stayed bridge is provided to allow ship passage. Courtesy: Norwegian Public Road Administration (NPRA)

bridges should be designed for easy fabrication and installation as well as control during use.

The development of technology for floating bridges has benefitted from transfer of technology from other sectors and implemented by proper consideration of the possible difference in serviceability and the target safety levels. In particular, the principles and methods developed for oil and gas platforms and ships as well as for floating airports in the Mega-Float project in Japan during 1995–2000 and the MOB project in USA, especially in the period 1997–2000 [23, 31, 52], have provided a basis for design and analysis of floating bridges. Earlier reviews on the development of floating bridges were conducted by Moan and Eidem [34] and Wang et al. [59].

In this chapter, recent advances and challenges in the design and analyses of floating surface bridges, with a special focus on the Bjørnafjord bridge, are addressed. The attention is particularly paid to research and developments of

- load effect analysis associated with ultimate limit state (ULS) and fatigue limit state (FLS) criteria—considering numerical simulations, and model and field testing.
- analysis of ship impacts in view of accidental collapse limit state (ALS) criteria and possible novel criteria based on a risk assessment.

2 General Considerations

2.1 Introduction

Bridges should satisfy serviceability and safety requirements during the operational life, of say 100 years of the bridge, with a minimum of life cycle costs. In the design of floating bridges the following loads should be considered: dead load, traffic loads, hydrostatic pressure (including buoyancy), live load, wind-, wave-, seismic-, temperature change and ocean current loads; effects of tidal change, seabed movement, movements of bearings; snow load, effects of storm surges, ship waves, possible effects of drift ice and ice pressure, and effects of marine growths, accidental loads—such as impact loads due to ship collisions, the effect of fires and explosions, abnormal traffic loads, as well as landslides possibly causing tsunami waves or direct damage to the mooring anchoring. Additionally, hazard due to fatigue, corrosion and other chemical attacks, should be considered. It is noted that the time variant loads can cause dynamic response in floating bridges, which have many natural frequencies in the frequency range of the dynamic loads.

Safety of floating bridges implies the absence of global structural failure and serious accidents. In Norway, floating bridges should be designed based on the highest consequence and reliability class (CC3, RC3) [37]. Since there is limited design experience with floating bridges, the Bridge Design Handbook [42] contains limited information about design criteria for floating bridges. A particular design basis is hence established for bridges for a given fjord crossing based on the standards of ocean engineering and bridge engineering (i.e. [23] and [38]). Eventually the design basis for various projects forms the basis for standardizing the design criteria in the NPRA Bridge Handbook. When using the information from other sectors, due account of differences in function and target safety levels, should be made. Criteria for the Bjørnafjord floating bridges are given in NPRA [41].

Serviceability and safety of the bridge is maintained during the life cycle by a proper design, fabrication and in service follow-up with respect to the criteria outlined subsequently.

2.2 Serviceability

The purpose of serviceability criteria is to ensure that the structure fulfils its function as specified by the owner; i.e. the comfort and safety of both drivers and pedestrians. Besides requirements to the width of the roadway, separation of traffic and pedestrians that use the bridge, the bridge should satisfy serviceability requirements with respect to deflections and motions under environmental loads, while accounting for the direct effect of the wind on vehicles in NPRA [41], such that the down-time due to environmental conditions (strong winds, wave-induced motions) is within acceptable values. For instance, the Bjørnafjord bridge will be closed for road traffic when the wind speed exceeds the annual maximum value. Moreover, it is noted that the maximum vertical angular change of the roadway for an annual effect of tide variation and waves, is 3.5° for speed limit of 70 km/h, while an angular change is not acceptable for larger vehicle speeds. This requirement especially applies to the ramp at the entrance of the floating bridge, while the bridge girder itself does not have abrupt changes.

A particular issue arises in connection with ship impacts as discussed later, in that a fundamental serviceability criterion is to ensure that the bridge is available for road traffic. The criterion for traffic disruption depends on the consequences of disruption, i.e. available alternative roads and the inconvenience to the society, including costs of using the alternative roadways. Both disruption due to accidents and planned disruption due to maintenance should be considered.

2.3 Safety

Current safety design criteria are formulated in terms of the ultimate limit state (ULS), fatigue limit state (FLS), and accidental limit state (ALS), where the load effects and resistance are specified by characteristic values and partial safety factors [34]. While structural reliability analysis methods are used to ensure that semi-probabilistic criteria used in practice for ULS an FLS criteria comply with target safety levels, quantitative risk analysis is applied to support ALS design criteria, as explained in Sect. 5.

The main failure mode of floating bridges is structural failure of components or, say, the bridge girder as a whole. For floating bodies, like the pontoons, loss of buoyancy is an important event that increases the loads on other parts, especially the bridge girder and can eventually cause structural failure. It is noted that sinking or capsizing of the pontoon, which are explicitly considered as failure modes of freely floating structures, do not represent a primary failure mode for the floating bridge in operational condition, because the pontoons are an integrated part of the bridge girder that is supported at its ends.

To ensure adequate safety, the design criteria should be balanced with operational monitoring or control. Hence, inspection should be carried out during fabrication and operation to detect possible cracks or corrosion and initiate maintenance or repair. Moreover, surveillance of the ship traffic and initiation of actions in case of hazardous situations, is important to control the ship collision risk besides choosing bridge layout and designing against certain ship impacts in the ALS check. As discussed later, the structural design, as one among mitigation measures, might especially imply introducing a subdivision of pontoons in compartments to limit the buoyancy loss, and hence the consequences of ship impacts.

In general, quality assurance and control to detect possible human errors and ameliorate their effect is particularly important for novel concepts such as floating bridges. To ensure the safety of floating bridges, it is necessary to conduct life-cycle structural integrity management including the design, inspection, monitoring, maintenance, repair, and replacement [34].

2.4 *Sustainability*

There has been concern about the sustainability of the activities of man on the earth for some decades. But now attention is paid to establishing and operationalizing sustainability requirements in the life cycle of floating bridges. Partly efforts are made with respect to the ocean ecology and partly energy use and emissions during fabrication as well as the possible reuse of the bridge components, material after end of the service life [25, 26].

3 Load Effect Analysis for Ultimate and Fatigue Limit States

3.1 Current Version of the Bjørnafjord Floating Pontoon Bridge

The preferred concept, based on an economic assessment, including a recognition of the uncertainties associated with novel technologies, is a low pontoon bridge, curved in the horizontal plane and with pontoons spaced at 100–200 m. The current idea is to make the box bridge girder and pontoons of steel—while in the initial stages of concept evaluation, reinforced concrete was also considered for the pontoons. In addition to the pontoons, the Bjørnafjord bridge consists of a fixed cable-stayed high bridge to facilitate a navigation channel to limit the risk of ship collisions. A curved bridge carries horizontal forces by "arch effect"—i.e. by axial forces, and with anchoring of the bridge ends. Yet, the intention is to equip the bridge with three clusters of mooring lines. In this way, additional damping is introduced, and the damage tolerance is increased, i.e. by preventing the bridge to drift off in case a collision event "breaks the girder in two".

For the Bjørnafjord floating bridge, the main hazards (besides wave-, current- and wind loads) are ship impacts as well as possible abnormal traffic loading—due to abnormal large and heavy vehicles. Semi-probabilistic design is based on a set of characteristic loads that correspond to a certain annual exceedance probability for each limit state. Moreover, load reduction factors are applied on the characteristic loads to account for the fact that extreme values of different loads (load effects) do not occur at the same time.

3.2 Self-weight and Traffic Loads

Self-weight and traffic loads induce vertical response. While road traffic causes "static" global effects, wheel loads could include local dynamic effects. A traffic flow simulation based upon recorded traffic data, normally serves as basis for assessing traffic loading for medium and long span structures, considering various flow scenarios for the structure, i.e. congested flow/accidental flow, free flow and/or a combination of both. A crucial issue is the assumed frequency of events, e.g. congestion events per time unit (e.g. year), as well as the influence of the assumed percentage of trucks in the traffic on characteristic load effects and how it might develop over time. The Eurocode LM1 load case was developed based on a 100% share of trucks, which is extremely conservative if a realistic share is 10–15%. The Eurocode model was developed and calibrated for application on influence lengths up to a maximum limit of 200 m, hence application to lengths greater than 200 m represents a particularly conservative assumption. Another important aspect is the assumed load distribution relative to the influence function for a given response. The comparison of fatigue loads in Eurocode LM5 and measured loads, made by Svangstu [51], also clearly shows the conservatism in the design codes.

3.3 Environmental Conditions

Wave, current and wind conditions are obtained by hindcasting based on meteorological and oceanographic data and direct field measurements at the bridge site, to characterize the short- and long-term variability of the environmental conditions. This is challenging due to the length of the bridge and the spatial variability of the environmental conditions due to the complex topography in fjords and Bjørnafjord in particular (Fig. 2). Hindcast predictions for Bjørnafjord have been compared with field measurements for a period of 19 months [6–8] to assess the uncertainty of hindcast data which are necessary for long-term estimates of extreme conditions, see estimate for the significant wave height H_s in Fig. 3. The model uncertainty is defined as the ratio between measured H_s and simulated H_s, given by $X_j = H_{s,Mj}/H_{s,Sj}$. For H_s (> 0.6 m) the model uncertainty has a mean value of 0.767 and a coefficient of variation of 0.228. The mean plus one standard deviation is 1.005.

The environmental conditions are given in Table 1 but might be updated as more information is collected. Fatigue analysis is based on long-term data specified by the joint probability density function of wave and wind conditions, as discussed in NPRA [41] Augustson [3] and Cheng et al. [7].

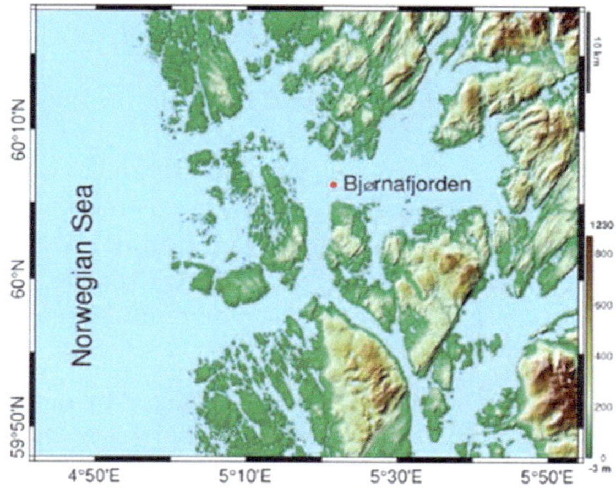

Fig. 2 Map of the Bjørnafjorden area showing topographical features with elevation indicated by color gradients. A red dot marks the location of Bjørnafjorden

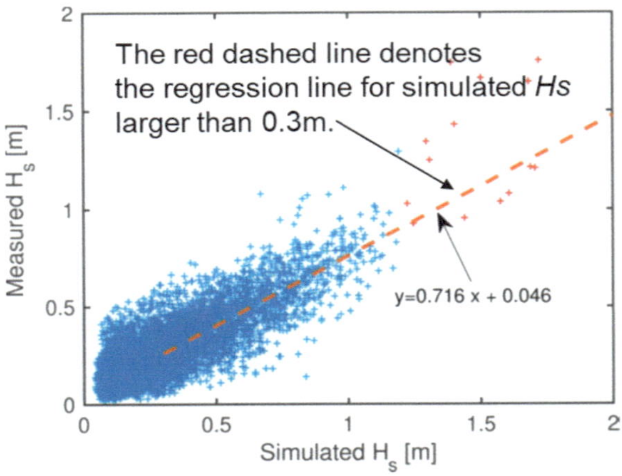

Fig. 3 Scatter plot of simulated and measured significant wave height, from February 2016 to June 2017. Red marks indicate events with simulated Hs larger than 1.2m [7]

Table 1 Environmental conditions for ULS criteria, initially specified for the Bjørnafjord site (not recognizing inhomogeneity in environmental conditions)

Waves		H_s	T_p	Short-crestedness function
	Wind sea	2.4 m	5.9 s	$Cos^4\theta$
	Swell	0.3 m	17.5 s	
Wind		Mean speed at 52 m height	Turbulence intensity, I	Mean wind profile U_{ref} $(z/10)^{\alpha}$, at height z
		39.3 m/s	I = 0.14	$\alpha = 0.12$

3.4 *Environmental Loads*

3.4.1 Wave and Current Loads

The hydrodynamic modelling of wave loads for the floating pontoon bridges has been comprehensively addressed e.g. in Cheng et al. [5–7], Fenerci et al. [21], Xu et al. [65, 66], based on the state-of-art in Faltinsen [20] and the standard [39]. The following wave load categories might be relevant:

- first order, wave frequency loads induced by surface waves
- second order slow drift loads—at a difference and sum frequency. Such forces are an order of magnitude less than the first order loads but can impose significant response if their frequencies coincide with natural frequencies, estimated either by a full second order theory or Newman's approximation.
- loads generated by internal gravity waves occurring in the sea due to a layer of fresh and salt water with different densities.

The hydrodynamic features of the pontoons in the global dynamic models are based on hydrodynamic coefficients, such as added masses, radiation damping, and transfer functions of wave excitation forces, estimated based on the potential flow theory. Challenging features in modelling sea loads are discussed in Sect. 3.6.3.

3.4.2 Wind Loads

Wind forces may stem from the mean pressure and the fluctuations in the incoming air flow (buffeting forces), oscillations of the structure itself (motion induced forces) as well as possible vortex shedding. The response to the first three types of loads are treated separately, because they occur in well separately wind velocity regions, however, without clear borders. It is assumed that the forced response in all cases is outside the critical wind velocity implying instability [50].

Aerodynamic loads on the column, tower and cables are normally modelled by viscous drag forces. The wind load acting on the bridge girder consists of three parts: the mean force due to mean wind velocity, the buffeting force due to fluctuating wind velocity in the two transverse directions, and the frequency-dependent force induced by girder motion [50]. For low bridges the mean force and buffeting force are considered using quasi-steady theory, based on instantaneous cross sectional drag and lift forces and moment per unit length [50] while the frequency-dependent motion-induced aerodynamic forces are neglected. A particular feature is the effect of waves on the turbulence and its effect on the bridge girder in the low part as described in Sect. 4.2.

3.5 Limit States

3.5.1 Ultimate Limit States

The steel bridge is made up by stiffened panels subjected to in-plane and partly lateral loads. Many alternative strength formulations for such components are available; e.g. in DNV-RP-C201 [17]. This implies that load effects for each stiffened panel involves in-plane forces/stress and possibly lateral pressure, due to self-weight, traffic and environmental loads. Traffic loads vary in time and space and to determine extreme values for ULS they need to be combined with environmental loads with allowance for their correlation. Characteristic extreme traffic loads for global load effects are based on a "static" load scenario—with the bridge densely populated with vehicles. However, lateral loads due to vehicle wheels involve dynamic effects.

A particular ultimate failure mode involves dynamic instability of the bridge girder, which will be subjected to compressive (and tensile) axial loads caused by the horizontal loads, steady and time-variant environmental loads (see Sect. 3.6.4).

3.5.2 Fatigue Limit States

The fatigue design check is based on resistance in terms of SN-curves and the Miner-Palmgren approach for the treatment of variable amplitude load effects and refers to the stress at the relevant joint [18]. The load effects for welded joints are normally expressed by the stress range, while the mean stress also is of importance especially for base material.

Fatigue design criteria are given in terms of characteristic fatigue damage, i.e. based on characteristic values of the SN-curve and expected load effect (stress range) histories and the acceptable fatigue damage is given by 1/DFF, where DFF is the design fatigue factor that depends on the consequences of failure and the effect of an inspection or monitoring program.

The main concern for fatigue in floating bridges is the bridge girder and support columns. These components are affected by the global load effects in terms of section forces and moments due to environmental loads, while lateral vehicle wheel loads on the girder deck also cause cyclic stresses in the plating and stiffeners in the upper girder deck. The fatigue load spectrum for the local loading depends on the road traffic pattern—vehicle number and sizes [51]. For a given structural detail affected by global and local stresses due to the are subjected to wave and wind loads and traffic loads. The fatigue loading then needs to be determined by combining the stress histories and then counting the stress ranges (cycles). It is noted that simplified ways to obtain the loading exist based on verification of the complete approach by cycle counting of the combined stress histories (e.g. [15]).

3.6 Load Effect Analysis

3.6.1 Overall Procedure

The load effect analysis under self-weight—, traffic—and environmental loads (Sect. 3.4) for ULS and FLS design checking is challenging due to the complex wave and wind environment and the large structure with multiple natural frequences causing resonances and the sensitivity to uncertain damping. The load effects in structural components are typically required in terms of multiple forces/moments or stresses. Extreme values (with a return period of 100 years) are required for ULS design checks while stress range time series are required for fatigue design checks.

While the girder in principle might be modelled by the Euler–Bernoulli beam theory, shear deformation, shear lag and warping torsional behavior might need the use of enhanced beam elements [4]. In particular areas like the girder-column connection (Fig. 4) shell elements and even solid elements (for fatigue analysis) are necessary to use. However, enhanced element models seem to exist only in "research" software and not in commercial software that integrate the load and structural models.

Shear lag (in beam flanges) affects the stiffness about the weak axis of the girder and hence moment distribution in the statically indeterminate girder as well as the

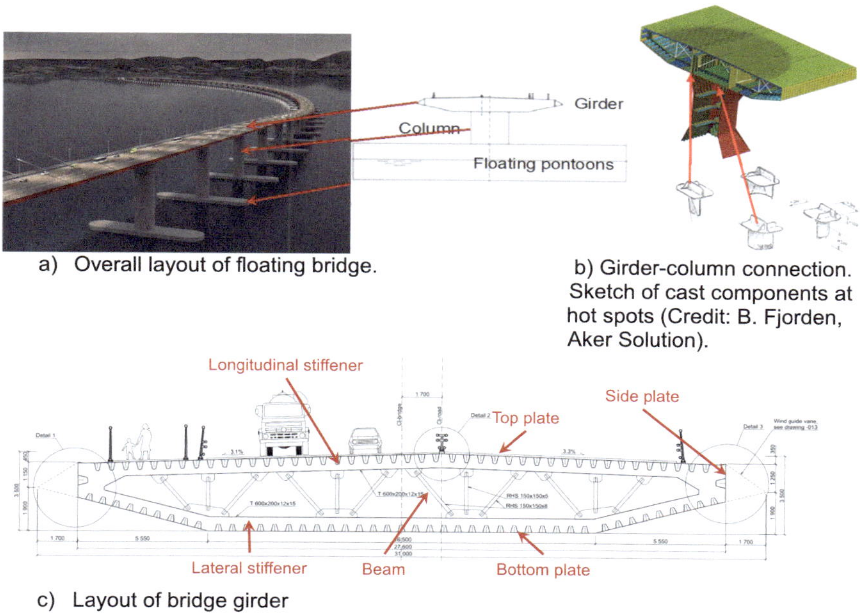

Fig. 4 Main components of the floating bridge structure. Panel **a**) shows the overall layout of the bridge with girder, columns and pontoons. Panel **b**) illustrates the girder-column connection, highlighting cast components at hot spots. Panel **c**) details the cross-section of the bridge girder

stress distribution in the cross section. The effect of shear lag depends on the ratio of width and span length and might be important for the weak axis bending moment [29]. For a dynamic problem, the effective span depends on the mode shapes, see e.g. Fig. 5. Luan et al. [29] found, as a rule of thumb, that the shear lag needs to be considered when the ratio of the effective width to the effective span length is 1/7 or more. In such a case, enhanced beam elements need to be used. Shell elements need to be employed in areas like the girder-column connection in the post-processing needed to obtain the local stresses due to the global stochastic dynamic response of the whole bridge in the time domain.

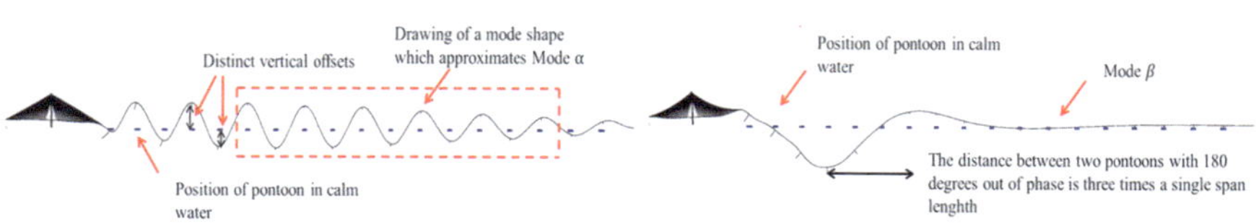

Fig. 5 Characteristics of two typical vertical mode shapes obtained by a beam element model [29]

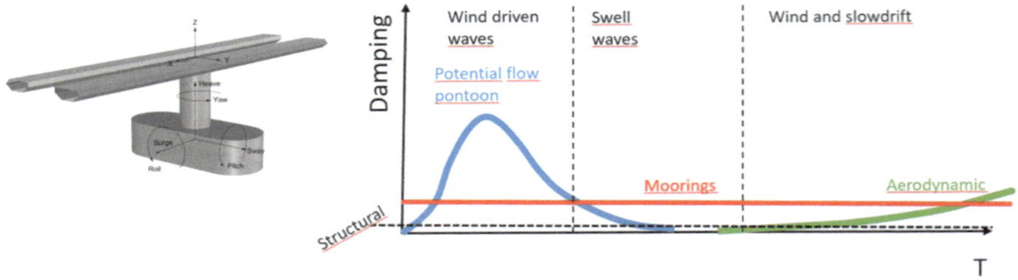

Fig. 6 Indication of sources of damping (due to potential flow, mooring and aerodynamics) on the pontoons as a function of excitation period and sources for dynamic response (wind driven waves, swell, wind and slow-drift wave forces). In addition, there will be viscous damping, which is important in the intermediate period range)

3.6.2 Challenging Modelling Features

Major challenges associated with the analysis of long floating bridges subjected to environmental loads, are:

- data and treatment of inhomogeneity of long-term environmental conditions and coherence in short-term conditions [5–10, 13, 60]
- hydrodynamic load interaction between pontoons [21, 63]—see Sect. 4.2.
- wave-current interaction [14, 57].
- damping includes aerodynamic damping, potential and viscous damping of pontoons [46], drag damping on mooring lines and structural damping. Viscous damping needs to be artificially introduced, preferably based on experiments, as mentioned in Sect. 4.,either by modifying the boundary conditions in the potential theory or simply through the quadratic viscous drag term in the Morison's equation (Fig. 6).

3.6.3 Strategy for Efficient Load Effect Analysis

Load effects should be determined with due account of relevant limit states. In principle, frequency or time domain (FD or TD) methods are applied in the dynamic analysis of environmental load effects. While FD methods are well developed for wave induced response analysis (e.g. [43]), FD for combined wave and wind loads are at an early stage of development [21, 65, 66]. FD analysis is exact and efficient for linear problems. Approximations are introduced e.g. by linearizing nonlinear effects. TD analysis is widely applicable but is time consuming for large MDOF systems with stochastic loading. Therefore, an appropriate balance of the stochastic time model and the models of the structural behavior and loads, should be used. The equations of motions in the time domain, can be expressed by Xu et al. [65, 66]:

$$\boldsymbol{M}_s\ddot{\boldsymbol{u}}(t) + \boldsymbol{C}_s\dot{\boldsymbol{u}}(t) + (\boldsymbol{K}_s + \boldsymbol{K}_h)\boldsymbol{u}(t)$$

$$= \underbrace{\boldsymbol{F}_{mean} + \boldsymbol{F}_{Buff}(t) + \boldsymbol{F}_{se}(t)}_{\boldsymbol{F}_{Aero}} + \underbrace{\boldsymbol{F}^{(1)}_{WA}(t) + \boldsymbol{F}^{(2)}_{WA}(t) - \boldsymbol{F}_{Rad}(t)}_{\boldsymbol{F}_{Hydro}} \quad (1)$$

where $\mathbf{M}_s$, $\mathbf{C}_s$ and $\mathbf{K}_s$ symbolize the still-air mass, damping and stiffness matrix, respectively, and $\mathbf{u}$ represents the degrees of freedom of the finite element model. $\mathbf{F}_{Aero}$ represents the wind actions, which consist of a time-invariant component $\mathbf{F}_{mean}$ due to the mean wind velocity, a dynamic component $\mathbf{F}_{Buff}$ due to turbulence in the wind field and self-excited forces $\mathbf{F}_{se}(t)$ generated by the motion of the girder. $\mathbf{F}_{Hydro}$ represents the wave actions, which consist of the radiation forces $\mathbf{F}_{Rad}$ induced by the motion of the submerged part of the pylons, the hydrostatic restoring stiffness $\mathbf{K}_h$, and the first-order and second order wave excitation forces $\mathbf{F}^{(1)}_{WA}$ and $\boldsymbol{F}^{(2)}_{WA}(t)$.

The wave loads consist of the excitation and radiation forces relating to the motions of the structure and include inertia (added mass) and potential damping forces. The radiation forces can be expressed by the inertia term at added mass for infinite frequency and a convolution term [12, 53]. Similarly, the aerodynamic self-excited forces involve a damping and stiffness term for harmonic motion at a single frequency and can be expressed by a convolution integral in the time domain to model self-excited forces for stochastic motion. An increased efficiency can be achieved by reformulating this equation as a state-space formulation—as a set of first order differential equations (e.g. [66]).

To achieve an efficient time domain analysis of many load conditions with combinations of all relevant loads and especially all wave and wind conditions with a structural model that provide the necessary load effects for limit state design checks, a global model based on (enhanced) beam elements is first used to estimate section forces and moments. The local load effects are obtained by

- using a finite element shell sub model based on input from the global analysis. This would involve solving the governing static equations of the sub model.
- establishing influence functions for local load effects expressed by section forces and moments for the different relevant response parameters by a static analysis. The joint time series of the (multiple) load effects at the various locations can then be obtained and used in the limit state check.

3.6.4 Static and Dynamic Instability Due to Axial Load

The curved bridge carries horizontal loads by axial forces, which, if compressive, can cause buckling of the slender structure. The dynamic character of the loads can on one hand imply that the buckling strength increases due to the stabilizing effect of inertia forces that are set up at the onset of buckling due to an axial load with short duration compared to the first natural period of the system. On the other hand, time varying axial force can result in parametric excitation or parametric resonance causing an instable response of the system. A generalized equation of motion can be expressed as

$$\tilde{M}\ddot{\mathbf{y}} + \tilde{C}\dot{\mathbf{y}} + (\tilde{K} - \tilde{K}_G N(t))\mathbf{y} = \tilde{F}(t) \tag{2}$$

where the product $\tilde{K}_G N(t)$ is the geometric stiffness matrix and N(t) the axial force and the excitation force $\tilde{F}(t)$ varying with time. For damped systems, the risk of parametric resonance is highest for a single harmonic axial force at twice the natural frequency of the mode considered. The parametric resonance could in principle be predicted by simulations including time varying geometric stiffness but in case of stochastic loading, it would require a large sample since parametric instability depends on both the intensity and the history of the stochastic axial load. Hence, an alternative strategy is to introduce a stability criterion to eliminate parametric resonance [64, 67].

$$\sigma_{\mathrm{N},m} < 0.5\mathrm{A}_{\mathrm{cr,harmonic,m}} \tag{3}$$

where $\sigma_{\mathrm{N,m}}$ is the standard deviation of the axial force in mode m, and $\mathrm{A}_{\mathrm{cr,harmonic,m}} = 4\zeta\, \tilde{K}/\tilde{K}_{G,}$ where ζ is the damping ratio, which has a significant effect on the dynamic stability of the bridge.

3.6.5 Load Effect Analysis of the Bridge with Damage or Under Repair

As shown in Sect. 5, to allow road traffic on the bridge in damaged and repair conditions, the bridge needs to fulfill design criteria, especially ULS criteria. If the damage/repair conditions involve a winter season, 100-year environmental conditions, would apply. A particularly demanding situation is when a dry dock, with a displacement about 4–5 times that of pontoons, is used for pontoon repair, especially the pontoon supporting the high column (see Fig. 1).

Comparison studies of the dynamic response under environmental conditions, of the bridge in intact and repair conditions were conducted by considering various combinations of environmental conditions [10, 11, 35]. Eigen-value analysis indicates that pendulum modes of the pontoon-columns exist in the repair condition due to the large mass of the floating drydock. These modes can be excited by swell waves and contribute significantly to the bending moment of the girder weak axis. It is found that the structural load effects at the damaged pontoon axis, can be considerably increased. Unless the excitation of the pendulum modes is reduced or additional damping is provided or the bridge is strengthened, road traffic disruption will occur. See also Sect. 5.

Figure 7 compares the response spectral density for intact and damaged bridge. The significant increase in the weak axis bending moment is noted, while the torsional moment M_x decreases at axis-A3 when compared to that of the intact bridge. The orders of the excited modes are also indicated.

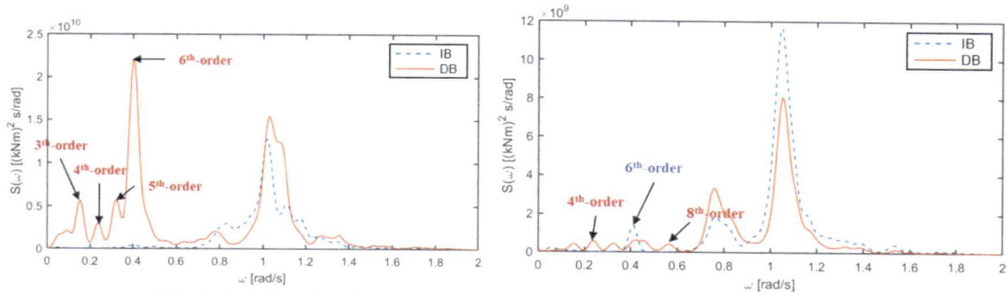

a) Weak Axis Bending Moment

b) Torsional Moment

Fig. 7 Power spectral density of the time series of the response at the axis A3 (with the tallest column). IB and DB signify intact and damaged bridge, respectively [10]

3.7 Stochastic Load Effect Analysis

ULS and ALS design checks are based on characteristic values of extreme load effects corresponding to specified annual exceedance probabilities. These load effects are calculated by accounting for the long - and short-term variations in environmental conditions. Each short- term condition is based on the assumed stationarity of wave and wind fluctuations, typically 3 h and 10 min, respectively. For practical reasons 1-h short term periods are considered. The long term variation long-term approach accounts for the probability of occurrence of the parameters describing the wave condition (H_s, T_p, θ_{wave}) and wind condition (mean wind, U_w, θ_{wind}) in the long-term period.

The long-term characteristic extreme value, corresponding to annual exceedance probabilities of 10^{-2} and 10^{-4}, respectively, for ULS and ALS design checks [38, 41] can be determined by the full long-term method (FLM) which essentially integrates short-term response statistics (i.e., distributions of all peaks, distributions of extreme values or mean upcrossing-rate) over all short-term environmental conditions [43]. It incorporates both the long-term variability of environmental conditions represented by a joint probability distribution of environmental parameters, $f(u_w, h_s, t_p)$ and the variability of short-term extreme values characterized by the conditional distribution of short-term responses for each environmental condition. The cumulative distribution function (CDF) of the extreme value, that is, the extreme value over a long-term period T can be expressed as the integral of the short-term response over all possible environmental conditions [43]:

$$F_{\widehat{X}}(\zeta) = exp\left\{ -T \int\limits_{V} \int\limits_{H_s} \int\limits_{T_p} \nu_X^+(\zeta|u, h_s|, t_p) f(u, h_s, t_p) du dh_s dt_p \right\} \quad (4)$$

where ν_X^+ denotes the average ζ-upcrossing rate for a short-term environmental condition.

While the FLM approach represents the long-term variability of environmental conditions accurately, it is very computationally demanding. Hence the following approximate approaches are commonly used:

- the simplified long-term method (SLM) which integrates short-term response statistics by including short-term environmental conditions that contribute to the exceedance probability of extreme events (e.g. [55, 66]).
- the inverse FORM (e.g. [61]).
- the environmental contour method (ECM), which is a simplification of the IFORM, decouples the uncertainty in the environmental conditions and the short-term extreme values [22]. Fundamentally, the ECM calculates the contour surface corresponding to a selected exceedance probability. It is further assumed that the most important combination of environmental parameters on the contour surface can be used to approximate the long-term extreme value. Neglecting the short-term variability in the extreme values can give non-conservative results. Thus, a higher percentile than the median or expected maximum is used as for each short-term characteristic value [22]. The ECM has been frequently applied in ocean engineering endeavors to search for the appropriate short-term design case.

So far, most research studies have focused on simplified methods for wave induced load effects in offshore structures. However, Xu et al. [65, 66] addressed a very complex structural response problem, i.e., the three-span suspension bridge with two floating pylons in Fig. 8, subjected to combined wind and wave loading, which both are important for this kind of structure.

The bridge model includes nonlinear effects due to geometrical effects and buffeting wind load components. Three environmental parameters were considered, namely, the mean wind velocity $U_{w,}$ the significant wave height H_s and the peak wave period T_p. Both (linearized) FD and TD approaches were used. The accuracies of the SLM, ECM and IFORM were validated through a comparison with the FLM. Table 2 shows that the simplified methods provide adequate results and they thus can be used for predicting the wind- and wave-induced extreme load effects in this bridge concept.

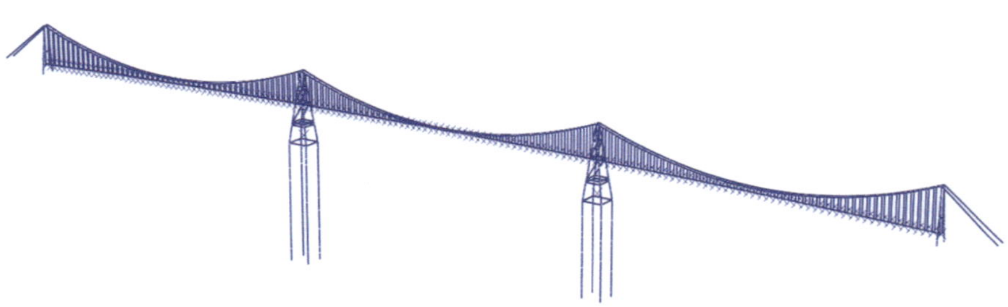

Fig. 8 Structural model of the three span suspension bridge with two floating pylons (towers) [65, 66]

Table 2 Comparison of 100 yr extreme values of girder bending moment based on different methods [66]

Method	Extreme response value based on FD results (MNm)	Extreme response value based on FD&TD results (MNm)
FLM	50.43	–
SLM	50.26	–
ECM	49.56 to 58.57	60.8[a]
IFORM	51.54	62[a]

[a]Only the environmental conditions around the design "point" were simulated in the time domain approach to account for nonlinearities (i.e., $28 < V < 30$ m/s, $4 < H_s < 6$ m and $7 < T_p < 9$ s)

The time domain simulations showed that nonlinearities imply an increase in extreme values of the bridge girder weak axis bending moment of up to 20%. However, time domain simulations can be very time consuming, even by applying the ECM or IFORM. Therefore, a computationally efficient approach was proposed to predict the long-term extreme response values based on the combined FD and TD simulation results and the use of IFORM and ECM. TD was used only in simulations close to the design "point". A similar approach might be used in combination with the SLM approach.

4 Role of Physical Testing to Assess Environmental Load Effects

4.1 General

Physical testing might be conducted to

- demonstrate feasibility of or document a product
- provide data to support design
- provide basis for assessing the uncertainty in numerical models and the associated partial safety factors

4.2 Model Tests

Small-scale experiments in controlled environments in wave tanks, (e.g. [30, 62]), and field measurements in natural environments, are commonly used to assess uncertainties of numerical predictions of the global behavior, whereas full-scale tests of components are commonly carried out in the laboratory to validate the strength or durability of the components.

Aerodynamic testing is typically made in wind tunnels to determine drag, lift and moment coefficients and aeroelastic properties (mass, stiffness and damping) based on 2D model testing. In the wind model test for the Bjørnafjord bridge [40], a free-surface boundary test was carried out in the wind tunnel, to estimate the influence of waves on the vertical turbulence component, and its effect on the bridge vertical response. It was found that the wind velocity standard deviation in the vertical direction could increase and that the turbulence increased with increasing velocity. Hence, the wave height and the distance between the mean water level and the bridge girder are important when evaluating the effect of waves on the wind turbulence on the bridge girder in low bridges.

Tests with combined wave and wind (and current) on very large floating bridges with dynamic behavior, is challenging due to the conflict between Froude, Reynolds and Cauchy scaling and limited size of laboratories. The solution can then be to use a proper scaling of the most notable features and use of numerical analyses to generalize the model test results. Simplifications might also involve using a truncated model to avoid too small scaling [47, 56]. An alternative to address the challenges would be to use a hybrid testing by combining a physical model, e.g. of wave loads and impose wind loads by mechanical equipment, as obtained by a numerical model, which is run in real-time with feedback from measurements of the physical model, e.g. Zasso et al. [68] and Rodrigues et al. [47].

In recent years, several model tests have been conducted to enhance the understanding and design of large floating bridges. These tests have focused on various aspects of hydrodynamic performance, structural responses, and dynamic behavior under different environmental conditions. Experimental campaigns can typically be grouped in two main categories:

- Tests to find hydrodynamic loads and corresponding hydrodynamic load coefficients.
- Tests to study global response in the bridge exposed to various environmental loads.

Hydrodynamic loads are normally studied for single components exposed to hydrodynamic loads (typically one bridge pontoon with vertical shaft, Fig. 9), but multiple pontoons may also be studied to investigate effects of interaction between pontoons, global dynamic behavior etc. (see Fig. 10). Tests were carried out either with a fixed model to study hydrodynamic excitation forces, or as forced oscillation tests to study added mass and damping from radiated waves and viscous effects. In such tests, requirements to stiffness/natural frequency of the test set-up are important.

For eigenmodes exposed to excitation from environmental loads, aero- and hydrodynamic, mooring—and structural damping is important. In particular, hydrodynamic damping of the pontoons has been investigated in several experimental campaigns.

Ravinthrakumar et al. [46] investigated damping properties of a pontoon forced to oscillate in surge at different KC numbers and oscillation periods. The hydrodynamic coefficients of the pontoon in forced surge motion were extracted and compared against numerical simulations. Added mass as well as damping coefficients were

Fig. 9 Forced oscillation test with single pontoon to estimate hydrodynamic damping [46]

Fig.10 Hydro-elastic, truncated model with 10 pontoons, of the Bjørnafjord bridge concept 5K12 in SINTEFs Ocean Basin [56]

found to be dependent on the KC number in the tested range of oscillation periods. Numerical simulations were performed using a hybrid potential flow-CFD solver. Simulations results were in fair agreement with those obtained from the model tests and this suggests that flow separation along the pontoon and at the sharp edges along the flat bottom provide the dominant hydrodynamic damping. Heave motions and hydrodynamic pontoon interaction were also studied.

Numerical studies (e.g. Xiang [63]) have shown that hydrodynamic interaction between pontoons with a distance of about 100 m, may influence the heave motion and the structural response of the bridge girder. Xiang [63] proposed an efficient, simplified method to account for hydrodynamic interaction between pontoons based on model tests with a single and three pontoons as well as a hydro elastic 10-pontoon model.

When waves and currents interact, the wave properties and corresponding hydrodynamic loads can be influenced. Dai et al. [14] investigated global structural responses of a 5 km long straight side-anchored floating bridge under the combined action of waves and current considering the wave-current interaction. It was concluded that wave-current interaction had a significant effect on a fjord-crossing floating bridge studied.

In model tests that aim at reproducing the global dynamic response of the bridge, it is important to apply a proper scaling and accurate modelling of structural stiffness and mass of the bridge. Normally a reasonably large scale (typically 1: 30–40) is selected to keep scaling effects as small as possible (since floating bridge is used in areas with moderate wave conditions). Long floating bridges with lengths of several km normally need to be truncated to due to limited size in most laboratories. It is vital that the truncation of the test model and the corresponding boundary conditions of the bridge is selected such that the properties of important global modes are representative for the full-scale structure.

Rodrigues et al. [47] carried out a comprehensive model test campaign with a truncated bridge in scale 1:33 representing a 4.75 km straight floating bridge with 4 mooring clusters. Various combinations of wave, current and wind were considered, including an inhomogeneous environment. Wind excitation was modelled by actuator winches applying wind force at distributed locations along the bridge.

Viuff et al. [56] made a comprehensive experimental study of a 1:31 scale truncated section of the Bjørnafjord concept. The truncated section has a double-curved girder and includes a simplified stay-cable tower and 10 pontoons. The tests showed a clear coupling between the transverse motion in the horizontal plane and rotation of the bridge girder due to the significant distance between the bridge girder centerline and the supporting pontoons. From the time series, the first two natural periods were found to agree with the numerical estimates. Tests with current were performed with two current velocities corresponding to the 1-year and 10-year currents at the Bjørnafjord site. The results from the above model test were analyzed further by Viuff et al. [57]. In particular, the effect of current on the response was studied. Collinear current was observed to have a significant influence on the responses from both 100-year wind- and 10,000-year swell wave conditions, especially due to amplified vertical response,suggesting that wave–current-interaction effects should be properly accounted for in the design. The response was also found to be sensitive to the direction of the environmental loads. The current uncertainty that seems to exist regarding predictions of wave-current interaction, show that further investigations are needed by numerical and experimental methods.

4.3 In-Situ Measurements—Monitoring

Monitoring of the performance of bridges might serve various purposes:

- Recording the bridge structural response under wave loads to identify the natural frequencies and modes.
- Recording the environmental load condition and the response to assess the uncertainty in the response analysis (of motions, internal forces/moments).
- Identify possible hazards, such as wind speed, ships on collision course etc. in advance to initiate risk mitigation effects like closing the bridge for road traffic.
- Identify damage which reduces the strength, or cause flooding that reduces the buoyancy and increased structural response.

It is necessary to differentiate between damage due to extreme loads or gradual deterioration. While events causing overload damage can be observed and trigger examination of the damage, gradual deterioration due to crack growth is more difficult to detect before it gets rather large, without significant instrumentation. Hence, identification of the damage based on inspection might be more efficient. Anyway, damage is only "acceptable" as long as there is sufficient residual strength. A particular aspect of floating bridges is its large size with many potential locations of damage.

There are limited systematic field testing of floating bridges under environmental loads. However, a comprehensive monitoring system was installed on the 933 m long Bergsøysund Bridge and was used for measurements of environmental conditions and response of the bridge. The instrumentation includes 5 anemometers, 6 wave radars, 14 triaxial accelerometers, and a global navigation satellite system (GNSS) sensor [27]. The wave action is the dominating process at this site. Owing to the complexity of the terrain, the correlation between the wind direction and the wave direction in the fjord is limited.

Kvåle and Øiseth [28] compared the numerically predicted and measured response of the Bergsøysund Bridge based on an initial and updated numerical method. The numerical method focused on waves based on linear wave theory while a simplified wind load model was used. A beam model of the bridge was applied, with assumed structural damping. A sensitivity-based finite element model updating technique [44] was used to calibrate the numerical model. The standard deviations and response spectra of the bridge displacements (accelerations), obtained by predictions and measurements, were compared with reasonable agreement. The model updating and the structural damping (for the vertical response) turned out to be important and the question is how the lessons learned can be used to predict the response for new concepts at the design stage, e.g. of a larger size like the Bjørnafjord bridge. Moreover, despite the comprehensive instrumentation, there are uncertainties in the measurements that should be considered.

5 Analysis for ALS Design Check Relating to Ship Impacts

5.1 General

Accidental loads are primarily handled by ALS criteria, which were introduced for offshore structures in 1984 to ensure more robust structures by a design check in two steps [32]:

- Estimate accidental damage (fabrication defects, deterioration or due to accidental loads) at an annual probability of 10^{-4} (and expected value of other loads).
- Check that the damaged structure survives gravity/payloads and environmental loads at a certain annual probability of occurrence (annual, 10-year or 100-year max, depending on the correlation between the damage situation and the environmental loads) and use of load factors equal to 1.0.

This approach is also adopted for bridges (e.g. [42]), considering relevant accidental loads, namely ship impact loads, for the Bjørnafjord bridge. The accidental load should be determined by a risk analysis.

Some standards for bridges [1, 16, 19] refer to risk-based design of bridges. The latter reference contains some guidance on risk-based design of bridges according to the Eurocode.

Terndrup Pedersen et al. [54] made a review of risk-based design of bridges and presented a risk-based design approach for ship impacts on bridge piers fixed to the seabed. As outlined by Sha et al. [48, 49], Jin et al. [24], Moan and Jin [35] and Moan [33] the existing procedures have limitations with respect to the impact analysis methodology and the link between impact scenario and ultimate consequences as applied to floating bridges. To establish a basis for risk management relating to ship impacts, a comprehensive risk analysis has been conducted for an early design of the Bjørnafjord floating pontoon bridge.

5.2 Ship Impact Risk Analysis Procedure

The purpose of the present risk analysis is to estimate the ultimate consequence of ship impacts, namely: fatalities to users of the bridge as well as road traffic disruption caused by the impact. The ship collision risk analysis [33, 35] is carried out by estimating the

- Frequency versus impact energy and—direction (traffic pattern (see e.g. Fig. 11), root causes of operational and technical faults, bridge layout, considering the mitigation measures) was estimated by Rambøll [45] using the methodology in Mølsted et al. [36].
- Consequences

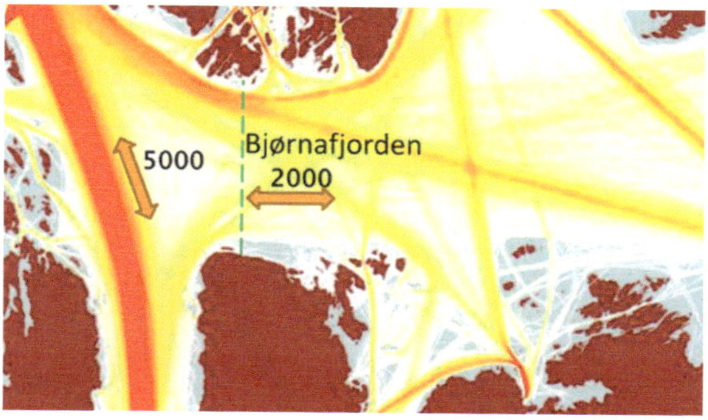

Fig. 11 Ship traffic pattern in Bjørnafjorden for the years 2015-2016. The traffic density is described by red-yellow lines. A dashed blue line indicates the bridge location. The background includes red landmasses and light blue water areas [34]

- Immediate consequences in terms of structural damage and possible buoyancy loss as well as girder motions
- Ultimate consequences in terms of fatalities/injuries and roads traffic disruption

It turns out [35] that the impacts on the pontoons make the largest contribution to the ship impact risk for the current bridge. Examples are therefore directed towards pontoon impacts. It is also found that the contribution to the fatality risk due to ship collisions is small when compared to the road traffic risk of fatalities on the bridge. Hence, the focus is on road traffic disruption.

A three-step analysis procedure is necessary to determine the immediate consequences in terms of internal forces, possible permanent damage and the girder motions (accelerations) [35, 49]:

Step (1) A local, nonlinear shell analysis is carried out to determine the force-displacement relationship in the impact region (see the top of Fig. 12 and the curves for different impact locations and directions in Fig. 14).

Step (2) A global impact analysis is carried out to: a) identify the energy distribution between the local structural damage and the global motion; b) determine the overall bridge response, including the indentation in the pontoon). As illustrated in Fig. 12, the ship is modeled by a mass-spring system connected to the bridge. The stiffness of the nonlinear spring is defined by the force-deformation curve obtained in step 1. The default global model of columns/girder is based on a linear beam model. If it is found that the column will suffer damage, a nonlinear FE shell model is introduced for the column (LSDYNA). Figure 13 shows an example of damage caused by a transverse impact on the pontoon of the tallest column. Figure 14 shows the force-indentation curves for ship-pontoon impacts.

Step (3) Based on the results from these first two steps, the damage/flooding are determined. First the structural damage is estimated based on impact locations/directions, force-indentation curve. The indentation in the pontoon together with the layout of the pontoon are used to determine possible damage that causes leaks and then the subsequent consequence of flooding of pontoon compartments.

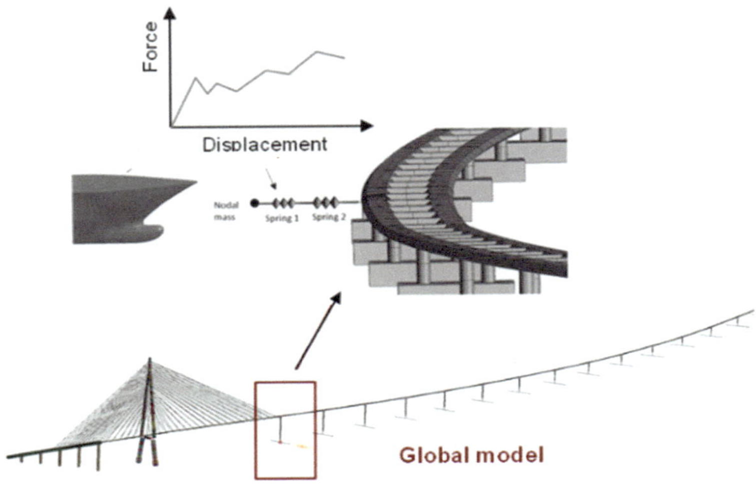

Fig. 12 Structural Impact analysis model based on force-displacement (indentation) relationship for the contact between the ship and the pontoon (Fig. 14) and a global model of the girder, columns and pontoons

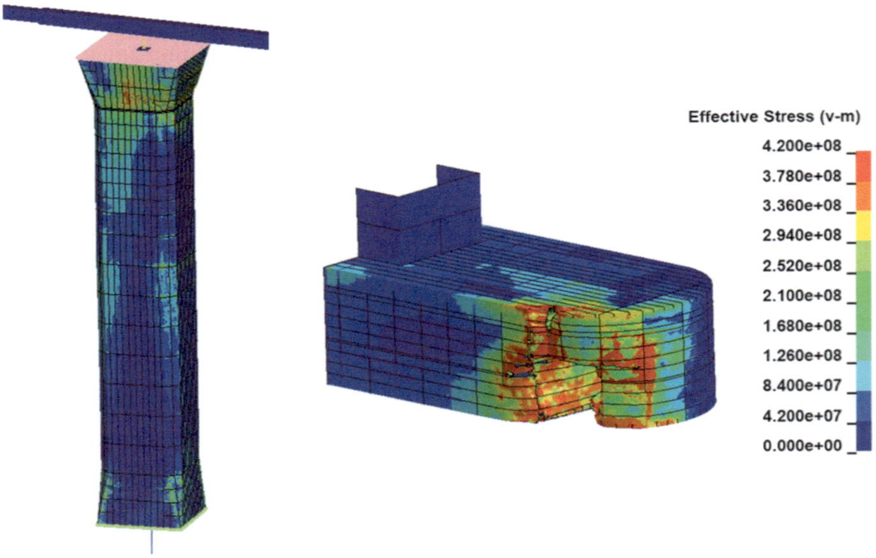

Fig. 13 Example of damage in a 125 MJ ship impact between bulkheads in the pontoon with tall column. The color plot represents the effective stress level (in Pascals)

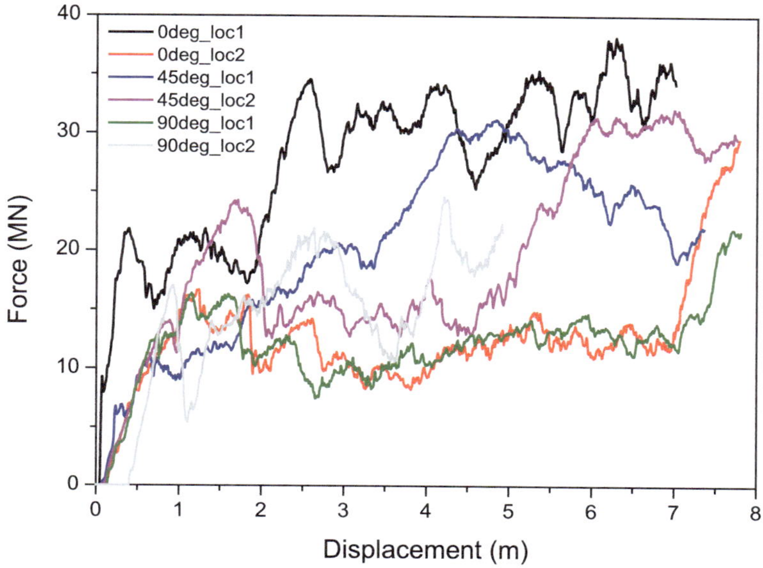

Fig. 14 Local force-indentation curve for ship bow-pontoon impact (for three directions and two locations). The zero deg. direction is along longitudinal pontoon axis

The repair time—including mobilization of the equipment—was estimated by Aker Solutions [2] to be between 2 and 20 months—where the longest time refers to taking the column-pontoon out and carrying out repair in a yard or replacement (Fig. 15c). The main question then is whether the bridge can be open for some road traffic or not. To allow road traffic, it was assumed that all criteria, especially ULS criteria, need to be satisfied for the bridge in damaged and repair conditions. Since the accident scenario could occur at any time during the year, ULS requirements should be fulfilled for the 100-year environmental conditions. Global analysis considering self-weight, traffic and environmental loads, including dynamic effects, was conducted.

(a) Use of cofferdam to provide a dry working space.

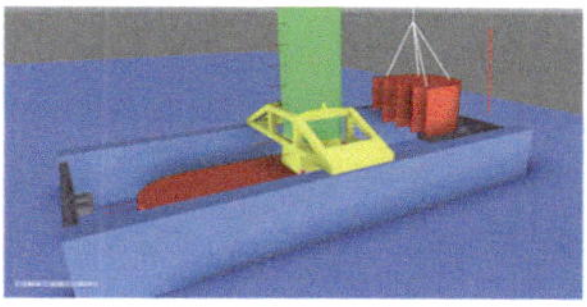

(b) Use of drydock to provide a dry work working space.

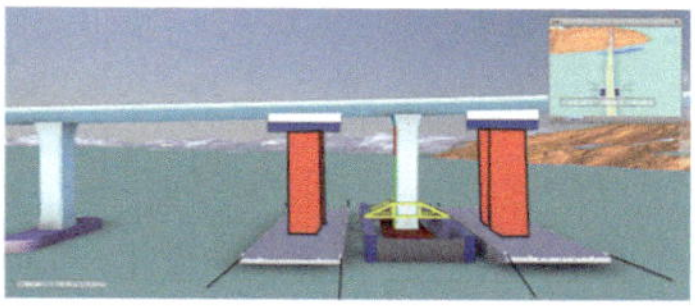

(c) Use of barges/temporary columns to support the girder during repair of column/pontoon.

Fig. 15 Alternative devices used for repair of damaged pontoon (figures are not to scale). The cofferdam and drydock provide a dry working space for part and the whole pontoon, respectively

The traffic disruption time was estimated by

- the annual frequency of each impact scenario (impact location, direction, energy discretized in ranges).
- the immediate consequence (structural damage, flooding), considering the uncertainties in the damage calculation, especially the uncertainty in the impact force-indentation (Fig. 14).
- estimated time for damage and repair conditions based on the damage condition and choice of repair method for conditions with no compliance with the ULS and SLS criteria.

5.3 *Risk Estimate and Mitigation*

It turned out that the expected annual downtime due to ship impacts was about 7 days with 90% of the downtime stemming from impacts on the pontoons with an energy up to 50 MJ. This result caused scrutinizing of the impact frequency analysis, which initially was focused on severe impacts in view of the ALS criterion referring to energies at an annual frequency of 10^{-4}. Moreover, further ship traffic control measures to reduce the impact probability, were pursued. Another option (proposed by Moan and Jin [35]) was to introduce a limit state in addition to the ALS to address the risk associated with frequent impacts [33], and strengthening of the areas of the girder and columns that did not fulfill the ULS or SLS criteria in damaged and repair conditions.

6 Concluding Remarks

Recent research and development activities to improve the design basis for floating bridges are described, with a focus on the planned 4.6 km Bjørnafjord pontoon bridge, as an extension of the Moan and Eidem paper in the first WCFS conference in 2019. Particular efforts are devoted to the analysis of the effects of sea loads and ship impacts for design check related to ultimate and accidental collapse limit states, respectively.

The challenges and advances associated with sea loads include handling inhomogeneity in environmental conditions, wave interaction between pontoons, wave-current interaction and viscous damping, global dynamic buckling of the bridge as well as proper and efficient account of the long-term variability in environmental conditions, are highlighted. The important role of laboratory experiments in this connection is noted. A risk analysis procedure to deal with the major accidental load—ship impact—for floating bridges, is presented. It involves an estimate of the frequency of impact scenarios, the immediate structural damage and possible flooding, the expected time for repair and ultimate consequences in terms of fatalities and road traffic downtime. Owing to the high traffic disruption rate, a scrutiny of the impact frequency for relatively frequent low energy events (instead of the initial focus on 10-4 events relevant for the ALS criterion), possible introduction of extended ship traffic control measures, as well as to possibly introduce a limit state in addition to the ALS to address the risk associated with frequent impacts.

References

1. ASHTO (2009) Guide specification and commentary for vessel collision design of highway bridges, 2nd edn. American Association of State Highway and Transportation Officials, Washington, DC
2. Aker Solution (2021) Repair after ship impact study on Bjørnafjorden floating bridge. Report 160025-34-AS-Z-0001, Ver 03, 08 Dec 2024
3. Ágústsson H (2018) E39—measurements, simulations and analysis of wind conditions for long bridges. Met-ocean Workshop, 12 Mar 2018 NTNU/SINTEF, Trondheim
4. Cambronero-Barrientos F, Díaz-del-Valle J, Martínez-Martínez J-A (2017) Beam element for thin-walled beams with torsion, distortion, and shear lag. Eng Struct 143(2017):571–588
5. Cheng Z, Gao Z, Moan T (2018) Wave load effect analysis of a floating bridge in a fjord considering inhomogeneous wave conditions. Eng Struct 163:197–214
6. Cheng Z, Svangstu E, Gao Z, Moan T (2019a) Field measurements of inhomogeneous wave conditions in Bjørnafjorden. J Waterway Port Coast Ocean Eng 145(1):05018008
7. Cheng Z, Svangstu E, Moan T, Gao Z (2019) Long-term joint distribution of environmental conditions in a Norwegian fjord for design of floating bridges. Ocean Eng 191(2019):106472
8. Cheng Z, Svangstu E, Moan T, Gao Z (2021) Assessment of inhomogeneity in environmental conditions in a Norwegian fjord for design of floating bridges. Ocean Eng 2021:220
9. Cui M, Cheng Z, Moan T (2022) A generic method for assessment of inhomogeneous wave load effects of very long floating bridges. Mar Struct 83:103186. https://doi.org/10.1016/j.marstruc.2022.103186
10. Cui M, Cheng Z, Moan T (2023) Dynamic response analysis of an end-anchored floating bridge with a damaged pontoon under repair operation, paper OMAE2023-101805. In: Proceddings of 42nd OMAE conference, Melbourne, Australia, 11–16 June 2023
11. Cui M, Cheng Z, Moan T (2023) Effects of inhomogeneous wave modeling on extreme responses of a very long floating bridge. Appl Ocean Res 134:103505
12. Cummins WE (1962) The impulse response function and ship motions. Schiffstechnik 9:101–109
13. Dai J, Stefanakos C, Leira BJ, Alsos HS (2021) Effect of modelling inhomogeneous wave conditions on structural responses of a very long floating bridge. J Mar Sci Eng 9(5):548
14. Dai J, Abrahamsen BC, Viuff T, Leira BJ (2022) Effect of wave-current interaction on a long fjord crossing floating pontoon bridge. Eng Struct 266:114549
15. Dirlik T, Benasciutti D (2021) Spectral Methods in vibration fatigue: a review with a historical perspective. Metals 11(9):1333. https://doi.org/10.3390/met11091333
16. DS/EN 1991-1-7 DK NA:2013 National annex to eurocode 1: actions on load carrying structures—part 1–7: general actions—accidental loads (by The Danish Road Directorate)
17. DNV-RP-C201 (2023) Buckling strength of plated structures, Oslo, Norway, 2023–9
18. DNV-RP-C203 (2021) Fatigue design of offshore steel structures, Oslo, Norway, Edition 2019–09—Amended 2021–09
19. EN 1991 (2010) Eurocode 1: actions on structures—part 1–7: general actions—accidental actions. July 2006 CEN (incorporated corrigendum February 2010)
20. Faltinsen OM (1993) Sea loads on ships and offshore structures. Cambridge University Press, Cambridge
21. Fenerci A, Kvåle KA, Xiang X, Øiseth O (2022) Hydrodynamic interaction of floating bridge pontoons and its effect on the bridge dynamic responses. Mar Struct 83:103174

22. Haver S, Winterstein SR (2009) Environmental contour lines: a method for estimating long term extremes by a short term analysis. Trans Soc Nav Arch Mar Eng 2009(116):116–27
23. ISO 19900 (2023) Petroleum and natural gas industries—offshore structures—part 1: general requirements and parts for different types of structures. International Standardization Organization, London
24. Jin Y, Moan T, Sha Y (2021) Numerical study of the structural consequences of ship impacts on a floating bridge in a risk analysis perspective. In: Marstructure conference, Trondheim
25. Kanyilmaz A, Birhane M, Fishwick R, del Castillo C (2023) Reuse of steel in the construction industry: challenges and opportunities. Int J Steel Struct 23(5):1399–1416. https://doi.org/10.1007/s13296-023-00778-4
26. Kanyilmaz A, Hoi Dang V, Kondratenko A (2023) How does conceptual design impact the cost and carbon footprint of structures? Structures 58:105102
27. Kvåle KA, Øiseth O (2017) Structural monitoring of an end-supported pontoon bridge. Mar Struct 52:188–207
28. Kvåle KA, Øiseth O (2021) Dynamic response of an end-supported pontoon bridge due to wave excitation: numerical predictions versus measurements. Shock Vib 2021:8842812
29. Luan C, Moan T, Cheng Z, Kvåle KA (2024) Shear lag effects on global bending moments in a long-span pontoon bridge under self-weight, traffic and environmental loads. Mar Struct 93:103518
30. Løken AE, Oftedal RA, Aarsnes JV (1990) Aspects of hydrodynamic loading and response in design of floating bridges. In: Second symposium on strait crossings. Trondheim, Norway. ISBN: 9061911184, pp 479–486
31. MOB (2000) Mobile offshore base science and technology program. Final report, technical report TR-2125-OCN, naval facilities engineering and expeditionary warfare center, Port Hueneme CA, Dec 2000
32. Moan T (2009) Development of accidental collapse limit state criteria for offshore structures. Struct Saf 31(2):124–135
33. Moan T (2024) Estimation and mitigation of ship collision risk of a floating bridge. Submitted for publication
34. Moan T, Eidem ME (2020) Floating bridges and submerged tunnels in Norway—the history and future outlook. In: Wang CM et al (eds) WCFS 2019 proceedings of the world conference on floating solutions, Springer Nature Singapore, lecture notes in civil engineering, vol 41
35. Moan T, Jin Y (2023) Risk analysis of a floating bridge subjected to ship collisions. NTNU report to NPRA, 15 Jan 2023
36. Mølsted RF, Kürstein GK, Kristina M, Gamborg HM, Koldborg JT, Tue LS, Søren RT (2012) Quantitative assessment of risk to ship traffic in the Fehmarnbelt Fixed Link project. J Pol Safety Reliab Assoc 3:123–134
37. Norwegian Standard (2016) NS-EN 1990:2002+A1:2005+NA:2016 basis of design (Norwegian version of EN 1990)
38. NORSOK N-001 (2024) Structural design (—and associated suite of standards). Norwegian Technology Standards, Oslo
39. NORSOK N-003 (2007) Actions and action effects. Norwegian Technology Standards Institution, Oslo
40. NPRA (2018) Wind model-tests floating bridge. Step 1: small scale testing, SBJ-32-C4-SOH-20-RE-001, revision A, Oslo, Noway, 6 June 2018
41. NPRA (2023) Design basis Bjørnafjorden: end anchored floating bridge. Norwegian Public Road Administration, Oslo
42. NPRA (2024) Handbook N400, design of bridge (in Norwegian). Norwegian Public Road Administration, Oslo
43. Næss A, Moan T (2013) Stochastic dynamic analysis of marine structures. Cambridge University Press
44. Petersen ØW, Øiseth O (2017) Sensitivity-based finite element model updating of a pontoon bridge. Eng Struct 150:573–584

45. Rambøll (2020) Bjørnafjorden—Skibsstødsanalyse for ende-forankret flydebro—detaljerede resultater, revision 2, 27 Oct 2020, Copenhagen, Denmark
46. Ravinthrakumar S, Alsos H, Økland OD, Lie H, Xiang X, Grytå OA, Kristiansen T (2024) Forced surge motion of a floating bridge pontoon to evaluate the hydrodynamic damping properties. Ocean Eng 303(2024):117696
47. Rodrigues JM, Viuff T, Økland OD (2022) Model tests of a hydroelastic truncated floating bridge. Appl Ocean Res125. Article 103247. https://doi.org/10.1016/j.apor.2022.103247
48. Sha Y, Amdahl J (2017) Ship collision analysis of a floating bridge in Ferry- free E39 project. Paper OMAE2017–62720. In: Proceedings of 36th OMAE conference, Trondheim
49. Sha Y, Amdahl J, Liu K (2019) Design of steel bridge girders against ship forecastle collisions. Eng Struct 196:109277
50. Strømmen E (2010) Theory of bridge aerodynamics. Springer Science & Business Media
51. Svangstu E (2019) Comparison of the Eurocode Fatigue Load Model 5 (FLM5), and measured traffic loads relevant for the E39 project crossing Bjørnafjorden. Priv Commun 20(09):2019
52. Suzuki H (2005) Overview of megafloat: concept, design criteria, analysis and design. Mar Struct 18:111–132
53. Taghipour R, Perez T, Moan T (2008) Hybrid frequency–time domain models for dynamic response analysis of marine structures. Ocean Eng 35:685–705
54. Terndrup Pedersen P, Chen J, Zhu L (2020) Design of bridges against ship collisions. Mar Struct 74:102810
55. Videiro PM, Moan T (1999) Efficient evaluation of long-term distributions. Paper OMAE 1999-6014. In: Proceedings of the 18th OMAE conference (ASME). https://doi.org/10.1115/OMAE2017-61444
56. Viuff T, Ravinthrakumar S, Økland OD, Grytå OA. Xu X (2023a) Model test of a hydroelastic truncated floating bridge with a stay-cable tower. Appl Ocean Res 135:103539
57. Viuff T, Ravinthrakumar S, Økland OD, Grytå OA, Xu X (2023b) Experimental study of floating bridge global response when subjected to waves and current. Appl Ocean Res 138:103588
58. Wang CM, Wang B (2015) Large floating structures. Technological advances. Springer, Singapore
59. Wang C, Cui M, Cheng Z, Moan T (2024) A review on design and analysis of floating bridges: numerical simulations, model tests and field measurements. Ocean Eng 306:118065
60. Wei W, Fu S, Moan T, Song C, Ren T (2018) A time-domain method for hydroelasticity of very large floating structures in inhomogeneous sea conditions. Mar Struct 57:180–92
61. Winterstein SR, Ude TC, Cornell CA, Bjerager P, Haver S (1993) Environmental parameters for extreme response: inverse FORM with omission factors. In: Proceedings of 6th international conference on structural safety and reliability, Innsbruck, Austria
62. Xiang X, Løken A (2019) Hydroelastic analysis and validation of an end-anchored floating bridge under wave and current loads, paper OMAE2019–95114. In: 38th OMAE conference, Glasgow
63. Xiang X (2024) Floating bridge global responses with hydrodynamic interaction. Ocean Eng 291:116420
64. Xie WC (2006) Dynamic stability of structures. Cambridge University Press. ISBN 9780521852661
65. Xu Y, Øiseth O, Moan T (2018) Time domain simulations of wind- and wave-induced load effects on a three-span suspension bridge with two floating pylons. Mar Struct 58:434–452
66. Xu Y, Øiseth O, Moan T, Næss A (2018) Prediction of long-term extreme load effects due to wave and wind actions for cable-supported bridges with floating pylons. Eng Struct 172(2018):321–333
67. Øiseth O et al (2018) Dynamic stability of elastic nonlinear systems subjected to random excitation. NTNU Report to NPRA
68. Zasso A, Argentini T, Bayati I, Belloli M, Rocchi D (2017) Super-long bridges with floating towers: the role of multi-box decks and Hardware-In-the-Loop technology for wind tunnel tests. IOP Conf Ser 276(1):012008

Torgeir Moan is a Professor Emeritus of Marine Technology at the Norwegian University of Science and Technology (NTNU), and former director of a research centre of excellence, CeSOS. Dr. Moan has served on the Royal Inquiry Commission of the Alexander Kielland accident and other accident inquiries and on various committees for the industry and standardization organisations for e.g. offshore structures (ISO); wind turbines (IEC) and floating bridges (NPRA). He has served as the inaugural adjunct Keppel professor at NUS in Singapore and has honorary academic positions in several leading Chinese universities. His main disciplines are structural analysis and design and, with a focus on safety assessment, as applied to ships, oil and gas platforms, floating bridges and offshore wind turbines. Professor Moan has authored or co-authored about 800 journal and peer-reviewed conference papers, supervised about 440 MSc candidates and 88 PhD candidates, respectively, and delivered about 50 keynote and award lectures at major conferences and received several international awards, e.g. including the Bruce Wallace award at MIT; the James W. Rice award of ASME; the Offshore Energy Center Hall of Fame Award, Houston and several best paper awards. He has been elected Fellow of five Academies, including the Royal Academy of Engineering, UK and the Chinese Academy of Engineering.

Ole David Økland is a Senior Researcher at Sintef Ocean. Dr. Økland's main research focus is dynamic structural response and hydrodynamic loads. He has worked with analysis, code development and experimental testing/full scale measurements related to marine structures in general as researcher, research manager and division manager. During the last decade his main focus has been model testing of dynamic response for marine structures exposed to strongly non-linear hydrodynamic forces, where he has been QA/Scientific adviser for model tests and also project leader for 10–15 commercial model tests. Dr. Økland has also participated in national and international research projects and projects/committees related to recommended practise and industry standards. Dr. Økland has authored or co-authored close to 100 technical reports and about 25 journal and peer-reviewed conference papers.

Ole Andre Øiseth is a professor of structural dynamics at the Department of Structural Engineering at the Norwegian University of Science and Technology and is the head of the structural mechanics research group. His main research areas are wind engineering, dynamics of structures subjected to environmental loading, structural monitoring and system identification, as especially applied to bridges. Dr. Øiseth has worked with numerical and experimental methods as well as field studies in his research. He has supervised more than 20 PhD candidates and 100 master's students and published more than 100 peer-reviewed papers within his research areas. He has served as consultant to the Norwegian Public Road Administration and many companies.

Modular Floating Structures for Offshore Urban Development in Real Sea Conditions

Gil Wang

Abstract This chapter outlines some of the fundamental stages of the design process specific to Modular Floating Structures (MFS), which encompass sustainable methodology and spatial optimization in open water conditions. MFS are multibody configurations consisting of several floating modules designed to provide a sustainable urban development space in the adjacent marine environment to coastal cities, which suffer from land scarcity. All modules considered herein are assumed to be rigid elements that comply with the acceptable box dimensions in the shipbuilding industry. They are arranged in proximity, moored on the seabed, and may or may not be interconnected. The specific configuration of each module depends largely on its urban function, which determines both the shape and geometry of the superstructure and the main particulars of the substructure, including the global hydrodynamic characteristics that influence human comfort. The design process of a MFS starts on a module level and continues to a multi-body arrangement in accordance with zoning, metocean conditions, and other physical and spatial considerations.

Keywords MFS design · Wave-structure interactions · Hydrostatics · Hydrodynamics · Regulations · Floating breakwaters

1 Expanding Coastal Cities into the Sea

This section offers a brief introduction, highlighting the key motivations for expanding coastal cities through large floating solutions. It also explores current available technologies, emerging floating innovations, and explains the decision to focus on Modular Floating Structures (MFS) as the primary configuration. From a design perspective, the section also addresses the gap between the available legal frameworks for residential building on land and the missing legal frameworks for

G. Wang (✉)
Coastal and Marine Engineering Research Institute (CAMERI), Technion – Israel Institute of Technology, Haifa, Israel
e-mail: wang@technion.ac.il

B. T. Wang et al. (eds.), *Large Floating Solutions*, Lecture Notes in Civil Engineering 636, https://doi.org/10.1007/978-981-96-5435-2_6

residential buildings offshore. It provides design guidelines intended to bridge this gap, which are demonstrated for this purpose.

1.1 The Marine Expansion Option

The world is experiencing the most significant phase of urban expansion recorded. In 2014, 54% of the global population resided in cities, with projections indicating an increase to 68% by 2050 [12]. Rapid expansion of urban areas presents major difficulties for global cities and megacities, making it increasingly challenging to oversee their sustainable management and regulate their spatial growth [29]. In coastal cities, which represent approximately 70% of the world's megacities, this issue is even more pronounced since their spatial expansion is restricted to at least one side by a body of water. The forecasts for the future indicate that by 2030, the land areas of megacities in developing nations are expected to increase three times, while in developed nations, the expansion is projected to be 2.5 times [20]. In addition, ocean thermal expansion and glacier melting, which have been the main causes of global sea level rise, are creating a growing danger to numerous coastal regions [4]. Considering that almost half of the industrialized population lives within a kilometer of the coast, the need for land and space is reaching a critical point as the global population continues to grow exponentially [2].

1.2 Current Practice and Emerging Technologies

The following sections provide a brief overview of the currently used non-buoyant and buoyant engineering method along with emerging floating solutions for extending urban spaces into the sea. It also provides a methodical evolution of the school of thought that led to the focus on the MFS configuration for both sheltered and open water conditions.

1.2.1 Land Reclamation

Land reclamation is an extremely well-proven technology for creating new land areas by filling designated plots with sand [27]. However, although this method has been effective for many centuries, it has recently faced significant environmental and technical challenges. From a technical perspective, bathymetric conditions play a significant role where mainly the depth of the water is considered a limiting factor. Therefore, countries that are not part of the continental shelf or are characterized by deep waters cannot apply this method without wasting a large amount of fill material, usually high-quality sand. According to the literature, land reclamation is considered uneconomical at depths greater than 20 m [31]. A different technical challenge is related to the quality of the seabed, if the seabed is not suitable, the land reclamation method becomes neither economically viable nor practical [7]. From an environmental point of view, fill-material-based land reclamation changes the natural landscape, damages marine habitats, and impacts ecosystems and biodiversity. In addition, it can harm the natural sedimental balance and coastal morphodynamics, which can have a very negative influence on the coastal environment in the near- and distant future. The natural aggregate used as the filling agent—composed of sand, gravel, clay, and bedrock—is an essential resource for various industries, whose largest consumers are the construction and land reclamation industries. The aggregate is currently being extracted much faster than it can be regenerated [23].

1.2.2 Maritime and Offshore Industries

In this Edited Collection, large floating solutions are presented in the context of sustainable use of sea space and annexation of the marine environment as a viable development alternative for land scarcity. However, this nascent multidisciplinary science field has strong foundations in other maritime and offshore practices. The shipping industry accounts for most large-scale floating structures, and the offshore industry, perhaps a closer relative, accounts for most stationary large floating solutions positioned offshore.

Research shows that the energy industry is responsible for most independent offshore structures designed and constructed over the years [11]. These structures represent established technology, which is operational, and are subject to ongoing advancements and studies focusing on various aspects of design and performance, interactions with waves, mooring systems, and seakeeping, among other factors. But like ships, independent offshore structures (e.g., mobile offshore drilling platform, drill barges, jack-up drilling platforms/rigs, and offshore wind turbine platforms) do not provide the requirements suited for expanding coastal cities—mainly in terms of spatial consideration and occupant comfort.

1.2.3 Emerging Floating Solutions

In recent years, considerable advances have been made in floating solutions specifically designed for coastal expansion, both in literature and in practice. In addition to the substance presented in this book, some of the notable work was published in Springer Nature's Lecture Notes in Civil Engineering, following the World Conference of Floating Solutions in Tokyo in 2023. Beyond the various configurations developed in the offshore industry, two principal configurations (archetypes) have been dominating the research of floating solutions. One is the Very Large Floating Structures (VLFS) which can be either rigid or can undergo hydroelastic deformation [8–10], and the other is the modular floating structures (MFS) which vary in both size and shape [22, 25, 26, 35]. This chapter focuses on MFS.

The MFS concept represents a floating design that combines practical implementation with a sustainable methodology. The practical aspect primarily concerns the dimensions of the modules, which fit within the size limits of standard vessels, allowing them to leverage the existing infrastructure and expertise of the widely established marine industry. This means that the proposed MFS modules could be constructed using existing shipyards and use the widely accepted regulatory framework of the marine industry. Additionally, size conforming means that the entire marine sector infrastructure can be used. This includes transport and handling by conventional tugboat and the use of existing mooring equipment, saving cost, energy, and resources. Another practical aspect concerns spatial arrangement. The MFS offers the versatility to add, remove, or reorganize models as needed. Owing to their relatively small dimensions, the modules can be arranged in various configurations, facilitating spatial efficiency and design flexibility. A significant ecological benefit of the MFS's grid arrangement is that it permits sunlight to penetrate between the modules, thereby preventing the detrimental effects of a constant shaded footprint.

1.3 The Common Rule: Reconciling the Legal Framework

Currently, there are no dedicated regulatory frameworks or building codes for floating offshore urban development, particularly when considering residential use as the primary urban function of the modules. However, the design and construction of structures for human use, whether on land or at sea, is probably best depicted under the umbrella of fields such as civil engineering and naval architecture, where the rooted common denominator includes both structural integrity (ultimate limit state) and human compliance (service limit state). Naturally, design objectives change according to the specific requirements denoted by each field in terms of function, geometry, and regulations, which can be reconciled by the common rule approach.

The maritime regulations, under the auspices of the International Maritime Organization (IMO), cover aspects of ship design, pollution prevention, collision prevention, and safety (www.imo.org). The regulations and guidelines for civil buildings, such as the Eurocode, cover mainly the structural integrity and the durability of buildings. Design elements outside the scope of the Eurocode, for example, serviceability, spatial requirements, and other nonstructural regulations, are dealt with 'case-to-case', according to the relevant local annexes.

Dividing the MFS module into substructures and superstructures, facilitates the way to match the relevant existing regulation to each part, fusing maritime requirements with building codes. The substructure is subjected to all IMO regulations as the minimum required standard, and the superstructure is designed in adherence to Eurocode regulations and local annexes. To achieve this, an integrated regulatory framework was created by combining residential building codes with the classification standards mandated by IMO.

Figure 1 shows the main requirements of civil engineering and naval architecture and their mutual regulations.

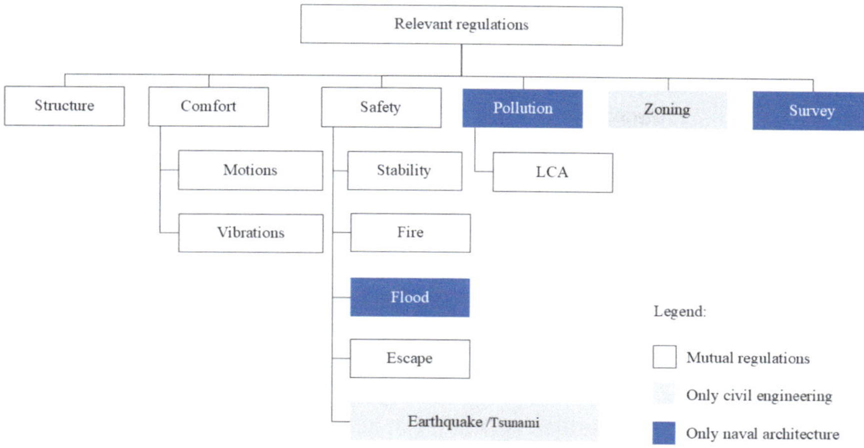

Fig. 1 The Common Rule flow chart; reconciling the legal framework of civil engineering and naval architecture building codes requirements [35]

1.3.1 Mutual Regulations

With reference to Fig. 1, Structure', Safety', and 'Comfort' are mutual building regulations when considering the design of onshore residential buildings or offshore structures. From a legal standpoint, the approval process for onshore buildings differs slightly from that of marine structures. Onshore building approvals are managed by local authorities, whereas marine structures are regulated through classification societies acting on behalf of the IMO and recognized by the local authority (flag state). Consequently, regulations for marine structures tend to be more universal. However, while the structural aspects of ultimate limit state design are similar in both areas, the definitions of loading differ significantly.

Building safety refers to the ability of a building to withstand abnormal impacts, such as fire, floods, and earthquakes, and maintain the required degree of protection for evacuation and escape. In the context of MFS, two main regulations have been reviewed, Eurocode (1 and 3), and the International Convention for the Safety of Life at Sea (SOLAS).

The dynamic behavior of a fully displaced buoyant structure, subjected to constant weather changes, presents significant challenges in maintaining minimal motion across all operational modes and six degrees of freedom. Consequently, the development of a novel Modular Floating Structure (MFS) for urban use must prioritize occupant comfort and incorporate design measures to minimize discomfort. When considering the MFS module as a residential unit, comfort standards should align with those of land-based buildings, rather than the motion sickness criteria typically used in the marine industry. This topic is explored in detail in Sect. 2 of this chapter.

1.3.2 Only Civil Engineering

Zoning consideration and earthquake engineering are considered exclusive to the field of civil engineering. The position along the coast for a potential MFS project is influenced by development needs and proximity to a coastal city. The offshore position is mainly driven by policy, technology, and costs. Zoning is neither an IMO regulation nor a class requirement when designing marine structures—especially the nonstationary structures such as marine vehicles. Offshore structures positioned in the Exclusive Economic Zone (EEZ) will need to be authorized by the local government prior to installation according to spatial local conditions (e.g., shipping routes, nature reserves, bathymetry, closed military areas, etc.). In addition, earthquake resistance is an important topic in building safety. However, floating structures do not share this problem as they are inherently base isolated [31]. In a Tsunami event, floating structures, positioned at a sufficient depth, can help mitigate, or even eliminate, the disastrous effects of a Tsunami, which accumulate into a destructive order only when approaching the shore.

1.3.3 Only Naval Architecture

With reference to Fig. 1, 'Stability', 'Flooding', 'Pollution', and 'Survey' are required regulations in naval architecture. When compared to shipbuilding or residential development, the MFS fabrication layout is rather unique. The MFS module is fabricated offsite, and once completed, it is towed to the site, where it is linked to a combined multibody MFS structure. Naturally, the transport period of a module is relatively short compared to its entire life span as a stationary component in an MFS unit. However, during this period, the MFS module must comply with all the related IMO requirements in terms of stability and safety. Flooding is also an issue of grave concern in shipping safety but is not a common risk for buildings situated on land. SOLAS convention addresses this issue rather comprehensively. It distinguishes between two main scenarios that can lead to flooding: 'Intact stability' and 'Damaged stability' with associated implications that influence almost every aspect of the design. It includes structural division (watertight bulkheads), benchmarked hydrostatic properties, openings (size and locations), minimum freeboard requirements, and reserved buoyancy.

The International Convention for the Prevention of Pollution from Ships (MARPOL) was first adopted by the IMO in 1973, as a response to 1976–1977 tanker accidents and their grave effect on the marine environment [6]. The MARPOL Convention includes regulations aimed at preventing and minimizing pollution from ships, both accidental pollution and routine operations. Currently, it includes six technical annexes [18]. The relevant input will be adequately implemented in the conceptual development phase of the MFS. So far, there are no specific regulations regarding pollution prevention in the Eurocode. However, based on local policy, guidelines for preventing pollution on construction sites can also be mandatory. In addition, for a vessel to maintain its classification approval, a periodic survey is mandatory. The survey is a visual examination that consists of an overall inspection of the items identified in the survey rules. The purpose of a survey is to ensure that the vessel and the components inspected are in adequate condition and that the voyage is allowed from safety and pollution perspectives. Survey and class renewal are normally held every five years. This requirement is not mandatory for residential buildings according to the Eurocode.

1.4 Demonstrator: MFS Module Ultimate Limit State Design

The previous sections examined related regulations for floating offshore structures for urban use, outlining the relevant input needed and the relevant input missing, in order to convey a coherent design guideline. This part offers a demonstration of the conceptual design framework mentioned above, focusing on both structural aspects and stability, viewed as the ultimate limit state.

1.4.1 Substructure Dimensions and Superstructure Configuration

The geometry of a single MFS module (length, beam, and depth) can be customized to support the desired superstructure design. However, constructing the module substructure within the size range of conventional ships or barges allows the use of existing technologies and shipyards. Most commercial shipyards are constrained by vessel size limits, primarily due to draft (linked to local water depth) and beam, which is the maximum width of the fabrication facility. These limitations stem from the Panama Canal's size restrictions—32 m in beam and 12 m in draft—known as Panamax, which has influenced shipbuilding for the past century. To ensure production and cost efficiency, the module's dimensions will not exceed Panamax specifications. The selected module, "Module9000," represents a single barge housing three individual buildings, each with 10 floors of 300 square meters (15×20 m). The name "Module9000" reflects the total building area available per barge. Figure 2 shows the principal geometry of Module9000 inside and plan views.

1.4.2 Substructure's Scanting Calculation

The continuous structure is the most dominant weight post and contributes ca. 60% of the total Light Ship Weight (LSW) [35]. The continuous structure consists of all the continuous longitudinal (parallel to keel) elements (e.g. outer shell, decks, double-bottom, stringers and girders), contributing to the cross-sectional properties of the module. The continuous structure is the main attribute to resist the global longitudinal bending moment—the Still Water Bending Moment (SWBM).

Longitudinal strength analysis was performed to define the continuous structure. Calculating the SWBM is an iterative process where LSW is initially assumed based

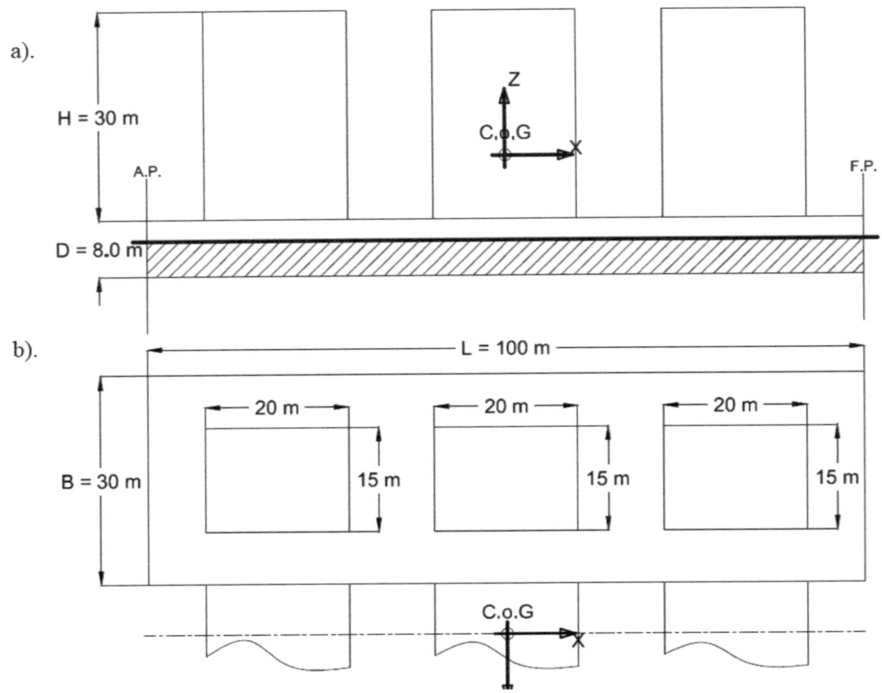

Fig. 2 Module9000 **a** side view and **b** plan view [36]

on the geometrical dimensions and properties chosen and the expected buildings' weight. The superstructure weight (including payload) taken is 1.1 ton per square meter, hence based on the geometric input given in the previous section the weight of each floor is 330 tons, and each building weighs 3,300 tons.

In addition to the physical SWBM, a stochastic wave moment was added according to: Lloyds Register Longitudinal Strength, Part 3, Chapter 4, in: Rules and regulations for the classification of ships. Figure 3 shows the superposition of the SWBM distribution on the barge length (x-axis) and the plotted Wave Moment. The Wave Moment is a stochastic Rule moment which is linked to the module's parameters: Length (L) and Beam (B). The Wave Moment is considered constant at 20% amidships, and then it tapers down linearly towards the ends (bow and stern). The sum of these two curves along every section of the hull is the total bending moment used to determine the continuous longitudinal structure. To validate the obtained results, a compatibility check was performed using DNV rules of ships Pt. 5 Ch. 7 Sec. 14—Page 43 (Hull strength), and the results showed good agreement.

The side structure includes the open frames (perpendicular to the keel), which form the internal frame grid of almost any vessel. The side structure supports the outer shell (the plate bending span decreases) subjected to hydrostatic pressure. The frame dimensions and spacing are obtained by Rule scantling calculations. In

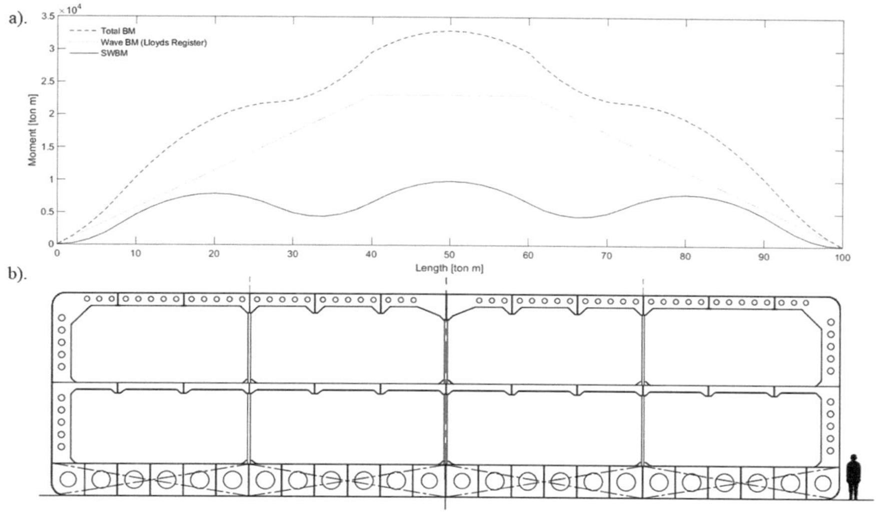

Fig. 3 **a** Module9000 Longitudinal Bending Moment diagram (SWBM + Wave Moment), **b** main frame scantling [35]

addition to the framing system, pillars are installed to reduce the bending span of the longitudinal girders that cross the open frames.

The watertight bulkheads are transverse division walls installed in the hull of the substructure. The watertight bulkheads divide the internal substructure space into watertight compartments to comply with the damage stability criteria. Module9000 has been divided into seven watertight compartments. In addition to damage stability, the watertight compartments together with the pillars act as support members to the longitudinal structure and as foundations to the superstructures. The dimensions of the bulkhead components (stiffeners and plates) are determined using Rule scantling calculations, similar to those used for the side structure.

1.4.3 Stability, Safety and Hydrostatic

Based on the structural scanting calculations, the LSW is 2800 tons and contains the structural components mentioned above, including the intercostal plate stiffeners and the unexpected weight factor of 5% from the LSW. The total weight of the three superstructures is 9900 tons. These inputs, together with the substructure geometry and buoyancy characteristics, allow the determination of the total weight and final displacement. Additionally, a weight calculation is performed to define the center of gravity. Module9000 dimensions and geometrical properties are given in Table 1.

The range of positive stability is determined by the relationship between the heel angle and the righting lever (GZ). Figure 4 shows three scenarios: (1) when the module is not subjected to side force, and thus there is no heeling angle, i.e. upright

Table 1 Module9000 dimensions and geometrical properties

Description	Dimension	Description	Dimension
Length (L)	100.0 m	Transverse metacentric radius (BM_T)	14.70 m
Beam (B)	30.0 m	Vertical center of buoyancy (KB)	2.55 m
Depth (D)	8.1 m	Vertical center of gravity (KG)	16.61 m
Design Draft (T)	5.1 m	Transverse metacentric height (GM_T)	1.47 m
Displacement (Δ)	15,688 tones	Block coefficient (C_b)	1.00

condition, (2) the module is heeled to the port side under a certain heel angle, and the righting lever is positive, (3) the module is heeled to portside under a larger heeling angle, and the righting lever is negative. When GZ is positive (the intersection of buoyancy line B with the initial centerline is above the C.o.G), the righting moment (RM) tends to restore the module to its upright position. When GZ is negative (the intersection of buoyancy line B with the initial centerline is below the C.o.G), the RM direction changes, and it tends to capsize the module.

In this hydrostatic analysis, two separate loading conditions are defined, in transit and when in stationary. During transit, Module9000 has no sufficient means of self-propulsion and is towed to the MFS site by another vessel. At this point Module9000 is unmanned and has no openings on deck. For this scenario Det Norske Veritas (DNV) Rules for Ships, January 2010, Section 14: Barges, was adopted. Accordingly, this load case must comply with the following regulations:

I. The minimum range of positive stability is 20 degrees.

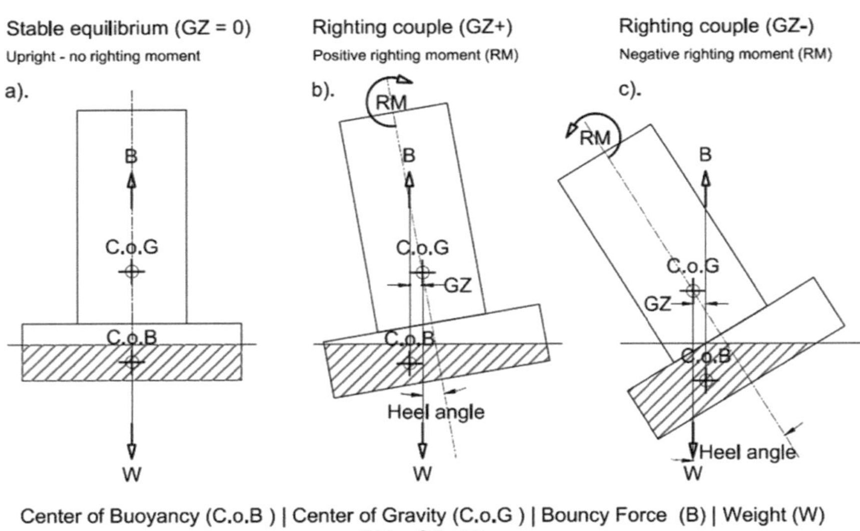

Fig. 4 Righting lever (GZ) schematics under heel angles [35]

II. The static heel angle due to the horizontal wind load of 0.54 kPa (wind velocity of 30 m/s) should not exceed an angle corresponding to the half of the freeboard for the relevant loading condition.
III. The area under a righting lever curve up to the angle of maximum righting lever (GZ) should not be less than 0.08 m-radians.

The stationary loading condition refers to the Module9000 lifespan as part of the consolidated MFS unit. This is a stationary condition where the module is no longer underway but is considered fully occupied. The loading condition must comply with the criteria stipulated above (I, II and III) as minimum required standard, also corresponding to IMO Res. A. 749(18) Ch. 4.7).

The transit load case with light displacement meets the stability regulation for the module delivery from the fabrication site to the MFS site. Upon arrival, the module is expected to increase its payload by 20%, and as a result it does not comply with the stability regulations. To improve the hydrostatic performance of the MFS module, a multibody configuration was examined. The twin hull configuration is accomplished by rigidly connecting the modules side by side with a 15-m gap between the hulls. This configuration met the stability requirements with large margins of safety. To conclude, Module9000 now has two basic configurations: monohull and twin hull. Retaining the monohull version, even after establishing the twin-hull as the primary configuration, facilitates module transportation and helps reduce costs. The 30-m beam (monohull) can pass through the Panama Canal and can be transported using conventional tugboats.

1.5 Section Summary

Adopting the twin-hull configuration for the Module9000 as a standard MFS module directly impacts the loads and design criteria. The governing load shifts from the longitudinal bending moment (SWBM + wave moment) to the transverse bending moment acting on the cross-structure that connects the two hulls. The transverse bending moment, along with the associated shear force, is calculated according to DNV Pt.3 Ch.1 for vessels over 50 m. The bending moments—Ms (still water transverse bending moment), Mp (pitch connecting moment), and Mt (twin-hull torsional moment)—and the shear forces are not of major concern due to the cross-structure's height and its non-buoyant nature. Thus, based on the given criteria, the Module9000 structure complies with structural regulations. The multibody effects of a consolidated MFS platform are not analyzed in this preliminary design phase but will be detailed in Sect. 3. Regardless of the chosen consolidated MFS layout, designing the module to meet hydrostatic benchmarks as a standalone unit ensures that, in any configuration, its stability properties remain sufficient. This guaran-tees the module's independent hydrostatic performance and positive stability throughout its entire life cycle, from seafaring transport to its stationary use in an MFS, and eventual removal or reuse.

2 Occupant Comfort for Offshore Dwelling

In line with the framework of the Common Rule described in Sect. 1.3, this part investigates the hydrodynamic behavior of an MFS module in relation to the accepted motion standards allowed in onshore structures. Providing an Occupant Comfort-Based Design for a typical MFS module in real-sea conditions. Although demonstrated for a specific module configuration, the proposed methodology is universal and can be applied to any module design, configuration, and environmental conditions.

2.1 From Seasickness to Homesickness

"People may be divided into three classes; the living, the dead, and the seasick"; this quote is attributed to Anacharsis, a Greek philosopher from the sixth century BC who described the feeling of motion sickness as a limbo between life and death. In the context of floating urban solutions, motion sickness should be unacceptable. Therefore, the design ethos of the MFS is to provide the same level of comfort, in terms of frequency and acceleration, acceptable in residential buildings as a threshold requirement. This section discusses the hydrodynamic properties of a MFS module and its associated occupant comfort performance in relation to the relevant standards.

2.1.1 Occupant Comfort Criteria for MFS

The excitation of the building is caused by internal and/or external sources. Internal sources can be the product of electromechanical systems (propulsion engines, pumps, air conditioning units, etc.) which are usually within the frequency range of 1–80 Hz and typically may cause noise and vibration [14]. External building excitations originate from wind or seismic activities. As this section examines the service limit states of buildings with respect to comfort, seismic activities such as earthquakes are not considered relevant. In the case of floating systems, earthquakes are even more insignificant, since they do not impact the structure. However, wave energy is a major source of external excitation in floating structures, occurring at frequencies below 1 Hz [36], which is comparable to the frequency range of wind load excitations in buildings. Moreover, the frequency range in which motion sickness and its related symptoms occur is below 1 Hz, commonly between 0.1 and 0.5 Hz [16]. Figure 5 shows the frame of reference for Module9000 (x, y, and z) and the six degrees of freedom. It also includes an illustration from ISO 2631, detailing human exposure to whole-body vibrations.

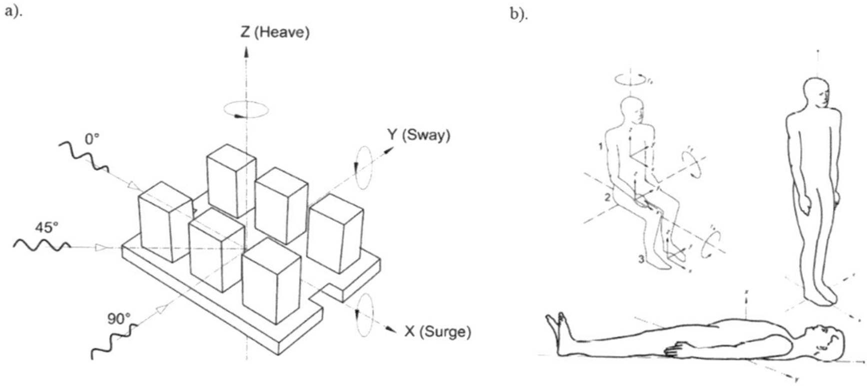

Fig. 5 **a** Frame of reference for Module9000 also showing six degrees of freedom and the wave headings calculated in this analysis [36]. **b** ISO 2631 reference to human exposure of whole-body vibrations [14]

2.2 *The Range of Comfort*

Motion can cause various forms of discomfort and disability in humans. Some of the most notable effects include Sopite syndrome, Motion-Induced Fatigue (MIF), Motion-Induced Interruptions (MII), Motion-Sickness Incidents (MSI), and a decline in cognitive performance. Each condition is influenced by different parameters, primarily frequency and acceleration. Sopite syndrome, a manifestation of motion sickness, arises from prolonged or brief exposure to oscillating movement [19]. MIF, though similar in nature, occurs only after extended exposure [36]. Common symptoms of Sopite syndrome include yawning, drowsiness, disinterest in work (physical or mental), and lack of participation in group activities. It represents a milder form of motion sickness, which can range from light drowsiness to nausea and vomiting, classified as MSI. MII, on the other hand, refers to a loss of balance and mobility disruptions due to movement. Since this study aims to align the dynamic performance of floating stationary offshore platforms with the comfort standards of residential buildings, MSI and MII are less critical within the service limit state category. Human sensitivity to whole-body vibration and motion in artificial environments is addressed by various standards and regulations. Many of these guidelines are based on foundational experimental research conducted by O'Hanlon and McCauley [21] and Lawther and Griffin [17], which underpin much of the empirical data in professional literature. A summary of the relevant motion-induced occupant comfort standards is provided in Table 2.

Table 2 Motion sickness standards

By	Title:	Year
ABS	Guide for Passenger Comfort on Ships	2015
BS 6472	Guide to evaluation of human exposure to vibration in buildings. Vibration sources other than blasting	2008
ISO-10137	Bases for design of structures—Serviceability of buildings and walkways against vibrations	2007
NATO-4154	Standardization Agreement (STANAG), Common procedures for seakeeping in the ship design process	2000
ISO-2631	Mechanical vibration and shock—Evaluation of human exposure to whole-body vibration	1997
BS-6841	Guide to measurement and evaluation of human exposure to whole-body mechanical vibration and repeated shock/	1987
NORDFORSK	Seakeeping Criteria	1987
BS 6611	Guidelines for the evaluation of the response of occupants of fixed structures, especially buildings and offshore structures, to low-frequency horizontal motion (0.063–1 Hz)	1985
ISO 6897	Guidelines for the evaluation of the response of occupants of fixed structures, especially buildings and offshore structures, to low-frequency horizontal motion (0.063–1 Hz)	1984

2.2.1 Limiting Curves and Residential Building Standards

One important aim of the current research is to evaluate the comfort of the occupant of an MFS module according to the acceptable standards in residential buildings, which are more stringent than their equivalent standards from the offshore and shipping industry. BS-6611 and the equivalent ISO-6897 limiting curves, together with ISO-10137 evaluation curves, are the most adequate criteria in the current context [16, 36]. The motion perception thresholds in BS 6472, given for the frequency range of 1–80 Hz, exceed the frequency magnitudes relevant for the current research, which are below 1 Hz. The same applies to the limit curve in BS 6841. Therefore, to evaluate module occupant comfort in the context of residential buildings, only ISO-6897 and 10137 are used. ISO-10137 specifies curves of acceleration limit that theoretically offer acceptable limits for habitability with respect to human response to horizontal accelerations on a one-year return period. The guideline applies for office and residential occupancy where the residential curve, relevant to current research, is 2/3 of the limit curve for office buildings [15]. The ISO-10137 residential curve is based on close to 90% of the perception probability for a one-year period event, while the ISO-6897 residential curve considers the worst ten minutes of an infrequent event in a period of five years, where not more than 2% of the occupants comment adversely about the motion [13].

2.2.2 Seakeeping Criteria

The American Bureau of Shipping (ABS) guide on passenger comfort offers a classification standard detailing the comfort criteria for passenger vessels, which accounts for human physiological and psychological responses to whole-body motions aboard ships [1]. Furthermore, ABS has introduced COMF and COMF + certifications for ships that meet their classification requirements. To achieve the COMF + certification within the frequency range of 0.1–0.5 Hz, the acceleration, expressed in Motion Sickness Dose Value (MSDVz) at the maximum root mean square (RMS) of vertical acceleration amplitudes, must not surpass 30 $m/s^{1.5}$ over a three-hour duration. Over extended periods, human sensitivity to motion diminishes as an adaptation process begins.

In the maritime industry, the term 'comfort' can also be related to absolute RMS accelerations, separated from the associated frequencies. The absolute comfort of the RMS accelerations varies according to the classification of the vessels and the performance of the task. In addition to the MSDVz and MSI outlined above, Table 3 shows the general comfort restrictions of ISO 2631, in absolute RMS values. NORDFORSK and NATO's STANAG 4154 also provide seakeeping benchmarks in absolute RMS values as given in Tables 4 and 5.

Table 3 ISO 2631 comfort restrictions

Comfort description	General accelerations
Not uncomfortable	From 0.315g[a]
Little uncomfortable	From 0.315 to 0.63g
Fairly uncomfortable	From 0.5 to 1.0g
Uncomfortable	From 0.8 to 1.6g
Very uncomfortable	From 1.25 to 2.5g
Extremely uncomfortable	Above 2g

[a] g *is the gravitational acceleration constant (taken as 9.81* m/s^2*)*

Table 4 NORDFORSK 1987: criteria for acceleration and Roll

Activity description	Vertical (heave) (g)	Lateral (surge/sway) (g)	Roll
1. Cruise liner	0.02	0.03	2°
2. Transit passengers	0.05	0.04	2°
3. Intellectual work	0.10	0.05	2°
4. Heavy manual work	0.15	0.07	2°
5. Light manual work	0.20	0.1	2°

Table 5 NATO STANAG 4154: Human performance criteria—seakeeping

Criteria set	MSI	Roll	Pitch	Vertical	Lateral
Default	N/A	2°	2°	0.20g	0.10g
Recommended	20% at 4h				

2.3 *Demonstrator: MFS Module in Eastern Mediterranean Sea*

This section provides a concise occupant comfort analysis for one free-floating MFS module in real-sea conditions. Module9000, described in Sect. 1.4 is used as a demonstrator to showcase this methodology. The first-order calculation is performed using a potential flow solver in the boundary element method to determine the response amplitudes of the module in six degrees of freedom. The specific environmental conditions for the demonstrator, in terms of wave statistics, are also provided. The calculated response amplitudes are then transformed into a real-sea representation using the JONSWAP spectrum, correlating the hydrodynamic calculation to the environmental conditions in irregular waves. To evaluate module performance according to the collected occupant comfort standards (see Sect. 2.2), the motion amplitudes are transformed into accelerations using the linear relationship between the unit wave amplitude and the structural response.

2.3.1 Compliant Mooring System

To maintain the module's position onsite, a compliant mooring system is used. The system consists of extended lengths of cables or chains, with one end attached to the module and the other secured to an anchor. The anchor is placed at a distance, usually a few module lengths away, in the x-direction and lies on the seabed at a specified depth in the y-direction. The line between the two dead ends, in catenary, forms a type of nonlinear spring. The objective of the compliant mooring application is to resist the horizontal mean forces (wind, current, wave drift) operating at the water plane (surge, sway, and yaw), which are typically considered as a second order relative to the linear wave forces. Therefore, due to the relatively flexible spring action the module is free to move in all six DOFs in the first order, and the mooring system contribution to occupant comfort is insignificant.

2.3.2 Environmental Conditions

Waves are the main energy source for inducing motion response in floating structures, and they serve as the only input for the occupant comfort analysis. Thus, in this demonstration, the environmental conditions refer to the wave climate at the eastern Mediterranean Sea, off the Israeli coastline. The waves were measured using the

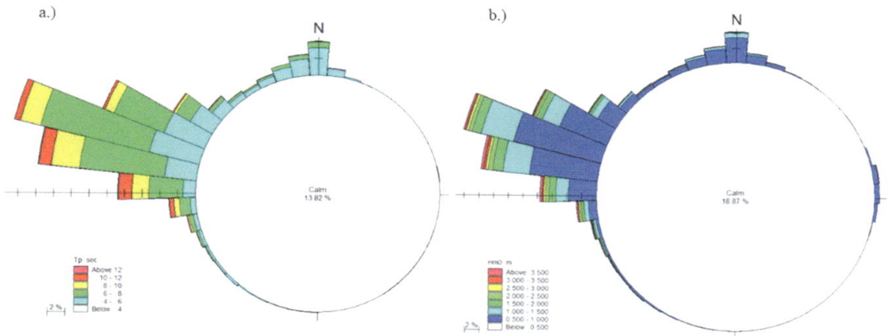

Fig. 6 Wave statistics from: 01/04/1994 to 31/03/2023, measured with Datawell's (Directional Wave rider) Buoys off the coast of Israel [5]. **a** Showing the heading and wave peak periods (T_p) in s. **b** Showing heading and wave significant wave height (H_{m0})

directional Datawell waverider buoy, during an extensive period of three decades at a water depth of 24 m.

Figure 6 shows the measured wave statistics, presenting both significant wave height and wave periods, and their heading (direction). It shows that ca. 20% of the time waves are lower than 0.5 m and with relatively short periods, which have insignificant influence with respect to wave-structure interaction. However, significant wave heights up to 3.5 m are within the scope of the annual characteristics, including wave periods greater than 12 s.

2.3.3 ANSYS-AQWA the Numerical Analysis Platform

To calculate the wave-structure interaction, a parametric boundary element model was programmed for the ANSYS-AQWA analysis system using parametric MATLAB code to generate the required preprocessing data files. The model was first verified to reconstruct the results presented by Ray [24] for a simple box, calculated by both WAMIT and MOSES—two widely used and well-proven 3D diffraction and radiation computer software. The hydrodynamic analysis focuses mainly on the Response Amplitude Operators (RAOs) which are determined by solving the following equation of motion in the frequency domain

$$M_s\ddot{X} + M_A\ddot{X} + C\dot{X} + KX = F \tag{1}$$

In Eq. (1), the RAO is indicated as X, and can be calculated when writing, $X = X_0e^{-i\omega t}$, and $F = F_0e^{-i\omega t}$, M_s is the structural mass matrix of the chosen module, M_A the hydrodynamic added mass matrix, C the linear damping matrix, K the stiffness matrix, and F is the external wave force as calculated from

$$\boldsymbol{H} = \left(K - [M_s + M_A]\omega^2 - iC\omega\right)^{-1} \tag{2}$$

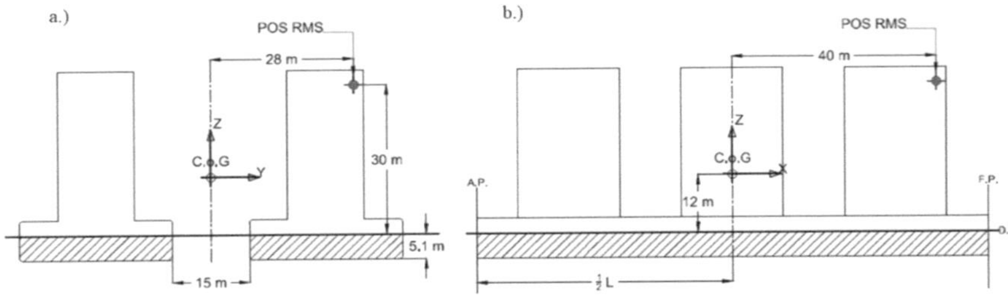

Fig. 7 Position of calculated Point RAO, **a** Module9000 looking forward and **b** profile view

where $\boldsymbol{H}$ is the transfer function which relates input forces to output response as given by

$$X_0 = \boldsymbol{H} F_0 \tag{3}$$

A detailed explanation of the first order analysis used by the solver can be found in the literature [32] and software's manual [3].

Figure 7 shows the location of the position RAO calculated for this demonstrator, and Fig. 8 shows the obtained results.

2.3.4 Real Sea and RMS Acceleration

This study adopts the JONSWAP spectrum to transfer the RAO calculated in monochromatic waves into real sea (irregular sea). In Fig. 8 the data set of the position RAO includes wave periods between 4 and 20 s in the three predefined wave headings. The Position Root Mean Square (RMS) amplitude of motion represents the average motion amplitude for each sea state, which is represented by the two parameters H_{m0} and T_P, that are substituted in the JONSWAP spectrum formula. Figure 9 shows the spectral density curves for the selected significant waves.

2.3.5 Occupant Comfort and Seakeeping Results

Figure 10 shows the accelerations of Module9000 measured at the specified point in three predefined wave headings 0°, 45°, and 90°. These accelerations are associated with the five significant wave heights (H_{m0}: 0.5, 1.0, 1.5, 2.0 and 2.5 m) under real sea conditions, using the JONSWAP spectrum. The results are presented on the limiting curve of ISO 6897 (equivalent to BS 6811) and ISO 10137.

- Curve 1 (ISO 6897) is the lower threshold of perception of horizontal motions by humans.
- Curve 2 (ISO 6897) is the average threshold of perception of horizontal motions by humans.

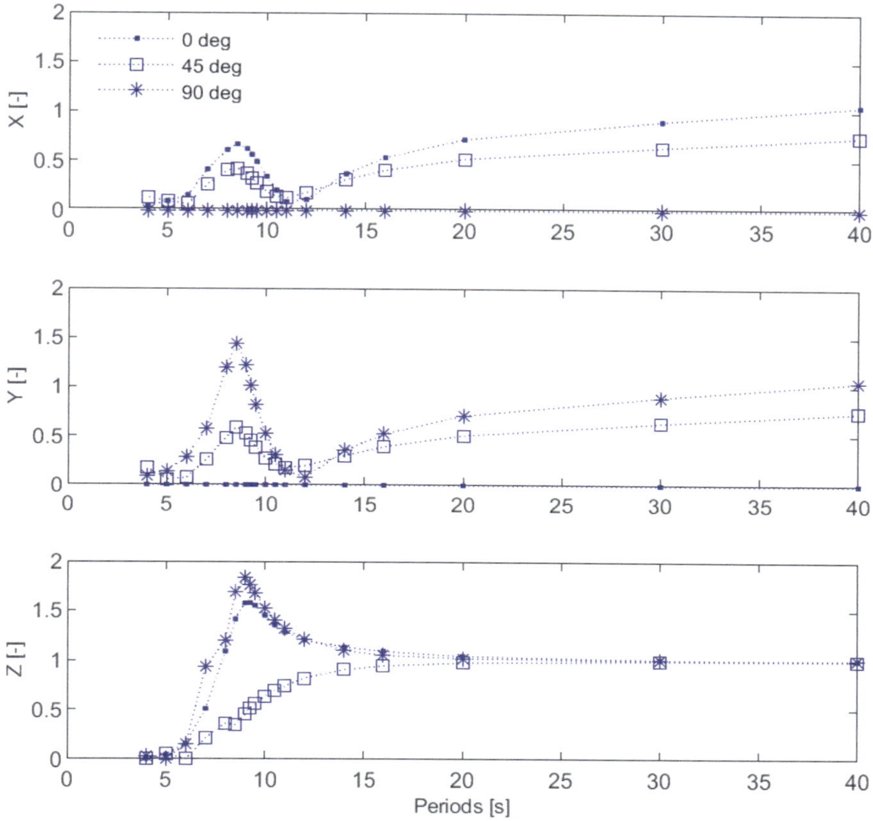

Fig. 8 Point RAO determined for a period sweep from 4 to 20 s

- Curve A (ISO 6897) is the suggested magnitudes of horizontal motion of buildings used for general purposes.
- Curve B (ISO 10137) is the evaluation curve for wind induce vibration for residential buildings for a one-year return period.
- Curve C (ISO 10137) is the evaluation curve for wind induce vibration for office building for a one-year return period.
- Curve D (ISO 6897) is the suggested magnitudes of horizontal motion for offshore structures (none floating; fixed structure).

Figure 11 shows the calculated results of the MSDV and Fig. 12 shows the calculated MSI values. Both results were calculated using the transformed RAO of Module9000.

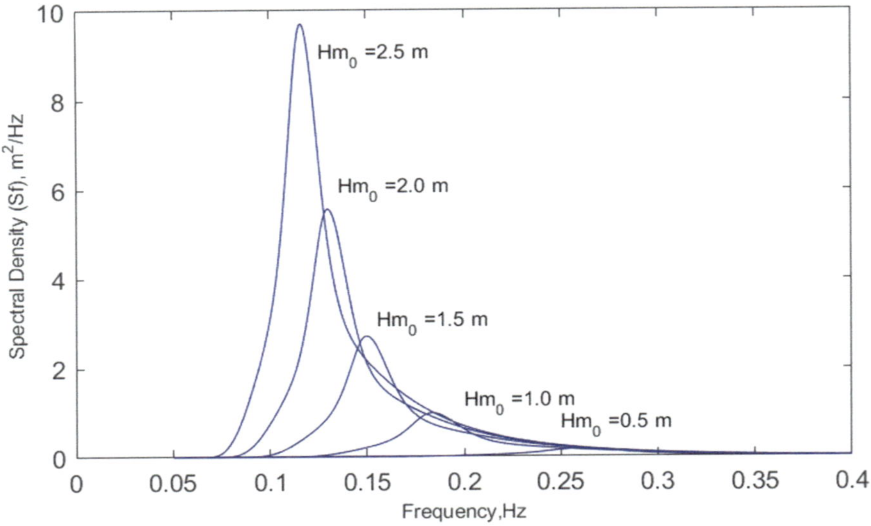

Fig. 9 JONSWAP based spectral density curves for the chosen significant wave heights of 0.5, 1.0, 1.5, 2.0, and 2.5 m

2.4 Section Summary

This section presented a new approach and methodology to examine the habitability characteristics of floating structures for residential purposes offshore. The evaluation concentrated on the response spectrum of the free-floating Module9000. The calculated structure response was constructed and evaluated according to current available comfort criteria used in both residential buildings and the marine industry. The aim was to determine whether the selected module configuration can provide the comfort required in residential buildings in a predefined environment and in a free-floating state.

With respect to occupant comfort, wave heading was found to play a significant role. Oblique waves caused much lower motion amplitudes than the bow/stern waves and the side heading waves. Regarding ship's comfort criteria, the suggested Module9000 complies with the accepted Motion Sickness Doze Value (MSDVz) with a large margin for the oblique wave heading (45°) up to a significant wave height of 2.5 m. At 0° wave heading the MSDVz passed with a very little margin at 2.5-m significant wave height, and at 90° wave heading the module fails to satisfy the criterion at above 1.75 m significant wave height. Therefore, when considering the location of the module and its relative position in the direction of the major waves, an angle of about 45° to the wave heading is recommended in terms of comfort performance.

In the absence of occupant comfort criteria for residential buildings offshore, this research acknowledges the ISO-6897 and ISO-10137 limiting curve suggested for

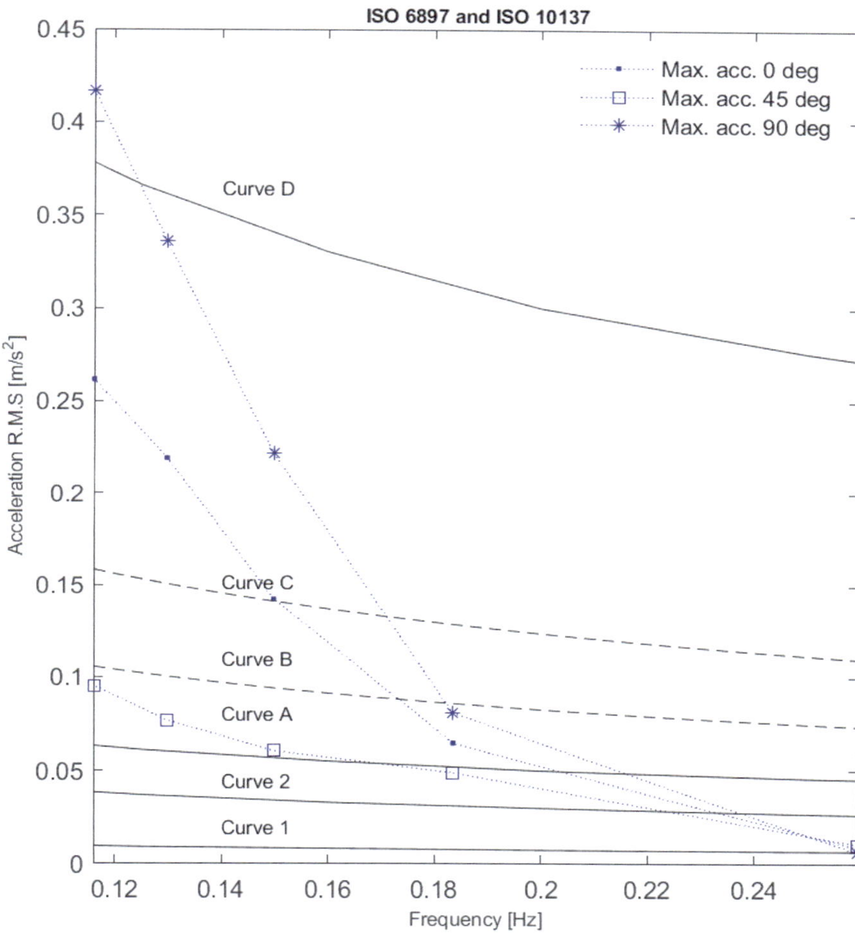

Fig. 10 Max. RMS. accelerations of Module9000 overlaid on the ISO 6897 and 10137 curves

residential buildings as the initial benchmark. This is a conservative approach to the available seakeeping or ship comfort criteria presented in this section, however also a realistic one, as ship motions (or any other vehicle) are generally accepted by the public (passengers), while building motions cause fear, anxiety, and alarm.

Figure 13 shows the vertical acceleration for Module9000 at different significant wave heights from 0.5 to 4 m. The results are overlaid on the relevant limitations for occupant comfort. Curves (d), ISO-6897 and (c) ISO-10137 are acceptable accelerations for residential purposes, curve (b) ISO-10137 provides the acceptable accelerations for office use, and curve (a) NORDFORSK shows the limiting acceleration criteria for cruise liners.

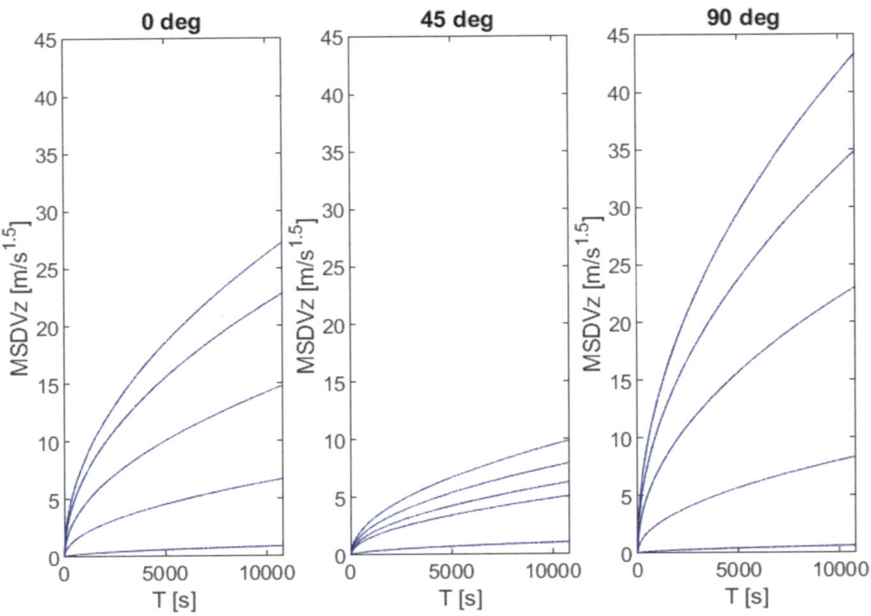

Fig. 11 Seakeeping performance based on ISO-2631, Motion Sickness Doze Value in the vertical direction (MSDVz) curves represent Hm0 of 0.5, 1.0, 1.5, 2.0, and 2.5 m

3 Multi-body Configuration and Circular Economy of Retired Ships

Using the design methodology presented in Sect. 1 and the comfort criteria analysis presented in Sect. 2, this section looks at the multibody MFS configuration to increase the performance limit of MFS in open water conditions. The occupant comfort performance limit of an MFS module, Module9000, was evaluated in the previous section, and was found suitable for residential purposes up to a significant wave height of ca. 2 m. Looking back at our demonstrator off the eastern Mediterranean coast, and the chosen environmental conditions, we can see that this floating solution does not provide the required occupant comfort all year round, as the significant wave heights are larger than 2 m for approximately 24 days a year [5].

The MFS concept is based on a multibody configuration, and the question examined in this section is whether this multibody arrangement could help mitigate wave motions and increase the performance limit of the individual modules composing it.

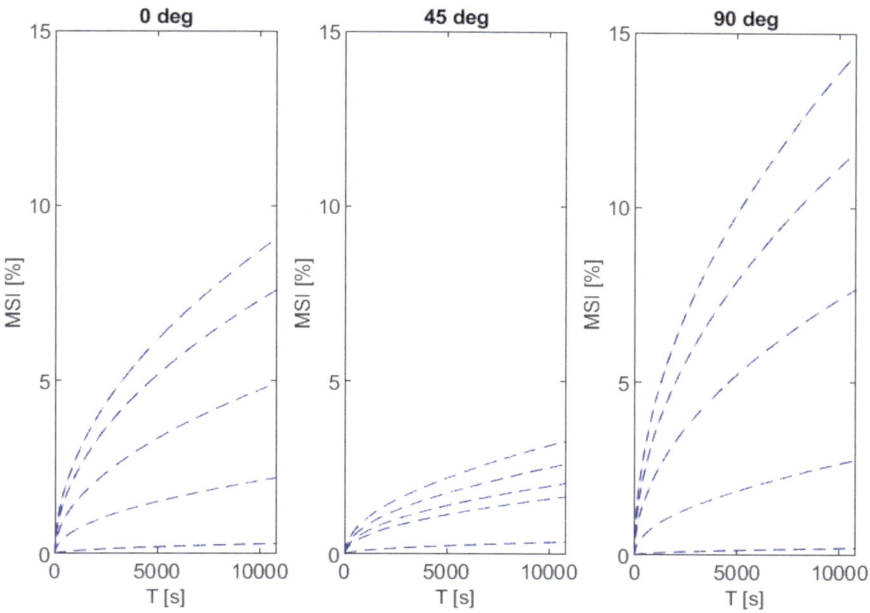

Fig. 12 Seakeeping performance based on ISO-2631 Motion Sickness Doze Value in the vertical direction (MSDVz) for a duration of three hours

3.1 From Urban Module to Urban Block

The proposed MFS "neighborhood unit" or Urban Block takes on the smart city principles of mixed used spaces, where each block encompasses residential, commerce, office space, education (e.g., preschool and primary education), and leisure. The idea is that the Urban Block is located near the center of coastal cities, and acts as an additional neighborhood as opposed to a standalone floating city.

In this study the principal module was designed for residential use, which has stringent comfort criteria, but it could also be used for other urban activities. Figure 14 shows a schematic representation of such Urban Block made from 52 equal Module9000, which can accommodate more than 20,000 inhabitants. Figure 15 shows the MFS Urban Block near Tel Aviv—Yafo, providing a floating offshore neighborhood near the core of the city. The position offshore was determined by the local bathymetry where water depth greater than 20 m was threshold requirement.

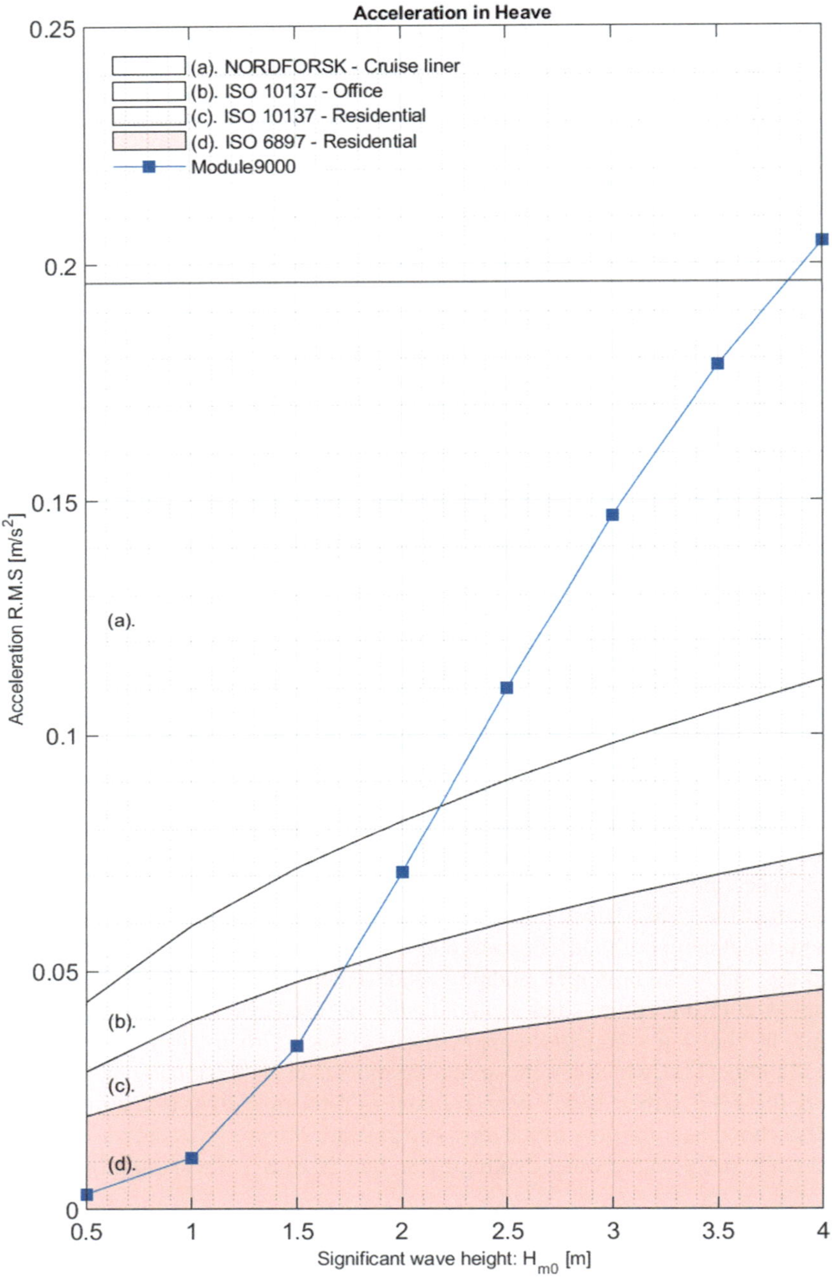

Fig. 13 Module9000 allowed accelerations versus significant wave height, the vertical acceleration is given for different significant wave heights from 0.5 to 4 m. The results are overlaid on the relevant limiting curves for occupant comfort

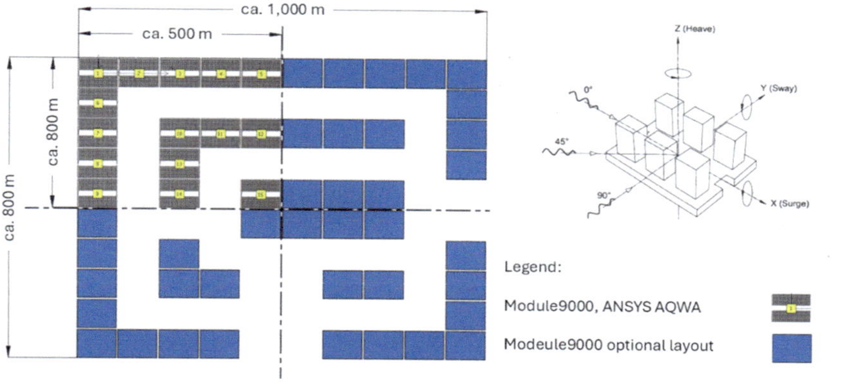

Fig. 14 Schematic representation of an MFS urban block, showing the modules distribution, and the dimensions of the urban block. The upper left side corner of the Urban Block was used in the hydrodynamic analysis outlined in Sect. 3.2 and in Fig. 12

Fig. 15 Overlay of MFS Urban Block on Google Maps, off the coast of Tel Aviv-Yafo in a prime real estate area location near city center. MFS Urban Block is positioned at an angle of 45° to the main wave direction, to reduce the vertical motion amplitude as discussed in Sects. 2.3 and 2.4

3.2 *Frequency Domain Wave-Structure Interaction*

Presented in Figs. 14 and 15, the MFS Urban Block consists of 52 modules, but the hydrodynamic (wave-structure interaction) analysis will concentrate at the upper left corner, at the modules directly facing the wave direction. The modules are all-in free-floating conditions and are not motion restricted in the first order, and hence the modules motions are linear to the wave unit amplitude. At this point there are no structural constraints nor inter-modules connections, and the analysis is performed in the frequency domain only. The reason that the connections of the modules are not included in this preliminary design phase relates directly to the outcome of the analysis. As each module, in its twin-hull arrangement, weighs over 30,000 tones, the shear forces loading the connections during vertical oscillations will be very high, especially if the connectors are rigid. Therefore, the reduction of motion amplitudes in the modules is not only a comfort objective but a structural one too. Figure 16 shows the upper left corner of the MFS Urban Block in four different wave periods. The wave periods 8, 9 and 10 s represent storms with an annual return, while 16 s wave period is assumed as a storm of 100-year return.

Examining Fig. 16, we can visually evaluate the influence of the modules on incident wave direction and amplitude. At an 8-s wave period (Fig. 16a), the wave diffraction process is more pronounced, with a notable 100% amplitude decrease on the leeward side of the third-tier module. In contrast, at a 16-s wave period (Fig. 16d), the incident wave amplitude passing through the corner of the Modular Floating Structure (MFS) remains unchanged, indicating that the MFS model is hydrodynamically transparent. This interaction among the selected modules effectively limits motion response by reducing incident wave amplitude for wave periods up to approximately 10 s. Consequently, we can conclude that the modules' effectiveness in attenuating incident waves increases as the wave period decreases, establishing a direct relationship between the dimensions of the modules and the wavelength.

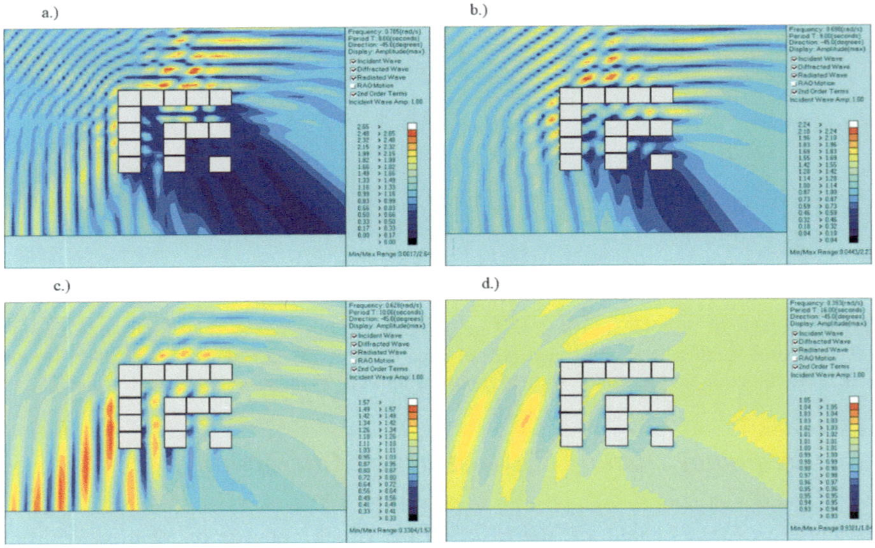

Fig. 16 Wave-structure interaction analysis in the frequency domain, displayed using ANSYS-AQWA's Graphical Supervisor (AGS) [34]. Only the upper left corner of the MFS Urban Block is calculated in the analysis at **a** 8 s wave period, **b** 9 s wave period, **c** 10 s wave period, and **d** 16 s wave period

3.3 *Zoning Consideration and Motion Amplitudes*

3.3.1 Service Limit State Design—Comfort

In terms of occupant comfort, Curve A of ISO 6897, which suggests acceptable magnitudes of horizontal motion for general-purpose buildings, serves as the benchmark for residential use. In contrast, Curve C of ISO 10137, which addresses wind-induced vibrations in office buildings over a one-year return period, establishes a much lower perception threshold than Curve A. Based on the results shown in Fig. 10 and in relation to Curve C, it is evident that the free-floating module (positioned at a 45° angle to the wave direction) comfortably meets the required threshold, even at significant wave heights greater than 2.5 m. Recognizing at least two relevant criteria for evaluating the module's comfort allows us to strategically plan their locations within the MFS arrangement according to their classification—residential or office, for instance. The differing motion thresholds also serve as an optimization tool, enabling us to position office modules closer to the wave impact area, thereby creating a protected space for residential modules. Naturally, additional classifications for other module types, such as commerce and leisure, need to be developed regarding allowable accelerations, which could potentially result in even lower perception thresholds than those indicated by ISO 10137 Curve C. The NORDFORSK 1987 criteria for acceleration and roll, provided in Table 4, could serve as a suitable starting point for establishing benchmarks for these additional module classifications.

3.3.2 Ultimate Limit State Design—Safety

In extreme weather conditions, such as a 100-year storm, modules are not expected to provide occupant comfort but rather to endure and maintain the required structural integrity, and the range of stability according to IMO regulations. A similar approach is also taken in buildings designed for rare high-magnitude earthquake events. The transparent hydrodynamic response of the modules in 16 s wave period (Fig. 16d) implies that their response amplitude will be very similar to the transmitted unit wave amplitude. The significant wave height during a 100-year storm in the eastern Mediterranean is 8.5 m, which means very high-load impacts on any connecting modules; even free-floating modules position in proximity could collide while risking both structure and occupant safety.

To ensure the safety of the MFS Urban Block a maximum wave height must be defined and controlled by a protective environment. In the ethos of this research, floating breakwaters can be used to attenuate the incident wave height, serving as protection seawall to enable the use of MFS in open water conditions. Understanding the inherent relationship between the dimensions of the modules and the wavelength, it is obvious that only very large floating structures will be useful for this purpose.

3.4 The Floating Seawall Concept

This section focuses on the transformation of retired ships into deep-water floating breakwaters and investigates gap resonance effects to increase energy dissipation and improve the transmission coefficient during critical wave periods. The suggested design approach of 'The Circular Economy' reduces building cost and fabrication time, together with material waste, and the environmental impact associated with shipbreaking and conventional breakwaters. The floating seawall is made from large rigid modules positioned in staggered formation to allow them to move safely in the boundary stiffness of their mooring system. Each module is made from two large scale ocean-going-vessel rigidly connected alongside with a methodical clearance between them. In this example, two Ultra Large Crude Carrier (ULCC) with the length (L) = 400, beam (B) = 60, and draft (T) = 40 m with clearance of 20 m between them to induce gap resonance at 16 s wave period. Figure 17 shows the principal action of the floating seawall concept, where all the wave amplitudes in the leeward side of the seawall are smaller than half of the incident wave amplitudes. This means that the wave climate behind the floating sea wall does not exceed 4.25 m significant wave height (where the incident wave height is 8.5 m).

3.4.1 Circular Economy of Retired Ships

Millions of tons of transported cargo are moving continuously by 100,000 ocean-going cargo ships around the globe—of which about 2% comes to their end-of-life annually [28]. The disposal of retired ships is one of the most dangerous industries regarding occupational health and safety, in addition to being a significant source of marine pollution [30, 37]. However, reusing them, as a stationary floating breakwater for deep-sea applications, will not only mitigate hazards related to their end-of-life, but it will also enable, and widen, some of the most sustainable developments

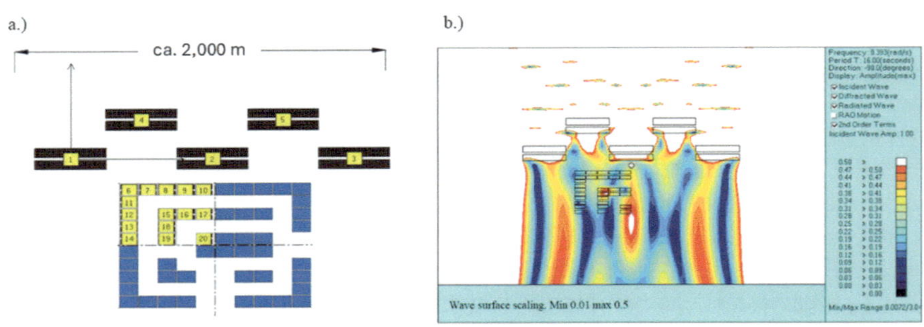

Fig. 17 **a** Typical floating seawall arrangement positioned at wave ward side of an MFS Urban block and **b** wave-structure interaction analysis in frequency domain, displayed using ANSYS-AQWA' AGS, where all wave amplitudes given are with respect to the incident wave amplitude 1.00. And all the amplitudes ≧ 0.5 are omitted [34]

Fig. 18 Main test configuration of the floating seawall concept tested [33]

of recent years (e.g., renewable energy farms, aquaculture, coastal protection, and additional useable surface that is not substituting green fields). This research focuses on the transformation of retired ships into deep-water floating breakwaters to control extreme waves, and to enable a safe environment for MFS application in open water.

3.4.2 Gap Resonance and Wave Attenuation

The influence of the gap resonance on wave attenuation, and the ability to forecast its behavior with numerical tool was investigated in research published in 2024 [33]. The research methodology included both numerical analysis (using ANSYS-AQWA) and physical experiments to test the hypothesis that gap resonance could contribute to wave attenuation of floating breakwaters. This research concluded that the numerical software predicts very accurately the gap resonance phenomena but tends to overshoot the structural response due to mathematical limitation associated with the viscosity and nonlinearity of the fluid in the gap resonance region. The physical experiment on the other hand, showed the contribution of the gap resonance, where 80% reduction of the incident wave height was registered. Figure 18 shows testing setup in the Ship Hydrodynamic Lab at Delft Technical University (TU Delft).

3.5 Section Summary

This section provides a concise description of the MFS configuration, with and without the use of a protective floating seawall, emphasizing the contribution of multimodule arrangement as a measure to reduce motion response amplitudes, for both structural and comfort perspectives. The results presented in Fig. 19 show great improvement in the module performance up to significant wave height of 4 m when it is part of an MFS arrangement, even in free floating conditions. The values are given with respect to the third-tier module shown in Fig. 16. The figure is based on Fig. 13, and the original results (Module9000) are shown in grey.

The section also includes storms with a return period of 100 years, which denotes the design wave (maximum significant wave height) for the ultimate limit state design with respect to structure and safety. Therefore, the use of a protective seawall in the open water conditions at the eastern Mediterranean Sea is required. The presented floating seawall concept, made entirely from end-of-life ships, provides a good starting point for setting a controlled environment in terms of wave height characteristics—enabling the use of MFS also in open water conditions with extreme storms.

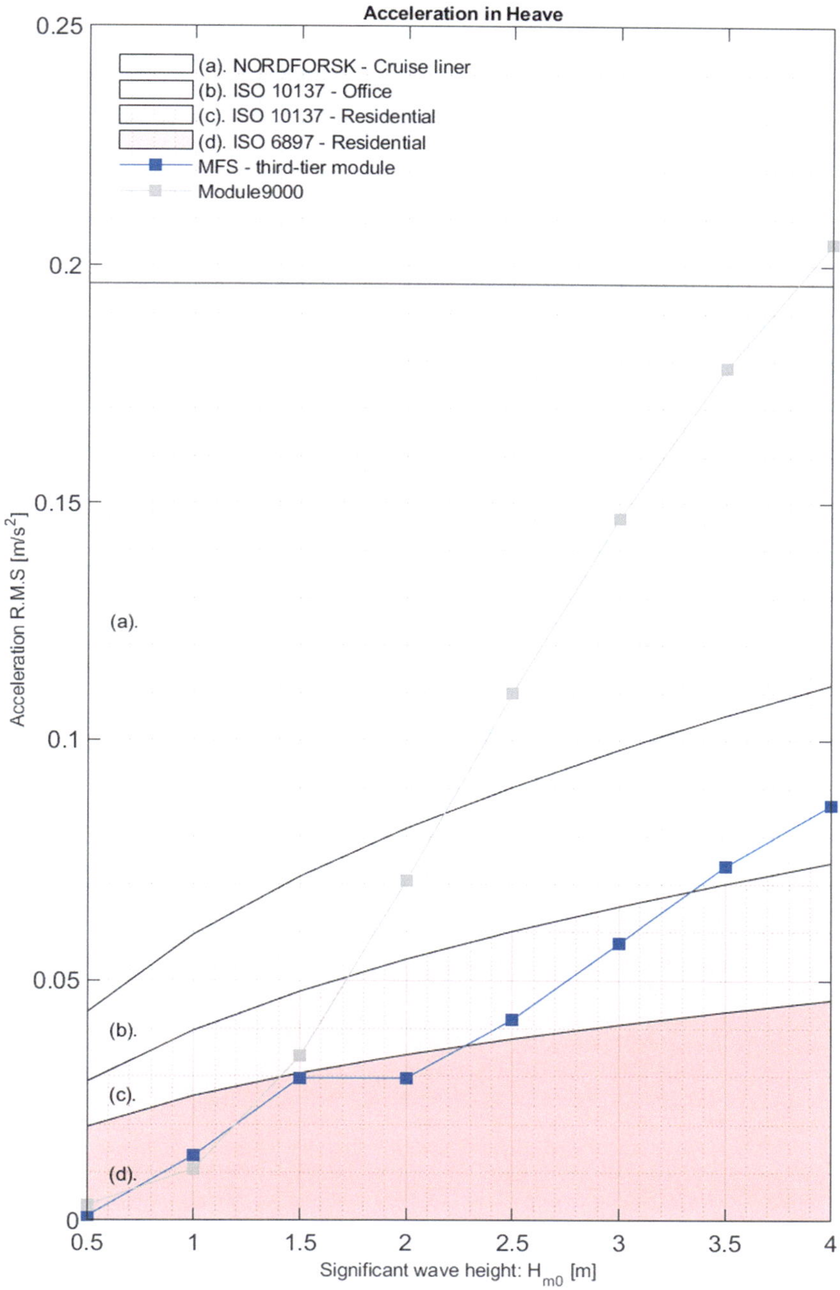

Fig. 19 Module9000 allowed accelerations versus significant wave height, with respect to the third-tier module in an MFS configuration. The vertical acceleration is given for different significant wave heights from 0.5 to 4 m. The results are overlaid on the relevant limiting curves for occupant comfort

References

1. ABS (2015) Guide for passenger comfort on ships. American Bureau of Shipping, ABS Plaza 16855 Northchase Drive Houston, TX 77060 USA
2. Ajibade I, Shah SH (2024) Can floating homes make coastal communities resilient to climate risks? Nature 632:733–736. https://doi.org/10.1038/d41586-024-02679-w
3. ANSYS (2022) Aqwa reference manual. Canonsburg USA
4. Bamber JL, Oppenheimer M, Kopp RE, Aspinall WP, Cooke RM (2019) Ice sheet contributions to future sea-level rise from structured expert judgment. Proc Natl Acad Sci 116:11195. https://doi.org/10.1073/pnas.1817205116
5. Bar D, Zavadski A, Trifonova E, Levin A, Sladkevich M (2023) Processing of hydrographic data for the Haifa region (No. P.N. 998/23). Technion City, Haifa
6. Bellefontaine N, Johansson T (2014) The role of the international maritime organization in the prevention of illegal oil pollution from ships: North Sea special status area. In: Carpenter A (ed) Oil pollution in the North Sea, the handbook of environmental chemistry. Springer International Publishing, pp 49–67. https://doi.org/10.1007/698_2014_294
7. Bo MW, Chang M-F, Arulrajah A, Choa V (2011) Ground investigations for Changi East reclamation Projects. Geotech Geol Eng 30:45–62. https://doi.org/10.1007/s10706-011-9448-3
8. Drimer N, Gafter R (2017) Delta-type VLFS—hydrodynamic aspects. Ships Offshore Struct 1–14. https://doi.org/10.1080/17445302.2017.1384440
9. Fujikubo M (2005) Structural analysis for the design of VLFS (very large floating structures). Mar Struct 18:201–226. https://doi.org/10.1016/j.marstruc.2005.07.005
10. Fujikubo M, Suzuki H (2015) Mega-float. In: Wang CM, Wang BT (eds) Large floating structures, ocean engineering & oceanography. Springer Singapore, pp 197–219. https://doi.org/10.1007/978-981-287-137-4_8
11. Haritos N (2007) Introduction to the analysis and design of offshore structures—an overview. Electron J Struct Eng 7:55–65
12. Huang K, Li X, Liu X, Seto KC (2019) Projecting global urban land expansion and heat island intensification through 2050. Environ Res Lett 14:114037. https://doi.org/10.1088/1748-9326/ab4b71
13. ISO-6897, 1984. 6897:1984, Guidelines for the evaluation of the response of occupants of fixed structures, especially buildings and off-shore structures, to low-frequency horizontal motion (0,063 to 1 Hz).
14. ISO-2631, 1997. 2631–1:1997 Mechanical vibration and shock: evaluation of human exposure to whole-body vibration
15. ISO-10137, 2007. 10137:2007. Bases for design of structures—serviceability of buildings and walkways against vibration
16. Lamb S, Macefield VG, Walton D, Kwok KCS (2016) Occupant response to wind-excited buildings: a multidisciplinary perspective. Proc Inst Civ Eng Struct Build 169:625–634. https://doi.org/10.1680/jstbu.15.00017
17. Lawther A, Griffin MJ (1987) Prediction of the incidence of motion sickness from the magnitude, frequency, and duration of vertical oscillation. J Acoust Soc Am 82:957–966. https://doi.org/10.1121/1.395295
18. MARPOL (2005) International convention for the prevention of pollution from ships (MARPOL) [www document]. URL http://www.imo.org/en/About/Conventions/ListOfConventions/Pages/International-Convention-for-the-Prevention-of-Pollution-from-Ships-(MARPOL).aspx. Accessed 19 Dec 2019
19. Matsangas P, McCauley ME, Becker W (2014) The effect of mild motion sickness and sopite syndrome on multitasking cognitive performance. Hum Factors 56:1124–1135. https://doi.org/10.1177/0018720814522484
20. McMichael C, Dasgupta S, Ayeb-Karlsson S, Kelman I (2020) A review of estimating population exposure to sea-level rise and the relevance for migration. Environ Res Lett 15:123005. https://doi.org/10.1088/1748-9326/abb398

21. O'Hanlon JF, McCauley ME (1974) Motion sickness incidence as a function of the frequency and acceleration of vertical sinusoidal motion. Aerosp Med 45:366–369
22. Palo P (2005) Mobile offshore base: hydrodynamic advancements and remaining challenges. Mar. Struct 18:133–147 (very large floating structures). https://doi.org/10.1016/j.marstruc.2005.07.007
23. Peduzzi P (2014) Sand, rarer than one thinks. Environ Dev 11:208–218
24. Ray J (2000) WAMIT-MOSES hydrodynamic analysis comparison study. McDermott Engineering, LLC, JRME
25. Riggs HR, Ertekin RC, Mills TRJ (2000) A comparative study of RMFC and FEA models for the wave-induced response of a MOB. Mar Struct 13:217–232 (very large floating structures (VLFS) part I). https://doi.org/10.1016/S0951-8339(00)00029-0
26. Rognaas, G., Xu J, Lindseth S, Rosendahl F (2001) Mobile offshore base concepts. Concrete hull and steel topsides. Mar Struct 14:5–23 (very large floating structures (VLFS) part II). https://doi.org/10.1016/S0951-8339(00)00019-8
27. Sengupta D, Chen R, Meadows ME (2018) Building beyond land: an overview of coastal land reclamation in 16 global megacities. Appl Geogr 90:229–238. https://doi.org/10.1016/j.apgeog.2017.12.015
28. Solakivi T, Kiiski T, Kuusinen T, Ojala L (2021) The European ship recycling regulation and its market implications: ship-recycling capacity and market potential. J Clean Prod 294:126235. https://doi.org/10.1016/j.jclepro.2021.126235
29. United Nations Human Settlements Programme (2022) World cities report 2022: envisaging the future of cities. United Nations, Erscheinungsort nicht ermittelbar
30. Wan Z, Wang L, Chen J, Sperling D (2021) Ship scrappage records reveal disturbing environmental injustice. Mar Policy 130:104542. https://doi.org/10.1016/j.marpol.2021.104542
31. Wang CM, Watanabe E, Utsunomiya T (eds) (2007) Very large floating structures. CRC Press, London ; New York
32. Wang G, Bar D, Hooimeijer F, Schreier S (2024a) Floating urban development—sustainable growth and affordable housing, lecture notes in civil engineering. Springer Nature Singapore, Singapore
33. Wang G, Bar D, Schreier S (2024) The potential of end-of-life ships as a floating seawall and the methodical use of gap resonance for wave attenuation. Ocean Eng 298:117246. https://doi.org/10.1016/j.oceaneng.2024.117246
34. Wang G, Drimer N, Goldfeld Y (2020) Modular floating structures (MFS) for offshore dwelling a hydrodynamic analysis in the frequency domain. Ocean Eng 216:107996. https://doi.org/10.1016/j.oceaneng.2020.107996
35. Wang G, Goldfeld Y, Drimer N (2019) Expanding coastal cities—proof of feasibility for modular floating structures (MFS). J Clean Prod. https://doi.org/10.1016/j.jclepro.2019.03.007
36. Wang G, Rosenfeld Y, Drimer N, Goldfeld Y (2020b) Occupant comfort analysis for rigid floating structures—methodology and design assessment for offshore dwelling module. Ships Offshore Struct 1–16. https://doi.org/10.1080/17445302.2020.1718267
37. Zhou Q, Du Z, Liu J, Liang J, Jiao Y (2021) Factors influencing green ship recycling: a conceptual framework and modeling. J Clean Prod 322:129155. https://doi.org/10.1016/j.jclepro.2021.129155

Gil Wang is a naval architect and marine engineer currently serving as the Chief Science and Technology Officer at the Israeli Coastal and Marine Engineering Research Institute (CAMERI), where he leads academic research and innovation. His career began at the Amsterdam-based superyacht design studio, Dykstra Naval Architects, where he worked for nearly a decade as a key member of the design team. He then pursued a Ph.D. in Civil and Environmental Engineering at the Technion – Israel Institute of Technology, followed by a two-year postdoctoral fellowship at the Ship Hydromechanics Laboratory at Delft University of Technology (TU Delft). His work integrates numerical simulations and physical testing methodologies to advance the design and performance of complex floating structures for urban development in open-water conditions. In addition to investigating the hydromechanics of modular floating structures, he incorporates industrial ecology methods, GIS tools, and ecological frameworks to enhance biodiversity through innovative floating solutions.

Environmental and Sustainability Considerations

Industrialization Challenges in Floating Wind

Carlos A. Garrido-Mendoza

Abstract The transition to floating wind technology introduces a unique set of industrialization challenges worldwide, each shaped by regional characteristics across Europe, Asia, and the Americas. This chapter examines the primary obstacles to scaling floating wind energy, including technological, economic, regulatory, social, and infrastructural barriers, and how these challenges differ by region due to variations in market maturity, regulatory frameworks, and geographic conditions. In Europe, key issues include supply chain limitations and grid integration. Asia contends with deep-water deployment and resilience to typhoons, while the Americas face regulatory complexities and financing challenges. Understanding these regional distinctions allows stakeholders to develop targeted strategies for more effective global scaling of floating wind technology.

Keywords Floating wind · Industrialization challenges · Renewable energy · Regulatory barriers · Technological innovation

1 Introduction

Floating wind technology represents a significant advancement in the renewable energy sector, offering the potential to harness wind energy in deeper waters where traditional fixed-bottom turbines are not feasible. However, the path to widespread industrialization of floating wind is fraught with challenges that vary significantly across different regions of the world. By 2030, an estimated 10.9 GW of floating wind will be built in the world, and by 2050, 2% of the world's energy supply is expected to come from floating wind. In Europe alone, which currently continues to be the leader in floating installations, the development of 250 GW is expected by 2050, which is equivalent to an investment of €800 bn.

The floating wind industry is still in its nascent stages, with only a few large-scale projects currently operational. The technology involves anchoring wind turbines on

C. A. Garrido-Mendoza (✉)
Saitec Offshore Technologies S.L., Leioa, Spain
e-mail: carlosgarrido@saitec.es

B. T. Wang et al. (eds.), *Large Floating Solutions*, Lecture Notes in Civil Engineering 636, https://doi.org/10.1007/978-981-96-5435-2_7

Fig. 1 Left: DemoSATH platform at bimep testing center. Right: TetraSpar moored at the test site in Norway. Courtesy: SAITEC and STIESDAL

floating structures that are moored to the seabed. This allows for the exploitation of wind resources in deeper waters, which often have stronger and more consistent wind speeds.

The development of floating wind technology has, however, accelerated in recent years, driven by advancements in materials science, engineering, and a growing emphasis on sustainable energy solutions. Key milestones include the deployment of pilot projects and the establishment of frameworks for larger commercial projects (DemoSATH project [1] and TetraSpar Demonstrator [2] as shown in Fig. 1). As the demand for clean energy rises and conventional offshore wind technologies reach their limits in deeper waters, floating platforms are emerging as a promising solution to harness wind potential in marine areas further from the coast.

In recent years, there has been rapid progress in the development and industrialisation of these floating platforms. Several countries, such as Norway, Portugal, and Scotland, are leading the research and deployment of large-scale pilot projects (Hywind Scotland and Tampen [3, 4] and Kincardine Offshore Wind Farm [5] as shown in Fig. 2). These initiatives have helped to improve the understanding of the technical and economic challenges associated with floating platforms, as well as to drive innovation in design and manufacturing.

One of the main challenges in the industrialisation of floating platforms for offshore wind energy is the reduction of costs. So far, floating platforms have been more expensive to construct and operate when compared to fixed structures. However, as production scales up and manufacturing processes are optimised, costs are expected to decrease significantly. Moreover, the development of more efficient anchoring technologies and advanced control systems has improved the stability and survivability of floating platforms, which also contributes to long-term cost reductions.

Fig. 2 Top: Hywind Scotland pilot park 30 MW. Bottom: Kincardine Offshore Wind Farm 50 MW. Courtesy: Equinor ASA and Principle Power

Another crucial aspect is the advancement in the fabrication of components and structures for floating platforms. New manufacturing techniques, such as additive manufacturingmodular construction, are being explored to streamline the construction process and reduce delivery times. Additionally, the adoption of lighter and more durable composite materials is enabling the construction of more efficient platforms that can withstand extreme marine environmental conditions.

Research into anchoring and mooring systems has also made significant progress. Innovative technologies, such as suction anchors and gravity anchors, are being developed to provide greater stability and ease of installation in deep waters. These advancements in anchoring technology are fundamental to expanding the reach of floating offshore wind energy and tapping into the potential of areas with challenging seabed conditions.

Furthermore, significant efforts are being made to standardise the designs and manufacturing processes of floating platforms. Standardisation can facilitate industrialisation by reducing development costs and improving efficiency in construction and maintenance. Collaboration between industry, standardisation bodies, and regulators plays a key role in promoting common standards and accelerating the deployment of floating platforms worldwide.

In summary, the state of the art of the industrialisation of floating platforms for offshore wind energy is experiencing rapid advancement. Progress in design, manufacturing, anchoring, and standardisation is paving the way for greater expansion of offshore wind energy in deeper waters. As costs decrease and technologies are refined, floating platforms are expected to play a significant role in providing clean energy and contributing to the transition towards a sustainable energy future.

2 Industrialization Challenges

The industrialization of floating wind faces several significant challenges, which can be broadly categorized into technological, economic, regulatory, and infrastructural.

2.1 *Technological Challenges*

Technological challenges in the floating wind sector are primarily centred on the development and deployment of reliable and cost-effective floating platforms, mooring systems, and subsea cables. The need for stability in harsh marine environments and the complexities associated with maintenance and grid integration further complicate these technological hurdles.

2.1.1 Floating Platforms and Mooring Systems

The design and construction of floating platforms are fundamental to the success of floating wind projects. These platforms must withstand the dynamic forces of waves, currents and wind. There are several types of floating platforms currently being tested, including spar-buoy, semi-submersible, barge-type, and tension-leg platforms (see Fig. 3).

Developed by Equinor, the Hywind Scotland project [3] utilizes spar-buoy platforms. This project has demonstrated the feasibility of deploying floating wind turbines in deep waters and has set a precedent for future developments. However, the cost and complexity of these platforms remain significant barriers to broader adoption.

Mooring systems, which anchor the floating platforms to the seabed, must be designed to handle the considerable forces exerted by the energetic offshore environmental condition. These systems must ensure the stability of the turbines while allowing for some movement to absorb energy from waves. The Kincardine Offshore Wind Farm [5], using semi-submersible platforms, has highlighted the need for advanced mooring technology to ensure both stability and flexibility.

2.1.2 Subsea Cables and Grid Integration

The integration of floating wind farms into existing power grids poses another significant technological challenge. Subsea cables that transmit electricity from the turbines to the shore must be robust enough to withstand harsh marine conditions while minimizing transmission losses.

Projects like WindFloat Atlantic [6] have illustrated the importance of reliable subsea cable technology. WindFloat Atlantic uses dynamic cables designed to move

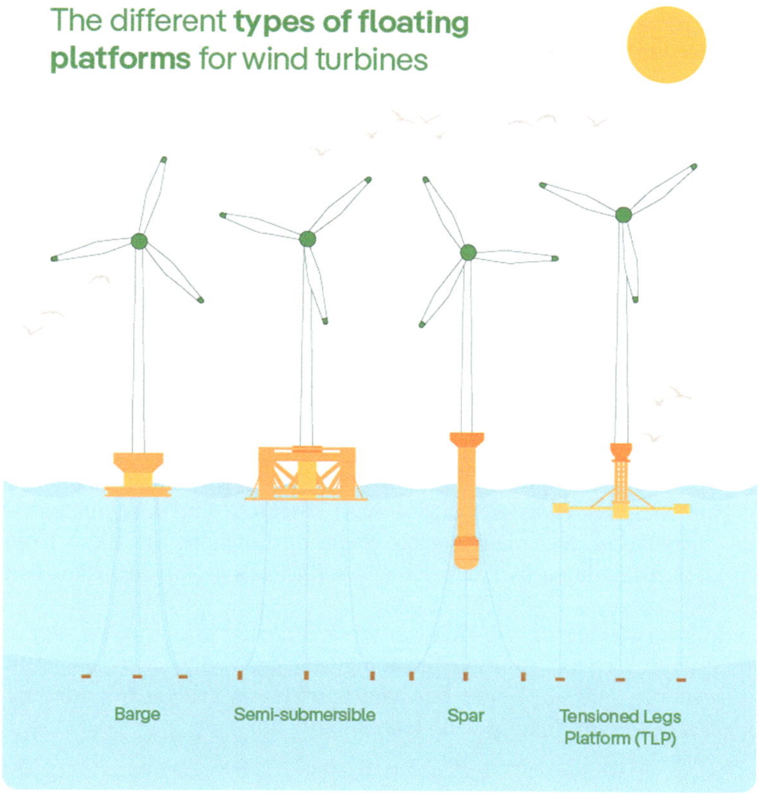

Fig. 3 Various types of floaters for wind turbines. Courtesy of Iberdrola

with the floating platforms, reducing the risk of damage and ensuring consistent power de-livery. Furthermore, the grid integration of these floating wind farms requires advanced technology to manage the variable output of wind energy and ensure stability in the power supply.

2.1.3 Maintenance and Reliability

The maintenance of floating wind turbines is more complex and costly when compared to fixed-bottom turbines due to their offshore locations and the dynamic nature of their platforms. Ensuring reliability and minimizing downtime are critical for the economic viability of floating wind projects. Innovations in autonomous drones and robotic technologies for maintenance are being explored to address these challenges. The DemoSATH [1] and Floatgen [7] projects off the coast of Biscay,

Spain, and France, respectively, are an example where such concrete-based technologies are being tested to perform inspections and minor repairs, potentially reducing maintenance costs and increasing operational efficiency. Depending on the technology and the size of the farm, it is estimated that the contribution of O&M costs may be between 30 and 50% of the LCoE (Levelized Cost of Energy), which gives an idea of the huge potential market that the maintenance of future farms will entail. However, floating wind presents unique challenges that must be addressed to ensure its long-term viability and efficiency, specifically in O&M, since the prototypes deployed to date have only been installed for a few years and it is only now that the problems that may arise are beginning to be known, for which, in many cases, There are still no technological solutions with sufficient maturity.

2.2 *Economic Challenges*

The high initial capital expenditure for floating wind projects is a major economic barrier. This includes costs associated with research and development, manufacturing, installation, and maintenance. Securing financing for these large-scale projects can be difficult, particularly in markets that lack established frameworks for floating wind energy.

Economic challenges in the floating wind industry primarily revolve around high initial capital costs, securing financing, and achieving cost-competitiveness with other energy sources. These financial barriers are critical to address for the widespread adoption of floating wind technology.

2.2.1 High Initial Capital Costs

The development of floating wind projects requires substantial upfront investment. The costs associated with research and development, manufacturing, installation, and maintenance of floating wind turbines are significantly higher than those for fixed-bottom turbines. The Hywind Scotland project, for instance, had a capital expenditure of approximately £210 million for a 30 MW capacity [8], translating to around £7 million per MW, which is higher compared to traditional offshore wind projects. These high costs are attributed to the need for advanced materials, specialized equipment, and complex installation processes.

2.2.2 Securing Financing

Securing financing for floating wind projects is challenging due to the perceived risks and uncertainties associated with the technology. Investors are often hesitant to fund projects that involve unproven or emerging technologies, especially when the return on investment is uncertain. Therefore, public–private partnerships and government incentives are crucial in mitigating these financial risks. The European Investment Bank's support for the WindFloat Atlantic project, which provided a €60 million loan, is an example of how public funding can catalyse private investment [9].

2.2.3 Achieving Cost-Competitiveness

To achieve cost-competitiveness with other renewable and non-renewable energy sources, floating wind technology must undergo significant cost reductions. This can be achieved through economies of scale, technological advancements, and streamlined manufacturing and installation processes.

The Carbon Trust's Floating Wind Joint Industry Project aims to reduce the levelized cost of energy (LCoE) for floating wind to below £100/MWh by 2030 through collaborative research and development efforts. Achieving such cost reductions is critical for the long-term viability of floating wind technology.

2.3 Regulatory Challenges

Regulatory challenges vary widely by region but generally include the need for clear policies and permitting processes. In many areas, existing regulations are tailored to fixed-bottom offshore wind and may not adequately address the specific needs of floating wind projects. Regulatory challenges encompass the need for clear and supportive policies, streamlined permitting processes, and international collaboration. Regulatory frameworks must evolve to accommodate the unique requirements of floating wind technology and ensure its integration into national energy strategies.

2.3.1 Clear and Supportive Policies

The absence of clear and supportive policies can hinder the development of floating wind projects. Regulatory uncertainty can deter investment and delay project timelines. Policies that provide long-term stability and incentives for renewable energy projects are essential. In Europe, the EU Offshore Renewable Energy Strategy aims to increase offshore wind capacity to 300 GW by 2050, with a significant portion expected to come from floating wind. This strategy provides a clear roadmap and support mechanisms for the industry.

2.3.2 Streamlined Permitting Processes

The permitting process for floating wind projects is often complex and time-consuming, involving multiple regulatory bodies and stakeholders. Streamlining these processes is crucial to accelerate the deployment of floating wind technology.

For example, the ScotWind Leasing round, managed by Crown Estate Scotland, has introduced a more streamlined approach to leasing seabed areas for offshore wind development. This initiative is expected to facilitate faster project approvals and reduce administrative burdens [10].

2.3.3 International Collaboration

International collaboration is essential for sharing best practices, standardizing regulations, and fostering innovation in the floating wind sector. Collaborative initiatives can help overcome regulatory barriers and promote the global adoption of floating wind technology.

In 2019, the Global Wind Energy Council (GWEC) and the World Bank launched the Offshore Wind Development Program, which aims to support the development of offshore wind projects, including floating wind, in emerging markets. This program promotes international cooperation and knowledge exchange. The programme will convene developing country governments, commercial developers, development partners, and wind energy experts to raise awareness around offshore wind opportunities in emerging markets and lay the groundwork for a pipeline of new projects that could be supported by World Bank or IFC financing. The World Bank and IFC will work with public and private sector partners to undertake technical studies and develop national strategies to facilitate the adoption of offshore wind [11].

2.4 Infrastructural Challenges

Infrastructural challenges encompass the development of ports, manufacturing facilities, service vessels, and supply chains capable of supporting large-scale floating wind projects. Additionally, grid integration and the construction of subsea cables are critical components that require substantial investment and coordination. Infrastructural challenges involve the development of ports, manufacturing facilities, supply chains, and grid integration systems necessary to support large-scale floating wind projects.

2.4.1 Port and Manufacturing Infrastructure

The successful deployment of floating wind projects requires significant port and manufacturing infrastructure capable of handling large components and supporting assembly and maintenance activities. Ports must be equipped with facilities to handle the assembly and deployment of large floating structures. For example, the Port of Rotterdam has been upgraded to support offshore wind projects, including floating wind, by providing extensive quay space, heavy-lift capabilities, and logistical support.

2.4.2 Supply Chain Development

Developing a robust supply chain for floating wind components is essential to ensure the timely and cost-effective delivery of materials and equipment. The supply chain must be able to scale with the growing demand for floating wind technology. The Floatgen project has highlighted the importance of developing local supply chains to reduce costs and support regional economies. By sourcing materials and components locally, the project has demonstrated cost savings and reduced lead times.

2.4.3 Grid Integration and Subsea Cables

Integrating floating wind farms into existing power grids requires advanced infrastructure, including subsea cables and grid management systems. These systems must be capable of handling the variable output of wind energy and ensuring stable power delivery. The Dogger Bank Wind Farm, although primarily a fixed-bottom project, is developing infrastructure that will also support future floating wind projects. The use of high-voltage direct current (HVDC) technology in this project enhances the efficiency of power transmission over long distances, which is crucial for offshore and floating wind farms [12].

3 Regional Perspectives

The challenges associated with the industrialization of floating wind vary significantly across Europe, Asia, and the Americas, reflecting differences in market maturity, regulatory environments, and geographic conditions.

3.1 Europe

Europe is currently the global leader in floating wind technology, with several large-scale projects underway. However, the region faces challenges related to supply chain constraints, grid integration, and the need for coordinated regulatory frameworks across different countries. Europe is currently the global leader in floating wind technology, with several large-scale projects underway and a robust regulatory framework to support its development. The region's commitment to renewable energy and offshore wind is evident through various tenders, regulatory initiatives, and future development plans.

3.1.1 Current Floating Wind Tenders

Europe has been at the forefront of floating wind development, with significant investments in tendering processes to support large-scale deployment. The United Kingdom's ScotWind Leasing round is a notable example, where the Crown Estate Scotland has offered seabed leases to developers for offshore wind projects, including floating wind. This initiative is expected to add up to 10 GW of new offshore wind capacity, demonstrating a strong commitment to the sector.

One of the most significant steps towards industrializing floating wind in Europe has been the implementation of tenders specifically for floating wind projects. The ScotWind Leasing round by Crown Estate Scotland is a prime example, offering around 8600 km^2 of seabed for offshore wind development, including significant areas allocated for floating wind. This tender has attracted substantial interest from major energy companies, highlighting the commercial viability and investor confidence in floating wind technology.

France is also making strides with its tendering processes. Figure 4 shows the identified areas for next competitive tendering procedure. The French government launched a tender in 2021 for a 250 MW floating wind farm off the coast of Brittany, expected to be operational by 2026. This project highlights France's ambition to lead in floating wind technology and its commitment to renewable energy targets. This project is part of France's broader strategy to deploy 6.2 GW of offshore wind capacity by the end of the decade, with a significant portion dedicated to floating wind due to the country's deep waters.

3.1.2 Regulatory Frameworks

Europe's regulatory frameworks are evolving to support the unique needs of floating wind technology. The European Union's Offshore Renewable Energy Strategy is a key policy document that outlines the goal of increasing offshore wind capacity to 300 GW by 2050, with a significant focus on floating wind. This strategy includes funding mechanisms, research and development support, and streamlined permitting processes to accelerate deployment.

Individual countries have also developed supportive regulatory frameworks. For in-stance, the United Kingdom has established a Contracts for Difference (CfD) scheme, which guarantees a fixed price for electricity generated by renewable sources, including floating wind. The CfD mechanism reduces financial risk for developers and has been instrumental in the growth of the offshore wind sector in the UK.

The regulatory environment in Europe is generally supportive of floating wind, but there are variations between countries. The European Union's Offshore Renewable Energy Strategy aims to enhance regulatory coordination across member states, facilitating smoother and faster project approvals. This strategy includes simplifying permitting processes and establishing clear guidelines for offshore wind development, which are critical for the industry's growth.

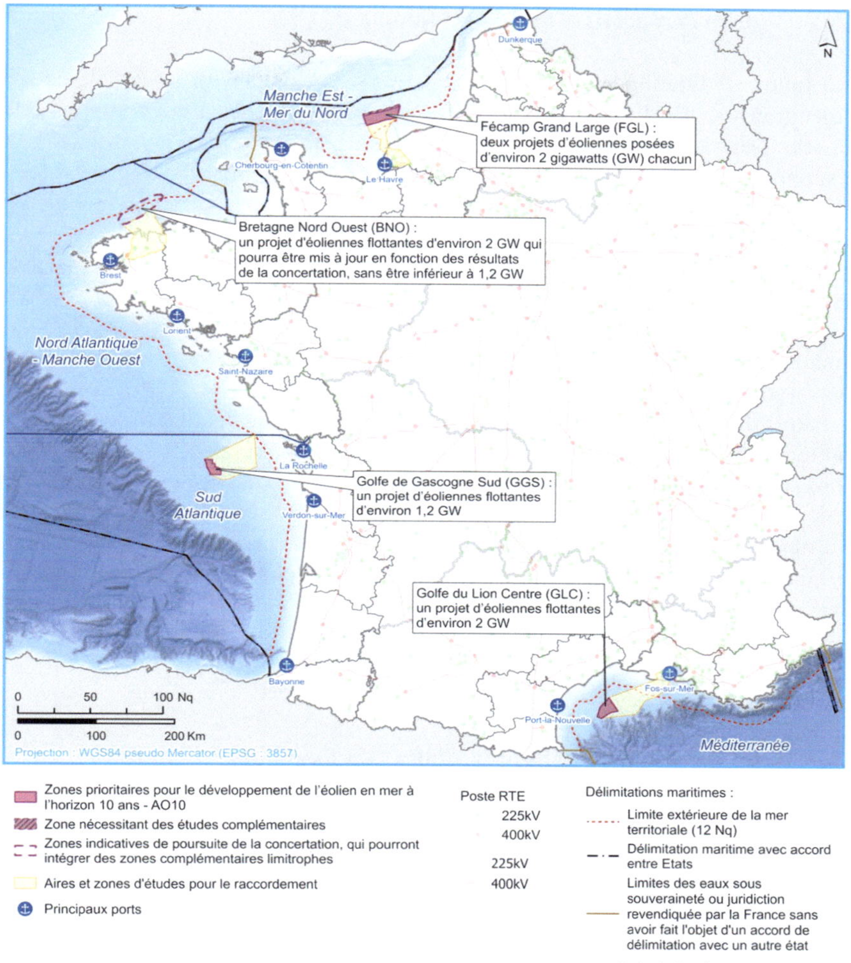

Fig. 4 Areas identified for the next competitive tendering procedure (AO10). *Source* RTE, IGN, SHOM, EMODnet, Gouvernement

In the UK, the Offshore Wind Sector Deal has set ambitious targets, aiming to generate 40 GW of offshore wind capacity by 2030, including significant contributions from floating wind. The deal provides a framework for collaboration between the government and industry, addressing regulatory and logistical challenges to expedite project development.

3.1.3 Future Perspectives and Developments

The future of floating wind in Europe looks promising, with several large-scale projects in the pipeline. The North Sea Wind Power Hub, a collaborative effort between Denmark, Germany, and the Netherlands, aims to create an artificial island to support up to 100 GW of offshore wind capacity, including floating wind farms. This ambitious project highlights Europe's commitment to integrating floating wind into its long-term energy strategy.

Additionally, advancements in technology and cost reductions are expected to drive further growth. The European Technology and Innovation Platform on Wind Energy forecasts that floating wind could achieve cost parity with fixed-bottom offshore wind by the mid-2030s, making it an economically competitive option for deep-water regions.

Future developments in Europe are focused on scaling up floating wind technology and reducing costs. The Norwegian government has announced plans for the Utsira Nord project, a large-scale floating wind farm with a capacity of up to 1.5 GW. This project is expected to be a testbed for new technologies and operational strategies, contributing to cost reductions and increased efficiency.

The European Investment Bank (EIB) continues to play a crucial role in financing floating wind projects. In 2021, the EIB provided a €300 million loan to the Kincardine Offshore Wind Farm, emphasizing the bank's commitment to supporting innovative renewable energy projects.

3.1.4 Europe: Navigating Supply Chain and Regulatory Complexity

Europe is the global leader in floating wind technology, driven by ambitious renewable energy targets and significant investments. However, the region faces considerable challenges in scaling up this technology:

Supply Chain Constraints

The supply chain for floating wind in Europe is still evolving. The production of specialized components such as floating platforms, mooring systems, and subsea cables requires significant investment and coordination. For instance, the Hywind Scotland project, developed by Equinor, highlighted the logistical complexity of sourcing and transporting components from various parts of Europe. As projects scale up, bottlenecks in the supply chain could delay deployment and increase costs.

Grid Integration

Integrating floating wind farms into the European grid poses technical challenges. Floating wind farms are often located far from shore, requiring advanced grid management systems and high-capacity subsea cables. The Dogger Bank Wind Farm, although primarily fixed-bottom, is developing infrastructure that will support future floating wind projects by using high-voltage direct current (HVDC) technology that is crucial for efficient power transmission over long distances.

Regulatory Frameworks

The regulatory environment in Europe varies significantly between countries, adding complexity to project development. The European Union's Offshore Renewable Energy Strategy aims to enhance regulatory coordination across member states, but differences in national policies and permitting processes remain a challenge. Streamlining these processes is essential for accelerating the deployment of floating wind.

3.2 Asia

Asia presents unique challenges for floating wind industrialization, including deep-water deployment and resilience to extreme weather events such as typhoons. Countries like Japan and South Korea are making significant investments, but they still face hurdles related to technology adaptation and market readiness.

Asia presents unique opportunities and challenges for the industrialization of floating wind. The region's deep waters, susceptibility to extreme weather events such as typhoons, and diverse regulatory environments require tailored approaches to develop floating wind technology effectively.

3.2.1 Current Floating Wind Tenders

In Asia, several countries have launched tenders to accelerate the deployment of floating wind projects. Japan, with its ambitious renewable energy targets, has initiated tenders for multiple floating wind projects. The Japanese government has launched multiple tenders for floating wind projects, with a notable example being the tender for a 16.8 MW floating wind farm off the coast of Fukushima. This project, supported by the Ministry of Economy, Trade, and Industry (METI), aims to demonstrate the viability of floating wind in the region's deep waters [13]. Another notable example is the Goto Floating Wind Farm project, which aims to install 16.8 MW of floating wind capacity off the coast of Goto City by 2024. This project is part of Japan's broader goal to achieve 10 GW of offshore wind capacity by 2030, with floating wind playing a crucial role due to the country's deep coastal waters.

South Korea is another major player in the Asian floating wind market. The government has announced plans to develop 12 GW of offshore wind capacity by 2030, including significant contributions from floating wind projects. The Ulsan

Floating Offshore Wind Farm, with a planned capacity of 1.4 GW, is a key project supported by government tenders and incentives. South Korea's government is also actively promoting floating wind through its Renewable Energy 3020 Plan, which includes significant investments in offshore wind. The Ulsan Floating Wind Farm project, currently in the tendering phase, is expected to add up to 1.5 GW of capacity by 2030. This project reflects South Korea's strategy to become a leader in floating wind technology and leverage on its extensive shipbuilding expertise [14].

3.2.2 Regulatory Frameworks

Asia's regulatory frameworks for floating wind vary significantly by country, reflecting different levels of market maturity and policy support. Japan has implemented the Renewable Energy Act, which includes provisions for offshore wind development and aims to streamline the permitting process for floating wind projects. This regulatory support is crucial for overcoming bureaucratic hurdles and accelerating project timelines.

In South Korea, the government has introduced the Renewable Portfolio Standard (RPS), which mandates that power producers generate a certain percentage of their electricity from renewable sources. The government has also introduced streamlined permitting processes for offshore wind projects, including floating wind, to reduce administrative burdens and accelerate development timelines. This policy, combined with financial incentives and support for research and development, has created a favourable environment for floating wind projects.

3.2.3 Future Perspectives and Developments

The future of floating wind in Asia is characterized by ambitious targets and significant investment. Japan's government has set a target of achieving 45 GW of offshore wind capacity by 2040, with floating wind expected to make up a substantial portion of this capacity. This target is supported by technological advancements and increased private sec-tor participation.

South Korea's floating wind market is also poised for growth, with several large-scale projects in the planning stages. The country's focus on innovation and collaboration with international partners is expected to drive technological advancements and cost reductions. The Korean government's commitment to achieving carbon neutrality by 2050 further underscores the importance of renewable energy, including floating wind, in the country's energy transition strategy.

Future developments in Asia are focused on overcoming technological and environmental challenges to enable large-scale deployment of floating wind. Japan's ambitious plans include the development of a 1 GW floating wind farm off the coast of Nagasaki by 2030. This project, part of the government's Green Growth Strategy, aims to demonstrate the scalability and economic viability of floating wind in the region.

South Korea's floating wind ambitions are underscored by the planned development of the Ulsan Floating Offshore Wind Farm, which is expected to be one of the largest floating wind farms in the world upon completion. This project will leverage South Korea's strengths in marine engineering and shipbuilding to achieve cost reductions and technological advancements.

China, while still in the early stages of floating wind development, is expected to play a significant role in the future. The Chinese government has announced plans to support pilot floating wind projects in the South China Sea, with a focus on deep-water deployment and typhoon resilience. These projects aim to leverage China's vast offshore wind potential and rapidly growing renewable energy sector.

3.2.4 Asia: Technological and Environmental Adaptation

Asia's floating wind market is burgeoning, with countries like Japan and South Korea leading the charge. However, the region must overcome several key challenges:

Deep-Water Deployment

Asia's geographic conditions necessitate the deployment of floating wind technology in deeper waters compared to Europe. Projects like the Fukushima Forward demonstrate the feasibility of deep-water floating wind, but scaling up such projects requires further technological advancements. Developing cost-effective and reliable floating platforms capable of withstanding harsh marine conditions is critical.

Extreme Weather Resilience

Asia's vulnerability to extreme weather events, such as typhoons, presents significant risks to offshore wind infrastructure. Designing floating wind farms capable of withstanding these conditions is crucial. For example, South Korea's Ulsan Floating Wind Farm integrates advanced engineering to enhance resilience against typhoons. Continued innovation in this area is essential to ensure the long-term viability of floating wind farms in such challenging environments.

Market Readiness

While there is growing interest in floating wind in Asia, market readiness varies. Countries like Japan have established supportive regulatory frameworks, such as the Renewable Energy Act, which simplifies the permitting process and offers financial incentives. However, other countries may need to develop similar policies to attract investment and accelerate project timelines.

3.3 Americas

In the Americas, key challenges for floating wind development include regulatory obstacles and securing adequate financing. While the United States has immense offshore wind potential, it contends with a fragmented regulatory landscape and intense competition for investment from other renewable energy sectors. The region presents a varied market landscape, with differing levels of readiness, regulatory frameworks, and geographic conditions that shape the sector's growth trajectory.

3.3.1 Current Floating Wind Tenders

In the United States, the Bureau of Ocean Energy Management (BOEM) has been actively involved in issuing leases for offshore wind development, including areas suitable for floating wind projects. The California Offshore Wind Energy Lease Sale, held in 2021, is a significant step.

The United States has seen growing interest in floating wind technology, with several states initiating tenders to explore its potential. California has emerged as a key player, with the California Public Utilities Commission (CPUC) launching a tender for a 150 MW floating wind project off the coast of Humboldt County. This project aims to harness the state's strong offshore wind resources and reduce dependence on fossil fuels [15].

Maine is also exploring floating wind opportunities through the Maine Aqua Ventus project, a 12 MW floating wind farm supported by the University of Maine and funded by the U.S. Department of Energy. This project aims to demonstrate the technical and economic viability of floating wind in the Gulf of Maine, with potential for future expansion [16].

3.3.2 Regulatory Environment

The regulatory environment in the United States is complex and fragmented, posing significant challenges to the development of floating wind projects. Federal and state agencies share jurisdiction over offshore wind development, leading to lengthy and often duplicative permitting processes. The Bureau of Ocean Energy Management (BOEM) plays a key role in leasing federal waters for offshore wind, but coordination with state and local agencies is critical [17].

To address these challenges, the Biden administration has introduced the Offshore Wind Executive Order, which aims to streamline permitting processes and set a target of deploying 30 GW of offshore wind capacity by 2030. This initiative includes specific provisions for floating wind development and emphasizes the importance of regulatory clarity and support.

In Latin America, Brazil has shown interest in offshore wind, including floating wind, through its National Energy Plan 2050. The plan identifies offshore wind as a key component of Brazil's future energy mix and outlines regulatory measures to support its development. However, significant challenges remain in terms of infrastructure and market readiness [18].

3.3.3 Future Developments

Future developments in the Americas are focused on unlocking the region's vast offshore wind potential and overcoming regulatory and financial barriers. The United States is expected to lead the way with large-scale floating wind projects off the coasts of California and Maine. These projects aim to demonstrate the commercial viability of floating wind and pave the way for broader adoption [15, 16]

Brazil's government has announced plans to launch pilot floating wind projects in the Northeast, capitalizing on the region's strong wind resources and the country's deep-water expertise. These initiatives aim to attract international investment and establish Brazil as a leader in offshore wind development within Latin America [18]

Canada is also exploring floating wind opportunities, particularly in the Atlantic provinces. The Government of Nova Scotia has initiated studies to assess the feasibility of floating wind projects, aiming to diversify the province's energy mix and reduce greenhouse gas emissions.

3.3.4 Americas: Regulatory Barriers and Financing Hurdles

The Americas, particularly the United States and Brazil, hold significant potential for floating wind development. However, several challenges must be addressed:

Regulatory Complexity

The regulatory landscape in the United States is fragmented, with federal and state agencies sharing jurisdiction over offshore wind development. This leads to lengthy and often duplicative permitting processes. The Biden administration's Offshore Wind Executive Order aims to streamline these processes and set ambitious deployment targets, but effective implementation is key.

Securing Financing

Floating wind projects require substantial upfront investment. In the United States, se-curing financing is challenging due to competition from other renewable energy sectors and the perceived risks associated with floating wind technology. The Maine Aqua Ventus project, supported by the U.S. Department of Energy, demonstrates the importance of government funding in de-risking projects and attracting private investment.

Infrastructure Development

The development of floating wind in the Americas also hinges on the availability of suitable infrastructure. This includes port facilities for assembling and launching floating platforms, as well as grid infrastructure for integrating offshore wind energy. The California Public Utilities Commission's tender for a 150 MW floating wind project off Humboldt County aims to leverage on existing infrastructure, but further investment is needed to support large-scale deployment.

3.4 *Australia*

Australia presents a unique case for the development of floating wind energy due to its vast coastlines, significant offshore wind potential, and existing renewable energy infrastructure. However, the country also faces specific industrialization challenges, including regulatory hurdles, the need for technological adaptation, and competition from other renewable energy sources.

3.4.1 Current Floating Wind Tenders

Australia is at the nascent stage of floating wind development, with few active tenders dedicated specifically to this technology. Nonetheless, the Australian government and private sector are showing increasing interest. One notable tender is the proposed Star of the South project, located off the coast of Victoria, which although primarily fixed-bottom, includes considerations for future floating wind expansions. This project aims to deliver up to 2 GW of power, providing a foundation for future floating wind developments.

In Tasmania, the state government has expressed interest in exploring offshore wind potential, including floating wind, as part of its Renewable Energy Action Plan. This initiative is still in the exploratory phase, but it highlights the growing recognition of offshore wind's potential in Australia.

3.4.2 Regulatory Environment

Australia's regulatory environment poses significant challenges for the development of floating wind. The current framework for offshore renewable energy is primarily tailored to oil and gas extraction, requiring substantial adjustments to accommodate floating wind projects. The Offshore Electricity Infrastructure Bill 2021 aims to create a new regulatory regime specifically for offshore electricity projects, which will be crucial for enabling floating wind development.

Permitting processes are another hurdle, often being lengthy and complex due to the involvement of multiple agencies and jurisdictions. The need for environmental impact assessments and stakeholder consultations further adds to the regulatory burden. Simplifying and streamlining these processes will be essential to accelerate floating wind projects.

3.4.3 Future Developments

Future developments in Australia will likely focus on overcoming regulatory and technological barriers to enable large-scale floating wind deployment. The government's commitment to reducing carbon emissions and transitioning to renewable energy sources provides a supportive backdrop for these efforts. The integration of floating wind into Australia's National Electricity Market (NEM) will be a critical factor, requiring enhancements in grid infrastructure and energy storage solutions to manage the intermittent nature of wind energy [19].

Private sector initiatives are also expected to drive future developments. Companies, like Oceanex Energy, are exploring potential sites for floating wind farms, leveraging Australia's extensive maritime industry and engineering expertise. Collaboration between the public and private sectors will be key to advancing floating wind technology and achieving cost reductions [20].

Australia's vast Exclusive Economic Zone (EEZ) offers significant opportunities for floating wind, particularly in areas where seabed conditions make fixed-bottom turbines impractical. The Bass Strait and areas off the coast of Western Australia are identified as prime locations due to their strong and consistent wind resources.

3.4.4 Australia: Overcoming Initial Barriers

Australia's vast coastlines and significant offshore wind potential position it well for floating wind development. However, the country faces specific industrialization challenges:

Regulatory Hurdles

Australia's current regulatory framework for offshore renewable energy is primarily de-signed for oil and gas extraction. The Offshore Electricity Infrastructure Bill 2021 aims to create a new regulatory regime for offshore electricity projects, but its effective implementation will be crucial for enabling floating wind development.

Technological Adaptation

Australia needs to adapt existing technologies to suit its unique marine conditions. The proposed Star of the South project off Victoria, which includes considerations for floating wind, highlights the need for technological adaptation. Developing platforms and mooring systems that can withstand the Australian marine environment is essential.

Market Competition

Floating wind in Australia faces competition from other renewable energy sources, particularly solar and onshore wind, which are already well-established. To attract investment, floating wind projects must demonstrate economic viability and cost-competitiveness. The Tasmanian government's Renewable Energy Action Plan, which includes offshore wind, aims to diversify the state's energy mix and reduce reliance on other renewables.

3.5 Synthesis and Future Outlook

The industrialization of floating wind technology is a multifaceted challenge that re-quires addressing region-specific barriers. Key strategies for overcoming these challenges include:

3.5.1 Enhancing Supply Chains and Infrastructure

Developing robust supply chains and infrastructure is critical for the large-scale deployment of floating wind. This involves investing in manufacturing capabilities for specialized components, expanding port facilities, and upgrading grid infrastructure. For example, Europe's focus on high-voltage direct current (HVDC) technology for long-distance power transmission is a step in the right direction.

3.5.2 Streamlining Regulatory Frameworks

Streamlining regulatory frameworks and simplifying permitting processes are essential to reduce project timelines and attract investment. The European Union's Offshore Renewable Energy Strategy and the Biden administration's Offshore Wind Executive Order are examples of efforts to harmonize regulations and support offshore wind development.

3.5.3 Fostering Technological Innovation

Technological innovation is key to addressing the unique challenges of floating wind. This includes developing cost-effective floating platforms, resilient mooring systems, and advanced grid integration technologies. Projects like the Ulsan Floating Wind Farm in South Korea and the Fukushima Forward in Japan are leading the way in technological advancements.

3.5.4 Securing Financing and Investment

Securing financing for floating wind projects requires a combination of government support, private investment, and innovative funding mechanisms. Government initiatives such as the U.S. Department of Energy's support for the Maine Aqua Ventus project play a crucial role in de-risking projects and attracting private capital.

3.5.5 Building Stakeholder Support

Gaining stakeholder support, including from local communities, environmental groups, and industry players, is vital for the successful deployment of floating wind projects. Transparent communication, environmental impact assessments, and stakeholder consultations can help build trust and address concerns.

4 Conclusion

The industrialization of floating wind technology holds great promise for the future of renewable energy, but it is not without its challenges. By understanding and addressing these challenges on a regional basis, stakeholders can develop more effective strategies for scaling up floating wind projects globally.

The industrialization of floating wind technology is a critical component of the global transition to renewable energy. As highlighted in the regional perspectives, Europe, Asia, the Americas, and Australia each face unique challenges and opportunities in the development of floating wind. This section synthesizes these challenges and explores how they shape the future trajectory of floating wind industrialization in these regions.

The industrialization of floating wind technology represents a significant opportunity for the global renewable energy sector. Each region faces unique challenges, from supply chain constraints and regulatory complexities in Europe to deep-water deployment and extreme weather resilience in Asia, regulatory barriers and financing hurdles in the Americas, and technological adaptation and market competition in Australia. By addressing these challenges through targeted strategies, stakeholders can unlock the full potential of floating wind and contribute to a sustainable energy future.

The collaborative efforts of governments, industry players, and research institutions will be crucial in overcoming these challenges. Continued investment in technological innovation, infrastructure development, regulatory streamlining, and stakeholder engagement will pave the way for the successful industrialization of floating wind. As regions learn from each other and share best practices, the global floating wind industry can achieve significant scale, driving down costs and accelerating the transition to renewable energy.

In conclusion, while the path to industrializing floating wind is fraught with challenges, the potential benefits in terms of renewable energy generation, economic growth, and environmental sustainability are immense. By addressing the specific barriers in each region and fostering a collaborative and innovative approach, the floating wind sector can play a pivotal role in the global energy transition.

References

1. DemoSATH demonstration project. First floating wind turbine connected to Spanish grid. https://saitec-offshore.com/en/projects/demosath/
2. TetraSPAR Demonstrator. The world's first fully industrialized offshore foundation. https://www.stiesdal.com/offshore/the-tetraspar-full-scale-demonstration-project/
3. Hywind Scotland. The world's first floating wind farm, the 30 MW Hywind Scotland pilot park. https://www.equinor.com/energy/hywind-scotland
4. Hywind Tampen. The world's first renewable power for offshore oil and gas. https://www.equinor.com/energy/hywind-tampen
5. Kincardine Offshore Wind Farm. World's largest floating offshore wind farm. https://www.principlepower.com/projects/kincardine-offshore-wind-farm
6. WindFloat Atlantic. World's first semi-submersible floating offshore wind farm. https://www.windfloat-atlantic.com/
7. Floatgen project. France's first offshore wind turbine and BW Ideol's first unit. https://www.bw-ideol.com/en/floatgen-demonstrator
8. Hywind Scotland first commercial-scale floating offshore wind farm capital cost. https://masdar.ae/en/renewables/our-projects/hywind-scotland
9. Chris Welch (2021) A wind power wonder off Portugal's stormy coast. https://www.eib.org/en/stories/floating-wind-farm
10. ScotWind leasing round. https://www.crownestatescotland.com/scotlands-property/offshore-wind/scotwind-leasing-round
11. Adnan Durakovic (2019) GWEC and world bank group target emerging offshore wind markets https://www.offshorewind.biz/2019/03/07/gwec-and-world-bank-group-target-emerging-offshore-wind-markets/
12. Dogger Bank Wind Farm. Building the world's largest offshore wind farm https://doggerbank.com/
13. METI (Ministry of Economy, Trade, and Industry) (2021) Japan Floating Wind Initiatives. https://www.meti.go.jp/
14. Korea Energy Agency (2021) Ulsan floating offshore wind project. https://www.energy.or.kr/
15. California Public Utilities Commission (CPUC) (2021) California offshore wind projects. https://www.cpuc.ca.gov/
16. University of Maine. Maine Aqua Ventus Project. https://umaine.edu/
17. Bureau of Ocean Energy Management (BOEM) (2021) Offshore Wind Development. https://www.boem.gov/

18. Brazilian Ministry of Mines and Energy (2021) National energy plan 2050. https://www.gov.br/mme
19. Australian Energy Market Operator (AEMO) (2021) Integrating renewables into the national electricity market. https://www.aemo.com.au/
20. Oceanex Energy (2021) Floating Wind Initiatives. https://oceanexenergy.com/

Dr. Carlos Ariel Garrido-Mendoza a distinguished Mechanical Engineer with an Executive MBA, brings over a decade of expertise in the renewable energy sector, specializing in offshore wind innovation. As Director of Innovation at Saitec Offshore Technologies, Dr. Garrido drives cutting-edge research and development initiatives to advance floating wind solutions, aligning with European and global market standards. His leadership has propelled transformative projects, which aim to reduce costs and accelerate industrialization in floating offshore wind. Dr. Garrido's academic foundation includes a Ph.D. in Maritime Technology, complemented by multidisciplinary master's degrees in aerospace and sustainable energy technologies, and a Executive Master's Degree in Business Administration.

Hydrodynamic Responses of a Modular Floating Solar Platform

Xuanlie Zhao, Lidong Zhang, Jia Wang, and Yao Zhou

Abstract Modular floating platforms have recently emerged as a promising solution for offshore renewable energy applications. This chapter explores the hydrodynamic behaviour of modular photovoltaic platforms and their arrays under wave action. The three-dimensional wave diffraction and radiation problems involving multiple floating bodies were analysed numerically using the boundary element method in the frequency domain. An in-house, parallelized code was developed to assess the hydrodynamic responses of multi-body systems with complex connections. Parametric studies were conducted to examine the performance of an isolated module, a modular solar platform, and an array of interconnected platforms under wave conditions. The results revealed that the motion response of an isolated module exhibited a single resonance peak. In contrast, modular solar platforms demonstrated multiple peaks due to the hydrodynamic interactions between individual modules. Furthermore, the vertical forces acting on the connectors of the weather-facing modules were significantly higher than those on other modules. For photovoltaic arrays, the trends in heave and pitch motion responses as functions of the wavenumber were consistent with those of a single modular platform. However, the loads on the connectors varied notably with changes in the incident wave angle. These findings provide valuable insights for designing and optimizing floating modular structures for offshore renewable energy applications.

Keywords Floating solar platform · Offshore photovoltaic farm · Modular design · Potential flow theory · Lagrange multiplier method · Motion response

---

X. Zhao (✉) · L. Zhang · J. Wang · Y. Zhou
College of Shipbuilding Engineering, Harbin Engineering University, Harbin, China
e-mail: xlzhao@hrbeu.edu.cn

L. Zhang
e-mail: zhanglidongonline@163.com

J. Wang
e-mail: Jia_Wang_heu@163.com

Y. Zhou
e-mail: yaozhou_heu@163.con

B. T. Wang et al. (eds.), *Large Floating Solutions*, Lecture Notes in Civil Engineering 636, https://doi.org/10.1007/978-981-96-5435-2_8

1 Introduction

The rapid growth of renewable energy technologies has positioned offshore energy solutions at the forefront of global energy research and development. Offshore renewable energy generally includes offshore wind, wave energy, offshore solar energy and tidal energy. There have been extensive investigations into the performance of wave energy devices, tidal energy devices, offshore wind systems, and hybrid renewable energy systems, employing methods such as hydrodynamic modeling, laboratory testing, and prototype testing [4, 5, 8, 10, 13, 15].

Among the abovementioned offshore renewable energy sources, floating solar photovoltaic (PV) platforms have garnered increasing attention due to their potential to harvest solar energy in marine environments while alleviating land-use constraints. Floating solar systems can generally be categorized into two primary structural types: membrane-based systems with lower freeboard and modular floating structures with elevated decks.

Dai et al. [1] described a modular floating PV platform designed specifically for reservoir conditions, and analysed the structural performance of the floating modules and the inter-modular connectors. Floating solar platforms, however, exhibit different structural characteristics that enable their deployment in offshore environments. Emami and Karimirad [2] presented the hydrodynamic and aerodynamic design requirements for floating solar platforms in marine environments. Li et al. [6] developed a modular floating structure for photovoltaic systems and performed a hydroelastic analysis of this type of platform. Zhang et al. [16] conducted a numerical analysis of offshore modular floating solar platforms, and the response of the platform was illustrated. Zhang et al. [17] performed a frequency-domain hydroelastic analysis of membrane-based floating PV platforms.

Several innovative concepts for offshore floating PV platforms with elevated decks have been proposed and developed, such as SolarDuck's floating solar platform and the "Huaneng Yellow Sea No. 1" floating solar platform. These platforms demonstrate the feasibility of combining modular and scalable designs with robust engineering solutions to operate in the challenging offshore environment.

López et al. [7] proposed a novel concept called "HelioSea", which consists of an array of tension leg platforms (TLPs) with solar panels mounted on elevated decks. Each TLP is individually moored to the seabed and arranged separately. Zhang et al. [18] developed a compliant modular floating photovoltaic farm designed for coastal areas, where full-scale experimental tests were carried out to validate the feasibility of the PV farm.

For modular offshore solar farms, one of the issues is the development of efficient numerical tools to evaluate the dynamic responses of the modular farm under wave conditions. This represents a complex multi-body hydrodynamic problem in terms of wave-structure interactions. Similar challenges arise in other fields, such as wave energy devices, hybrid wind-wave energy systems, and multifunctional offshore structures. Several researchers have conducted pioneering studies on multi-body hydrodynamic problems in these areas (e.g., [3, 11, 12, 14, 15]).

Motivated by SolarDuck floating PV platform, this chapter presents a detailed hydrodynamic analysis of modular floating solar platforms. The study employs the boundary element method (BEM) to solve 3D wave diffraction and radiation problems for multiple floating bodies and develops an in-house parallelized code to evaluate the hydrodynamic responses of multi-body systems with complex connections. Parametric investigations are conducted to assess the performance of an isolated module, a modular solar platform, and an array of platforms under different wave conditions.

2 Description of Floating Solar Platform

2.1 Floating Solar Platform and Photovoltaic Farm

Referring to the SolarDuck photovoltaic platform, the modular solar platform comprises four identical modules with hinged connections as shown in Fig. 1a. The module is composed of a deck which is raised above the water surface by three columns as shown in Fig. 1b. Their dimensions and structural properties are shown in Table 1. The photovoltaic farm consists of three platforms (i.e., 12 modules, with 4 modules for each platform) arranged in a line as illustrated in Fig. 2. The spacing between two adjacent units, denoted as s, is 32 m. The incident wave angle is denoted by β, and the water depth is assumed to be 20 m.

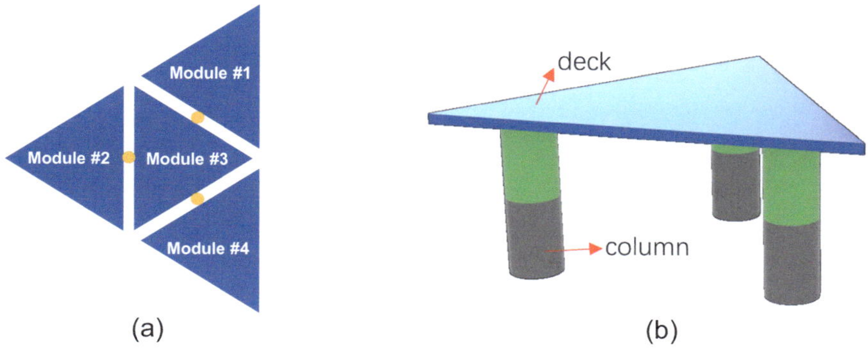

Fig. 1 Sketch of modular platform (**a**) and an isolated module (**b**)

Table 1 Dimensions and structural properties of a module

Term	Dimensions and properties
Side length of triangular platform	16 m
Deck thickness	0.4 m
Column height	6 m
Column draft	3 m
Column diameter	2 m
Center-to-center distance between columns	10 m
Roll moment of inertia of the platform	5.327×10^5 kg·m^2
Pitch moment of inertia of the platform	5.327×10^5 kg·m^2
Yaw moment of inertia of the platform	8.213×10^5 kg·m^2

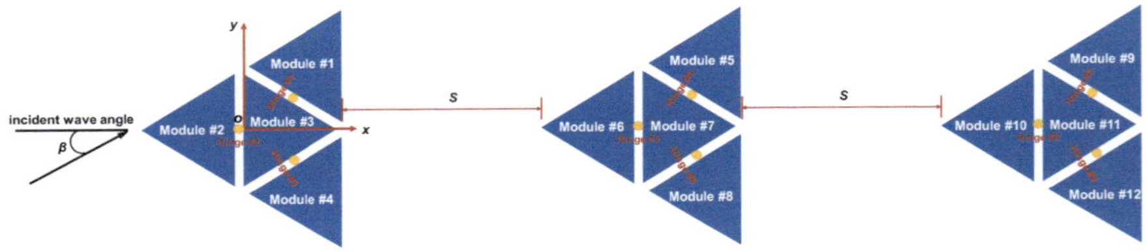

Fig. 2 Entire photovoltaic farm (spatial arrangement of platforms)

3 Numerical Methods

Under the framework of the linear potential flow theory, the boundary integral equation for the wave diffraction and radiation potentials of wave interacting with a multi-body system may be expressed as

$$2\pi\phi_m(\overrightarrow{x})+\phi_m(\overrightarrow{x})\iint\limits_{S_B}\frac{\partial G(\overrightarrow{\xi},\overrightarrow{x})}{\partial \mathbf{n}}\mathrm{d}s=-\iint\limits_{S_B}G(\vec{\xi},\vec{x})\frac{\partial\phi_0}{\partial\mathbf{n}(\overrightarrow{\xi_j})}\mathrm{d}s,\ m=6n+1 \tag{1}$$

and

$$2\pi\phi_m(\overrightarrow{x})+\phi_m(\overrightarrow{x})\iint\limits_{S_B}\frac{\partial G(\overrightarrow{\xi},\overrightarrow{x})}{\partial \mathbf{n}}\mathrm{d}s=\iint_{S_B}G(\vec{\xi},\vec{x})\mathbf{n_m}\mathrm{d}s,\ m=1,\ldots,6n \tag{2}$$

where ϕ_0 represents the incident wave potential, and ϕ_m ($m = 1, \ldots, 6n$) the radiation potentials, ϕ_{6n+1} the diffraction potential, $\mathbf{n_m}$ the normal vector along the wet surface (S_B) of the $6n$-th degree of freedom (DoF) in the multi-body system, respectively. Note that the multi-body system under consideration consists of n floating bodies, each with 6 degrees of freedom. So, the entire system has a total of $6n$ degrees of freedom. $G(\vec{\xi}, \vec{x})$ is the free surface Green function [9]. $\vec{\xi}$ and $\vec{x}$ represent the coordinate vectors of the source point and the field point, respectively. These boundary integral equations may be discretised using the boundary element method (BEM), where the wet surface of the floating body is divided into several surface elements. The velocity potentials on the surface element can be solved numerically. Based on the computed diffraction and radiation potentials, the wave exciting forces, added mass, and radiation damping coefficients of the floating bodies can be obtained. In this study, we developed a parallelized BEM solver (in-house), implemented in Python, capable of computing the wave excitation force vector ($\boldsymbol{f}$), the added mass matrix (**a**), and the radiation damping matrix (**b**) for multi-body systems.

Considering the hydrodynamic coupling of a multi-module system, the 6 n-DoF motion equation of a modular platform without the connection is given by

$$\left(-\omega^2(\mathbf{M}+\mathbf{a}) - \mathrm{i}\omega\mathbf{b} + \mathbf{C}\right)\boldsymbol{\varepsilon} = \mathbf{f} \tag{3}$$

where i is the imaginary unit, ω the wave angular frequency, **M** the mass matrix, **a** the added mass matrix, **b** the radiation damping matrix, **C** the restoring force matrix, $\boldsymbol{\varepsilon}$ the motion response vector and **f** the wave excitation force vector.

The motion Eq. (3) may be compactly written as $\mathbf{K}\varepsilon = \mathbf{f}$, where $\mathbf{K} = -\omega^2(\mathbf{M} + \mathbf{a}) - \mathrm{i}\omega\mathbf{b} + \mathbf{C}$, $\boldsymbol{\varepsilon} = [\boldsymbol{\varepsilon}_1, \boldsymbol{\varepsilon}_2, \ldots, \boldsymbol{\varepsilon}_n]^{\mathrm{T}}$, $\boldsymbol{\varepsilon}_i = \left\{\varepsilon_i^{(1)}, \varepsilon_i^{(2)}, \ldots, \varepsilon_i^{(6)}\right\}$, $i\ (= 1, 2, \ldots, n)$ represents the index of the modules, $\varepsilon_i^{(j)}$ corresponds to the complex response of the i-th module in the j-th DoF. $j = 1, 2, \ldots 6$ corresponds sequentially to the surge, sway, heave, roll, pitch and yaw DoF.

The planar view of the modular platform is presented in Fig. 1a. The central module (Module #3) is surrounded by three adjacent modules (Modules #1, #2, and #4). Hinge #1 represents the connector between Module #1 and Module #3. The layout of Modules #1, #2, …, #12, along with Hinges #1, #2, …, #9, in the photovoltaic farm is illustrated in Fig. 2. The connectors considered are column hinges. Based on the Lagrange multiplier method [12, 15, 16], the corresponding motion equation is derived as follows:

$$\begin{bmatrix} \mathbf{K} & \mathbf{D}^{\mathrm{T}} \\ \mathbf{D} & \mathbf{O} \end{bmatrix} \begin{bmatrix} \boldsymbol{\varepsilon} \\ \boldsymbol{\lambda} \end{bmatrix} = \begin{bmatrix} \mathbf{f} \\ \mathbf{0} \end{bmatrix} \tag{4}$$

In Eq. (4), **D** denotes the constraint matrix, and the force vector of the hinges is expressed as $\boldsymbol{\lambda} = [\boldsymbol{\lambda}_1, \boldsymbol{\lambda}_2, \ldots, \boldsymbol{\lambda}_m]^{\mathrm{T}}$, where $\boldsymbol{\lambda}_i = \left\{\lambda_i^{(1)}, \lambda_i^{(2)}, \ldots, \lambda_i^{(5)}\right\}$, $i = 1, 2, \ldots, m$ denotes the hinge number, $\lambda_i^{(j)}$ corresponds to the load of the i-th hinge in the j-th DoF. For instance, $\lambda_1^{(j)}, j = 1, 2, \ldots, 5$ sequentially represents the forces in the x'-, y'-, and z'-directions, and the torques in the x'- and z'-directions, within the local coordinate system $o' - x'y'z'$ acting on Hinge #1.

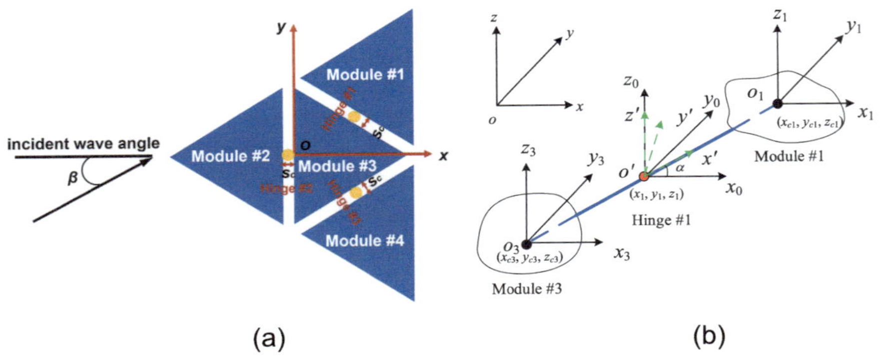

Fig. 3 Top-view of modular platform (**a**), coordinates of two adjacent modules and corresponding connector (**b**). β is incident wave angle, s_c denotes spacing between adjacent modules

Figure 3b presents the gravitational centers of Module #1, Module #3 and the corresponding coordinate of the connector Hinge #1. The coordinate system of o'-$x_0y_0z_0$ is parallel to the global system o-xyz. The o'-x' axis in the o'-$x'y'z'$ system is aligned with the connecting line between the two gravitational centers. In this instance, the angle, α, is formed by the o'-x_0 and o'-x' axes. o-z, o'-z_0 and o'-z' axes point upwardly.

In the o'-$x'y'z'$ coordinate system, the relative pitch motion is released due to the action of the connectors. The constraint matrix **D**, which characterizes such kind of motion, can be derived as:

$$\mathbf{D} = \Xi \begin{bmatrix} 1 & 0 & 0 & 0 & -(z_{c1}-z_1) & (y_{c1}-y_1) & -1 & 0 & 0 & 0 & (z_{c3}-z_1) & -(y_{c3}-y_1) \\ 0 & 1 & 0 & (z_{c1}-z_1) & 0 & -(x_{c1}-x_1) & 0 & -1 & 0 & -(z_{c3}-z_1) & 0 & (x_{c3}-x_1) \\ 0 & 0 & 1 & -(y_{c1}-y_1) & (x_{c1}-x_1) & 0 & 0 & 0 & -1 & (y_{c3}-y_1) & -(x_{c3}-x_1) & 0 \\ 0 & 0 & 0 & 1 & 0 & 0 & 0 & 0 & 0 & -1 & 0 & 0 \\ 0 & 0 & 0 & 0 & 1 & 0 & 0 & 0 & 0 & 0 & -1 & 0 \\ 0 & 0 & 0 & 0 & 0 & 1 & 0 & 0 & 0 & 0 & 0 & -1 \end{bmatrix} \tag{5}$$

and

$$\Xi = \begin{bmatrix} \cos\alpha & \sin\alpha & & & & \\ -\sin\alpha & \cos\alpha & & & & \\ & & 1 & & & \\ & & & \cos\alpha & \sin\alpha & \\ & & & & & 1 \end{bmatrix} \tag{6}$$

where (x_{c1}, y_{c1}, z_{c1}) and (x_{c3}, y_{c3}, z_{c3}) denote the gravitational center coordinates of Module #1 and Module #3, respectively, while the (x_1, y_1, z_1) represents the hinge point coordinate of the connector (Hinge #1).

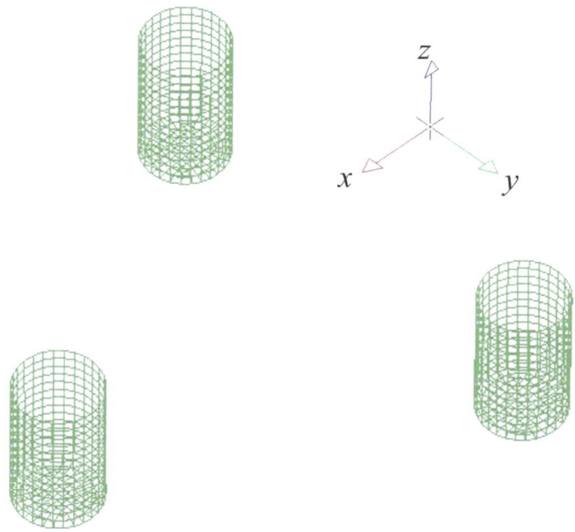

Fig. 4 Boundary mesh of a single module (wet surface) with a total of 1488 elements

4 Numerical Results and Discussions

4.1 Hydrodynamic Responses of an Isolated Module

Figure 4 shows the adopted boundary mesh of a single module (wet surface) with a total of 1,488 elements. Figure 5 presents the heave and pitch motion responses of an isolated module for the case of $\beta = 0°$. The numerical results reveal that both the heave and pitch motions exhibit a parabolic trend with respect to increasing angular wave frequency. The curves of $\left|\varepsilon_1^{(3)}\right| / A$ and $\left|\varepsilon_1^{(5)}\right| / A$ are notably narrow, indicating sharp response peaks. The peak values are observed at 1.65 rad/s and 1.00 rad/s, with corresponding response amplitude operators (RAOs) of 12.94 and 1.81 rad/m, respectively. As depicted in Fig. 5a, the heave motion response amplitude remains below 3.00 across the tested frequency range, except at its peak value. Similarly, the pitch motion response amplitude remains below 0.50 rad/m, except for a significant peak of 1.81 rad/m, as shown in Fig. 5b.

4.2 Hydrodynamic Responses of a Modular Solar Platform

The mesh of the modular solar platform used for the BEM computation is presented in Fig. 6. Figure 7 illustrates the heave and pitch motion responses of a modular solar platform (a 4-module unit) for $\beta = 0°$. When compared to the isolated module, the

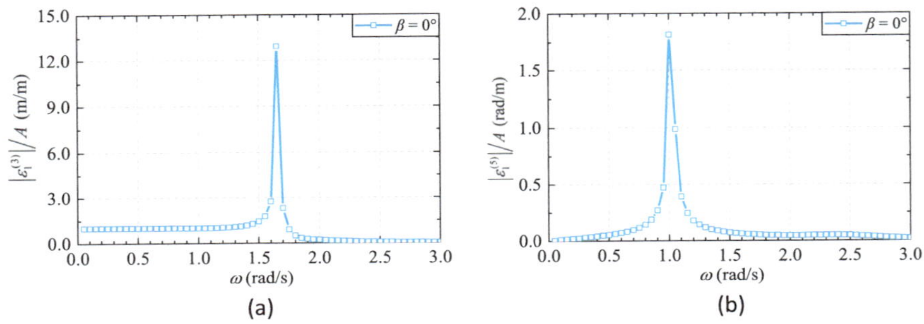

Fig. 5 Response amplitude operators (RAOs) of an isolated module in heave (**a**) and pitch (**b**) DoF, denoted as $\left|\varepsilon_1^{(3)}\right| / A$ and $\left|\varepsilon_1^{(5)}\right| / A$, respectively

results indicate that the number of peaks differs due to the hydrodynamic interactions among the modules.

Figure 7a shows that the heave motion amplitudes of Module #1 and its symmetrical counterpart (Module #4) exhibit the same variation trend, with two distinct peaks observed in both modules. In addition to the peak frequency corresponding to the isolated module at 1.65 rad/s, another peak appears at 1.50 rad/s, which is identified as the primary peak. Two peak values are observed in the heave motion responses of Module #2 on the weather side, with peak frequencies consistent with those of Module #1 and Module #4. However, the leeside Module #3 exhibits peak

Fig. 6 Boundary mesh of a four-module solar platform (wet surface), with each module consisting of 1488 elements

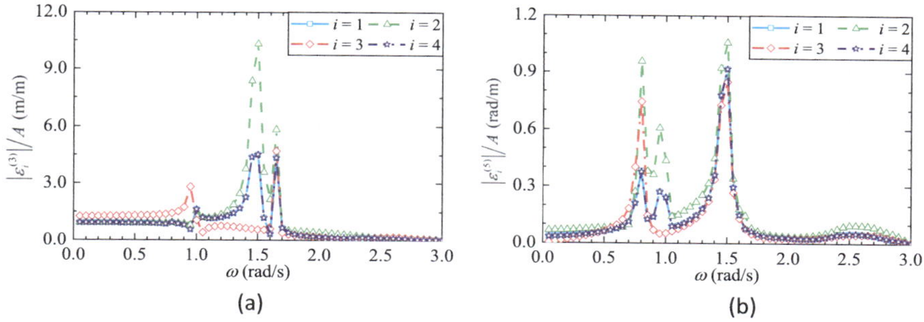

Fig. 7 Motion responses of modular solar platform (a 4-module unit) for $\beta = 0°$

values at 0.95 rad/s and 1.65 rad/s. The maximum heave RAO is found in Module #2 within the tested frequency range for $\beta = 0°$. The peak value of 10.32 occurs at $\omega = 1.50$ rad/s, which is lower than that of the isolated module.

Figure 7b illustrates the trend of the pitch motion response of the modular solar platform. Three peaks are observed in Module #1, Module #4, and Module #2, occurring at frequencies of 0.80 rad/s, 0.95 rad/s, and 1.50 rad/s, respectively, with the frequency (0.95rad/s) close to the peak frequency of the isolated module (see Fig. 5b). In contrast, the leeside Module #3 exhibits two peak values, occurring at 0.80 rad/s and 1.50 rad/s.

Similarly, the maximum pitch motion response amplitude is observed in Module #2, with a value of 1.06 rad/m, which is significantly lower than that of the isolated module.

Figure 8 presents the vertical forces acting on the connectors between the four hinged modules. The numerical results indicate that Hinge #1 and its symmetrical counterpart, Hinge #3, experience identical vertical forces, both exhibiting two peak values. These peaks occur at 1.00 rad/s and 1.50 rad/s, with the primary peak force of 0.11 MN observed at $\omega = 1.50$ rad/s. For the connector on the weather side, Hinge #2, two peak values are also observed, with corresponding frequencies identical to those of Hinge #1 and Hinge #3. However, the primary peak force for Hinge #2 reaches 0.22 MN, which is approximately twice the maximum vertical force experienced by Hinge #1 (or Hinge #3).

To examine the influence of the incident wave angle on the response of the 4-module solar platform, the heave and pitch motion responses of Module #2, as well as the vertical forces acting on Hinge #2, are presented in Fig. 9 and 10, respectively, for $\beta = 0°$, 30°, and 60°.

For the heave and pitch motion responses of Module #2, the numerical results in Fig. 9 indicate that the variation in wave angle has a slight effect on the overall trend of the motion responses. Within the tested frequency range, the heave and pitch motion responses exhibit two and three peak values, respectively, with identical peak frequencies observed for $\beta = 0°$, 30° and 60°. When the incident wave angle is held constant, Fig. 9a and b show that the peaked motion amplitudes of the heave and pitch motion gradually decrease as the wave angle increases.

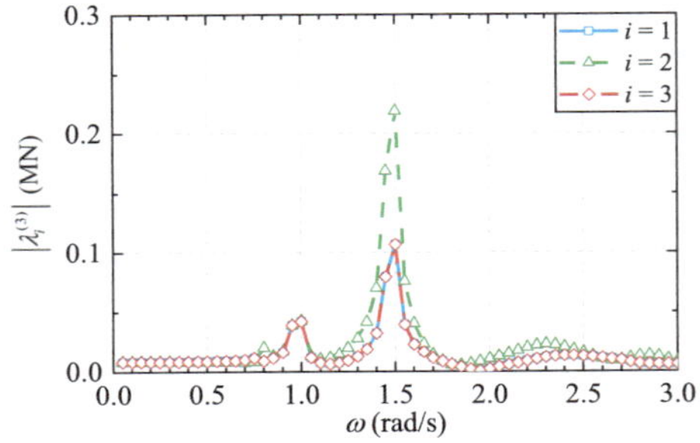

Fig. 8 Vertical forces acting on connectors for $\beta = 0°$

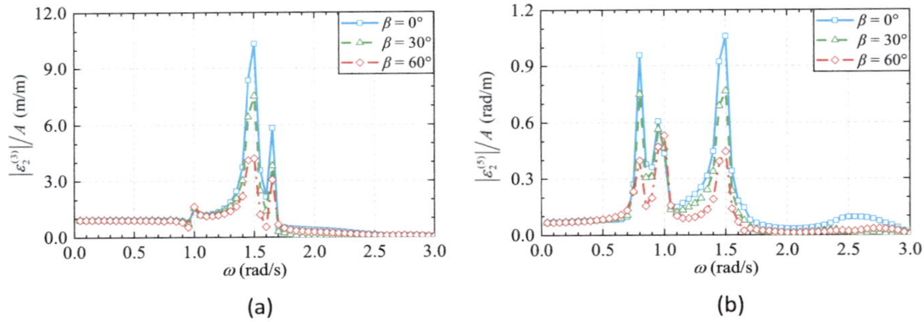

Fig. 9 Heave (**a**) and pitch (**b**) motion response of Module #2 for different wave angles

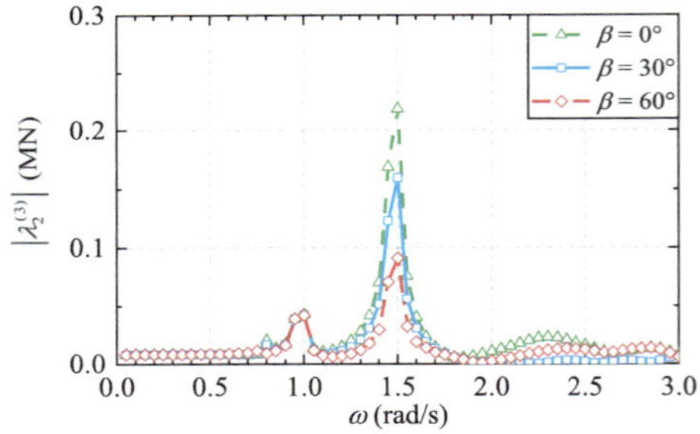

Fig. 10 Vertical force acting on Hinge #2 for different wave angles

Similar to the variation trend observed in the motion responses, Fig. 10 shows that the vertical forces acting on Hinge #2 are slightly affected by changes in the wave angle. As the wave angle increases, the peak vertical force decreases, while the location of primary peak remains unchanged. However, for $\omega > 2$ rad/s, the vertical force acting on Hinge #2 at $\beta = 60°$ is slightly greater than that at $\beta = 30°$.

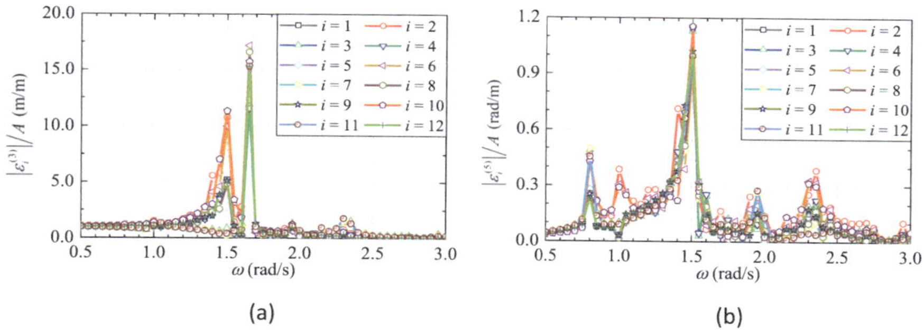

Fig. 11 Heave (**a**) and pitch (**b**) motion responses of each module in photovoltaic farm ($\beta = 0°$)

4.3 Hydrodynamic Performance of Photovoltaic Farm with Three Modular Solar Platforms

Considering the spatial arrangement of the solar platform, this section examines the hydrodynamic performance of the photovoltaic farm, consisting of three 4-module units. The modular platforms are aligned along the x-axis with a spacing of $s =$ 32.0 m, and the heave and pitch motion responses of all 12 modules are presented in Fig. 11 for $\beta = 0°$.

The results in Fig. 11 indicate that the heave and pitch motion responses of Module #1 and #4, Module #5 and #8, and Module #9 and #12, which exhibit symmetrical characteristics, follow the same variation trend and share identical amplitudes for $\beta = 0°$. For modules at the same position across different units (e.g., Module #2, #6, and #10), the results in Fig. 11a show that Module #6, located in the middle unit, exhibits the largest heave motion response amplitude. However, it is evident that the pitch motion response of modules at the weather side and leeside tends to be larger than those at other locations. Specifically, $\left|\varepsilon_1^{(5)}\right|$ is larger than $\left|\varepsilon_5^{(5)}\right|$ and $\left|\varepsilon_9^{(5)}\right|$, $\left|\varepsilon_{10}^{(5)}\right|$ is larger than $\left|\varepsilon_2^{(5)}\right|$ and $\left|\varepsilon_6^{(5)}\right|$, and $\left|\varepsilon_3^{(5)}\right|$ is larger than $\left|\varepsilon_7^{(5)}\right|$ and $\left|\varepsilon_{11}^{(5)}\right|$. This contrasts with the behavior observed for heave motion.

Figure 12 presents the numerical results of the vertical forces acting on the connectors of the photovoltaic farm when the incident wave angle is $\beta = 0°$. Since the three solar platform units are arranged along the x-axis, the vertical forces acting on Hinge #1 and #3, Hinge #4 and #6, and Hinge #7 and #9, which exhibit symmetrical characteristics, are identical for $\beta = 0°$ (i.e., $|\lambda_1^{(3)}| = |\lambda_3^{(3)}|$, $|\lambda_4^{(3)}| = |\lambda_6^{(3)}|$, $|\lambda_7^{(3)}| = |\lambda_9^{(3)}|$). Similar to the numerical results for an isolated 4-module solar platform, the vertical forces acting on the connectors located on the weather side (Hinge #2, #5, and #8) are relatively larger compared to those on other connectors within the same modular platform. Furthermore, it is observed that the primary peak of the vertical force on the connectors is located at $\omega = 1.50$ rad/s, consistent with the results for the isolated modular solar platform (see Fig. 10). By recalling Fig. 11, it is evident that the primary peaks of both the heave and pitch motion responses also occur at $\omega =$

Fig. 12 Vertical force acting on connectors of each modular solar platform in the photovoltaic farm ($\beta = 0°$)

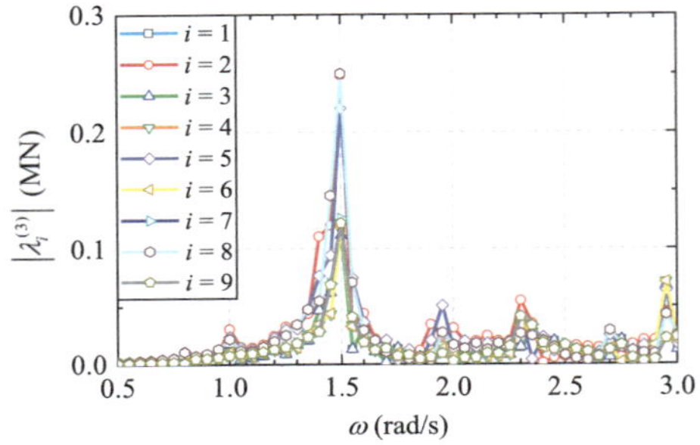

Fig. 13 Vertical force acting on connectors of photovoltaic platform for different incident wave angles

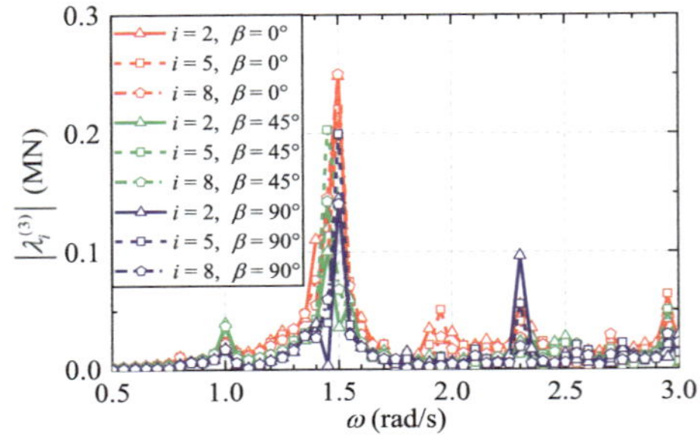

1.50 rad/s. Therefore, the forces acting on the connectors are closely associated with the motion responses of the modules.

The variation of incident wave angle may affect the loads acting on the connectors. To investigate this phenomenon, Fig. 13 presents the vertical forces acting on the connectors (Hinge #2, #5, and #8, see Fig. 4) under different incident wave angles. It can be observed that, as the incident wave angle increases, the connector loads are generally larger in head seas ($\beta = 0°$) and smaller in beam seas ($\beta = 90°$) across most of the tested frequency range. Notably, at $\omega = 2.30$ rad/s, the connector load for $\beta = 90°$ is relatively high. The primary peak of the vertical connector loads is located near $\omega = 1.50$ rad/s, which is consistent with the peak location observed in Fig. 11. This indicates that the mechanical characteristics of the modular solar platform arrays are similar to those of an isolated modular solar platform. The variation in loads on connections under different incident wave angles should be considered during the design process.

5 Conclusions and Prospect

This chapter explored the applications and research frontiers of offshore solar platforms. Inspired by SolarDuck's photovoltaic platform, a hydrodynamic analysis of an offshore solar platform designed based on a modular floating body concept was conducted. An in-house code was developed to compute the hydrodynamic coefficients, wave exciting forces, and system responses of a multi-body system with complex connections. The hydrodynamic responses of the modular platform were calculated and analyzed. The following main conclusions may be drawn for this study:

(1) Isolated Module versus Modular Solar Platform

For an isolated module under head seas ($\beta = 0°$), the heave and pitch motion responses exhibit a parabolic trend. However, the response curves are very narrow, indicating that the motion amplitudes are significantly reduced except near the peak frequency. For a modular solar platform (a 4-module unit), multiple peaks are observed in the motion response curves due to hydrodynamic interactions between the modules. When $\beta = 0°$, the peak values of heave and pitch motion responses for each module in the modular platform are smaller than those of an isolated module. Additionally, vertical loads on the hinges were computed for the modular platform, and it was observed that peaks in the vertical loads correspond to the frequencies where response peaks occur. This highlights the need to avoid such scenarios during the design of modular platforms.

(2) Photovoltaic Farm with Multiple Modular Platforms

The hydrodynamic performance of a photovoltaic farm consisting of three modular solar platforms was also investigated numerically. The results indicate that the responses of each platform in the array are like those of an isolated solar platform. Furthermore, when the array is exposed to beam seas ($\beta = 90°$), the vertical loads on the hinges are relatively smaller compared to other wave angles.

Floating modular solar platforms differ from traditional offshore structures, such as offshore oil and gas platforms or wind turbine foundation, due to their complex structural features. These systems are typically composed of arrays of modular floating units interconnected by connectors, with solar panels installed on elevated decks. Such configurations present engineering challenges requiring innovative and efficient solutions. Firstly, it is imperative to develop robust and accurate methodologies for analyzing the motion responses of these multi-body systems, particularly given their complex connections and nonlinear low-frequency motion characteristics. Secondly, protecting the upper deck from wave impacts is crucial to prevent damage to solar panels. While raising the deck height can reduce wave slamming risks, excessively high decks increase construction costs and compromise platform stability. It is necessary to balance them. Lastly, the mooring system is a critical component for ensuring platform survivability in extreme sea conditions. Designing efficient mooring systems that can withstand harsh marine environments is essential for the long-term operational safety of floating PV platforms. Addressing these challenges will require advancements in hydrodynamic modeling, structural optimization, and integrated system design, enabling the successful deployment of floating modular PV platforms in diverse offshore conditions.

References

1. Dai J, Zhang C, Han VL, Ang KK, Qian X, Wong JLH, Tan ST, Wang CL (2020) Design and construction of floating modular photovoltaic system for water reservoirs. Energy 191:116549
2. Emami A, Karimirad M (2025) Further development of offshore floating solar and its design requirements. Mar Struct 100:103730
3. Hallak TS, Gaspar JF, Guedes Soares C (2024) A mathematical model for the dynamic analysis of multi-body floating platforms with complex mechanical constraints. Ocean Eng 314(part 2):119640
4. Legaz MJ, Coronil D, Mayorga P, Fernández J (2018) Study of a hybrid renewable energy platform: W2Power. OMAE2018–77690, V11AT12A040
5. Li Y, Ong MC, Wang K, Li L, Cheng Z (2022) Power performance and dynamic responses of an integrated system with a semi-submersible wind turbine and four torus-shaped wave energy converters. Ocean Eng 259:111810
6. Li Z, Chen D, Feng X (2023) Hydroelastic and expansibility analysis of a modular floating photovoltaic system with multi-directional hinge connections. Ocean Eng 289(Part 1):116218
7. López M, Claus R, Soto F, Hernández-Garrastacho ZA, Cebada-Relea A, Simancas O (2024) Advancing offshore solar energy generation: the HelioSea concept. Appl Energy 359:122710
8. Marino E, Gkantou M, Malekjafarian A, Bali S, Baniotopoulos C, van Beeck J, Borg RP, Bruschi N, Cardiff P, Chatzi E, Čudina I, Dinu F, Efthymiou E, Ferri G, Gervásio H, Heng J, Jiang Z, Lenci S, Lukačević I, Manuel L, Meyer A, Méndez-Morales M, Osmanović A, Pakrashi V, Pandit A, Rega G, Skejić D, Tesch L, Ungureanu V, Uzunović T, Verma AS (2024) Offshore renewable energies: a review towards floating modular energy Islands—monitoring, loads, modelling and control. Ocean Eng 313(Part 1):119251
9. Newman JN (1985) Algorithms for the free-surface Green function. J Eng Math 19(1):57–67
10. Roy A, Auger F, Dupriez-Robin F, Bourguet S, Tran QT (2018) Electrical power supply of remote maritime areas: a review of hybrid systems based on marine renewable energies. Energies 11(7):1904
11. Stansby P, Li G (2024) A wind semi-sub platform with hinged floats for omnidirectional swell wave energy conversion. J Ocean Eng Mar Energy 10(2):433–448
12. Sun L, Taylor RE, Choo YS (2011) Responses of interconnected floating bodies. IES J Part A Civil Struct Eng 4(3):143–156
13. Zhao H, Stansby P, Liao Z, Li G (2024) Multi-objective optimal control of a hybrid offshore wind turbine platform integrated with multi-float wave energy converters. Energy 312:133547
14. Zhao X, Wang J, Geng J, Zhou S, Guedes Soares C (2024) Hydrodynamic investigation of a multi-module floating platform with a wave energy device array. China Ocean Eng 38(6):1034–1046
15. Zhao X, Xue F, Chen L, Göteman M, Han D, Geng J, Sun S (2023) Hydrodynamic analysis of a floating platform coupled with an array of oscillating bodies. Ocean Eng 287(Part 1):115439
16. Zhang L, Wang J, Zhao X, Geng J, Hu Z, Guedes Soares C (2024) Numerical investigation on motion response of modular semi-submersible floating platform in waves. In: The 4th world conference on floating solutions
17. Zhang Y, Zhang X, Chen Y, Tian X, Li X (2024) A frequency-domain hydroelastic analysis of a membrane-based offshore floating photovoltaic platform in regular waves. J Fluids Struct 127:104125
18. Zhang C, Dai J, Ang KK, Lim HV (2024) Development of compliant modular floating photovoltaic farm for coastal conditions. Renew Sustain Energy Rev 190(part A):114084

Xuanlie Zhao is an Associate Professor in Harbin Engineering University, China, and the Vice Director of the Ministerial-Level Key Laboratory of Deep-Sea Engineering Equipment and Technology. His research interests are in the areas of wave-structure interactions, offshore renewable energy systems, and multi-purpose marine structures. He has led over 10 research projects as a principal investigator in these areas. He serves as a member of the editorial boards and scientific committees for several journals and international conferences.

Lidong Zhang is a Ph.D. candidate in Naval Architecture and Ocean Engineering at Harbin Engineering University. His research focuses on the performance evaluation of modular floating systems and floating offshore wind turbines.

Jia Wang is a Ph.D. candidate at the Sanya Nanhai Innovation and Development Base, Harbin Engineering University. His research focuses on multi-body hydrodynamics and hybrid floating structures.

Yao Zhou is a Ph.D. candidate in Naval Architecture and Ocean Engineering at Harbin Engineering University. His research focuses on the hydrodynamic analysis of multi-use marine structures and the development of potential-flow solvers.

Innovative Aquaculture "Penghu": the First Semi-submersible Wave-Energy-Powered Aquaculture Platform in China

Wanzhen Yue, Songwei Sheng, and Zhenxin Huang

Abstract This chapter presents the analysis, design, and construction of Penghu, China's first semi-submersible, wave-energy-powered aquaculture platform. Penghu is a notable example of an integrated offshore aquaculture farm and ocean renewable energy systems, powered by a combination of 60 kW Sharp Eagle wave energy converters (WECs) and 60 kW solar photovoltaic units. The platform features an internal pool with a volume of approximately 11,000 m^3 for finfish farming, along with a spacious deck for aquaculture equipment. Based on Penghu, the interactions between Sharp Eagle WEC and semi-submersible aquaculture platform are analyzed to enhance understanding of the potential and challenges of co-locating aquaculture and renewable energy systems. The Penghu project offers valuable insights and serves as a reference for advancing offshore aquaculture and expanding wave energy generation technology.

Keywords Offshore aquaculture · Sharp Eagle wave energy · Multi-use platform · Co-location

1 Introduction

The marine aquaculture industry has experienced significant global growth in recent decades. However, in coastal waters, this expansion has led to challenges such as competing space usage, stakeholder conflicts, and environmental impacts, including water contamination from eutrophication [1] (Fig. 1). These issues have prompted the industry to shift from coastal waters to more expansive offshore areas. The offshore

W. Yue · S. Sheng · Z. Huang (✉)
Guangzhou Institute of Energy Conversion, Chinese Academy of Sciences, Guangzhou, China
e-mail: huangzx@ms.giec.ac.cn

W. Yue
e-mail: yuewz@ms.giec.ac.cn

S. Sheng
e-mail: shengsw@ms.giec.ac.cn

B. T. Wang et al. (eds.), *Large Floating Solutions*, Lecture Notes in Civil Engineering 636, https://doi.org/10.1007/978-981-96-5435-2_9

Fig. 1 Marine aquaculture in coastal waters (retrieved from https://ishare.ifeng.com/c/s/7mdiO8WxMKo)

environment offers advantages, such as reducing the accumulation of fish waste and minimizing the spread of parasites and diseases [4]. As a result, offshore aquaculture is emerging as a promising approach to increase marine aquaculture production while alleviating pressure on coastal ecosystems [2].

There is no agreed definition of "offshore" aquaculture [9]. However, Chu et al. [4] proposed a definition that attempts to reflect the general consensus on the parameters defining offshore fish farming:

> Offshore fish farming involves (i) unsheltered sites defined by the sea space outside a straight line joining two major capes or promontories, at least about 3 km distance from shoreline but within the Exclusive Economic Zones (ii) water depth that is greater than 50 m or larger than 3 times the cage height and no less than 15 m between the cage bottom and the seabed, (iii) current speed within 0.1 m/s to 1 m/s and (iv) wave height exceeding 3m."

In offshore aquaculture, space is no longer a primary constraint; instead, the focus shifts to developing sustainable infrastructure, large-scale system designs, reliable power supplies, and addressing the economic challenges posed by harsh and exposed environments. Offshore conditions are characterized by more energetic waves, requiring aquaculture platforms' moorings and structures to withstand and operate in extremely rough seas. Additionally, aquaculture systems need stable power supplies and wave protection for efficient operations. Energy is required for a range of applications, including lighting, feeding, pumps, aeration, desalination, cleaning, and monitoring. According to Møller [17], the energy demand for the growth phase of finfish aquaculture is less than 700 kWh daily for a fish farm with a production capacity of 3120 tonnes, with peak power demands of 100–120 kW during daytime feeding.

As the aquaculture industry moves marine operations farther offshore, there is significant potential for co-locating aquaculture with ocean renewable energy (ORE) resources [15]. ORE offers a sustainable, on-site power solution for offshore aquaculture and can also provide wave protection for aquaculture operations [24]. Co-locating offshore aquaculture with ORE brings several advantages, such as shared use of space, integrated systems like mooring and electrical infrastructure, reduced reliance on diesel [8], increased autonomy for aquaculture equipment, and the potential to label seafood as being produced using renewable energy, thereby boosting its commercial value [21]. In support of these efforts, the European Union's Horizon 2020 research and innovation program launched the "Multi-use in European Seas" (MUSES) project in November 2016. This two-year initiative aims to combine compatible marine activities and promote the shared use of valuable ocean space [18] (Fig. 2).

Ocean renewable energy (ORE) resources are typically harnessed from waves, tides, ocean currents, and thermal or salinity gradients. Each of these sources presents unique advantages and challenges for powering aquaculture. Wave energy, in particular, stands out for its reliability, predictability, and higher energy density (3 kW/m^2) when compared to wind (500 W/m^2) and solar (170 W/m^2) [5]. Garavelli et al. [10] compiled a comprehensive database on the energy requirements of offshore aquaculture and conducted spatial analyses for co-location. Their findings suggest that wave energy could be a viable power source for aquaculture operations. As the aquaculture industry shifts further offshore, platforms will encounter increasingly energetic wave conditions, which not only benefit wave energy devices but also make wave energy an ideal solution for sustaining offshore aquaculture.

Fig. 2 Project MUSES (Multi-Use in European Seas) (retrieved from https://muses-project.com/)

The wave energy converter (WEC) industry has been exploring the deployment of small-scale WECs to generate power from low-wave energy resources, particularly for offshore aquaculture applications [6, 8, 19]. In 2014, two Albatern WaveNET arrays were deployed off the Isle of Muck in Scotland through a collaboration between Albatern and Marine Harvest Scotland [3]. The WaveNET arrays use interconnected SQUID-generating units, each with a capacity of 7.5 kW. A SQUID unit consists of a hollow central riser tube connected to three floaters via linking arms (Fig. 3). Articulated pumping modules connect these components, while the floaters house the power take-off (PTO) system and components for communication and hydraulic operations. This 14-week project, conducted during the summer of 2014, demonstrated the units' high survivability in rough sea conditions.

In 2018, a single MANTA WEC, with a capacity of 4.6 kW, was installed at Scotland's Teisti Geo finfish farm [7, 8] (Fig. 4). Developed by Aqua Power Technologies, the MANTA WEC is a submersible wave energy device designed specifically for the aquaculture industry. It features large hydrodynamic "wings" that move vertically, capturing wave energy to generate electricity and recharge battery banks.

Arrays of small-scale wave energy converters (WECs) have demonstrated improved economics in terms of average annual energy production, capacity factor, and costs [6], as well as enhanced maintainability. However, one of the challenges is the limited space for installing energy storage systems on aquaculture platforms. As a result, the power generated by WEC arrays often needs to be transmitted first to a

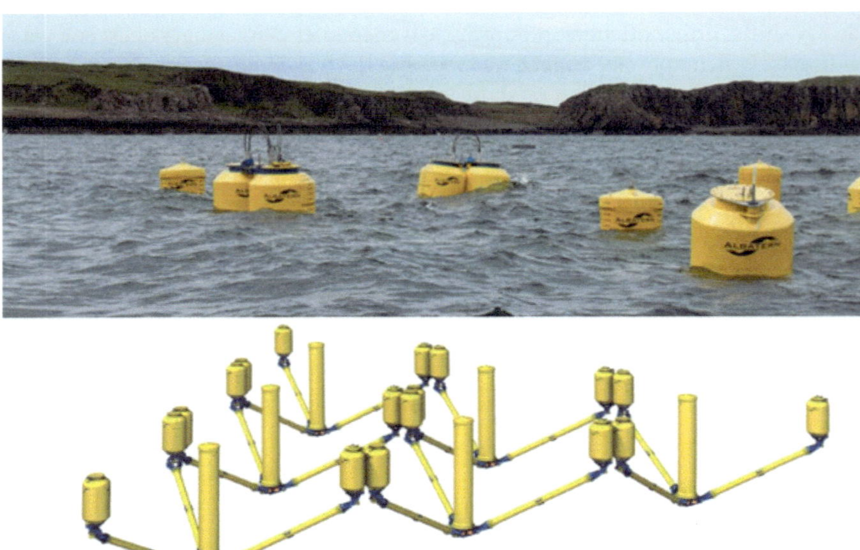

Fig. 3 WaveNET—a floating, flexible, modular and massively scalable wave power generation idea under testing in Scotland (retrieved from https://newatlas.com/albatern-wavenet-squid-floating-wave-energy/34942/)

Fig. 4 Aqua Power Technologies MANTA wave energy device installed at a Scottish salmon farm (image courtesy of Aqua Power Technologies)

shore station and then back to the offshore aquaculture site via underwater cables. In locations where installing submarine cables is not feasible, integrating WECs directly with aquaculture platforms and sharing mooring or foundation systems offers a practical alternative. This approach not only reduces infrastructure costs but also enhances the efficiency of co-locating energy generation with aquaculture operations.

Penghu exemplifies the integration of wave energy with offshore aquaculture. Designed by Guangzhou Institute of Energy Conversion (GIEC) of the Chinese Academy of Sciences, Penghu is China's first semi-submersible wave-energy-powered aquaculture platform [11] (see Fig. 5). It combines 60 kW Sharp Eagle wave energy converters (WECs) with 60 kW solar photovoltaic (PV) units to supply power for aquaculture operations. The platform includes an internal pool with a volume of approximately 11,000 m^3 for finfish farming and ample space for aquaculture equipment. Building on Penghu's success as a prototype, GIEC has developed a series of similar semi-submersible aquaculture platforms, including Fujian Mintou and Hainan Pusheng [25]. These platforms demonstrate the feasibility and benefits of integrating renewable energy systems with offshore aquaculture, providing both energy independence and operational efficiency.

2 Description of Penghu

The concept of integrating the Sharp Eagle wave energy device with aquaculture originated from aquaculture enterprises *operating near the site at where the Sharp Eagle wave energy device "Wanshan" was deployed.* These enterprises recognized that wave-generated electricity would be well-suited for powering aquaculture facilities located far from the mainland power grid. In response, Guangzhou Institute of Energy Conversion (GIEC) of the Chinese Academy of Sciences (CAS) launched

Fig. 5 Penghu platform for wave energy-powered offshore aquaculture

a three-year project, supported by the Chinese Ocean Renewable Energy Special Fund and the Provincial Special Fund for Promoting Economic Development of Guangdong. The project aimed to tackle key challenges, including providing a reliable energy supply and ensuring the structural resistance of aquaculture cages in energetic wave environments.

The design of Penghu is mainly based on the Sharp Eagle wave energy generation technology and semi-submersible rigid cage technology. The Sharp Eagle wave energy generation technology is a novel technology developed by GIEC. This technology is like Oyster [20], both of which are terminators of WEC. However, unlike Oyster, the curve of the absorbing floater in Sharp Eagle technology has been specially optimized for maximum captured energy,and the Sharp Eagle floaters are hinged on a semi-submersible platform, which has a large mass and moment of inertia to reduce the motion. Compared to Oyster, the design of this semi-submersible platform makes it easier to capture wave energy near the sea surface.

Penghu was deployed in 2019. Prior to 2019, GIEC had patented [22] and sea-tested [23] a series of semi-submersible platforms with the Sharp Eagle wave energy conversion technology. Penghu is a semi- submersible platform integrating wave and solar energy with offshore aquaculture. The basic idea of this hybrid platform is to provide green power generated by WECs for aquaculture operations and support all aquaculture systems with the column-stabilized platform. Figure 6 shows a 3D

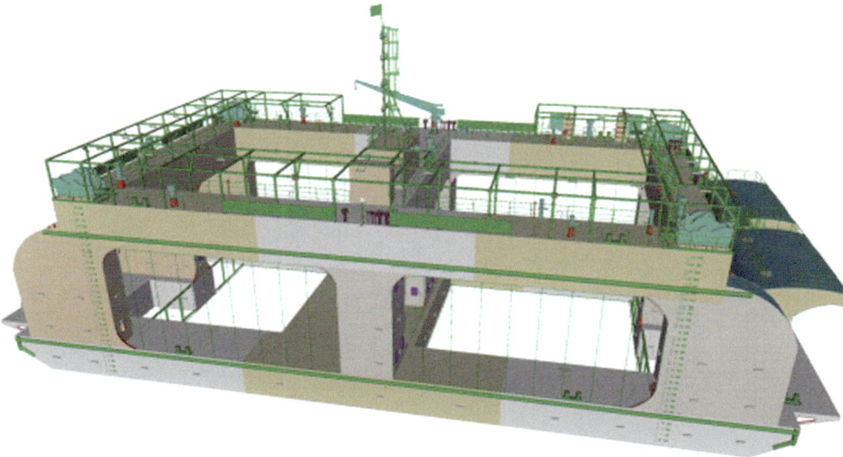

Fig. 6 3D view of the general arrangement of Penghu

view of the general arrangement of Penghu. The Penghu platform consists of two rectangular pontoons with nine columns, two sets of Sharp Eagle WEC at the front end, and solar panels installed on the roof. The Sharp Eagle floaters and hull are connected using hinges.

Penghu has three operating modes: (1) transit condition, (2) operational condition, and (3) self-protection condition. When waves reach a significant height of 0.5 m, the WEC system begins to operate. When the significant wave height exceeds 4 m, Penghu switches to the self-protection condition, where the Sharp Eagle floaters are fixed to the hull by locking devices to ensure that the hydraulic system is safe and dives to a 12.2 m draft. Table 1 summarizes the main parameters of the Penghu platform.

Penghu is regarded as a multifunctional platform, both addressing the energy needs of offshore aquaculture and providing a model for sustainable aquaculture operations in remote areas. Additionally, it offers tourists a unique experience to view an offshore fish farm powered by renewable energy.

3 Design, Analysis and Construction

3.1 Power Supply System

Penghu integrates Sharp Eagle WECs (60 kW) and solar PV units (60 kW) with the offshore aquaculture. The electricity generated by wave energy and solar energy is stored in the battery pack. Figure 7 shows the hydraulic and electrical equipment compartments inside Penghu. The storage systems' capacity is 500 kWh, while the

Table 1 Main parameters of the Penghu platform [26]

Parameters	Value
Platform size (sharp eagle floaters included) (m)	66 * 28 * 15.5
Material	AH36 steel
Draft (m)	Operational condition: 11.3 Self-protection condition: 12.2
Displaced mass (tone)	Operational condition: 3900.2 Self-protection condition: 4165.7
Center of gravity (CoG) (m)	Operational condition: (32.987, 0.0, 3.206) Self-protection condition: (33.021, 0.0, 3.291)
Radius of gyration of the lightship (m)	Roll: 11.023; Pitch: 21.296; Yaw: 22.873
Transversal frame spacing (m)	2.0
Longitudinal stiffener spacing (m)	0.5
Thickness of plate (mm)	Bottom~3 m ABL.: 10; 3 m ABL. ~ 9.25 m ABL.: 9; 9.25 m ABL. ~ 15.5 m ABL.: 8
Length of mooring lines (m)	600
Diameter of mooring lines (mm)	78

inverter has a rated output power of 200 kW and a voltage of 380 V (Fig. 7). In addition, the platform is equipped with a diesel generator set for redundancy.

The Sharp Eagle wave energy conversion system mainly consists of three parts: the Sharp Eagle floaters (wave-absorbing floaters resembling eagle beaks), the semi-submersible platform, and the power take-off system (PTO) located within the platform. The Sharp Eagle floaters are connected to the semi-submersible platform through hinges. By adjusting ballast waters, the semi-submersible platform can perform both surfacing and submerging, allowing it to adapt to various sea conditions. The platform provides large mass and moment of inertia, which offer a relatively stable foundation for the motions of the Sharp Eagle floaters. The schematic diagram of Penghu WEC is shown in Fig. 8, where an incident wave propagates towards the Sharp Eagle floaters.

Penghu uses a hydraulic energy conversion system to accommodate large force and low velocity of waves. It comprises a hydraulic cylinder, accumulator, control valve, tank, and hydraulic motor. The energy conversion process is shown in Fig. 9. First, the Sharp Eagle floaters are driven by waves. These floaters' motion then drives the hydraulic rod to convert the mechanical energy into stable hydraulic energy. Finally, a hydraulic motor and an electrical generator convert the hydraulic energy into electrical energy, powering aquaculture operations, seawater desalination system (Product flow of 250L/h; power consumption of 2.5 kW), and monitoring system.

(a) Hydraulic system (accumulator, control valve, tank, and hydraulic motor)

(b) Electrical system (e.g., rectifier, inverter, photovoltaic controller)

Fig. 7 Hydraulic and electrical systems inside Penghu

3.2 Aquaculture System

Penghu has two aquaculture cages, front and rear, which are separated by netting in the middle to meet the requirements of independent operation and breeding. A crane with lifting capacity of 1 ton and span of 8 meters, an automatic feeding machine (Fig. 10), and a live fish transfer system were equipped on Penghu for convenient breeding operations. Furthermore, Penghu possesses a internal storage capacity of

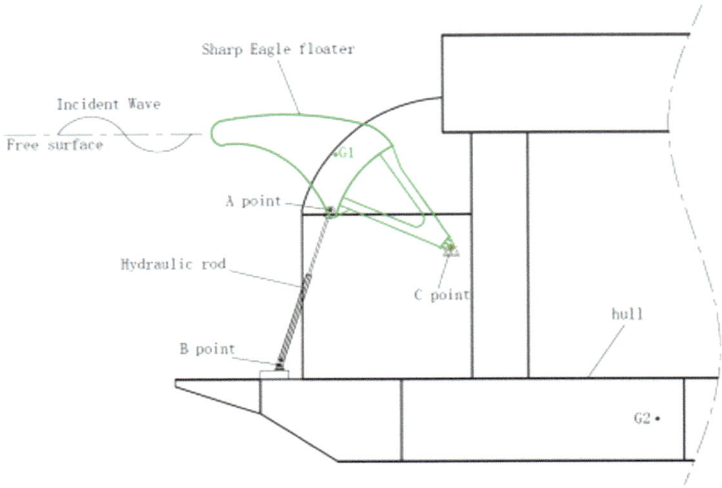

Fig. 8 Schematic diagram of the Penghu platform at the center line plane [26]

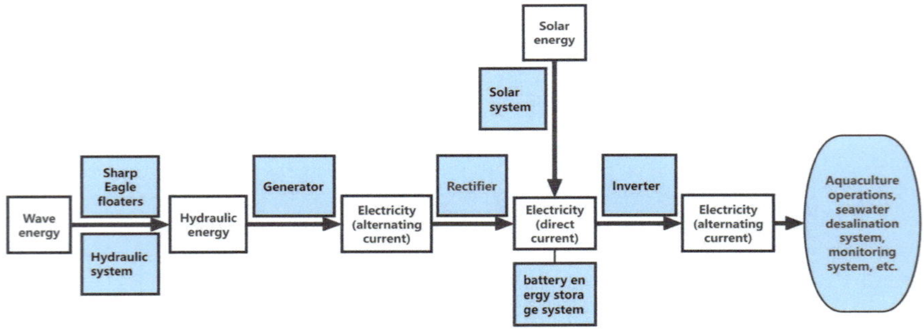

Fig. 9 Energy conversion process of Penghu

approximately 400 m^3 and a theoretical weight capacity of 2400 tons, and it also provides living space for not less than 13 aquaculture workers. In August, 2023, the netting structure was upgraded to double-layer design by adding a steel netting onto the outer layer of the nylon netting (Fig. 11).

To better implement aquaculture operations, supporting systems such as data collecting system, transmitting and communicating system, monitoring system were installed on Penghu. The data transmitting and communicating system employs three methods, respectively local data storage, 4G communication, and radio communication, to ensure a stable and reliable data transmission. Over 130 monitoring equipment enable Penghu to collect ample data. In view of 17 video monitoring channels, real-time monitoring of critical components' operational status became available. In addition, Penghu has satellite positioning capabilities.

(a) Crane (b) Automatic feeding system

Fig. 10 Aquaculture equipment on Penghu

Fig. 11 Netting structure was upgraded to double-layer design in August, 2023

3.3 Typhoon Defense Measures

Offshore aquaculture platforms in South China Sea face a much harsher sea state than nearshore aquaculture platforms because of the frequent occurrence of tropical cyclones and typhoons [12, 13]. To ensure that the aquaculture platform can survive and operate long-term in the typhoon-prone South China Sea, the main structure and mooring system of Penghu were specifically designed to withstand a once-in-50-years extreme sea conditions with a significant wave height of 9.07 m. Besides, Penghu has a large draft capacity to submerge to a depth of 12.2 m before typhoon

Fig. 12 Wave tank tests with a 1:15 scaled model in July 2017

arrives, this maneuver reduces the wave loads on the platform and prevents excessive movement that could harm the fish larvae. Additionally, a locking and breach mechanism was set at the lower limit of the Sharp Eagle floaters in case of extreme conditions. For example, right before typhoon arrives, the Sharp Eagle floaters can actively take in ballast water to submerge to their lower limit, at which point they are locked with the locking mechanism to prevent further movement and stop absorbing wave energy.

Before officially manufacturing Penghu, wave tank tests with a 1:15 scaled model were conducted in July 2017 at Tianjin Waterborne Transport Research Institute under the Ministry of Transport. During the tests, mooring forces under regular and irregular waves of the once-in-50-years extreme state were tested at different water depths of 30, 50, and 100 m. Through the tests, rationality of mooring scheme of Penghu was validated (see Fig. 12). Far from 2019 by now, though experienced over ten typhoons including Typhoon Saola, Penghu's main structure and mooring system are still performing normally.

3.4 Construction and Equipment Preparation

Penghu, with official construction started on December 20, 2018, sectional closing and equipment installation and debugging started on March 16, 2019, was fully constructed and delivered by June 30, 2019. Excluding the Spring Festival holiday, the whole construction and equipment commissioning of Penghu took no more than six months (Fig. 13).

Fig. 13 Photo of the launch of the Penghu ship on July 8, 2019

Penghu's length of 66 m had made it a large-scale structure. Therefore, a modular construction process was adopted by dividing it into 30 sections. Equipment was progressively hoisted in during the assembly process after all sections were completed, allowing simultaneous performing for both structure closure and equipment installation/debugging. Based on the above, speeds of structure construction, equipment installation and system debugging were significantly accelerated.

3.5 Open Sea Test and Commercial Operation

Penghu was deployed at sea in June 2019. After one month of river trials (including the ballast system test, submergence/emergence operation test and watertight tests), Penghu departed from Youlian Shipyard (Shekou, Shenzhen) (Fig. 14) on August 6, and successfully arrived at Zhuhai Dazhizhou Island Fish Farm after a 9-h journey.

The open sea test lasted for over a month, and the test results showed that, Penghu could maintain a proper posture, its anchoring system was securely positioned, and the power generation system, distribution system, and monitoring system operated normally.

Aquaculture and tourism stand for Penghu's commercial values. The first batch of golden pomfret farmed in Penghu that was brought to market on December 28, 2019 had made an important milestone. As OES [11] wrote in their report: "The at-sea trials have been successful and have shown that a 50–100 kW WEC can fulfill the energy needs of a 10,000–20,000 m^3 offshore cage." In the following five years, Penghu experienced its aquaculture journey by cultivating different kinds of fish

Fig. 14 Towing trajectory of Penghu on August 6, 2019

such as golden pomfret and grouper (Fig. 15). Meanwhile, Penghu expanded its tourism offerings by conducting activities such as sea fishing, educational outreach and moreover, providing on-sea platform and power support for marine scientific experiments conducted by universities and research institutions (Fig. 16).

(a) Golden pomfret (b) Grouper

Fig. 15 Penghu harvested golden pomfret and grouper: **a** retrieved from Guangzhou Daily, **b** retrieved from https://new.qq.com/rain/a/20210123A0D8PD00

(a) Tourism (b) Educational outreach

Fig. 16 Tourism function of Penghu

4 Interactions Between WECs and Aquaculture Platform

Unknown consequences of the interactions between the two uses is one of the main challenge in the development of integrated WEC and aquaculture platform [8]. These interactions may reflect in many aspects such as power generation, movement, structural response, and construction costs, which can be positive and beneficial for both parties, or negative and disadvantageous for one party.

Energy supply. The integration between WECs and aquaculture platform provides a rational and economical way to consume the electricity generated by WECs. In this way, the aquaculture platform obtains necessary power for its operations from the WECs instead of receiving electricity from the onshore grid via underwater cables. On the other hand, power-consuming resistors are no longer needed for WECs, which helps save costs to a certain extent.

Motion of platform. On one hand, WECs could reduce the motion of the aquaculture platform, providing a stable foundational condition for aquaculture operations and decreasing the risk of damage to fish larvae. The PTO system of WEC transfers part of the wave energy out of the flow field, the energy of the platform's incident force will be reduced [26]. As shown in Fig. 17, the PTO system decreases the platform's surge, heave, and pitch motions within almost the full range. Though the movement of the Sharp Eagle floaters can generate waves outwards, as mentioned before, the platform's large waterline area and high damping make it less susceptible to the movements of the Sharp Eagle floaters. On the other hand, in view of attachment of marine organisms to nets in aquaculture activities, scholars [14, 16] have not reached a consensus on the effect of net solidity ratio on motion response, as it is influenced by multiple factors including the platform's specific design and wave frequency. Lastly, considering that wave energy-powered aquaculture platforms often operate in areas with higher wave energy, the platform experiences greater motion under more intense wave actions. This significant movement can cause sloshing within the open pool of the platform, increasing the risk of fish collisions and injuries.

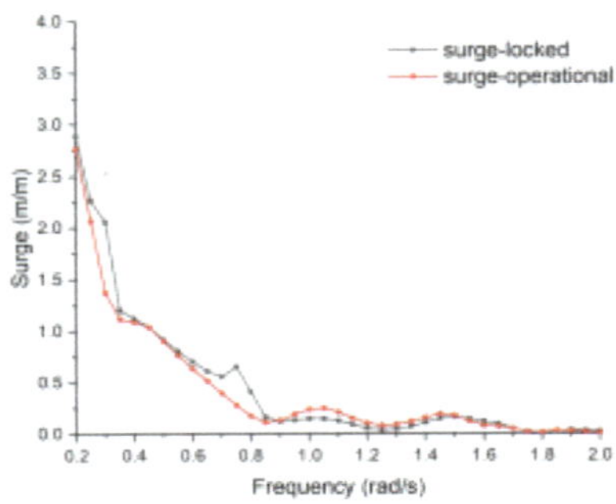

(a) Surge

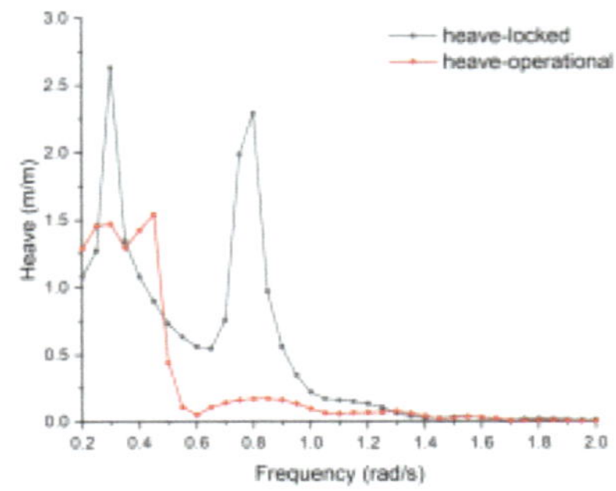

(b) Heave

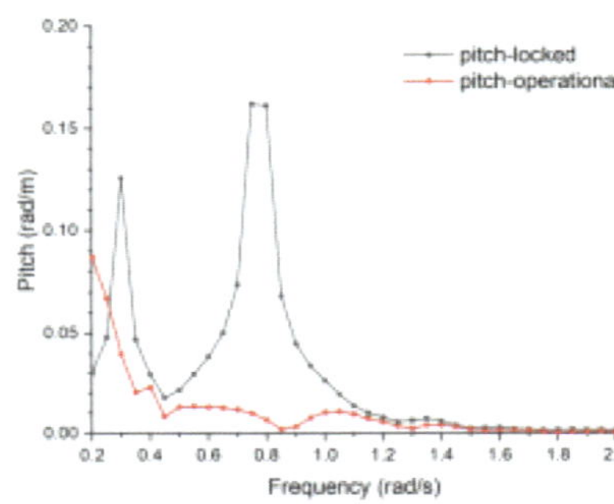

(c) Pitch

Fig. 17 RAOs of motion responses of Penghu platform

Structural response. For the platform itself, the impact on the platform motion of nets and attachment of marine organisms have not yet reached a consistent conclusion, making it temporarily challenging to determine the impact of aquaculture activities on the global strength of the platform. However, the wave loads acting on the nets are transmitted to the hull of the platform through the connection structures, which means that the local strength at the attachment areas between the nets and the platform is indeed affected by the attachment of marine organisms. On the other hand, adding wave energy absorbing floating bodies to aquaculture platform would bring more structural safety risk locations, such as the hinge and locking structure of the absorbing floating body, both of which would suffer fatigue fracture due to a high cyclic loading. Therefore, dedicated local strength analysis of yielding and fatigue strength are necessary to avoid water penetration caused by structural failure.

5 Concluding Remarks

Integrating aquaculture and wave energy has the potential to provide a multi-use platform and space. The first semi-submersible wave-energy-powered aquaculture platform in China, Penghu, which has been running for more than five years, is considered a case study by marine researchers and engineers. Penghu platform presents an innovative solution of providing green energy for offshore aquaculture by integrating wave energy and solar power generation technologies. Through the new design, as reliance on traditional energy sources was reduced, sustainability of aquaculture operations was enhanced, and autonomy of the aquaculture platform was increased. Penghu's successful deployment and operation demonstrate the feasibility of integrating offshore aquaculture and wave energy generation.

Unknown consequences of the interactions between the two uses are one of the main challenges for the development of co-located WEC and aquaculture platform. The interactions between Sharp Eagle WECs and semi-submersible aquaculture platform are analyzed to improve the understanding of the co-locating potential and challenge. The co-location of wave energy and offshore aquaculture platforms play a positive impact on energy supply. Nevertheless, challenges remain in the areas of platform motions and structural responses. Further research should be carried out on these aspects to better understand and address the issues involved.

In conclusion, Penghu represents a successful and groundbreaking attempt at integrating/co-locating wave energy converters (WEC) and aquaculture platforms. This innovation provides valuable experience and reference for the development of offshore aquaculture and the expansion of wave energy generation technology applications. As a mini integrated platform powered by offshore renewable energy, Penghu serves as a prototype of future ocean cities. The reliable Sharp Eagle WEC technology offers vast marine space and abundant renewable energy for various offshore production and living activities. This opens limitless possibilities for the construction of future ocean cities.

References

1. Bjelland HV, Føre M, Lader P, Kristiansen D, Holmen IM, Fredheim A, Grøtli EI, Fathi DE, Oppedal F, Utne IB, Schjølberg I (2015) Exposed aquaculture in Norway. Presented at the OCEANS 2015—MTS/IEEE Washington. IEEE, Washington, DC
2. Buck BH, Troell MF, Krause G, Angel DL, Grote B, Chopin T (2018) State-of-the-art and challenges for offshore integrated multi-trophic aquaculture (IMTA). Front Mar Sci 5:165
3. Campbell D (2017) Albatern WaveNet device Isle of Muck deployment. Northern Periphery and Arctic Programme
4. Chu YI, Wang CM, Park JC, Lader PF (2020) Review of cage and containment tank designs for offshore fish farming. Aquaculture 519:734928
5. Clemente D, Rosa-Santos P, Taveira-Pinto F, Martins P (2021) Influence of platform design and power take-off characteristics on the performance of the E-motions wave energy converter. Energy Convers Manage 244:114481

6. Coe RG, Ahn S, Neary VS, Kobos PH, Bacelli G (2021) Maybe less is more: considering capacity factor, saturation, variability, and filtering effects of wave energy devices. Appl Energy 291:116763
7. Cope C (2018) Aquaculture's bid to harvest marine energy. Shetland News
8. Freeman MC, Garavelli L, Wilson E, Hemer M, Abundo ML, Travis LE (2022) Offshore aquaculture: a market for ocean renewable energy. Ocean Energy Syst
9. Froehlich HE, Smith A, Gentry RR, Halpern BS (2017) Offshore aquaculture: I know it when I see it. Front Mar Sci 4
10. Garavelli L, Freeman MC, Tugade LG, Greene D, McNally J (2022) A feasibility assessment for co-locating and powering offshore aquaculture with wave energy in the United States. Ocean Coastal Manage 225:106242
11. International Energy Agency Ocean Energy Systems (OES) (2019) Annual report: an overview of ocean energy activities in 2019. Int Energy Agency Ocean Energy Syst
12. Jiang C, Cao R, Lao Q, Chen F, Zhang S, Bian P (2020) Typhoon Merbok induced upwelling impact on material transport in the coastal northern South China Sea. PLoS ONE 15:e0228220
13. Li C, Wang X, Zhu L, Liu K, Zong C, Wei N, Li D (2022) Enhanced impacts evaluation of Typhoon Sinlaku (2020) on atmospheric microplastics in South China Sea during the East Asian Summer Monsoon. Sci Total Environ 806:150767
14. Liu H, Lyu Y, Huang S, Chen J, Sheng S (2022) Influence of netting biofouling on hydrodynamic characteristics of wave energy offshore aquaculture platform. Ship Eng 44
15. LiVecchi A, Jenne D, Preus R, Green R, Copping A, Gorton A, Gill G, Geerlofs S, Gore S, Hume D, Schmaus C, Spence H, McShane W (2019) Powering the blue economy: exploring opportunities for marine renewable energy in maritime markets: U.S. Department of Energy (DOE), Energy Efficiency & Renewable Energy (EERE). Pacific Northwest National Laboratory, United States Department of Energy, Washington, DC
16. Miao Y, Ding J, Tian C, Chen X, Fan Y (2021) Experimental and numerical study of a semi-submersible offshore fish farm under waves. Ocean Eng 225:108794
17. Møller S (2019) Reduction of CO_2 emissions in the salmon farming industry: the potential for energy efficiency measures and electrification, Master thesis, NTNU, Norway
18. MUSES-PROJECT (2016) About MUSES (www document). https://muses-project.com/?page_id=45
19. Oikonomou CLG, Gomes RPF, Gato LMC (2021) Unveiling the potential of using a spar-buoy oscillating-water-column wave energy converter for low-power stand-alone applications. Appl Energy 292:116835
20. Renzi E, Doherty K, Henry A, Dias F (2014) How does Oyster work? The simple interpretation of Oyster mathematics. Eur J Mech B/Fluids Enok Palm Mem 47:124–131
21. Schultz-Zehden A, Lukic I, Ansong JO, Altvater S, Bamlett R (2018) Ocean multi-use action plan. MUSES (Multi-Use in European Seas) Project. Edinburgh
22. Sheng S, Wang K, Lin H, Zhang Y, You Y, Wang Z (2016) Semi-submersible open sea aquaculture cage with combined wave and solar power generation. PCT/CN2016/087091
23. Sheng S, Wang K, Lin H, Zhang Y, You Y, Wang Z, Chen A, Jiang J, Wang W, Ye Y (2017) Model research and open sea tests of 100 kW wave energy converter Sharp Eagle Wanshan. Renew Energy 113:587–595
24. Silva D, Rusu E, Guedes Soares C (2018) The effect of a wave energy farm protecting an aquaculture installation. Energies 11:2109
25. Yue W, Wang W, Sheng S, Ye Y, Hong T (2022) Analysis of the wave load and dynamic response of a new semi-submersible wave-energy-powered aquaculture platform. Ocean Eng 248:110346
26. Yue W, Wang Z, Ding W, Sheng S, Zhang Y, Huang Z, Wang W (2023) Feasibility of co-locating wave energy converters with offshore aquaculture: the pioneering case study of China's Penghu platform. Ocean Eng 288:116039

Wanzhen Yue a senior engineer of Guangzhou Institute of Energy Conversion, Chinese Academy of Sciences (CAS), is responsible for the structural security of the wave-powered aquaculture platform "Penghu". Her expertise focuses on wave-structure interactions in wave energy converters (WEC), as well as modeling and control design for these systems. Yue has directed and participated in several WEC projects supported by the Ministry of Science and Technology of the People's Republic of China. She received the Science and Technology Promotion Award from the CAS in 2022 for her significant contributions to the application of WEC technology in fish farming.

Songwei Sheng a research fellow, is the chief of the Ocean Energy Research Laboratory at the Guangzhou Institute of Energy Conversion, Chinese Academy of Sciences. He is the Vice Chairman of the Marine Resources Branch of the China Society of Natural Resources and the Director of the China Renewable Energy Society. Since 2002, Sheng has been engaged in the research and development of ocean wave energy technologies. He has managed several national science and technology projects and successfully developed the Sharp Eagle wave energy converting technology, which has been granted invention patents in China, the United States, the UK, and Australia, and has been certified by Bureau Veritas. He has led the development of Sharp Eagle wave energy converters with capacities of 10 kW, 100 kW, 500 kW, and 1 MW, as well as a large semi-submersible wave energy farming and leisure platform called Penghu. In 2022, Sheng won the Science and Technology Promotion Development Award from the Chinese Academy of Sciences and the first prize for technical invention in Guangdong Province in 2019.

Zhenxin Huang an assistant researcher of Guangzhou Institute of Energy Conversion, Chinese Academy of Sciences, focuses on wave energy generation research. He is a member of the China Renewable Energy Society and a member of the international expert group for TC114: Marine Energy—Wave Energy, Tidal Energy, and Other Water Current Energy Conversion Devices. Huang has been engaged in research on wave energy generation technology and the application of large-scale wave energy equipment technology for over 10 years. He has participated as a key member in several major research and development projects on wave energy technology. Huang has been involved in the feasibility study and validation of the Penghu project since 2016, responsible for the preliminary design and model testing, and he participated throughout the construction of the Penghu device. Till now, he conducted a series of initiatives to expand the application of marine renewable energy equipment and has participated in the implementation planning of the construction of Marine pasture in many places in China.

Modelling and Testing of Hinged Connected VLFS with Applicability to Offshore Renewables

Abel Arredondo-Galeana, Shen Li, Saishuai Dai, and Maurizio Collu

Abstract This chapter explores the role of very large floating structures (VLFS) in advancing offshore renewable energy solutions. Initially, the focus is on VLFS as foundations for wind turbines, highlighting their potential to support multiple turbines, thereby reducing installation costs and the number of mooring lines. Additionally, other applications of large floating structures in offshore renewable energy are discussed. While VLFS face challenges such as significant wave loading, the incorporation of hinge connections is proposed as an effective strategy to mitigate these effects. The chapter also provides an overview of numerical modelling and experimental testing methodologies for VLFS. Finally, the outlook for VLFS in offshore renewables is examined, emphasizing their potential to lower the levelized cost of energy for offshore arrays while maintaining structural reliability—key factors driving the innovative concepts presented in this chapter.

Keywords VLFS · Hinges · Floating platforms · Hydroelasticity · Offshore renewables

1 Introduction

In view of net zero, significant advancements in the development of decarbonised energy systems must be achieved in the immediate future to secure societal stability and sustainability for many years to come. Amongst others, the use of offshore renewable energy including wind, wave and tidal will be the key contributor.

Offshore wind is currently the major offshore renewable energy, and its growth has been tremendously successful with a total installed capacity of 14.7 GW in the UK [35] and 75.2 GW worldwide [15]. The early-stage offshore wind design employs bottom-fixed support structures such as monopile and jacket, ideal for shallow water deployments (< 60 m). As the exploitation of offshore wind moves to deep seas with

---

A. Arredondo-Galeana (✉) · S. Li · S. Dai · M. Collu
Department of Naval Architecture, Ocean and Marine Engineering, University of Strathclyde, Glasgow, UK
e-mail: abel.arredondo-galeana@strath.ac.uk

B. T. Wang et al. (eds.), *Large Floating Solutions*, Lecture Notes in Civil Engineering 636, https://doi.org/10.1007/978-981-96-5435-2_10

water depths greater than 60 m, bottom-fixed design is problematic, and the use of floating platforms such as semi-submersible, spar, barge and tension-leg becomes more relevant. Currently, a one-turbine-one-platform design philosophy is implemented in the design of offshore wind installations. Considering a floating offshore wind farm with 1 GW capacity, more than 60 floating platforms and hundreds of mooring lines will be required if each platform houses a single 15 MW turbine and is moored by 3–4 mooring lines. This could represent a major challenge from installation and maintenance perspectives. For this reason, multi-turbine shared platform approaches with the view to reduce the number of mooring lines is investigated by various stakeholders. To mitigate the turbine interference which potentially have an impact on power generation efficiency, these floating platforms are deemed to be of unconventional size, e.g., Very Large Floating Structures (VLFS). See for example the comprehensive review on VLFS by Wang et al. [40]. The VLFS concept was originally proposed in the 1920s as sea stations or floating docks to be used as aeroplane supply and navigating stations for Cross-Atlantic routes. This concept was further developed in Japan through the Mega-Float project, in which extensive research was conducted. For instance, two kinds of idealised structural models, namely a plane grillage model and a sandwich grillage model, were considered as the structural model for global response analysis of VLFS by Fujikubo and Yao [11]. Their applicability was examined through a series of elastic response analyses of the pontoon-type VLFS in waves. Structural modelling techniques for global response analysis and a two-step approach for stress analysis of detailed structures, based on the results of global response analysis, were outlined by Fujikubo [12]. As a result of the VLFS massive size and relatively small depth, it will deform under wave action due to its structural flexibility. A rigid-body assumption that is normally applied in hydrodynamic analysis of floating structures is arguably not applicable. To this end, the use of hydroelasticity theory is required [5]. Yang et al. [37] developed a three-step time-domain hydroelastic method for VLFS on depth-varying seabed and inhomogeneous waves. A discrete-module-based hydroelasticity method for floating structures was formulated by Wei et al. [41] in frequency domain. By applying different wave excitation forces to different modules, the effect of inhomogeneous waves was analysed, showing 2–3 times increase of the force responses in a specific wave frequency due to wave inhomogeneity. Building upon the same discretisation philosophy, a time-domain formulation was proposed by Wei et al. [42]. The time-domain analysis was also presented by Zhang et al. [43] to investigate the effect of transient loads. Apart from hydroelastic analysis, there has been growing interest in the recent years on hydro-elastoplasticity [33]. The behaviours of VLFS under extreme vertical bending was studied by Iijima and Fujikubo [16] using a segmented beam hydro-elastoplasticity method, in which the interaction between the surrounding fluid field and the deformation of VLFS, including both elastic and plastic, was taken into consideration. For validation, a scale-model test in extreme wave condition was reported by Iijima et al. [18].

VLFSs have been considered an alternative to create artificial lands on the sea in coastline cities that have scarce land resources. In fact, manmade islands where people have their residence exist now around the world. Other alternatives have been explored in terms of floating bridges and airports, where bottom based foundations would be prohibitively expensive, and therefore, floating structures provides a reasonable approach to construction. The scale of these constructions is in the order of 100 m, and therefore referred to as large floating structures.

In recent years, the race to net zero has propelled the offshore renewable industry to move to deep waters. Floating platforms in the form of semi-submersibles, spar buoys and tensioned legged platforms have been considered as floating foundations. In contrast, barge-type of structures have not been predominantly explored towards hosting offshore renewable devices. The main reason being that the large planform areas expose the structure to large loading and fatigue damage. In fact, the hydroelastic effects of such large structures are a crucial aspect, when designing bridges or airports, where local deformations in the platform can occur due to point loads or extreme wave events. It has been hypothesised that multiple renewable devices could be installed into large floating structures, reducing the costs of installation, maintenance and mooring lines [17, 21]. However, to ease the internal loading of the structures, these large planforms are segmented into modules and connected through hinges, that help the floating structure cope with large peak wave loads and hydroelastic effects [13, 14, 29, 39]. Figure 1 shows for example, the schematic of a novel hinged connected floating platform for multiple wind turbines. Although Fig. 1 shows a conceptual design, it synthetises the ability of a very large platform to host multiple devices, while minimising structural loads through hinge connections.

This chapter presents modelling and testing techniques used to characterise the motion dynamics of such hinged connected very large floating structures. Note that, although the chapter focuses only on the floating platform, the concepts can readily be developed and coupled to external loading, such as wind turbine loads. To further

Fig. 1 Hinged VLFS schematic with multiple wind turbines mounted on deck (Image courtesy of University of Strathclyde)

highlight the relevance of the topics discussed herein, the next section presents real life examples of large floating structures that have been deployed as foundation of offshore renewables.

2 Examples of Floating Platforms for Offshore Renewables

The previous chapter introduced the concept of a very large floating platforms as a foundation for offshore renewables (see for example Fig. 1). The concept depicted in Fig. 1 is one of many alternatives in which large floating structures can be used to deploy offshore renewables (ORE). It is therefore useful to review some large floating structures that have been deployed for ORE. By studying real-life examples, the practical implementation of very large structures for ORE seems within reach over the next years. Specifically, we discuss three examples: (1) the floating rigid hull of Orbital Marine Power as shown in Fig. 2a, (2) the hinged-liked connected wave energy converter Pelamis as shown in Fig. 2b and (3) the hybrid wind wave platform of Floating Power plant as shown in Fig. 3.

The rigid hull of Orbital Marine Power carries two tidal rotors alongside the main hull, as shown in Fig. 2a. The main hull is a tubular section of 74 m [27]. For rigid connected long structures subject to short waves, hydroelastic deformations might occur. These deformations could expose renewable devices to unexpected accelerations and additional loads that might need to be considered in design stages. Therefore, we introduce modelling techniques utilised to account for hydroelastic effects in large floating structures in the next section.

In contrast, hinge connected floating structures experience less hydroelastic deformation due to the reduced length of the blocks that constitute the floating platform. However, the platform motion due to the presence of the hinges needs to be considered

Fig. 2 **a** Orbital Marine Power—Launch 1 shared by S.clarkorbital under license CC-BY-SA-4.0 and **b** The Pelamis P2-001 and P2-002 wave energy convertors moored at Lyness, Hoy shared by Drnoble under license CC BY-SA 4.0

Fig. 3 Floating Power Plant platform with three wind turbines and wave energy converters distributed across the platform (Photo courtesy of Floating Power Plant [10])

to identify suitable locations that satisfy performance requirements of ORE. For some ORE applications, motion will be preferred whilst for others it will not. For example, a hinged connected device that required motion to provide power is the Pelamis wave energy converter (see Fig. 2b). Although the company is not active anymore, the Pelamis device extracted energy through hinged-like motion. In comparison, the hinged connected platform shown in Fig. 1 should be designed to minimise motion in the vicinity of the wind turbines and therefore, prevent undesirable accelerations in the turbines. Lastly, an important consideration in the use of hinged connected floating structures is mitigation of structural loading, which is one of the highlights of such types of floating structures [2].

Another example of a large floating structure deployed for offshore renewables is the hybrid wind wave floating platform by Floating Power Plant [10]. The platform is shown in Fig. 3, and has three wind turbines mounted on deck rated at 11 kW. The platform also has 10 hinged floats for wave energy conversion. The length and breadth of the platform are 25 m and 37 m, respectively. In this case, although the wave energy converters are hinged, the deformation of the platform is likely to be similar to that of a rigid connected structure. This is because the hinge axis does not decouple the hydroelastic motion of the platform, but instead, is used individually in each floater to extract wave energy.

In the following sections, we introduce modelling and experimental techniques used to study rigid and hinged connected very large floating structures, to provide guidance in implementing these techniques. Subsequently, we discuss the outlook of such type of structures in the context of offshore renewables and also, for different types of applications.

3 Numerical Modelling

The governing equation of motion of a floating structure is given by

$$-\omega^2(\mathbf{M} + \mathbf{A}(\boldsymbol{\omega}))\boldsymbol{\zeta} - i\omega\mathbf{B}(\boldsymbol{\omega})\boldsymbol{\zeta} + \mathbf{C}\boldsymbol{\zeta} = \mathbf{F_E} \tag{1}$$

where $\omega^2(\mathbf{M})\boldsymbol{\zeta}$, $\omega^2(\mathbf{A}(\boldsymbol{\omega}))\boldsymbol{\zeta}$, $i\omega\mathbf{B}(\boldsymbol{\omega})\boldsymbol{\zeta}$, $-\mathbf{C}\boldsymbol{\zeta}$ and $\mathbf{F_E}$ are the inertia, added mass, radiation damping, hydrostatic and wave excitation forces, respectively. Equation (1) is used to model the displacement of the centre of gravity of large floating platforms without the consideration of elastic effects [43]. Floating structures modelled by using Eq. (1) can be for example short length hulls, semi-subs or spar buoys of reduced planform areas. The hydrodynamic problem involves modelling the incident, scattered and radiated wave potentials around the floating structure. Subsequently, the forces acting on the structure, namely the added mass, radiation damping, and wave excitation forces are obtained through integration of the potentials around the wetted surface of the floating body [26].

Alternatively, the solution of the hydrodynamic forces can be obtained through potential flow solvers that are readily available in the literature. Such codes are for example open source or commercial solvers. For interested readers, Fig. 4 summarises some example of potential flow solvers. The list is, however, not exhaustive. Note that in Fig. 4, software Wadam is a module of software Sesam from DNV, while software Hydrostar is commercialised by Bureau Veritas and WAMIT by WAMIT Inc. In contrast, Nemoh was developed in Ecole Centrale de Nantes, Capytaine is a Python implementation of solver Nemoh, and HAMS was developed in Fortran [23]. In this chapter, software Capytaine version 2.0 was used to resolve the potential flow problem of the floating bodies analysed in Sect. 4.

Fig. 4 Summary or commercial and open-source potential flow solvers

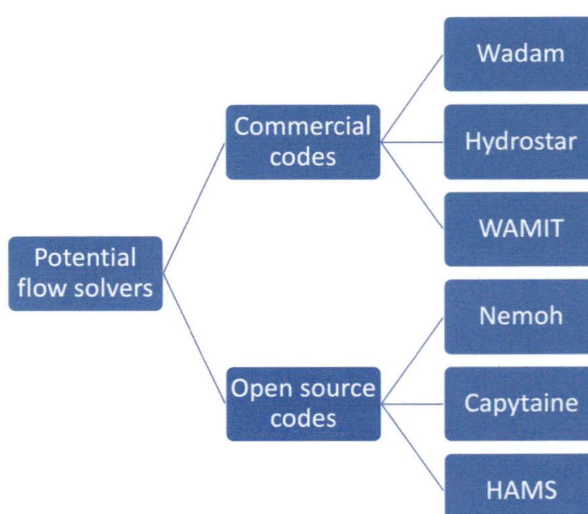

3.1 Elastic VLFS

The solution of fully coupled hydro structural problems involves not only hydrodynamic potential flow solvers, but also finite element models of the structure. Such coupled methodologies are numerically expensive and incur high computational costs. Hence, alternative methods that consider elastic deformations of very large structures are needed.

In recent years, low order methods have been developed to simplify the coupled problem and reproduce the structural response of the VLFS in a satisfactory manner. Therefore, as a complementary tool to hydrodynamics solvers, discrete module beam (DMB) methods, in time and frequency domain, have been developed to model the structural response of VLFS [43]. In this chapter, we introduced the main characteristics of such methods.

Specifically, the floating structure is discretised into a finite number of modules. Each of the modules is represented with a lump mass. The lump mass of each module is interconnected to the other masses through elastic beam elements of stiffness $\boldsymbol{k}$. A schematic of a VLFS discretised with N number of modules is shown in Fig. 5a, where the beam connections are in black and the lump masses in blue. Figure 5b, depicts the local frame of reference of two of the modules shown in Fig. 5a. In Fig. 5b, x, y and z represent the local reference frame, and u, v, z, θ_x, θ_y and θ_z represent the six degrees of freedom of each module in surge, sway, heave, roll, pitch and yaw, respectively. In Fig. 5b, the subindices 1 and 2 refer to module 1 or module 2, respectively.

The stiffness matrix ($\boldsymbol{k}$) of the single beam elements depicted as black connections in Fig. 5a, is given by [44]:

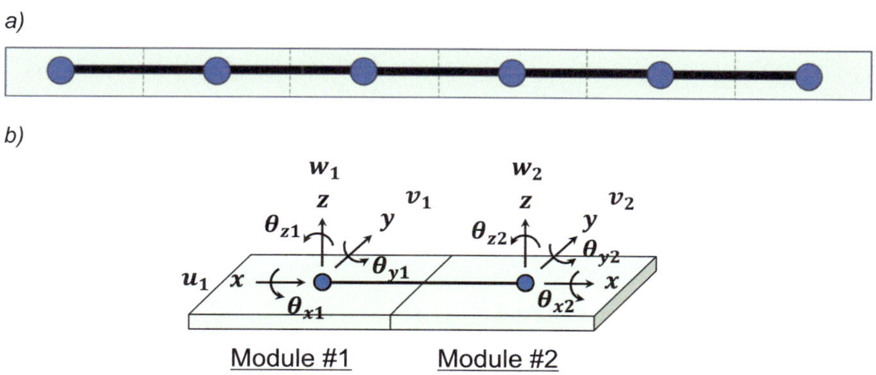

Fig. 5 **a** Model of multi-segmented VLFS with elastic beams connecting discrete modules. Each module is represented by a lump mass. **b** Two modules of the multi-segmented VLFS with the local frame of reference and 6 degrees of freedom depicted on each module

$$
\boldsymbol{k} = \begin{bmatrix}
\frac{EA}{l} & 0 & 0 & 0 & 0 & 0 & \frac{-EA}{l} & 0 & 0 & 0 & 0 & 0 \\
0 & \frac{12EI_z}{l^3} & 0 & 0 & 0 & \frac{6EI_z}{l^2} & 0 & \frac{-12EI_z}{l^3} & 0 & 0 & 0 & \frac{6EI_z}{l^2} \\
0 & 0 & \frac{12EI_y}{l^3} & 0 & \frac{-6EI_y}{l^2} & 0 & 0 & 0 & \frac{-12EI_y}{l^3} & 0 & \frac{-6EI_y}{l^2} & 0 \\
0 & 0 & 0 & \frac{GJ}{l} & 0 & 0 & 0 & 0 & 0 & \frac{-GJ}{l} & 0 & 0 \\
0 & 0 & \frac{-6EI_y}{l^2} & 0 & \frac{4EI_y}{l^2} & 0 & 0 & 0 & \frac{6EI_y}{l^2} & 0 & \frac{2EI_y}{l} & 0 \\
0 & \frac{6EI_z}{l^2} & 0 & 0 & 0 & \frac{4EI_z}{l^2} & 0 & \frac{-6EI_y}{l^2} & 0 & 0 & 0 & \frac{2EI_z}{l} \\
\frac{-EA}{l} & 0 & 0 & 0 & 0 & 0 & \frac{EA}{l} & 0 & 0 & 0 & 0 & 0 \\
0 & \frac{-12EI_z}{l^3} & 0 & 0 & 0 & \frac{-6EI_z}{l^2} & 0 & \frac{12EI_z}{l^3} & 0 & 0 & 0 & \frac{-6EI_z}{l^2} \\
0 & 0 & \frac{-12EI_y}{l^3} & 0 & \frac{6EI_y}{l^2} & 0 & 0 & 0 & \frac{12EI_y}{l^3} & 0 & \frac{6EI_y}{l^2} & 0 \\
0 & 0 & 0 & \frac{-GJ}{l} & 0 & 0 & 0 & 0 & 0 & \frac{GJ}{l} & 0 & 0 \\
0 & 0 & \frac{-6EI_y}{l^2} & 0 & \frac{2EI_y}{l} & 0 & 0 & 0 & \frac{6EI_y}{l^2} & 0 & \frac{4EI_y}{l} & 0 \\
0 & \frac{6EI_z}{l^2} & 0 & 0 & 0 & \frac{2EI_z}{l} & 0 & \frac{-6EI_z}{l^2} & 0 & 0 & 0 & \frac{4EI_z}{l}
\end{bmatrix}
$$

where I_y and I_z denote the second moment of area of the cross section with respect to y-axis and z-axis, respectively, E the Young's modulus, G the shear modulus, l the length of the beam element and J the torsional constant.

Considering the stiffness matrix of a beam element, the stiffness matrix of the full VLFS is concatenated following standard finite element procedures [43], and Eq. (1) is rewritten such that

$$
-\omega^2(\mathbf{M} + \mathbf{A}(\boldsymbol{\omega}))\boldsymbol{\zeta} - i\omega\mathbf{B}(\boldsymbol{\omega})\boldsymbol{\zeta} + (\mathbf{K} + \mathbf{C})\boldsymbol{\zeta} = \mathbf{F_E} \tag{2}
$$

where $\mathbf{K}$ is the stiffness matrix of the full structure. The consideration of structural deformation and the solution of Eq. (2) in terms of the motion displacement vector $\boldsymbol{\zeta}$ of each lump mass allows the computation of the internal forces in each lump mass, by assuming Hooke's law, such that $\boldsymbol{F} = \boldsymbol{k\zeta}$. Further details on the internal force computations may be obtained in the literature (for example, Sun et al. [31]).

3.2 Hinged VLFS

Noteworthy, VLFS ($L > 100$ m) are subject to high wave loading. To alleviate the internal loading of VLFS, mechanical hinges have been proposed to reduce the length of water plane areas that are exposed to wave loading. Structural loading can be alleviated either by elastic deformation as well [2, 20]. However, because mechanical hinges have been deployed in the offshore environment successfully [6] and the know-how of hinge construction is more advanced than that of elastic materials, we focus on the concept of a hinged connected VLFS.

Furthermore, hinge motion can also be exploited to extract wave energy directly. Therefore, the hinged VLFS dynamics are explored in this section. Figure 6 shows the schematic of a hinged VLFS. The hydrodynamic modelling of such system can also be found in the literature [43]. The main steps are briefly summarised as follows.

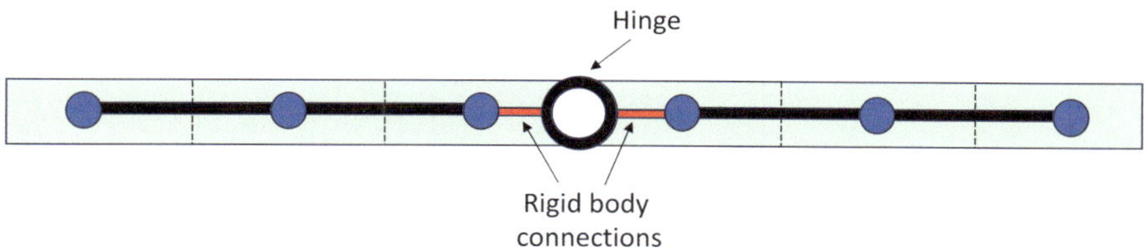

Fig. 6 A hinged connected VLFS with two pontoons, which are located to either side of the hinge and are identified in green colour. Each pontoon is discretised into finite segments with elastic beam connections

Similarly to the elastic VLFS, the pontoons of the VLFS are discretised into a finite number of modules. The modules are represented by lump masses and connected to each other through elastic beam elements, to account for elastic effects. The lump masses that are adjacent to the hinge are connected through rigid connectors. Assuming low stiffness mooring lines and therefore, negligible external forces due to moorings, the system of equations to be solved is

$$\begin{bmatrix} -\omega^2(\mathbf{M}+\mathbf{A}(\omega)) - i\omega\mathbf{B}(\omega) + (\mathbf{K}+\mathbf{C}) & \boldsymbol{\Xi}^T \\ \boldsymbol{\Xi} & 0 \end{bmatrix} \begin{bmatrix} \boldsymbol{\zeta} \\ \boldsymbol{F_J} \end{bmatrix} = \begin{bmatrix} \mathbf{F_E} \\ 0 \end{bmatrix} \tag{3}$$

where $\boldsymbol{\Xi}$ is the motion constrain matrix that specifies the transfer of motion from the lump masses to the hinge. Importantly, only the adjacent masses to the hinge affect the motion of the hinge. This influence is illustrated in Fig. 7 through the rigid beam connections. Hence, in Eq. (3) $\boldsymbol{\Xi}\boldsymbol{\zeta} = 0$ represents the motion transfer between the adjacent and the non-adjacent lump-masses of the VLFS to the hinge.

By assuming that the connection between hinge and adjacent lump mass is rigid, the motion transfer between point A (lump mass) and point a at the hinge is determined such that

$$\zeta_a = \mathbf{L}_\mathrm{a}\zeta_A, \tag{4}$$

where ζ_a is the motion vector of a, ζ_A is the motion vector of A, and $\mathbf{L_a}$ is the Lagrangian motion transfer matrix between A and a. The Lagrangian matrix is defined as

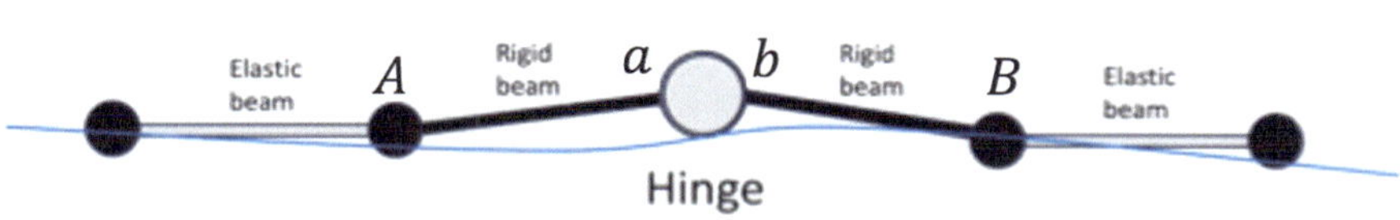

Fig. 7 Hinge connection with rigid connections to adjacent lump mass highlighted in black and elastic beam connections to non-adjacent lump masses

$$\mathbf{L_a} = \begin{bmatrix} 1 & 0 & 0 & 0 & y_a & -z_a \\ 0 & 1 & 0 & -y_a & 0 & x_a \\ 0 & 0 & 1 & z_a & -x_a & 0 \\ 0 & 0 & 0 & 1 & 0 & 0 \\ 0 & 0 & 0 & 0 & 1 & 0 \\ 0 & 0 & 0 & 0 & 0 & 1 \end{bmatrix} \tag{5}$$

where x_a, y_a and z_a are the horizontal, vertical and in-plane distances between points a and A, at rest conditions. Lastly, in Eq. (3) the force on the hinge F_J is translated to the adjacent lump masses and included in the force balance equation through $\Xi^T F_J$.

Lastly, the Langrangian matrix $\mathbf{L_a}$ is utilised to construct $\boldsymbol{\Xi}$, as illustrated for example in Zhang et al. [44]. Note that in Zhang et al. [44], $\mathbf{L_a}$ is 5×6, because the matrix is simplified on the observation that a point on the hinge has only 5 degrees of freedom due to the presence of the hinge.

For the system depicted in Fig. 7, the motion constrains matrix condition states that

$$\Xi\zeta = [\mathbf{0} \quad \mathbf{L}_\mathrm{a} - \mathbf{L}_\mathrm{b} \quad \mathbf{0}]\zeta = 0, \tag{6}$$

where $\mathbf{L}_\mathrm{a} - \mathbf{L}_\mathrm{b}\epsilon = 0$, arises from the fact that the motion of points a and b at the hinge in Fig. 7 are equal, and where $\mathbf{0}$ is the influence matrix of the non-adjacent lump masses to the hinge, which essentially is none.

Note that to construct $\mathbf{L}_\mathrm{b}$ in Eq. (6), the horizontal component x_b has opposite sign to the x_a elements shown in Eq. (5) in $\mathbf{L}_\mathrm{a}$, because point b is located to the right hand side of the hinge, rather than to the left, as indicated in Fig. 7.

4 Experimental Testing

Experimental testings of rigid and hinged connected VLFS were carried out in recent years for several applications. See for example, the experimental set ups of Yoon et al. [38], Bispo et al. [4], Arredondo-Galeana et al. [2]. In this chapter, we discuss some of the recent observations on testings carried out at the University of Strathclyde, for a 1:100th prototype scale VLFS design, as part of the Supergen ORE Hub program [3].

The scale model of VLFS is shown in Fig. 8a. Each pontoon shown in Fig. 8a was fabricated with a thin top layer of carbon fibre (2 mm) and a bottom layer of foam (50 mm). The dimensions of each pontoon were 580 mm × 52 mm. Three pontoons were joined to each other by off-the-shelf hinges, referred to as "piano hinges". In the case of a rigid connected VLFS, the hinges were replaced with rigid connectors. The hinge and rigid connectors were mounted on aluminium blocks to instrument the platform with strain gauges at the hinge locations. Additionally, the platform was instrumented with motion detection spheres. Because the strain gauges are connected

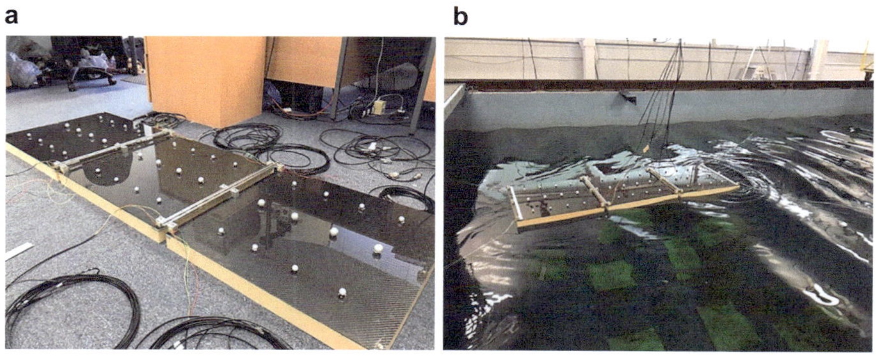

Fig. 8 **a** Experimental model of VLFS and **b** testing of rigid and hinged connected VLFS at Kelvin Hydrodynamics Laboratory, University of Strathclyde

to the VLFS via wires, enough slack should be given to the wire connections to avoid motion interference with the platform. In the case of this experiment, soft mooring lines were utilised to restrict the motion of the platform in surge, sway, roll and yaw, allowing platform motion only in heave and pitch.

The VLFS model was tested in regular waves and its relevance towards the application of offshore renewables was discussed in a conference paper presented at the 15th European Wave and Tidal Energy Conference by Arredondo-Galeana [2]. The experimental setup of the VLFS tested under regular waves at the Kelvin Hydrodynamics Laboratory at University of Strathclyde is shown in Fig. 8b.

The quality of the wave in the facility where the VLFS was tested needed to be carefully monitored during the experiments. Considering d to be the width of the VLFS and W the wave of the basin, and assuming shallow draft, blockage ratios of $d/W > 0.15$ could affect the physics of the flow around static objects [30], and therefore, could induce wall reflections after the wave's first arrival. Wave gauges were mounted upstream of the platform to monitor the wave quality.

A quality check of a regular incident wave measured at the Kelvin Hydrodynamics Laboratory at the University of Strathclyde is shown in Fig. 9. The wave probe is approximately 3 platform lengths upstream of the platform. In this example $d/W \approx 0.1$, and therefore, wave reflections are not expected to be a significant issue before the wave has reached the VLFS platform. Figure 9 shows a reasonably quality in the wave amplitude for all the wave amplitude which were $a_5 = 5, a_{10} = 10$ and $a_{20} = 20$ mm.

As part of the VLFS experimental campaign of the Supergen ORE Hub programme [32], tests were carried out at the COAST laboratory, University of Plymouth. Heave motion responses of the rigid and hinged connected platforms are shown in Fig. 10a and b for the case when the incident wave has a similar length to the full length of the platform. In Fig. 10a and b, the measured normalised heave amplitudes Δz were plotted with scatter points along 12 points of the platform for both rigid and hinged connected VLFS models. Numerical results were obtained by assuming 12

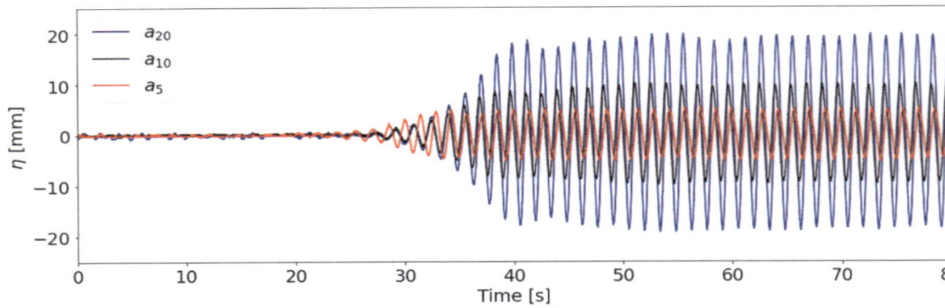

Fig. 9 Surface elevation (η) and wave amplitude (a) quality check at Kelvin Hydrodynamics Laboratory showing three wave amplitudes ($a_5 = 5$ mm, $a_{10} = 10$ mm and $a_{20} = 20$ mm)

discrete modules, and hydrodynamic data was computed in Capytaine version 2.0 [1]. The numerical solutions were interpolated with a Fourier fit for both rigid and hinged connected cases. Error bars, computed with the standard deviation of repeated measurements, were included in both Fig. 10a and b.

For this particular case when the wavelength is of similar length to the full length of the VLFS structure, Fig. 10a shows a single sagging shape for rigid connected structure. In contrast, Fig. 10b shows a triple sagging shape for hinged connected structure. In the context of offshore renewables, it is necessary to identify high motion points along the platform that can be detrimental or beneficial to the offshore renewable performance. For instance, for the case of a wind turbine, locations of reduced motion would be preferred. In contrast, if hinge motion is required for wave energy conversion, then high amplitude motions would be beneficial. In terms of strain gauge measurements, results indicated that the strain was significantly higher in the rigid connected platform when compared to the hinged connected case [2]. Hence, hinge connections can effectively reduce the stress in large floating structures while sustaining small motion response in certain areas of the platform.

5 Outlook for VLFS and Multipurpose Platforms

Although we have studied VLFS platforms within an applicability context of offshore renewables, the application of VLFS is vast. In fact, an offshore multi-purpose platform can be defined as an offshore structure designed to cater multiple industries. Many sectors have transitioned and/or are transitioning from land to sea, adapting to challenging environments to harness conventional (e.g. oil and gas) or renewable energy, to extract valuable materials through deep-sea mining, to produce food, to produce innovative nutraceuticals and pharmaceuticals. Also, there is a strong interest in this shift to support urban expansion and tourism in the nearshore and offshore areas. This transition resembles an "agricultural/industrial revolution", since it is

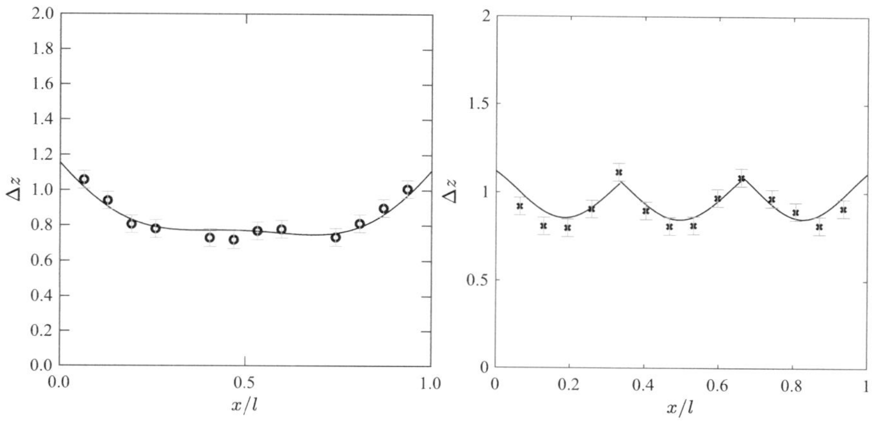

Fig. 10 Heave displacement (Δz) for **a** rigid and **b** hinged connected VLFS modelled numerically with Capytaine version 2.0 and measured experimentally in COAST Laboratory, Plymouth as part of the Supergen ORE Hub program [32]

marked by moving from resource gathering to systematic ocean resource utilization, and transitioning from isolated offshore facilities to interconnected industrial ecosystems—all these spurring technological advancements.

VLFSs offer the possibility to have a tightly integrated, rather than co-located, multipurpose platforms, i.e., the offshore structures serving the purpose of multiple industries can be accommodated on a single very large floating structure. Examples of these are: the aquaculture-wind-wave multipurpose platform proposed in the EU FP7 project H2Ocean [9], the Chinese Penghu Platform (described in one of the chapters of this book) and more recently in "The Blue Growth Farm" EU H2020 project [34], the wind-wave renewable energy devices analysed in the EU projects MERMAID [25] and MARINA [24] and the three concepts developed in the EU FP7 TROPOS project [36], namely, (1) the service hub for local maintenance of far-offshore wind farms, (2) the "green and blue" concept marrying offshore logistics and renewable energy, and (3) the "leisure island", looking at sustainable offshore tourism.

Furthermore, the concept of "seasteads" or "offshore floating complexes"—theoretical floating settlements, on VLFS—is starting to attract the attention of governments around the world. For example, in the UK [7], a series of workshops have been organised, involving a group of multidisciplinary experts, fundamentally to answer the questions: "*What happens if large-scale offshore floating complexes (OFCs) become a reality? What would that look like? What issues might the government need to respond to? How could—and should—the government respond to those issues?*". This reinforces the idea that VLFS as future settlements opportunities are on the horizon.

In summary, since the days of the Kamigoto National Oil Stockpiling Site (1988) and the "Megafloat" floating airport (1995–2001), VLFS have been adopted, and now multipurpose VLFS are considered as one of the key solutions to implement the Blue Economy agenda, creating economic growth and responding to the urgent energy and food needs of an ever-increasing population.

6 Concluding Remarks

In the next decade and beyond, we shall witness the construction of very large and sustainable floating platforms near congested coastal cities to create space on the adjacent water bodies for urban expansion. The implementation of such large platforms involves several engineering, economic and environmental challenges. However, large platforms, offer also great opportunities for innovation and novel developments.

This chapter summarises two commonly used approaches to model and test the hydrodynamic response of the type of floating structures. The numerical method consists in the discrete-beam module method [43] and it accounts for elastic effects through modelling the internal connections of the structure with beam elements. The experimental method involves measuring the displacement and shear response of the platform in controlled laboratory environments to validate numerical models and to improve design methods [2].

An overview and outlook of the applicability of VLFS as multi-purpose platforms is presented in Sect. 5. Owing to the numerous opportunities to deploy VLFS for ocean colonisation, it is envisioned that the future years will witness real applications grow and expand globally. Floating islands and hybrid offshore renewable platforms are, in the opinion of the authors, some of the promising applications that will be deployed with high priority across the globe.

Acknowledgements The work was part funded by the UK Engineering and Physical Sciences Research Council (EPSRC) as part of the 'Supergen ORE Impact Hub 2023' [EP/Y016297/1].

References

1. Ancellin M, Dias F (2019) Capytaine: a Python-based linear potential flow solver. J Open Source Softw 4(36):1341
2. Arredondo-Galeana A, Dai S, Chen Y, Zhang X, Brennan F (2023) Understanding the force motion trade off of rigid and hinged floating platforms for marine renewables. In: Proceedings of the European Wave and Tidal Energy Conference, vol 15, pp 1–10
3. Arredondo-Galeana A, Collu M, Brennan F (2024) Hinged very large floating structures for wave energy conversion and wind turbine foundation. Poster session presented at Supergen ORE Hub Seventh Annual Assembly, Plymouth , United Kingdom
4. Bispo IBS, Amouzadrad P, Mohapatra SC, Soares CG (2023) Motion analysis of a floating horizontal set of interconnected plates based on computer vision target tracking technique. In: Advances in the analysis and design of marine structures. CRC Press, Boca Raton, FL, pp 153–159

5. Bishop RED, Price WG, Wu YS (1986) A general linear hydroelasticity theory of floating structures moving in a seaway. Philos Trans R Soc Lond Ser A Math Phys Sci 316:375–426
6. Caio A, Davey T, McNatt JC (2021) Preliminary hydrodynamic assessment of Mocean energy's Blue Star WEC via fast turnaround physical model testing. In: Proceedings of the 14th European Wave and Tidal Energy Conference, Plymouth, UK, pp 5–9
7. Centre for Science and Policy (2021) Offshore floating complexes could be key to climate action. University of Cambridge. https://www.csap.cam.ac.uk/news/article-offshore-floating-complexes/
8. de Arcos FZ, Vogel CR, Willden RH (2023) A numerical study on the hydrodynamics of a floating tidal rotor under the combined effects of currents and waves. Ocean Eng 286:115612
9. H2OCEAN. Final Report Summary. Development of a wind-wave power open-sea platform equipped for hydrogen generation with support for multiple users of energy—Project ID: 288145. https://cordis.europa.eu/project/id/288145/reporting
10. Floating Power Plant. Accessed 1 Nov 2024. https://www.floatingpowerplant.com
11. Fujikubo M, Yao Y (2001) Structural modeling for global response analysis of VLFS. Mar Struct 14(3):295–310
12. Fujikubo M (2005) Structural analysis for the design of VLFS. Mar Struct 18(2):201–226
13. Gao RP, Tay ZY, Wang CM, Koh CG (2011) Hydroelastic response of very large floating structure with a flexible line connection. Ocean Eng 38(17–18):1957–1966
14. Gao RP, Wang CM, Koh CG (2013) Reducing hydroelastic response of pontoon-type very large floating structures using flexible connector and gill cells. Eng Struct 52:372–383
15. Global Wind Energy Council (2024) Global Offshore Wind Report 2024. Brussels, Belgium. https://gwec.net/global-offshore-wind-report-2024/. Accessed 20 Nov 2024
16. Iijima K, Fujikubo M (2018) Hydro-elastoplastic behaviour of VLFS under extreme vertical bending moment by segmented beam approach. Mar Struct 57:1–17
17. Ioannou A, Liang Y, Jalón ML, Brennan FP (2020) A preliminary parametric techno-economic study of offshore wind floater concepts. Ocean Eng 197:106937
18. Iijima K, Suzaki Y, Fujikubo M (2015) Scaled model tests for the post-ultimate strength collapse behaviour of a ship's hull girder under whipping loads. Ships Offshore Struct 10(1):31–38
19. Koh HS, Lim YB (2015) Floating performance stage at the Marina Bay, Singapore. In: Wang CM, Wang BT (eds) Large floating structures, ocean engineering and oceanography 3. Springer, Singapore, pp 37–60
20. Li L (2022) Full-coupled analysis of offshore floating wind turbine supported by very large floating structure with consideration of hydroelasticity. Renew Energy 189:790–799
21. Li S, Brennan F (2023) Feasibility assessment of very large floating structures for offshore wind: structural perspective. https://doi.org/10.2139/ssrn.4417894
22. Lim JEM (2019) Repurposing jack-ups, semi-submersibles and superbarges into offshore and nearshore settlements. In: Wang CM et al (eds) WCFS2019, Proceedings of the World Conference on Floating Solutions, Lecture Notes in Civil Engineering, vol 41, pp 329–357. Springer Nature, Singapore
23. Liu Y (2021) Introduction of the open-source boundary element method solver HAMS to the ocean renewable energy community. In: Proceedings of the 14th European Wave and Tidal Energy Conference, Plymouth, UK, Sep. 5–9, 2021
24. MARINA. Final report summary. Marine Renewable Integrated Application Platform—Project ID: 241402. https://cordis.europa.eu/project/id/241402
25. MERMAID. Final report summary. Innovative Multi-purpose off-shore platforms: planning, design and operation—Project ID: 288710. https://cordis.europa.eu/project/id/288710
26. Newman JN (2018) Marine hydrodynamics. The MIT Press, p 448
27. Orbital Marine Power. https://www.orbitalmarine.com. Accessed 1 Nov 2024
28. Ren NX, Zhang C, Magee AR, Hellan O, Dai J, Ang KK (2019) Hydrodynamic analysis of a modular multi-purpose floating structure system with different outermost connector types. Ocean Eng 176:158–168
29. Riyansyah M, Wang CM, Choo YS (2010) Connection design for two-floating beam system for minimum hydroelastic response. Mar Struct 23(1):67–87

30. Souppez JBR, Viola IM (2022) High-blockage corrections for circular arcs at transitional Reynolds numbers. J Wind Eng Ind Aerodyn 229:105139
31. Sun Y, Lu D, Xu J, Zhang X (2018) A study of hydroelastic behavior of hinged VLFS. Int J Naval Architect Ocean Eng 10(2):170–179
32. Supergen ORE Hub (2024) Supergen offshore renewable energy (ORE) Hub. https://www.supergen-ore.net. Accessed 18 Nov 2024
33. Tatsumi A, Li S, Benson S (2023) Collapse response analysis of a ship's hull girder in cyclic focused waves using a hydro-elasto-plastic beam model. In: Advances in the analysis and design of marine structures. CRC Press, Boca Raton, FL, pp 429–434
34. The Blue Growth Farm. Final report summary. Development and demonstration of an automated, modular and environmentally friendly multi-functional platform for open sea farm installations of the Blue Growth Industry—Project ID: 774426. https://doi.org/10.3030/774426
35. The Crown Estate (2023) UK Offshore Wind Report 2023. The Crown Estate, London. https://www.thecrownestate.co.uk/our-business/marine/offshore-wind-report-2023. Accessed 20 Nov 2024
36. TROPOS. Final report summary. Modular multi-use deep water offshore platform harnessing and servicing mediterranean, subtropical and tropical marine and maritime resources—Project ID: 288192. https://cordis.europa.eu/project/id/288192
37. Yang P, Li ZW, Wu YS, Wen W, Ding J, Zhang ZW (2019) Boussinesq-hydroelasticity coupled model to investigate hydroelastic responses and connector loads of an eight-module VLFS near islands in time domain. Ocean Eng 190:106418
38. Yoon JS, Cho SP, Jiwinangun RG, Lee PS (2014) Hydroelastic analysis of floating plates with multiple hinge connections in regular waves. Mar Struct 36:65–87
39. Wang CM, Riyansyah M, Choo YS (2009) Reducing hydroelastic response of interconnected floating beams using semi-rigid connections. In: Proceedings of the ASME 28th International Conference on Offshore Mechanics and Arctic Engineering, May 31–June 5, 2009, Honolulu, Hawaii, USA, OMAE2009-79692
40. Wang CM, Tay ZY, Utsunomiya T, Takagi K (2010) Literature review on methods in mitigating hydroelastic responses of VLFSs under wave action. Appl Mech Rev 63(3):030802
41. Wei W, Fu SX, Moan T, Lu ZQ, Deng S (2017) A discrete-modules-based frequency domain hydroelasticity method for floating structures in inhomogeneous sea condition. J Fluids Struct 74:321–339
42. Wei W, Fu SX, Moan T, Song CH, Ren TX (2018) A time-domain method for hydroelasticity of very large floating structures in inhomogeneous sea condition. Mar Struct 57:180–192
43. Zhang XT, Lu D, Gao Y, Chen LF (2018) A time domain discrete-module-beam-bending-based hydroelasticity method for the transient response of very large floating structures under unsteady external loads. Ocean Eng 164:332–349
44. Zhang X, Lu D, Liang Y, Brennan F (2021) Feasibility of very large floating structure as offshore wind foundation: effects of hinge numbers on wave loads and induced responses. J Waterw Port Coast Ocean Eng 147(3):04021002

Abel Arredondo-Galeana is a Research Fellow at the University of Strathclyde. He specialises in the design of offshore renewables with a focus in wave and tidal technologies. In addition to his work in marine renewables, Dr Arredondo-Galeana has worked in the field of hydroelastic floating structures, with international leading universities, such as Shanghai Jiao Tong University and as part of the UK EPSRC Supergen ORE Impact Hub programme. Dr Arredondo-Galeana has received funding from different entities, such as the EPSRC in the UK and the APS DFD in the US. The quality of his research is reflected with a Field-Weighted Citation Impact of 1.52 (Scopus), an *h*-index of 10 (Google Scholar) and the best paper award, as co-author, at the 15th European Wave and Tidal Energy Conference.

Shen Li is a Chancellor's fellow (lecturer) in the Department of Naval Architecture, Ocean and Marine Engineering at the University of Strathclyde, UK. His research focuses on the evaluation of ultimate and fatigue limit states, structural monitoring through digital twin technology, and structural optimization, particularly in relation to marine and offshore structures. He has participated in several research and knowledge exchange projects funded by EPSRC, Innovate UK and Scottish Enterprise etc., contributing as technical team member and project manager. He has also received two esteemed awards from the Royal Institution of Naval Architects, acknowledging his outstanding research contributions in the area of marine structural engineering.

Saishuai Dai is a Senior Lecturer in the Department of Naval Architecture, Ocean and Marine Engineering at the University of Strathclyde and serves as the Director of the Kelvin Hydrodynamics Laboratory. With over 10 years of experience in tank testing of marine structures—including ships, renewable energy systems, and conventional offshore floating structures—Dr Dai is a recognized expert in his field. He is currently a member of the ITTC Ocean Engineering Committee. Dr Dai has authored more than 70 research publications and actively collaborates with both industry and academia. His work provides valuable solutions and insights in hydrodynamics, tank testing, ocean engineering, and renewable energy systems.

Maurizio Collu is Professor in Offshore Renewable Energy Engineering, specializing in applied mechanics for offshore renewable energy systems, focusing on floating offshore wind turbines. With over 120 publications, an h-index of 30, and a Field-Weighted Citation Impact of 1.99, he is a leading researcher in his field. Prof. Collu has managed £4.5 M in research funding and contributed to £28.5 M worth of projects. He is involved in major research initiatives like Ocean RE-Fuel, The Blue Growth Farm, INNO-MPP, and HOME-Offshore. As chair of the ITTC Ocean Renewable Energy committee and editorial board member for prominent journals, his expertise is widely recognized. Prof. Collu collaborates extensively with both academia and industry, providing valuable insights and solutions in offshore wind energy.

Large Offshore Floating Fish Pens: Ocean Farm 1 and Havfarm 1

Per Arild Åland, Anders Fjellheim, and Torgeir Torgersen

Abstract As demand for sustainable food sources rises, particularly in the finfish sector, aquaculture is shifting from sheltered, coastal locations to offshore environments. This transition is driven by the need for increased output, improved fish welfare, and reduced environmental impact. Offshore fish farming presents opportunities for large-scale production in open waters, but it also introduces challenges such as harsh marine conditions, technological complexity, and sustainability requirements. Various innovative solutions, including both moored and mobile units, are being developed to address these challenges. This chapter explores the novel technologies being tested, with a focus on the operational parameters critical to the design and success of offshore aquaculture systems. We will also examine key factors like location, manning, and operational concepts that must be optimized to ensure efficient and environmentally responsible aquaculture. Finally essential factors for offshore fish farming including main concepts, design and fabrication as well as in-operations specifics are illustrated by an in-depth description of the two pioneering ocean fish farming installations in successful operation.

Keywords Ocean space · Aquaculture · Offshore fish farming · Exposed operations · Offshore installations

P. A. Åland (✉)
DNV, Høvik, Norway
e-mail: per.arild.aland@dnv.com

A. Fjellheim
Salmar Aker Ocean AS, Trondheim, Norway
e-mail: anders.fjellheim@salmarakerocean.no

T. Torgersen
NSK Ship Design AS, Harstad, Norway
e-mail: torgeri.torgersen@nsk.as

B. T. Wang et al. (eds.), *Large Floating Solutions*, Lecture Notes in Civil Engineering 636, https://doi.org/10.1007/978-981-96-5435-2_11

1 Introduction

All aquaculture market projections uncover a dramatic supply–demand imbalance in the next two to three decades (DNV [6] Aquaculture forecast 2050). Focusing on fin fish supply specifically reveals that output level is close to a saturation point for traditional areas with sheltered aquaculture [5, 10]. The environmental impact of fish farming activity in existing aquaculture coastal areas are in several areas close to sustainability limit.

Fish farming is expanding beyond sheltered, nearshore areas and moving to offshore locations, accompanied by innovative high-tech solutions. Additionally, new aquaculture facilities are being established worldwide. Traditional sheltered net pens are illustrated in Fig. 1. Several innovative solutions are being tried out in offshore fish farming in the process of industrialization of the aquaculture segment. The objective is a higher output volume of fin fish combined with improved fish welfare. Additionally, offshore farming has less conflicts with interests in the ocean space as well as to enable the scaling up of output volume without compromising environmental sustainability. As the world's population grows, the options for producing enough healthy, sustainable food are becoming limited. The oceans offer a vast opportunity to meet this demand with sustainable, safe and efficient offshore fish farming. Several interesting solutions for sustainable, large-scaled fish farming in offshore/exposed environment are in the pipeline. The industry needs to ensure that facilities for aquaculture can handle harsh environment and still maintain fish security.

Fig. 1 Fish farming is growing out of sheltered nearshore areas

In this chapter, we will investigate the various novel solutions being developed and piloted. We will also address critical factors that needs to be optimized to design and build an efficient offshore fish farm.

2 Ocean Aquaculture Concepts

Settling on the operational design and characteristics of an offshore aquaculture unit brings up a wide range of options to be considered [2]. While most new aquaculture solutions currently being tried out are permanently moored to a specific site of operation, we also see a few concepts on the drawing table meant to be operated with steering and propulsion.

2.1 Propulsion or Moored

Operational mode for these units is usually split in two. During calm weather, the fish farm is drifting or being temporarily moored on one or two predefined locations. Incoming rough weather or unfavorable seawater conditions such as increased temperature, high content of parasites, etc. will result in steaming to a different location with optimized conditions for marine aquaculture.

2.2 Operational Parameters

Operational parameters, as listed in Table 1, may be split into three main groups. For a permanently moored fish farming installation the first item is location of operation. Second item is manning on board. Finally, there is a question of operational concept and technology utilized on the fish farm. These three items are main factors essential for design and tuning safety (health, safety and environment—HSE) scheme on the fish farming facility [4]. The parameters more closely described below also play a crucial part in the arrangement and operation of fish farming essentials addressed later in this chapter.

Table 1 Essential operational parameters for exposed fish farming

Location	Manning	Concept
- Nearshore - Exposed - Offshore	- Permanently - Day-manned - Unmanned	- Remotely operated - Land power - Hybrid power concept - Green fuel (LNG/H_2) - Autonomy—full/partly

Locations

The location of the aquaculture site is normally spilt into tree main categories: Nearshore, exposed and offshore. The main focus here needs to be energy in the sea and significant wave height combined with extreme directions of wind and current. Often the operator is confronted with severe environmental conditions at a site categorized as "exposed". It often turns out that environmental loads are on a level expected for offshore location on a few days of the years cycle when weather sets in strong from an unfavorable direction.

Manning

The higher complexity technology applied in operations onboard the more personnel is required on Board. In some cased manning on board is replaced by Remote operation Center (ROC) on land, but this often result in increased personnel risk as daily visits to installation is required. As Marine personnel transfer is a high-risk operation representing significant challenges at rough weather, the ROC solution does not necessarily reduce the operational risk of the fish farm.

Operational concept

Operational concepts utilized in the new-tech ocean fish farming facilities are often a combination of digitally advanced technology and low emission power profile to minimize carbon footprint. As this industrial segment rapidly emerged into an more industrial advanced level, it immediately stood out quite clearly that green operation, fish welfare and profit only could be achieved by applying advanced digital technology. Main elements in this green smart-digital approach are:

- Remotely operated: Remote operations in aquaculture involve using advanced digital solutions to monitor, control, and manage fish farms from off-site locations. Usually the remote operating center (ROC) is located onshore. This technology includes sensors, cameras, and feeders linked to a central control unit. Remote operations allow fish farming operators to track water quality, fish behavior, feeding schedules, and overall health; thereby reducing the need for presence of personnel on-site. The main benefits include improved efficiency, reduced labor costs, less personnel safety risk due to challenging marine working conditions and in addition enhanced biosecurity.
- Land power: The use of land-cable, particularly when electricity is sourced from renewable energy (like wind, solar, or hydroelectric), can significantly lower the carbon emissions compared to using diesel generators on the installation. This aligns well with sustainability goals and reduces the environmental impact of aquaculture. Moreover, by eliminating or reducing the reliance on diesel generators, fish farms can minimize the wear and tear associated with fuel-based power generation, reducing the frequency of maintenance and repair. This leads to lower operational downtime and partly removes fuelling from the logistics operation.
- Hybrid power concept: Hybrid systems provide backup power, ensuring continuous operation of essential systems (feeding, monitoring, aeration) when renewable energy sources alone cannot meet the demand, or during periods of low

sunlight or wind. Hybrid power concepts in aquaculture involve combining multiple energy sources, such as renewable energy (solar, wind, or hydro) with conventional power sources (diesel generators or grid power), to provide a stable and reliable energy supply for fish farming operations. Hybrid power setups are becoming more common in advanced ocean fish farming since this enhances sustainability, flexibility, and cost-efficiency making hybrid concepts ideal for remote, offshore aquaculture facilities.

- Green fuel (LNG/H_2): Green fuel refers to environmentally friendly energy sources that results in little or zero carbon emissions and have a lower environmental impact compared to traditional fossil fuels. Green fuels are a key component in the global effort working toward more sustainable energy systems and reduce carbon footprints across the value chain of aquaculture industry.
- Autonomy—full/partly: Autonomy in the fish farming segment is a rapidly growing approach. Especially related to remote operations, certain functions are performed utilizing autonomy principles where robotics, machine learning and remote sensors are employed.

2.3 *Main Concepts in Ocean Aquaculture Development*

Open trusswork

Open truss aquaculture installations is a further development of the traditional sheltered net pen usually with rigid floating collar in steel or concrete. Additional structural arrangements for additional buoyancy, ballasting/de-ballasting and distention of net panels are common features. This category may be split in two sub-sets. One includes operating at the ocean surface only while an increasing number of designs have the option of submerged position to avoid splash-zone dynamics at rough sea states. The offshore fish farming installation Ocean Farm 1 in Fig. 2 is open trusswork-designed.

Semi-closed units

Main purpose of this solution is to separate fish environment from sea for the first few meters of the water column. This separation result in less exposure of parasites (such as sea lice) that lived in the first few meters of the water column. Figure 3 illustrates this concept with semi-closed fish farm Aquatraz G4 with fish pen lifted out of the water for cleaning and maintenance. Less environmental impact to the fish stock is also considered an advantage due to solid barrier between fish and surrounding water.

Submerged

Submerged fish farming installations are becoming a common solution. Environmental impact from rough sea is avoided for structure and net as splash-zone dynamics at rough sea states are prevented. The fish is exposed to less harmful parasites as the upper part of the water column is evaded. Several of these concepts are permanently submerged during the complete cycle of fish farming while others

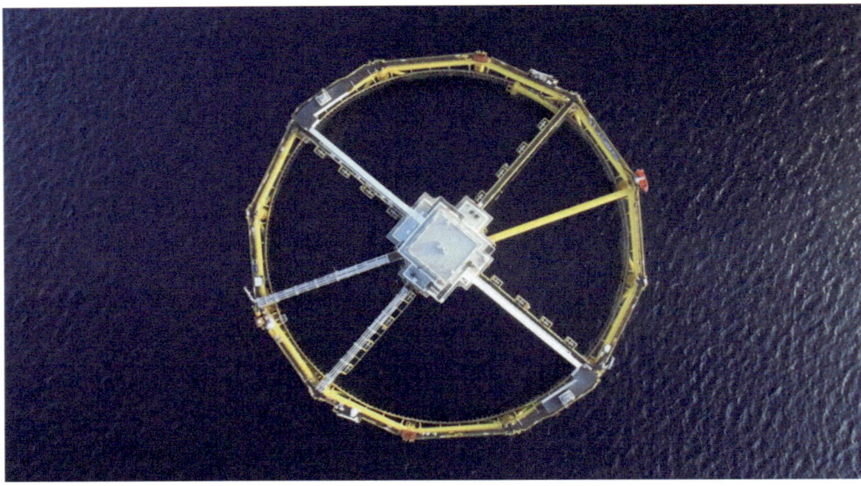

Fig. 2 Open trusswork aquaculture installation Ocean Farm 1 seen from above

Fig. 3 Semi-closed fish farm Aquatraz G4 being installed on site

are only during periods of high sea states. Hence, a wide variety of operational modes are used depending on species, sea conditions and level of technology applied.

Closed fish farming units

This concept is based on total separation between fish farming environment and marine environment at sea. This solution gives good control of what comes into the fish pen and what is discharged. There is a possibility to collect fresh water with favorable temperature containing a minimum of parasites at relevant depth to

Fig. 4 Modular closed fish farming installation Hexacage—computer graphics

improve wish welfare. One of several modular closed fish farming designs is shown in Fig. 4. Closed fish farming units enables operator to filter and clean return water from fish environment to reduce, or even totally remove, environmental impact to marine environment. A wide variety of solutions are currently being explored in this category.

3 Design and Fabrication

A few main categories are identified considering essential functional requirements of the fish farming facility. In this section, a brief description including how this influence the design specification of installations follows.

3.1 Fish Farming Essential Functions

Prevent escape

Fish control and prevention of fish escape is one of the main functions of a marine fish farm. A fundamental concept for sustainability in aquaculture is to restrict farmed fish from escaping into uncontrolled, wild marine environment where it may interact with wild spices and spread decease. This is a particularly dramatic issue for Atlantic salmon in northern part of Europe and Americas. Monitoring and maintaining net integrity are some of the most important issues operating a fish farm. Detecting wear and tear of the net at an early stage before sufficient opening resulting in escape

Fig. 5 Periodical inspection of net by divers or ROV to confirm integrity

is crucial. Visual periodical inspection is mainly performed by divers as shown in Fig. 5. However, more ROVs (Remote operated vehicles) and robots will be applied for future advanced aquaculture facilities.

In addition to the environmental aspect of maintaining fish control, there is a substantial financial incentive to keep this function intact. Escaped fish represents a dramatic loss of business revenue.

Feeding

Feeding is a critical aspect in the offshore fish farming operation and is arranged to ensure optimal growth rate, fish condition, and production efficiency. Fish feeding involves several components, each tailored to meet the nutritional needs of the species being farmed, while minimizing environmental impact.

Automated feed systems, combined with precision feeding strategies, are being applied in both operational and newly engineered advanced ocean fish farming installations. This increasingly being used to optimize aquaculture process as well as turning fish farming more productive and environmentally responsible.

Adopted feeding solutions are a critical success factor for ensuring the efficiency of offshore fish farms. Feeding costs typically account for around 60% of the total operational expenses (OPEX) of an aquaculture facility, making the ability to prevent overfeeding and feed waste essential for business profitability. Moreover, excess feed released into the marine environment has been shown to negatively impact ocean biodiversity.

For submerged fish farming installations novel techniques must be used to achieve efficient feeding. The so called "wet feeding" involves mixing fodder pellets with water before distribution into fish habitat. This approach very much relies on automated systems, specialized feed types, sensors, and remote monitoring technologies to ensure precise and efficient feed delivery.

Automated feeding technology consists of two main elements:

- Acoustic or camera-based sensors: This technology is installed both in surface- and submerged cages to monitor fish behavior, appetite, and feeding intensity.
- Underwater cameras: Cameras allow fish farmers to visually monitor the feeding process in real time. Operators can observe whether the fish are consuming the feed and adjust feed supply accordingly to prevent overfeeding.

Crowding and live transfer of fish

These processes cause significant stress and physical strain on the fish, reducing their resistance to diseases and potentially resulting in substantial losses and compromised fish welfare. Offshore fish farming often relies on advanced mechanical systems to manage these operations, but system failures can lead to operational disruptions and fish escapes. For example, Ocean Farm 1's approach to efficient fish crowding, as shown in Fig. 6, highlights the importance of reliable systems. While several innovative solutions for advanced ocean farming have shown promise, they frequently require major revisions or are entirely discarded due to the inability to integrate effective and dependable crowding or fish transfer systems on board.

Fig. 6 Sliding bulkhead at Ocean Farm 1 for fish crowding

Dead fish collection and handling

Dead Fish handling and ensilage processing. Typically, automated processes that involves much less manual handling compared to traditional sheltered methodology where these tasks are not carried out at the pen, but rather on the feed barge.

Water circulation to maintain oxygen level

Essential for fish welfare is the ability to keep continuous oxygen level above lower limit in all parts of the fish farming installation. In-depth analysis must be performed to ensure no "dead pockets" where the fish density at periods of low water circulation may be too high.

Cleaning and hygienic measures

In all aquaculture and offshore fish farming, cleaning and hygienic measures are crucial for maintaining fish health, protecting the environment, and finally to ensure operations according to regulatory compliance. By improving water quality and reducing the risk of disease, good hygiene practices contribute to the overall sustainability and efficiency in the fish farming industry [9]. More specifically the aqua-hygienic approach may be limited to two essential areas:

- *Minimizing environmental impact* because waste from aquaculture operations can negatively affect surrounding ecosystems. Excess organic matter from fish feces or feed surplus may pollute waters and damage marine habitats. Cleaning and sufficient waste control contribute to mitigate these effects, making the fish farming operations more sustainable.
- *Preventing disease spreading to wild populations* is mainly ensured by maintaining good hygienic performance. The objective is for the fish farm to reduce the risk of transmitting diseases to wild populations. This is a major concern, particularly for open fish farming installations where there are no separations between farmed fish and native species.

For these fish farming installations, the net is usually monitored daily and cleaned about once a week. Cleaning frequency increase as the seawater temperature rises. The cleaning scheme and intensity need to take material of the net together with any antifouling applied into account to minimize release of hazardous substances to the marine environment.

In most cases, the cleaning method used to remove biofouling is controlled to limit release levels of chemical compounds. These values need to be kept below threshold limit values for chemical substances in sediment and seawater. These limits are usually defined by local authorities of the operational area.

3.2 Personnel Safety—HSE

Personnel safety is mainly addressing arrangement for emergency escape and fire safety on board. This included lifesaving appliances, launching equipment and similar as well as fire detection and -extinguishing. In addition, other common safety items such as falling objects, working at heights and pinching/crushing must be considered as design and arrangement on board is concluded [8].

Other crucial operational issues include ability to perform hygiene and cleaning as well as feed handling and control may also be included considering the total operational performance of a high-volume fish farming installation.

To address and mitigate issues mentioned above there is a significant potential to utilize competence from traditional offshore and maritime industry to support operators of exposed fish farming units in design and tuning operational modes for efficient production. The industry is currently exploring possibilities and collecting experience from the various concepts being planned and put in operations. This learning process is an important phase in the development of offshore fish farming to supply future generations with sustainable and healthy marine proteins.

3.3 Standards and Quality Assurance

The main areas of concern when it comes to ensuring safe and reliable offshore fish farming units may be categorized into: Asset integrity, personnel safety, fish welfare and prevention of fish escape—see Table 2 [3].

Asset integrity includes structural strength, stability, mooring, technical arrangement, and solutions on board together with reliability of essential equipment installed.

Personnel safety primarily focuses on emergency escape procedures and fire protection. This includes lifesaving appliances, launching equipment, fire detection systems, and extinguishing measures. Maritime codes, such as SOLAS (Safety of Life at Sea), are commonly used as safety benchmarks, with local flag or coastal states providing their own interpretations of the code's requirements.

Fish welfare and requirements related to this vary depending on local authorities. It is essential to verify the reliability of technology utilized to monitor environment of the fish. Instrumentation indicating oxygen level, temperature, salinity, turbidity is

Table 2 Risk perspectives for safe and reliable offshore fish farming

Installation/unit	Personnel safety	Prevent fish escape
- Structural integrity - Stability - Mooring - Electrical and machinery	- Fire safety - Life saving - Emergency evacuation - General HSE	- Net pen integrity - Auxiliary equipment

subject for special attention. Maximum acceptable level of biomass is also a crucial parameter that needs to be monitored.

Fish containment and preventing escape are key functions of any fish farming unit. The structural integrity of the net systems, ropes, and the reliability of fish transfer systems are critical components for effective fish control. In rigid, high-volume steel fish farming installations, flexible net systems have been found vulnerable to fatigue, requiring special attention. Additionally, wear and tear from cleaning and handling pose further challenges. A significant number of reported fish escape incidents occur during fish handling, such as crowding for de-licing or transfer. Equipment involved in these operations must be carefully monitored and maintained to prevent such incidents.

The four different areas of concern are considered equally important for safe and sustainable fish farming offshore. They are also closely interconnected where integrity of one may support several others.

Competence from traditional offshore and maritime industry should be applied to help operators of exposed fish farming units to identify operational risks by applying technical rules and requirements from classification. Classification is often mentioned as the maritime executions model. As opposed to the offshore oil and gas industry, classification may not be obligatory in aquaculture, but it turns out that many developers and operators nonetheless choose to follow class requirements and recommendations.

Combining the well-known classification concept from maritime industry with balanced aquaculture-based requirements provides a robust and cost-efficient solutions to reducing risk in operation of offshore fish farming installations [1].

4 In-Operations Specifics

4.1 Logistics Routines

Managing the logistics of supplies and goods to and from offshore fish farming installations involves more complex marine operations compared to traditional sheltered farms. Higher sea states and increased vessel dynamics pose significant risks, especially during supply deliveries. These factors need to be carefully addressed when planning for emergency preparedness to ensure safety and smooth operations.

As in traditional offshore oil and gas installations, efficient logistics are essential for the smooth operation of a high-capacity fish farming facility. Continuous logistics streams, such as delivering fish feed and transporting dead fish silage, must often be carried out in challenging sea states and weather conditions. One of the most demanding logistics operations is the careful transfer of live fish without risking escape. This includes the placement of post-smolt into the pens and the retrieval of fully grown fish for transfer to live fish carriers during harvesting.

Personnel transfer is another key logistical challenge in offshore fish farming. The primary concern is to ensure the safe and efficient movement of workers from vessels to the fish farming installation, minimizing the risk of accidents and maintaining uninterrupted operations in difficult marine environments. Traditional methods such as crew transfer vessels (CTVs), basket transfers using onboard cranes, or helicopters are commonly used. However, the walk-to-work approach, utilizing an offshore gangway, has proven highly effective in the offshore wind sector and is now being considered for new, complex offshore fish farming installations. This is largely due to the limitations of traditional methods in rough sea conditions or in areas with restricted landing zones on fish farms.

The importance of the Walk-to-Work (W2W) solution lies in its ability to improve both personnel safety and operational efficiency. W2W systems, which typically apply motion-compensated offshore gangways, provide a stable and direct means of transferring industrial personnel—even in challenging sea states. This significantly reduces risks related to personnel safety and minimizes operational delays caused by waiting-on-weather.

4.2 Emission Control

Emission control of waste to sea because of fish farming is essential to maintain sustainable operation of the aquaculture facility. To consistently maintain emission control, procedures, equipment and systems that affect and regulate releases from aquaculture activity to the marine environment are often implemented on board.

Attending to these issues particular attention needs to be paid towards functionality to monitor and limit pollution from fish waste, feed surplus, chemical substances from net cleaning and emissions from medical treatment of feed stock.

Arrangements and design of these novel fish farming installations are commonly tuned to ensure that discharge to maritime environment because of aquaculture activity is kept at sustainable levels. Implemented procedures and systems on board are also applied to prevent spread of diseases and parasites originating from the fish stock.

The key focus areas for maintaining environmental integrity in modern high-tech fish farming installations typically include biosecurity, preventing fish escape, ensuring fish health, managing mortality, handling dead fish, controlling the use of chemical compounds, managing hazardous waste, and mitigating nutrient discharge.

4.3 Digital Monitoring

Digitally driven condition-based maintenance enhances reliability by continuously assessing the status of fish farming installations and monitoring the fish within them

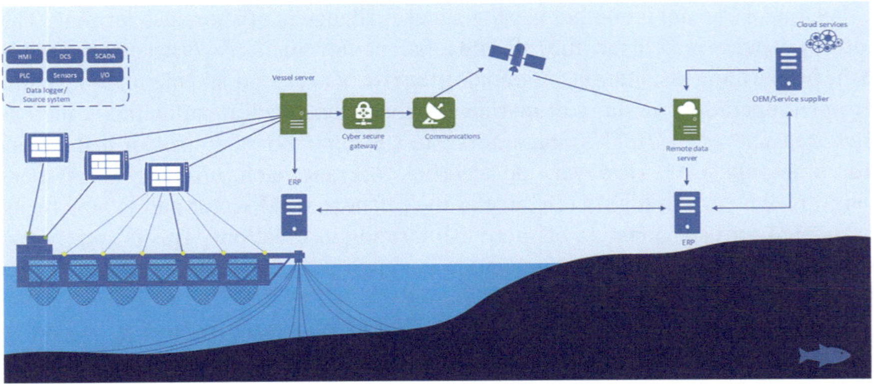

Fig. 7 Aqua digital asset management—descriptive model

[7]. The goal is to identify potential issues before they escalate into major problems, allowing for proactive measures that prevent failures and unplanned downtime. Implementing digital asset management can significantly improve efficiency and reduce operational costs [4]. For instance, by pinpointing equipment that requires maintenance or replacement, fish farmers can avoid costly breakdowns and minimize downtime. Figure 7 shows a conceptual model of data streaming infrastructure.

Digitally driven operational monitoring involves capturing information, analyzing collected data, synthesizing insights, and taking appropriate actions. The effectiveness of these solutions largely depends on the reliability and strategic placement of sensors throughout the fish farm. Onboard operational servers, data transfer solutions, and intermediate storage are also critical components of a digital asset management infrastructure. The stability and performance of the software used for processing and analyzing collected data pose significant risks and must undergo a robust assurance process to verify integrity. Additionally, safe data transfer and cybersecurity are essential for overall digital performance. While digital asset management offers numerous benefits, it also presents potential risks and challenges. Alongside data security threats and technical issues, maintaining data quality is a major concern, as the monitoring system relies on high-quality, accurate data to deliver meaningful insights into asset condition and performance Fig. 8 illustrates solution with cloud-based digital monitoring of aquaculture site.

Fig. 8 Aqua digital monitoring based on cloud solutions

5 Ocean Aquaculture Installations in Operation

5.1 Havfarm 1 (HF1)—Jostein Albert

General description

Nordlaks Havfarm 1 is a novel construction, where the main purpose is to provide a platform for safe and optimal fish farming in more exposed waters. Havfarm 1 has the outer shape and outline of a ship but is actually a framework structure in combination with semisubmersible platform when it comes to behavior and structural capacities. Reference is made to Fig. 9 for in-water details.

It is permanently moored with a turret solution in the bow to weathervane. It receives 100% power supply from shore. In addition, Havfarm 1 has 100% redundancy of power supply in case of long-term power failures on the shore power supply.

The fish is contained in 6 net pens, in a single row. Each net pen is operated by 9 winches for net and net weights to keep the net stretched and also for operating and changing of net volume during live fishing handling when well boats are harvesting from the Havfarm 1.

Fig. 9 Havfarm 1—outer shape and outline of a ship

On upper deck Havfarm 1 is equipped with fish farming systems, rail mounted service unit (RMSU), cranes and winches for integrated operations. Delivery of feed is from feed pellets carrier. The delivery/shipment of fish is from live fish carrier. The delivery of provision, fuel oil and fresh water, and shipment of ensilage is performed by aqua service vessels. Havfarm 1 even required new live fish carriers (well boats) to be developed for easy hook up of discharging hoses to each net pen for delivery of live fish to the well boats. These vessels have advanced dynamic positioning system which enable the well boat to follow the rotation of the Havfarm 1 as it is rotating on the turret system while fish transfer is performed.

A digital system for monitoring of the main structure is integrated into the condition-based monitoring scheme on board. The system registers and log the deflections, deformations and/or the stresses to give a more precise picture of the remaining fatigue life and if the structure has been exposed to stress levels close to design criteria. The system is applied in combination with traditional in-operation inspection services. This system is essential to pinpoint areas which need to be more thoroughly examined and is integrated into current inspection services reducing the manhours needed for inspection.

The control system installed on-board maintains safe operations for personnel and optimize the fish farming performance. The control system is split between maritime/safety systems and aqua production systems.

Ballast and bilge system will secure stability under both operational conditions and damage conditions. The ballast system handle operation at different levels of draft; operational draft, 30.7 m, survival draft 25.7 m, maintenance draft at 21.7 m and light ship condition for transfer from dock at lowest possible draft.

Fig. 10 Havfarm 1—operated in northern Norway by Nordlaks AS

The control system scope includes control of mooring tension, draft, position control within the maritime area. Further within safety system there are also fire detection and control of automatic/watertight doors.

Aquaculture production systems include fish feeding, dead fish handling and ensilage systems. The feed silos have approximately 1200 tons of feed pellets. The dead fish handling and ensilage systems can process and store up to 130 cubic meters of ensilage. A water replacement system with 6 stationary thrusters (one in each of the large columns) can move oxygen rich water into the net pens area and ensure sufficient water circulation. The thrusters are each 350 kW. All aquaculture productions systems are integrated into the main structure of Havfarm 1. Figure 10 shows Havfarm 1 on-site at operational draft in the northern part of Norway.

Power and IT network are closely integrated parts of the Havfarm 1 systems. The day-to-day operation use power from shore, along with signals to and from shore through the umbilical cable in the turret/swivel arrangement. The power cable and swivel is able to handle up to 3.5 MW at high voltage (17.5 kV.)

Three generator sets on-board can produce up to 3180 kW of power in case of loss of power from shore. An emergency generator is also fitted for safety functions such ballasting and firefighting.

The IT network operates with fiber optic cable, and it is possible to operate essential systems like fish feeding and general surveillance on Havfarm 1 from a remote operations center onshore. Logs of systems is sent to shore for storage and/or processing to optimize operations and maintenance.

A superstructure containing living quarters for 12 operators and guests with 8 single- and 2 double cabins. In addition, there is provision storage, control room, laboratories, workshops and garage for the RMSU is placed at the aft area of Havfarm 1.

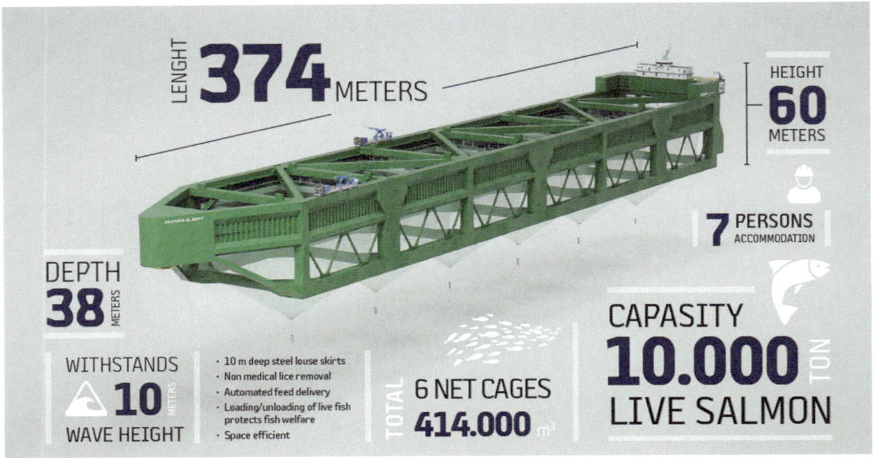

Fig. 11 Havfarm 1—overall information

The RMSU is equipped with a crane, winch and personnel basket dedicated to one of the cranes equipped for personnel transfer. The RMSU is used during fish handling operations and for inspection of the main structure in the splash zone both in and outside the corrugated bulkheads. The RMSU rails run along the Havfarm 1 "ship sides" from superstructure to the bow area.

Havfarm 1 is equipped with a remote operated vessel (ROV) for subsurface operations, such as inspection, maintenance, cleaning and repair. Both structure and nets are parts of the work scope for the ROV. Figure 11 shows a graphic compilation of vessel specifics.

Hydrostatics and stability

Loading conditions are identified as: lightship draft with lowest possible draft, maintenance draft 21.7 m, survival draft 25.7 m, and operational draft 30.7 m. Other loading conditions are included in analysis to account for temporary stages between the identified conditions.

Havfarm 1 is an open framework structure, with nonlinear change in waterplane area if the draft is varying. This makes both the hydrostatics and stability calculations more challenging.

The ballast arrangement is in the columns, the lower pontoons and transverse braces of the vessel. Stability calculations are performed for both intact and damaged condition at several drafts.

Main structure

Havfarm 1 is built as a framework structure. The main structure consists of two parallel lower pontoons from baseline up to a level which secure sufficient light ship draft to float out of a dock during construction. The lower pontoons have square cross section. They are connected to each other by transverse braces, one between each

Fig. 12 Havfarm 1—baptized "Jostein Albert" being connected to mooring system

of the six net pens, one in front of the net pen areas and one aft of the net pen area. The transverse braces have rectangular cross-section. Figure 12 shows forward part of structural members for Havfarm 1.

Transverse trusses are part of the global torsional stiffness. They run from joints between middle pontoons and columns up to upper transverse girders. Longitudinal trusses run between lower and middle pontoon to provide global vertical shear stiffness.

The stern is fitted with the aft columns and aft transverse brace, trusses and upper transverse girders. In the bow area, the lower pontoon is arranged at an angle in order to point inwards in the direction of turret center point. The lower pontoon has a horizontal orientation. Both middle pontoons and upper longitudinal girders will be inclined in a similar way to point to turret center point.

The mooring system is connected to the turret. The mooring system consists of several mooring lines to secure redundancy and sufficient capacity. The mooring lines is segmented for optimization of weight, capacity and cost.

Fish containment system

The six net cages are in the range of 400,000 cubic meters combined. The net pens are designed for 10 years lifetime and loads from hydrodynamic analyses is the input basis for design of the nets. Each pen is square shaped (seen from above) and slightly conical down to the lower pontoons. From the lower pontoon the net pens have the shape of upside-down pyramids. The cages are designed to secure sufficient distance from net to steel structures. This to avoid damages to the net.

The net pen is designed to optimize the available volume for fish inside the Havfarm 1 structures. The slope of the net at the bottom is ensuring that dead fish is sliding down to the Lift Up system for dead fish handling.

The lower point of the nets is connected to a clump weight hanging from the main structure. This clump weight is ensuring the net is stretched down towards the sea bottom. A centrally control and instrumented system at the working platform operates all ropes and winches.

Mooring and station keeping

The mooring system keeps Havfarm 1 (turret center) in position within a limited radius. The excursion is limited by the flexibility of the power and communication cable from shore.

Havfarm 1 weathervanes freely around a single point mooring system in the bow. The single point mooring consists of a turret solution with a swivel for transfer of power and communication from shore. The mooring system is designed according to the redundancy principle, meaning a mooring line failure shall not endanger the station keeping ability.

Mooring lines consist of several segments: top segment of chain, middle segment of fiber rope and bottom segment of chain. The mooring line is connected to the turret table at Havfarm 1 and to suitable anchors on the seabed.

5.2 *Ocean Farm 1 (OF1)*

General description

Ocean Farm 1 is the first marine aquaculture installation of its kind. The operator company Ocean Farming AS was established in 2010 and was at that time a subsidiary of SalMar ASA, one of the largest and most efficient producers of farmed salmon (Norway, Scotland and Iceland). Ocean Farming is now a subsidiary of SalMar Aker Ocean. SalMar Aker Ocean was established in 2021, owned 85% by SalMar and 15% by Aker. For SalMar Aker Ocean sustainable growth in the aquaculture industry depends on its ability to make use of new exposed locations currently not available for "traditional" fish farming, in areas that offer good biological conditions for the farming of fish stocks Ocean Farm 1 (OF1) was built in 2016–17 in China and is located at "Frohavet", a coastal area northwest of Trondheim, Norway. The location "Håbranden" has been thoroughly analyzed in relation to current, waves and other weather conditions.

As illustrated in Fig. 13 OF1 may look like some kind of novel oil rig design but is a fish cage of 110 m in diameter, rising 20 m above the surface and extending 43 m below it. OF1 has a crew of four trained mariners and fish keepers that stay onboard 24–7.

Fig. 13 Ocean Farm 1 operated in Norway by Salmar Aker Ocean

Ocean Farm 1 is built utilizing technical knowledge acquired from traditional aquaculture, ships, and oil and gas industry. Total steel weight adds up to 7300 tons. It houses a complex fish farming process. The salmon stocked here are 500–700 g, this enables annual production cycles. Running at full capacity 250,000 cubic meters of fish volume support the production of approximately 6300 (HOG) tons salmon every year. HOG stands for Head-On, Gutted. It refers to a common way of processing salmon, after harvest. When salmon are processed as HOG, heads remain on, but the fish are gutted—i.e., internal organs are removed.

The fish feed and other supplies such as water and diesel for the generator is delivered with regular intervals. Feed is stored in 16 silos with a total capacity of 640 tons.

Collection of dead fish is performed daily by applying a remotely operated vehicle (ROV) which also enables the fish farmer to keep an eye on what else is happening below the water surface.

The main dimensions of the offshore fish farm are as follows: CC Column diameter is 110.0 m, overall height (without LQ) is 50.0 m, diameter of periphery tanks is 12.6 m, diameter of center tank is 17.0 m, height of periphery tanks is 13.0 m, height of center tank is 7.0 m.

This fish farming installation have three different operational drafts depending on specific phase of operation: Inspection draft is 8.8 m, Survival draft is 38 m and Operational draft is 43 m.

Main structure and operations

The main geometry of the twelve angled (dodecahedron) fish farming cage comprises a top ring, a center ring and a bottom ring following the perimeter of the cage. Every second angle is equipped with six secondary columns. Each of the twelve angles

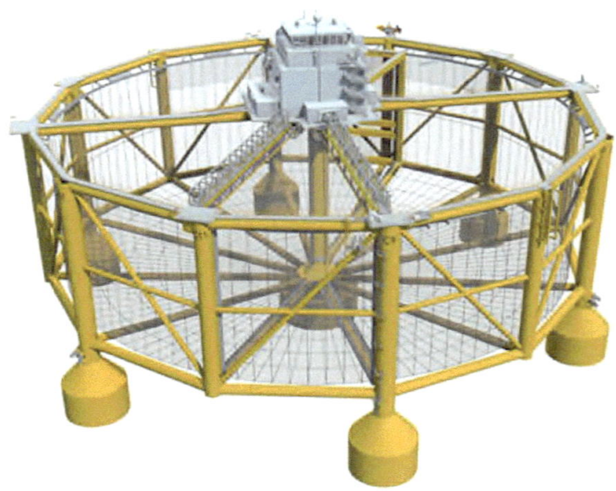

Fig. 14 Digital model of Ocean Farm 1

is connected to the center located main column with bottom beams in the radial direction. Each top cross member connecting circumference with main column is fitted with a walkway and in addition there are four fish feed distributors hanging on the side (i.e. 16 in total). The structural elements of the offshore fish farm are designed to sustain loads which occur during all temporary, operating and damaged conditions. Figure 14 shows an overview of structural members.

The design of Ocean Farm 1 is tuned to maintain acceptable safety for personnel and environment and have adequate durability against deterioration during the entire operational life of the steel structure. It is also worth mentioning that design and arrangements are founded on a concept where overall structural safety is settled based on preventive measures against structural failure put into design, fabrication and in-service inspection as well as residual strength against total collapse.

Examining the structural connections, it is noted that they are in general designed with the aim to minimize stress concentrations and reduce complex stress propagation through the details.

Marine personnel transfer is considered an operational risk and is specially addressed at OF1 to maintain safety of both regular personnel on board as well as many outside visitors that want to experience what the industrial shift in marine aquaculture looks like. The access system to the rig consists of a gangway that may be lowered or elevated depending on how high the fish cage is in the water. The movable gangway system has proven its value in waves when people must transfer between boat and cage.

Fish containment system

One of the main means to ensure sufficient water circulation is to prevent marine growth on the net. Automated cleaning robots brush the net on the cage with regular intervals ensuring hygienic standard is maintained. These units are mounted on trolleys that move automatically.

Continuous inspection and of net condition are carried out while cleaning is performed. A digital camera surveillance system detects anomalies of the net. The camera monitors that all the parts of the cage's main net have the right shape and weight load. This solution is in addition to the regular visual inspection program carried out by personnel on board.

For crowding and transfer of fish a complex radial sliding-net system is used. A large boom runs between the center and the edge of the cage. On it hangs a net which is dropped all the way down to the bottom of the main fish containment volume. The boom is then slowly moved around from one side of the circular cage, until the salmon are crowded into one single narrow sector of the cage. The fish is then pumped on board live fish carrier for transport to facility for further processing.

Mooring and station keeping

The mooring system is designed to include dynamic effects and non-linear effects such as slowly varying drift forces and the dynamic effects of the mooring lines. A complicating factor affecting dimensioning of mooring is verifying he biomass of the fish farm and its effect on the mooring lines. It is assumed, based upon current research that the increase in the static load will be approximately 15% due to the presence of the biomass.

Another interesting aspect worth mentioning for this kind of ocean installation compared to for example a drill rig is the fact that the fish farm will not have the possibility to stop the operation when adverse weather or other risk incident occurs. Ocean Farm 1 must therefore always be in operation condition. Hence, the mooring system must fulfil safety factors for consequence class 3 (operational condition) for extreme weather, i.e. 100-year condition. Figure 15 illustrates well utilization of offshore, oil and gas equipment and technology to succeed in the rapidly growing offshore fish farming industry.

Fig. 15 Ocean Farm 1 being towed into position—offshore vessels utilized in new industry

Biological results

The fourth production cycle of Atlantic Salmon will be harvested from Ocean Farm 1 in 2025. The biological results in all production cycles have been excellent with higher growth, lower mortality and close to no need for lice treatments, as compared to traditional salmon farming.

Further development

Ocean Farm 1 has been a development project that has furnished great learning and valuable experience that will be implemented in future ocean farms. This applies to all areas described above; HSE for people, biological production in extreme locations, technical operation in harsh conditions and logistics for people, fish and equipment. Salmar Aker Ocean is planning to build several new ocean farms both in Norwegian and international areas.

6 Concluding Remarks

Offshore fish farming is undergoing a period of rapid development, with disruptive solutions being tested and implemented. The future of offshore fish farming could evolve in multiple directions, depending on factors such as operational mode, location, sea conditions, species being farmed, and regulatory requirements. A common factor among all concepts is the benefit of experience gained from decades of designing and operating advanced offshore installations, which will play a crucial role in achieving success in this field.

Advanced offshore aquaculture systems hold significant potential for the sustainable global expansion of fish farming. By leveraging robust and complex installations, fish farming can be conducted in more open waters, closer to major markets.

This industrial shift presents an opportunity for the aquaculture industry to improve fish welfare and reduce environmental impact, enhancing its sustainability credentials while maintaining profitability. From a sustainability perspective, this development can contribute to protecting ocean health by minimizing discharge from aquaculture operations into the marine environment, thereby preserving biodiversity.

References

1. Åland PA (2024) Overall risk perspective for ocean aquaculture. In: WAS–EAS AQUA2024-conference, Copenhagen, Denmark, August 26–30, 2024
2. Åland PA (2024) Industrialization of ocean fish farming contributing to boost supply of marine proteins—solutions and challenges. In: 43rd International Conference on Ocean, Offshore and Arctic Engineering—OMAE 2024, June 9–14, 2024
3. Åland PA (2024) Reducing risk in design and operation of ocean aquaculture. In: 43rd International Conference on Ocean, Offshore and Arctic Engineering—OMAE 2024, June 9–14, 2024

4. Åland PA (2024) Unlocking the power of digital asset management: navigating the possibilities and overcoming the challenges. In: 43rd International Conference on Ocean, Offshore and Arctic Engineering—OMAE 2024, June 9–14, 2024
5. Carroll ML, Cochrane S, Fieler R, Velvin R, White P (2003) Environmental impacts of marine aquaculture: effects on benthic fauna and organic enrichment
6. DNV (2023) Aquaculture forecast 2050
7. Hassan W, Føre M, Urke HA, Kristensen T, Ulvund JB, Alfredsen JA (2019) System for real-time positioning and monitoring of fish in commercial marine farms based on acoustic telemetry and internet of fish (IoF). In: The 29th International Ocean and Polar Engineering Conference. ISOPE
8. Holmen IM, Utne IB, Haugen S, Ratvik I (2017) The status of risk assessments in Norwegian fish farming. Saf Reliab Theory Appl
9. Hvas M, Folkedal O, Oppedal F (2021) Fish welfare in offshore salmon aquaculture. Rev Aquacult 13(2):836–852
10. Wang O (2024) Monitoring regional benthic environment of Norwegian salmon cage farms

Per Arild Åland is a seasoned expert in emerging ocean space technologies, specializing in aquaculture. At DNV, he leads service and business development initiatives, focusing on innovative solutions that drive the sustainable growth of the marine aquaculture industry. With a Master of Science in Engineering from the University of Surrey, England, complemented by business and economics studies at leading Scandinavian universities, Mr. Åland combines technical expertise with strategic business insight. He brings over 30 years of experience in the maritime and offshore industries, with a track record in supply chain management, supplier and shipyard capability assessments, CE marking, risk assessments, and system/software integrity. Throughout his career, he has held key roles, including Senior Principal Engineer, Auditor, and Head of Section, contributing to offshore technology, business development, technology qualification, and management. In recent years, Mr. Åland has been at the forefront of marine aquaculture assurance services, helping to de-risk advanced, high-capacity offshore fish farming solutions. His work ensures the safety of installations, personnel, and marine life, with a focus on preventing fish escape and maintaining operational excellence. He has played a essential role in the safe and efficient design of several offshore aquaculture projects worldwide. A recognized capacity in his field, Mr. Åland is a frequent speaker and lecturer at international conferences and symposiums, where he shares insights on offshore aquaculture and ocean space technology.

Dr. Anders Fjellheim is COO in SalMar Aker Ocean (SAO). SAO is a pioneer within one of the new growth frontiers within sustainable salmon farming, offshore aquaculture, and is currently operating Ocean Farm 1 and Arctic Offshore Farming. Fjellheim graduated at NTNU in Trondheim with and M.Sc. (1999) and Ph.D. (2007). Since 1999 he has been working in different companies and positions in the Norwegian aquaculture industry.

Torgeir Torgersen is a naval architect with a keen interest in ship design and fish farming constructions. He has a mechanical degree from Gothenburg Technical Institute, Sweden, and a degree in Naval Architecture from University of Newcastle-upon-Tyne, UK. Torgeir has been design manager for NSK Ship Design since 2016 and has been the project manager for the world's largest exposed fish farm, Havfarm 1, the development project manager for Havfarm 2 and is currently the Sponsor for the Hydra fish farm project under construction in Turkey as well as CTO for Agalas AS. Torgeir has experience as a specialist engineer for structural design of subsea oil & gas pipelines for JP Kenny AS and Rolls-Royce Marine AS from 1998 to 2016 as Senior project engineer structures, Designer (concept) and Chief Designer (concept). He has lead a large range of newbuilding projects and headed development and design projects for approx. 50 vessels being built internationally and for clients like COSL (China), PaxOcean (China-Singapore), EZRA (Singapore), Farstad Shipping, Island Offshore, Bourbon, SIEM Offshore (Norway), Secunda Offshore (Canada) and Agalas as well as Eidesvik Agalas.

Floating Leisure Infrastructure: Integrating the World's Largest Floating Infinity Pool in Lake Como

Kimmo Kristian Saharinen and Tytti Sirola

Abstract Floating leisure infrastructure represents an evolving frontier in architectural and engineering innovation, offering sustainable and adaptable solutions for waterfront developments. This chapter examines the design, construction, and implementation of the world's largest floating infinity pool, developed for the Mandarin Oriental Lago di Como Hotel. The 600 m^2 floating structure, featuring a 40 m infinity edge presented unique engineering and architectural challenges, requiring bespoke technical solutions to address stability, comfort and environmental integration. Drawing on historical precedents of floating leisure infrastructure, this chapter contextualises contemporary floating pools within a broader architectural lineage, tracing developments from early Victorian floating baths to modern floating leisure facilities. The project's engineering solutions were informed by extensive hydrodynamic studies, including in-situ data collection and wave tank testing at Finland's VTT Technical Research Centre. A catamaran-based structural system, damping plates and a precision-designed mooring system were deployed to enhance stability and mitigate motion-induced discomfort. The chapter also explores the complex integration of an infinity-edge pool into a floating platform, a feature that demanded innovative hydraulic and structural strategies to maintain aesthetic and functional performance. As floating leisure solutions gain prominence, they present a compelling model of sustainable waterfront development, balancing luxury, environmental considerations and engineering excellence.

Keywords Floating pool · Floating baths · Floating structures · Waterfront development · Buoyant urbanism

K. K. Saharinen (✉) · T. Sirola
Bluet, Vantaa, Finland
e-mail: kimmo.saharinen@bluet.fi

B. T. Wang et al. (eds.), *Large Floating Solutions*, Lecture Notes in Civil Engineering 636, https://doi.org/10.1007/978-981-96-5435-2_12

1 Introduction

Our propensity to build our cities and settlements beside water bodies has provided us with a long, storied history of using our adjacent water bodies to host a range of leisure activities—one of which are ocean pools or tidal pools. Given the unpredictability of sea conditions, we design and construct these ocean pools to provide a safe space for swimmers to enjoy the sea under a wide spectrum of sea and weather conditions [10]. These pools emerge out of a long global history of lidos (outdoor public pools) going back to Roman baths and even further [12], with sea bathing rising in popularity in the mid 1700s due to the recommendation of physicians such as Dr Richard Russell [10].

Out of this long history of sea bathing emerged the concept of floating pools. From Victorian times to the 1930s, there were 'floating baths, temporary lidos, pontoons' along the River Thames [2] and floating baths that were built throughout the mid-1800s in Australia. But these early floating baths were vulnerable to weather. In Brisbane, the floating baths were repeatedly swept away by floods. Likewise, the Ramsden Dock floating swimming pool had a short lifespan, built in 1908 and then damaged and sunk in a storm in 1911. Floating pools then had a resurgence with Madelon Vriesendorp and Rem Koolhaas' 1978 floating pool design (see Chap. 15), Berlin's Arena Badeschiff (built in 2004), and Romer-Lee's proposal for a floating lido in the Thames that responded to an open call from the Architecture Foundation in 2013.

However, with the resurgence of floating pool proposals came a growing consciousness of how these installations in our adjacent bodies can be made eco-friendly. The Thames Estuary Growth Board [15] noted that the proposed floating lido for the Thames would incorporate recycled river water using chemical-free filters, tidal energy to meet some of the power requirements of the pool; geothermal energy or waste heat from surrounding offices to heat the pool. Other floating pools incorporate the construction of a compact wetland system using Forced Bed Aeration, designed to treat wastewater from show facilities, toilets and ancillary services [17]. Likewise, the floating pool studied in this chapter focused on enhancing the relationship of users with Lake Como through an infinity pool design while ensuring that design measures were put in place to prevent cross-contamination of the lake with the pool water.

This chapter provides a brief introduction to floating leisure infrastructure before examining the traditional and emerging construction practices that apply to these structures. It then presents a case study of the world's largest floating pool that was designed and built for the Mandarin Oriental Lago di Como hotel on Lake Como.

2 Traditional and Emerging Construction of Floating Leisure Infrastructure

Historically, most floating constructions have been made for transportation, such as ships. In the modern era, floating industrial applications have rapidly advanced, particularly due to the offshore industry. These include floating structures for the drilling, production, storage and offloading of hydrocarbons located in marine environments. More recently, floating structures have been adapted for floating wind energy production, solar energy production and various aquacultural applications.

Floating leisure constructions, in contrast, have been relatively undeveloped. They are, however, not a novel concept. The Roman Emperor Caligula designed elaborate two floating palaces (the Nemi Ships) that were deployed in Lake Nemi in the 1st Century AD and first discovered in 1446 by Cardinal Prospero Colonna and Leon Battista Alberti [9]. The floating palaces were described as having been 'filled with ample baths, galleries, and saloons, and supplied with a great variety of vines and fruit trees.' Interestingly, they had a relatively short lifespan having been stripped and sunk following the assassination of the Emperor [9].

However, the demand for floating leisure structures has never approached the volumes of their counterparts in transportation and industrial applications. In recent times, this has begun to change. New technologies, materials and techniques have made floating solutions more viable. Coastal areas are extremely desirable for leisure-oriented development, but economic and sustainability factors make shoreline construction sometimes difficult. A logical solution is thus the deployment of floating platforms, which can be developed faster, cheaper and more sustainably than similar-sized projects on land. These structures are also less disruptive to the environment, and they can be efficiently manufactured and assembled elsewhere, and then transported and towed to their eventual location (Fig. 1).

As these floating leisure structures are still relatively uncommon, there is a novelty factor which attracts owners, developers, local officials and the public. Generally, property owners tend to build their marketing approach around the buoyancy aspect of the structure as part of a strategy to engage public interests, attract investment and to distinguish such projects from other leisure infrastructure. These structures tend to have a wide variety of programmatic requirements as they are used across a range of purposes, including pools, spas, social areas, sports fields and restaurants, and they are critical amenities for the viability of floating residential areas. Bluet has participated in a varied number of these projects including (among others): a natural water pool, thermal spas, a bathing pavilion, a padel court (see Fig. 2), homes, a pool for aquatic zoo animals (Fig. 3) and an outdoor terrace for a restaurant. As these programmatic requirements are varied, they require bespoke solutions to meet the specific needs of the project and thus there is limited opportunity to streamline construction methodologies for economies of scale.

Further, floating leisure structures tend to be much smaller and located in more sheltered areas compared to industrial structures. For example, some of the largest

Fig. 1 Allas Sea Pool. Waterfront space is limited in downtown Helsinki, Finland, so developers chose a floating solution which Bluet helped implement. Allas Sea Pool has become a popular destination for locals and tourists

Fig. 2 A floating padel court. Originating in Mexico, padel is a racket sport similar to tennis

Fig. 3 The Korkeasaari Zoo in Finland is located on an island and has limited space. Bluet provided a floating aquatic habitat for the endangered ringed seal

offshore platforms could weigh up to 200,000 tonnes and need to withstand hurricane-strength winds and crushing ice [1, 19]. On the other hand, leisure structures, such as Singapore's Marina Bay Floating Platform (built in 2007, demolished 2023), which at the time was the largest floating stage in the world and hosted the opening ceremony of the Youth Olympic Games 2010, was just larger than the size of a football field and anchored in a sheltered body of water (the Marina Bay) [19]. As such, most of these installations are made of pontoon structures for inshore deployment with less use of the semi-submersible structures used in the oil and gas industry for open water deployments. In fact, many floating pools have been constructed from repurposed or refurbished vessels, such as the *Badboot* ('pool boat') in Atwerp, Belgium, a 'recycled ferry boat that serves as a restaurant and next to... a newly built ship that harbours an Olympic size swimming pool, sun decks, locker rooms with showers, and a party space' (Orischot et al. 2014).

Floating leisure infrastructure also face challenges that offshore industrial applications generally do not. These include aesthetics requirements that often require extensive community consultations and additional safety concerns, given that the users of such facilities are the general public instead of professionals trained to operate in offshore floating environments. As a consequence, many of these leisure infrastructure also contribute to the overall architectural iconography of the city, such as the swimming facility in Jubileumpsparken, Gothenburg, Sweden (Fig. 4).

There are existing offshore construction methodologies that are suited for use in the construction of floating leisure infrastructure. For example, the construction of modules and connectors allows for flexible modular construction that can then be assembled to meet the functional and operational requirements of the facility.

Fig. 4 The swimming facility in Jubileumsparken, Gothenburg, Sweden, is dragonfly-shaped. It has a freshwater training pool and two saltwater pools, one for diving and one for recreational swimming. *Photo* City of Gothenburg

These modules can be assembled offsite in multiple locations and then transported and assembled on site. This pre-fabrication enhances the efficiency of construction and can enable shorter construction times. Likewise, connection to land, provision of services, and connectors between modules that have been developed for offshore applications can be adopted in the use of floating leisure facilities.

However, due the smaller nature of floating leisure infrastructure, the pontoon structure that these facilities often take form as, and the need to meet the aesthetic demands of the general public, there is often a need to adapt and modify these construction solutions. For example, Bluet has employed a catamaran-like structure for many of its floating leisure infrastructure as it is lightweight, provides greater stability for user comfort and can still provide flexibility for arrangement and spatial programming. Separately, it is essential to also consider the mooring selection for these leisure infrastructure as it has to take into consideration the acoustic requirements as seen in our presented case study.

Critically, as the floating leisure infrastructure industry is underdeveloped, there are currently no industry certifications or standards. Unlike offshore industrial structures, which are subject to stringent international standards such as those set by the International Maritime Organisation (IMO) and the American Bureau of Shipping (ABS), floating pools and recreational platforms are often regulated under a patchwork of national and local laws. This regulatory ambiguity creates compliance challenges, particularly in jurisdictions where floating infrastructure does not neatly fit within existing legal classifications (see discussions in Chaps. 21 and 22). As such, there is a pressing need to engage with the design governance, policies, regulation and laws that apply to these floating structures.

In the case study set out below, the floating pool project on Lake Como had to adhere to both Italian building regulations and maritime safety codes, as well as local environmental protection laws governing water quality and ecosystem preservation.

In the absence of dedicated floating infrastructure standards, the project team relied on a combination of shipbuilding standards, offshore engineering best practices, and local leisure facility regulations to guide design and construction. The reliance on adapted regulatory frameworks underscores the need for dedicated guidelines that address the unique structural and operational characteristics of floating leisure infrastructure.

3 Case Study

We now turn to a case study on the world's largest floating pool, located on Lake Como. Mandarin Oriental Lago di Como is a 5-star luxury resort located in Blevio, a small commune on the shores of Lake Como in northern Italy. The heart of the resort is Roccabruna Villa, a neo-classical building once the haunt of opera singers and composers; Vincenzo Bellini wrote *Norma* there in 1831. Mandarin Oriental Hotel Group (Mandarin Oriental) purchased and renovated the property, opening it in 2019.

The Mandarin Oriental, which operates 38 hotels, 10 residences and a chain of properties across 25 countries and territories engaged Swiss architecture firm Herzog & de Meuron and Bluet to design and deliver an extensive floating leisure infrastructure project to serve hotel guests. Herzog & de Meuron was founded by Jacques Herzog and Pierre de Meuron in 1978 and has carried out over 600 projects in 40 countries, including the Tate Museum of Modern Art in London, FC Barcelona's stadium Camp Nou and the Prada Head Office in New York. Bluet is a Finnish company specialising in floating construction solutions, founded in 2016 by managing director Tytti Sirola, project director Petteri Houmo and technical director Kimmo Saharinen. A hydrological team, Finnish state-owned VTT, a research and technology centre, was also provided to the project. The core team includes engineers, designers and architects who have implemented floating leisure structures from the Nordic region to Australia.

While the resort had an existing floating swimming pool in the lake, the Mandarin Oriental was unsatisfied with it. The pool would move considerably due to wind and waves, to the extent guests could become nauseous. These comfort considerations of floating structures are considered in Chap. 2. Another issue was the design of the pool. The rectangular pool was surrounded by a sundeck, which limited the sublime views of Lake Como. Mandarin Oriental worked with the architect and the floating structure specialists Bluet to meet the following brief for a floating pool:

- Strict safety for guests and boaters.
- Less movement of the structure to improve comfort for guests.
- An impressive architectural design factor.
- Unprecedented size and scope.
- Infinity effect on long edge of pool to improve view of lake.
- Adequate size of pool and sundeck for normal number of guests.

- Design should not clash with the architectural style of property.
- Budget considerations.
- Deadline considerations.

4 Design Challenges

4.1 Stability

One of the main goals of the project was to provide a more comfortable experience to the hotel guests. The previous pool moved to an extent that some guests became seasick. Lake Como can be particularly rough due to the winds, boat traffic and the lack of natural beaches which dampen waves. The rock faces along the shore instead amplify waves, and the waterfront directly next to the pool consisted of stone arches, which caused even more volatility in the water.

To make the new pool more comfortable meant improving the stability and reducing movement. However, there is no single standard for comfort, which made finding a concrete goal to aim for difficult. A careful study of scientific literature revealed that reactions to different motions and accelerations generally depend on individuals [3–7, 11, 14]. There is limited guidance for floating structures and no industry standard. As benchmarks, the project used studies which found the average threshold, lower threshold and a "slightly nauseogenic" limit when humans become nauseous due to motion. For angular motions roll and pitch were used, as well as longitudinal, lateral and vertical accelerations. The classification of motion characteristics considered in this chapter are set out in Fig. 5.

In addition to short-term instability caused from waves, the team also needed to consider seasonal variations of water level. Lake Como is regulated with inflow and outflow dams, but rainfall, snowmelt and agriculture uses can vary considerably, contributing to seasonal variations of several meters. This consideration impacted how the pool could be stabilised and connected to the shore. The area can also be prone to earthquakes, which resulted in additional stability concerns.

4.2 Infinity Effect

A related problem was how to maintain the infinity effect on a floating platform. An infinity pool is designed so the water flows over one edge, producing a visual effect of water with no boundary (Fig. 6). The overflowing water is collected in a gutter which is pumped back into the tank which supplies the pool. The movement of the floating platform could cause sloshing or ruin the infinity effect [13].

Infinity pools require extensive structural, mechanical and architectural solutions not found in traditional pools. The infinity effect is normally used on the narrow end of a pool, minimising the extra costs for construction and maintenance. However, the

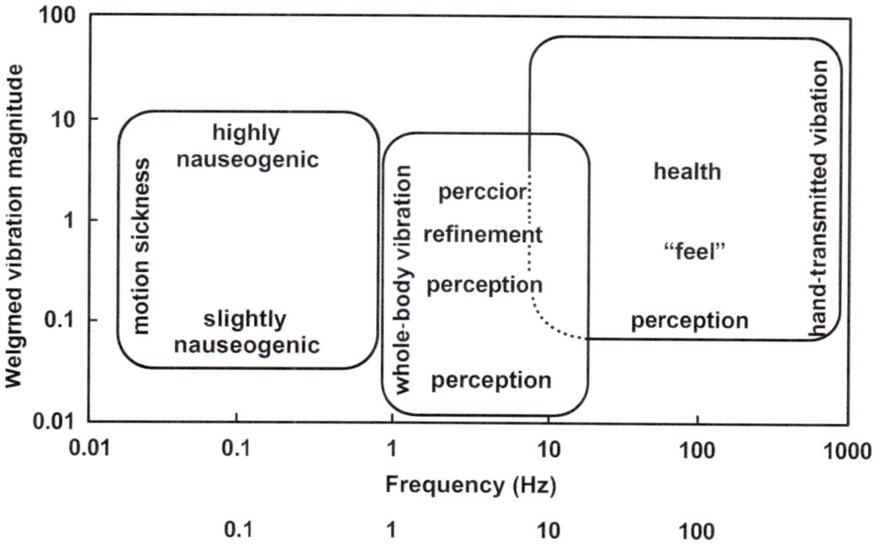

Fig. 5 Classification of motion characteristics [8]

Fig. 6 Infinity effect of new pool. The infinity effect is technically and architecturally challenging even on land

client and architect wanted a more elaborate infinity edge for a unique experience at the luxury resort.

In order to achieve an almost seamless edge between the pool and the lake surface to create the visual 'infinity' effect of the pool, the architect wanted the overflow gutter as low as possible for aesthetic reasons. However, to maintain the infinity effect of the pool, the volume of water spilling over the infinity edge was increased by boosting the pumping to about three times that of a regular infinity pool. If the infinity effect was ever lost due to the movement of the structure, it would quickly be regained with the higher pumping volume. This combination of the desired low overflow gutter and increased pumping raised concerns that there could be cross-contamination between the pool and the lake. Any cross-contamination would lead to non-compliance with water treatment rules given the inherent risks in pool water. In order to mitigate this risk, the design of the floating pool incorporated a gutter high enough to avoid a 1-m-high wave, which was higher than any observed wave height during data gathering at Lake Como (see next section) and which was used as a significant wave scenario during testing. In addition, the gutter angle was also slightly changed, with a larger gutter size specified that would be typical for a similar-sized infinity pool due to the increased water pumping volume.

4.3 Environmental and Water Quality Considerations

Floating structures inevitably interact with sensitive aquatic ecosystems, raising concerns about environmental sustainability, water quality management, and habitat disruption. Regulatory bodies overseeing inland and coastal waters typically impose strict environmental requirements on floating structures, particularly regarding wastewater discharge, chemical use, and potential impacts on biodiversity. The floating infinity pool had to comply with stringent water quality regulations to prevent contamination of the lake, a concern heightened by the pool's expansive infinity edge. The design team implemented a high-capacity filtration and recirculation system to ensure the pool water remained separate from the natural lake environment.

4.4 Public Safety

As described above, floating leisure structures, particularly those accessible to the public, must adhere to safety regulations that address structural integrity, buoyance, load capacity, and emergency preparedness. Unlike offshore industrial platforms, which are designed for professional use and governed by rigorous safety codes, floating leisure structures cater to untrained users, necessitating additional risk mitigation measures.

For the Lake Como floating pool project, compliance with European safety standards for swimming facilities, combined with maritime safety principles, guided the implementation of stabilisation features, including its mooring systems. However, the absence of a dedicated classification system for floating pools meant that safety assessments had to be drawn from a combination of land-based pool regulations, marine engineering guidelines, and floating structure performance metrics. The lack of industry-wide benchmarks for motion tolerances and human comfort levels on floating leisure platforms further complicates regulatory compliance, underscoring the need for dedicated research and standard-setting in this field.

4.5 Architectural Challenges

The design of the floating infinity pool presented several architectural challenges that required innovative solutions to harmonise the new structure with the historic and natural context of Lake Como.

The primary architectural challenge was to ensure that the floating pool complemented the 19th Century Villa Roccabruna and its surrounding structures, including the historic vaulted stone colonnade. The Swiss architectural firm aimed to create a design that blended seamless with the existing architecture without overshadowing it. To achieve this, the pool was finished in dark local Cardoso sandstone, allowing it to mimic the colours of the lake and maintain visual coherence with the villa's neoclassical aesthetics. The long infinity edge facing the lake was its key defining feature to create the visual illusion, making the pool surface erge with Lake Como, thus preserving the site's historic ambiance.

The material selection was also a significant consideration for the architects. These materials needed to withstand the aquatic site while maintaining aesthetic harmony. The use of prefabricated steel structures clad in glossy deep-green colour to reflect the water surface was intended to enhance the pool's integration with its surroundings. The sundeck, bordered by wooden planking, introduces natural and warm materials that echo the classic wooden motorboats of the region's nautical traditions. Further, the old pool had pile moorings, which were visible and noisy in the waves. The architect wanted a mooring system that was invisible and soundless for the guests.

5 Data Gathering

One of the considerations Bluet raised was that the depth to the lake floor and angles from the stone arches created unusual wave patterns which did not appear in general lake data collected and published by Italian authorities. The lakefront where the floating structure was to be placed created a unique "microclimate" which was not accurately reflected in public data.

Even more importantly, extensive, long-term hydrologic data has been collected for Lake Como, but this was found to be of limited use. Lake Como hydrological data is a statistical system built for wind-generated waves. Typical lake wave data removed outliers, which can be normal statistical procedures. This process treated abnormal waves—particularly those caused by boats—as measurement or processing errors. However, these were precisely the waves the project needed to consider. It was critical for the project's success to collect bespoke data and use that alongside traditional public hydrologic data.

The first task was to accurately measure the wave conditions. This information was used for computer modelling and controlled testing in a wave tank to evaluate different possible solutions. The goal was to find a solution which provided acceptable stability to the floating platform. The scientific literature proved that it was impossible to guarantee no one would ever become nauseous from the motion of the pool, yet it was possible to limit motion below thresholds published by the scientists.

In December 2021, the team visited the site on Lake Como. Sensors were placed on the old pool to measure:

- Longitudinal acceleration (mg).
- Lateral acceleration (mg).
- Vertical acceleration (mg).
- Pitch motion (°).
- Roll motion (°).

A buoy was installed by Italian partners at the site to measure the direction and strength of waves while video recordings were made of the motions of the existing pool. Also, detailed scans were made of the lake bottom and waterfront where the pool would be installed (Fig. 7). The lakefront extended down about 5 m in water depth and 3 m above the typical waterline.

Most waves were generated by passing boats, which was verified comparing the video with the elevated wave spectral density at certain frequencies, which correspond to boat-generated waves. Wave winds and swell were also recorded but were not a serious concern. Wave heights up to 0.32 m were recorded. Most of the existing

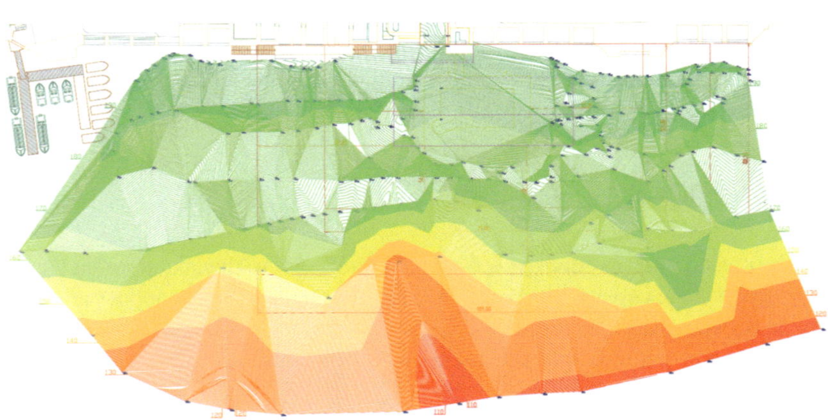

Fig. 7 Bathymetry of the site

pool's motions were on a relatively narrow frequency range from about 0.1–0.4 Hz. The largest amplitudes were typically between 0.25 and 0.30 Hz. Pitch motion was observed to be more significant than roll motion.

The data gathered on site was complemented by general lake data collected and published by the Italian authorities as a public service. The data-gathering period turned out to have more waves and adverse weather conditions than normal, but this was judged to be a good thing, as any proposed solution would be measured against unusually harsh waves.

6 Design and Testing

The new floating infinity pool was designed with CAD. Three possible new designs were created. The existing pool design was compared against the three possible new designs using theoretical methods. The three possible new designs were analysed and one was picked as most likely to meet the client's goals. The first choice was a rectangular shape, much longer than the existing pool (Table 1).

Typically, leisure floating structures have a barge structure, meaning a simple flat-bottomed monohull shape. However, other designs were also considered since one of the main goals of the new infinity pool was to increase stability (Fig. 8). An obvious possibility was a catamaran structure. A catamaran-like design can add stability due to having less contact with the water and hence less wave impact.

The testing partner was VTT Technical Research Centre of Finland, a research institution owned by the Finnish state. One of their specialties is the maritime sector, so they maintain modern software, a wave pool and hydrology experts. Ansys Aqwa, commercial software for hydrodynamic assessments, was used for the analysis (Figs. 9 and 10).

The selected experimental wave angles of 134° and 90° were chosen as both the typical and worst wave conditions the floating structure will encounter. As expected,

Table 1 Main dimensions of the floating pools

	Existing pool	New pool
Floating platform		
Length (m)	22.8	40
Width (m)	14.6	15
Draught (m)	1.4	1.72
Estimated weight (t)	445	457
Swimming pool		
Length (m)	15.9	39.3
Width (m)	8.4	6.3
Water depth (m)	1.35	0.23–1.20
Volume (m^3)	160	173

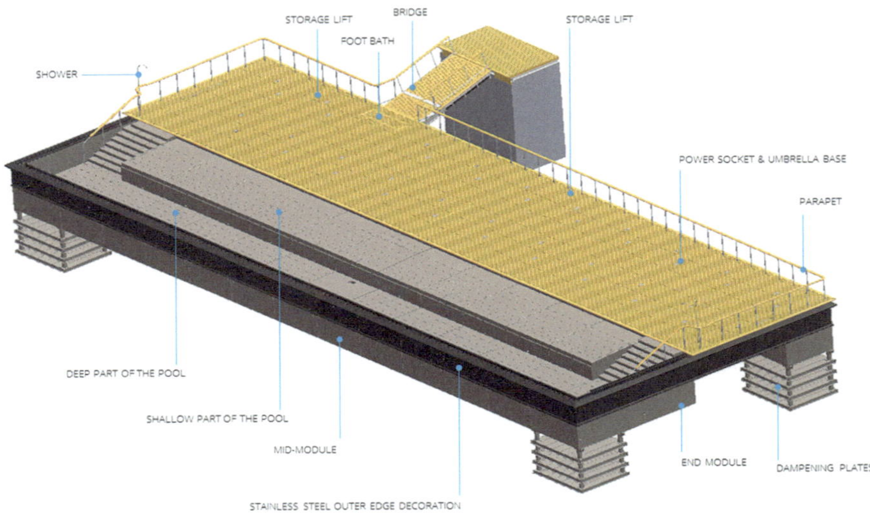

Fig. 8 New pool design. The pool is the grey area and the sundeck is brown. Note the infinity edge encompasses three sides of the pool

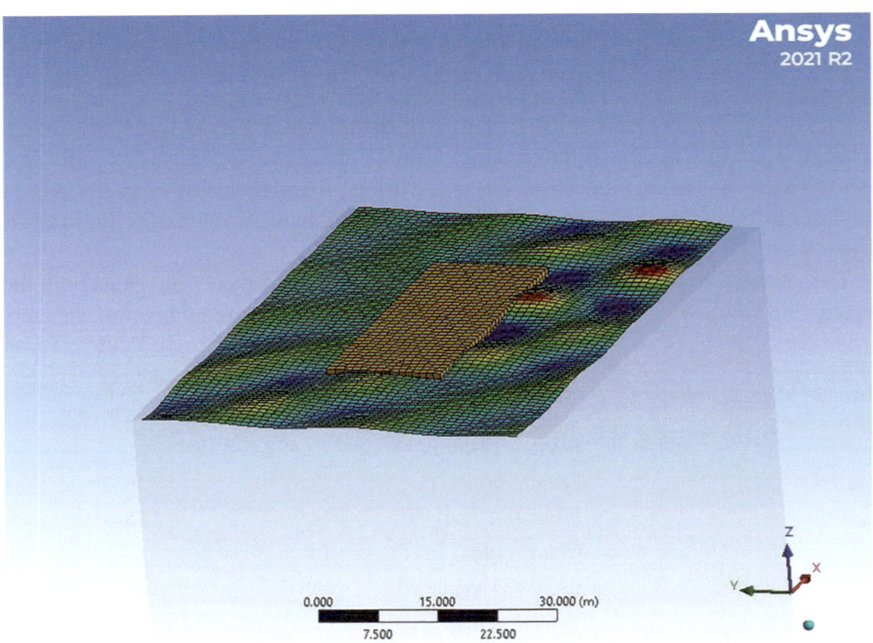

Fig. 9 Wave field around the floating pool at wave angle of 45°, the original pool geometry

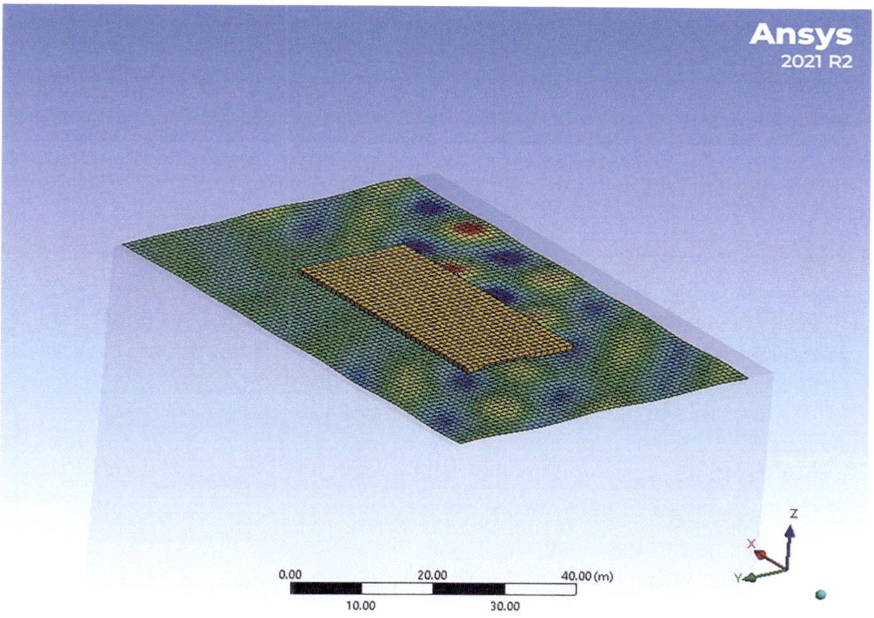

Fig. 10 Wave field around the floating pool at wave angle 45°, the 20% wider geometry

the peaks of roll motion occurred with beam waves at about 90° angles. Otherwise, magnitudes of roll were much lower.

6.1 Testing for Stability and Human Comfort

A main goal of the new infinity pool was to improve the comfort of the hotel's guests. Studies have shown vertical accelerations have a greater impact on people than lateral accelerations. A few studies [8, 14] show the greatest sensitivity to low-frequency motion is about 0.125–0.4 Hz.

The standard ISO 2631, Mechanical vibration and shock, gives some guidance to limiting vertical accelerations as a function of frequency and exposure. ISO 6897, Guidelines for the evaluation of the response of occupants of fixed structures, was also used to evaluate horizontal accelerations. As mentioned above, we note that ISO 6897 explicitly excludes floating structures, but it was believed that the standard's guidelines could be referenced in the project.

A very stable floating object, such as a heavy bulk carrier, has a very short roll and can quickly recover. Ironically, the quick recovery of stable ships is more likely to cause human discomfort. Generally, a slow 10-s roll is more comfortable than a quick 1-s roll. Cruise ships are therefore designed for slow rolls.

Criteria for human comfort for floating constructions is not well-established. Standards in the shipping and cruise industry typically use whole-body vibration limits, [18] which have frequencies beyond seen on a floating pool, or use a "susceptibility of vomiting" criteria, which is an unacceptably low standard for a luxury infinity pool.

A survey of the literature and relevant standards resulted in several thresholds to use:

- Suggested satisfactory magnitudes of horizontal motion for offshore fixed structures.
- Suggested satisfactory magnitudes of horizontal motion for buildings used for general purposes.
- Average threshold of perception of horizontal motion.
- Lower threshold of perception of horizontal motion.
- Average threshold of perception of vertical motion.
- Lower threshold of perception of vertical motion.
- Mansfield's "slightly nauseogenic" limit.

In confirmation to the client's concerns, the motions of the existing pool were found to exceed some thresholds, which explains how some guests experienced motion sickness.

It was determined that the best way to affect the stability of a floating platform were the main dimensions (length, width, draft) and mass distribution. As such, traditional naval architecture tools and techniques were adopted as these engineering solutions were found to be appropriate for floating, stationary structures. Damping solutions, such as the damping plates and moorings, were also important. Attempts to affect the waves, such as with breakwaters, was not feasible given the architectural objectives of the project.

6.2 Testing in Wave Tank

VTT's tank in Espoo, Finland, is 130 m long, 11 m wide and 5.5 m deep. It often uses a towing rig to test ship designs, but it also has a wave-maker for the offshore industry. The wave-maker was used to test the infinity pool. A 1:10 scale model of the proposed new design was constructed of stainless steel and the lakeshore model was built of wood (Fig. 11). The shoreline was approximately three times the width of the new infinity pool.

In total, 100 tests were performed with regular beam waves (90°), regular oblique waves (134°), and irregular waves (134°) (Fig. 12). The waves were based on the most demanding boat and wind waves as observed with the Lake Como buoy as well as computer modelling. In addition, decay tests were performed to determine the natural period of roll and the roll damping coefficient. Variations included:

Fig. 11 The scale model of the Lake Como shore used in the wave tank tests. The model accurately replicates all the arches and angles of the real shore

- Dampening plates.
- Gutters with different volumes.
- Elastic-type moorings.
- Hydraulic-type moorings.
- Wave dampers in the swimming pool.
- Buoyancy.

Testing was also performed on the infinity effect for the swimming pool. These tests included variations in the infinity overflow gutter and how it would perform in reaction to the waves. One possibility was to make the infinity edge a comb-like structure, with the water flowing through the grooves, which would break the surface tension and maintain the infinity edge effect. This would have lowered the power usage and made irregular sloshing less likely but would have negatively impacted the infinity illusion effect. Ultimately, the comb structure was found not to be necessary. Buoyancy additions under the pool and wave dampening plates in the swimming pool were also tested to see their impact on the infinity effect.

Fig. 12 Scale model test. The stainless steel infinity pool model is in front of the wooden shoreline model

6.3 Testing Conclusions

The roll damping plates, adapted from their use on boating piers, were found to be particularly effective. They decreased the roll about 40% when three plates were used on each corner. The moorings arms did not fit this project. The testing concluded that the existing pool exhibits motions that exceed the motion sickness lower threshold, but the new pool design remains mostly below this threshold, except for a very narrow region near 0.25 Hz. Therefore, the new pool was found to be much less likely to cause motion sickness, but it was impossible to guarantee no such incidents would ever occur (Table 2).

Table 2 Comparison of the movements of the existing pool and the new pool

	Existing pool	New pool	Difference
Longitudinal acceleration (mg)	3.7	2.4	37% improved
Lateral acceleration (mg)	3.3	2.7	18% improved
Vertical acceleration (mg)	6.6	3.5	47% improved
Pitch motion (deg)	0.134	0.094	30% improved
Roll motion (deg)	0.077	0.082	– 6% not improved

7 System Description

The new infinity pool solution is 600 square metres, 40 m by 15 m with a 1.72 m draft. It includes the pool, deck, pool infrastructure under the deck, catamaran hulls, weights, four 1,000-kg dampening plates on each corner and mooring system. It also includes the walkway leading to the pool structure from shore. The pool area is 39.3 m by 6.3 m, with a water depth of 0.23–1.20 m. The total volume is 173 cubic metres. The infinity effect runs along the entire 39.3 m of the long edge of the pool, around the corners, and back along the 6.3 m-short edges to the sundeck. The result is a pool with an infinity edge along three of the four sides.

The pool was constructed of steel with a coating technique similar to what is used on boats. The paint was the same as used on icebreaker vessels. Aluminium was used in the mooring system. Many of the architectural features were made with stainless steel or wood. The walkway and deck are covered with wood, while the inside surface and infinity edge of the pool is dressed in stone. Cardoso sandstone was selected because its colours closely complemented the colours of the lake. The sundeck is 352 square metres. The architects paid meticulous attention to details, such as requiring the electric sockets to be perfectly flush and the deck hatches completely hidden. Two open showers are installed at opposite ends of the sundeck.

The pool infrastructure is mostly located under the sundeck. A great deal of technology is packed into a tight space, including pumps, filters, electrical cabinets and wastewater tanks. The ballast tanks are operated manually. The pool technology is fully automated and can be controlled remotely. For example, the pool can be raised during the winter season, or it can be lowered to give a better view of the stone arches from the lake.

The mooring system turned out to be one of the biggest challenges. The architect wanted the moorings to be quiet and invisible, unlike the pile moorings in the old pool. Bluet arrived at a hybrid solution using existing underwater structures and 13-m underwater damped moorings typical to houseboats (Fig. 13).

8 Construction

The floating infinity pool was manufactured in Pori, Finland, where it was also preassembled. The floating structure was built in modules, which made it easier to transport to the busy, high-density area of Lake Como. There was some concern regarding the stresses the modules would face during assembly. The modules would have different weight and buoyancy distributions, which could cause flexing in the waves. It was decided to use factory welding instead of on-site welding where possible, because the environment was more controlled (Figs. 14 and 15).

The inland side of the site at Lake Como is densely built-up with narrow streets. The client preferred to avoid clogging the area with construction equipment, so the pool was assembled at a nearby location (Fig. 16) and towed to the hotel for

Fig. 13 Wave pool testing setup with the three dampening units in each corner of the pool model

Fig. 14 The pool structure being fabricated

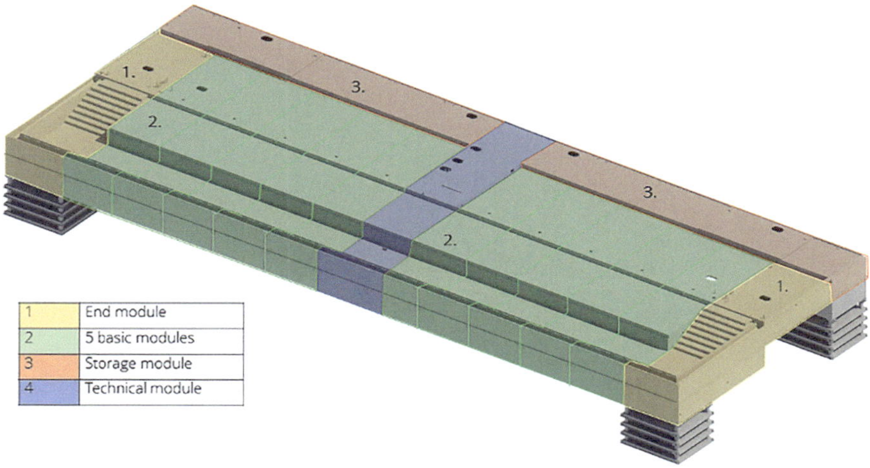

Fig. 15 The modules of the floating infinity pool

final assembly on the water (Fig. 17). The modules were transported to truck to the waterside. They were hoisted into the water and assembled (Fig. 18). Assembly was challenging due to buoyancy being different in different module types. This difference in buoyancy was resolved using separate weights or water filled tanks (Fig. 19, white tanks seen in the picture).

The logistical aspects of constructing and assembling the floating infinity pool added another layer of complexity. The pool was designed in Switzerland, constructed in Finland, and assembled in Italy, necessitating meticulous coordination across different countries and compliance with various regulatory standards. Transporting the prefabricated components and assembling them on-site required careful planning to ensure the integrity of the materials and the precision of the final structure.

Once installed, in the summer of 2023 the team once again installed video recorders and sensors on the pool and the wave buoy to verify if the infinity floating pool performed as anticipated. The results indicated that the structure was performing as expected and to design. The world's largest floating infinity pool opened early in 2024. It has turned into a tourist attraction, with tour guides pointing out the pool to guests on tour boats (Fig. 20).

Fig. 16 Como module installation. The suspended piece is part of a "basic module" (see Fig. 10)

Fig. 17 Como pool being towed. Once the floating infinity pool structure was assembled, it was towed to its location in front of the hotel

Fig. 18 Como assembly. After the main structure was safely moored the final assembly was completed

Fig. 19 During the assembly the difference in buoyancy between modules was resolved by using separate weights or water filled tanks (white tanks seen above)

Fig. 20 The new pool in use by hotel guests

9 Conclusion

The development of the world's largest floating infinity pool at the Mandarin Oriental Lago di Como underscores the growing viability of floating leisure infrastructure as an architectural and engineering solution. The project successfully integrated cutting-edge design, innovative engineering, and sensitive environmental considerations to deliver a luxury floating structure that harmonises the historic and natural setting of Lake Como. However, its execution revealed the unique challenges associated with floating leisure infrastructure, from regulatory ambiguity and structural stability to material selection and site-specific constraints.

A key lesson from this project is the necessity for early engagement with floating solutions experts, such as Bluet, to navigate the complex intersection of maritime engineering, architectural vision, and regulatory compliance. The successful outcome of the floating infinity pool was made possible by a collaborative approach, where designers, engineers and regulatory authorities worked together from the conceptual phase to final installation. Bluet's expertise in modular floating solutions, hydrodynamic assessments, and stability-enhancing technologies played a crucial role in mitigating risks, optimising structural integrity, and ensuring a seamless integration with the lake environment.

Ultimately, the floating infinity pool represents a milestone in floating leisure design, paving the way for further advancements in this emerging industry. With increased investment in research, regulatory clarity, and interdisciplinary collaboration, floating leisure infrastructure has the potential to become a mainstream solution in coastal waterfront developments worldwide. Early engagement with floating structure specialists, comprehensive site assessments, and a commitment to sustainability will be critical to ensuring the long-term success of these projects, bridging the gap between engineering feasibility, architectural ambition and regulatory acceptance.

References

1. Chandrasekaran S (2020) Offshore semi-submersible platform engineering. CRC Press, Boca Raton, FL
2. Davies C (2015) A history and celebration of swimming the River Thames. Aurum, London
3. International Organization for Standardization (1984) Guidelines for the evaluation of the response of occupants of fixed structures, especially buildings and off-shore structures, to low-frequency horizontal motion (0.063–1 Hz). International Standard, ISO 6897
4. International Organization for Standardization (1997) Mechanical vibration and shock—evaluation of human exposure to whole-body vibration—Part 1: General requirements, International Standard, ISO 2631-1
5. ITTC Recommended Procedures 7.5-02-05-04.1 (1999) Testing and extrapolation method: high speed marine vehicles—excerpt of ISO 2631, seasickness and fatigue
6. Kwok KC, Hitchcock PA, Burton MD (2009) Perception of vibration and occupant comfort in wind-excited tall buildings. J Wind Eng Ind Aerodyn 97(7–8):368–380
7. Lloyd ARJM (1998) Seakeeping: ship behaviour in rough weather
8. Mansfield N (2004) Human response to vibration. CRC Press, Boca Raton, FL
9. Milligan M (2020) The Nemi Ships were elaborate floating palaces built by Caligular in the 1st Century AD. Heritage. https://www.heritagedaily.com/2020/03/did-you-know-the-nemi-ships-were-massive-elaborate-floating-palaces-built-by-caligula-in-the-1st-century-ad/. Accessed 29 Mar 2020
10. Romer-Lee C (2023) Sea pools: 66 saltwater sanctuaries from around the world. Batsford Books, London
11. Scheu M, Matha D, Schwarzkopf MA, Kolios A (2018) Human exposure to motion during maintenance on floating offshore wind turbines. Ocean Eng 165:293–306
12. Slessor C (2019) A bigger splash: Britain's love affair with the swimming pool. The Guardian. https://www.theguardian.com/artanddesign/2019/jul/13/into-the-blue-history-swimming-pools-lidos-baths-britain-riba-v-and-a. Accessed 14 Jul 2019
13. Starr C (2024) Infinity pool by Herzog & de Meuron aims to "blend into the waters" at Lake Como. Dezeen. https://www.dezeen.com/2024/02/02/infinity-pool-herzog-de-meuron-lake-como/. Accessed 2 Feb 2024
14. Stevens SC, Parsons MG (2002) Effects of motion at sea on crew performance: a survey. Mar Technol SNAME News 39(01):29–47
15. Thames Estuary Growth Board (2021) Thames Baths (by Studio Octopi). https://thamesestuary.org.uk/thames-estuary-visionaries/thames-baths-by-studio-octopi/. Accessed 10 Feb 2021
16. Todini E (2022) On the benefits of collaboration between decision makers and scientists: the case of Lake Como. *Hydrology* 9(11):187
17. Van Oirschot D, Wallace S, Van Deun R (2015) Wastewater treatment in a compact intensified wetland system at the Badboot: a floating swimming pool in Belgium. Environ Sci Pollut Res 22(17):12870–12878

18. Vergassola G, Pais T, Boote D (2019) Low-frequency analysis of super yacht free vibrations. Ocean Eng 176:199–210
19. Wang CM, Wang BT (2006) Colonization of the Ocean and VLFS technology. Very Large Floating Structures, CRC Press, Abingdon, Oxon UK

Kimmo Kristian Saharinen is a Chief Engineer and Co-founder of Bluet Floating Solutions, having wide and unique, over two decades of experience in floating construction projects. He holds a Master of Science in Mechanical Engineering from the University of Oulu and a Bachelor Engineering degree from Wärtsilä Technical School. Kimmo's career highlights include technical design, engineering and supervision of floating construction projects on an international scale. Kimmo also leads research, and development works in terms of floating solutions, working with energy-efficiency and sustainability. Prior to Bluet, he worked as a Technical Design Engineer at Marina Housing and managed projects as CEO of his own company—Sahaki. Throughout his career, Kimmo has been a key player in iconic projects, such as the Mandarin Oriental floating heated pool at Lake Como, public floating pools in Sweden, geothermal floating pools in Iceland. His work reflects a strong commitment to innovation, quality coordination, and sustainable development in marine and urban environments. In the future Kimmo is looking forward to expanding his and Bluet's expertise and involvement in groundbreaking projects using floating megastructures, such as self-sustaining and independent floating islands.

Tytti Sirola with a visionary outlook on the niche but evolving industry of floating construction, co-founded Bluet Floating Solutions in 2016. Recognizing the unique potential of floating technologies already in 2013, she has led the company to specialize exclusively in the design and delivery of innovative floating solutions. Acting also as Bluet's CEO, her leadership has pioneered projects that blend functionality with sustainability, technical innovations and environmental consciousness, enabling executing both, public and private projects with high-profile end-solutions such as the Lake Como floating infinity pool, known for its uniqueness and seamless integration into the picturesque landscape with cutting-edge design. Throughout the years, Tytti has gathered experience from business and real estate development to project management, R&D and she also acts as a mentor to new entrepreneurs and growing startups in their global growth. Tytti's expertise and approach consistently demonstrates how collaboration over the academic and business world can coexist and with it is possible to develop innovations and creative solutions for the growing need of urban waterfront development. Her target is to be leading the industry towards a more dynamic and sustainable floating future, together with people sharing the same enthusiasm towards floating solutions.

Submersible Platforms for Offshore Marine Permaculture

Brian von Herzen, Huu Phu Nguyen, Chien Ming Wang, Chenxuan Huang, Perfecto Tubal, Eric Smith, Joseph Rauch, Sam Donohue, Lucy Schlueter, Sergey Ingram, and Theresa Theuretzbacher

Abstract This study presents a transformative approach to cultivate seaweed offshore through the development and deployment of advanced Marine Permaculture platforms. Emphasizing sustainability and ecological restoration, the research develops deep cycling architecture, which optimizes nutrient distribution by subtly transporting deep nutrients to cultivation zones through seaweed transport, thereby

B. von Herzen (✉)
The Climate Foundation, Woodford, QLD, Australia
e-mail: brian@climatefoundation.org

H. P. Nguyen · C. M. Wang · C. Huang
School of Civil Engineering, The University of Queensland, St Lucia, QLD, Australia

Blue Economy Cooperative Research Centre, Launceston, TAS, Australia

H. P. Nguyen
e-mail: huuphu.nguyen@uq.edu.au

C. M. Wang
e-mail: cm.wang@uq.edu.au

C. Huang
e-mail: chenxuan.huang@uq.edu.au

P. Tubal · E. Smith · J. Rauch · S. Donohue
The Climate Foundation, Compostela, Cebu, Philippines
e-mail: pepe@climatefoundation.org

E. Smith
e-mail: eric.smith@climatefoundation.org

J. Rauch
e-mail: rauch@climatefoundation.org

S. Donohue
e-mail: sam.donohue@climatefoundation.org

L. Schlueter
The Climate Foundation, Cincinnati, OH, USA
e-mail: LucyS@climateFoundation.org

S. Ingram
The Climate Foundation, Boulder, CO, USA
e-mail: Sergey@climatefoundation.org

B. T. Wang et al. (eds.), *Large Floating Solutions*, Lecture Notes in Civil Engineering 636, https://doi.org/10.1007/978-981-96-5435-2_13

enhancing seaweed growth and marine biodiversity. This novel Marine Permaculture design utilizes high-density polyethylene (HDPE) structures supported by ultra-high molecular weight polyethylene (UHMWPE) ropes, ensuring durability and flexibility in harsh marine environments. For improved resilience against extreme weather events such as hurricanes, the platforms use multi-point mooring systems, with modular and scalable SubTractor traction buoys that submerge temporarily in heavy waves. Field trials conducted in the Philippines demonstrate a substantial increase in seaweed productivity, enhanced carbon sequestration, and positive impacts on local marine ecosystems. Additionally, the integration of advanced monitoring technologies and renewable energy sources, like marine solar modules, underpins the operational efficiency and sustainability of these systems. The study also addresses critical regulatory challenges and explores future opportunities, including the synergistic co-location of seaweed farms with offshore wind turbines and the development of multitrophic aquaculture systems. By leveraging cutting-edge engineering and ecological principles, this research not only advances the state of offshore mariculture but also contributes to global efforts in climate mitigation, food security, and marine conservation. The findings underscore the potential of Marine Permaculture approaches for viable and scalable sustainable ocean management and economic development.

Keywords Mariculture · Seaweed · Platform · Offshore · Marine Permaculture · Seaweed cultivation · Submersible · Traction buoy · Deep cycling · Depth cycling

1 Introduction

Marine Permaculture is a sustainable and self-sufficient form of mariculture that mimics nature to create productive seascapes, regenerating seaweed forest habitat and other ecosystems in nearshore and offshore environments. Using forms of irrigation with deep water, essential nutrients can be absorbed at depth, enabling exemplary seaweed growth in conditions where seaweeds ordinarily struggle. Doing so enables large-scale, sustainable and continuous harvesting of seaweeds, while regenerating life in the ocean. Marine Permaculture addresses critical gaps in marine nutrient-value chains and supporting processes, primarily with forms of deepwater irrigation called deep cycling and regenerative upwelling. These can aid in restoring nutrients to nutrient-deprived surface waters destabilized from rising ocean temperatures, acidification, and deoxygenation.

As ocean temperatures rise, the stratification of the water column intensifies. Warmer surface waters form an energy barrier that inhibits the natural mixing of

T. Theuretzbacher
The Climate Foundation, Bergen, Norway
e-mail: theresa@climatefoundation.org

layers, preventing the nutrient-rich, deep waters necessary for the growth of phytoplankton, the foundation of the marine food web, from reaching the surface. The decline in phytoplankton growth cascades through the food web, affecting all marine organisms. Moreover, phytoplankton play a crucial role in sequestering carbon dioxide through photosynthesis and transport into the deep ocean when they die and sink. The decline in phytoplankton productivity weakens this carbon sequestration mechanism, reducing the ocean's effectiveness as a carbon sink and exacerbating the accumulation of carbon dioxide in the atmosphere and global warming.

The precipitous loss of over 2000 km^2 of kelp forests in Australia and North America alone over the past 100 years signals a considerable ecological crisis [31]. Climate change has wreaked havoc on the commercial seaweed industry. In the Philippines, the industry supports an estimated 200,000 families across the country, but it is increasingly fragile due to the rising temperature of coastal waters. In Indonesia, the sector is even more important. Indonesia produces more than a quarter of the world's seaweed, behind only China [13]. Seaweed forests also help regenerate biodiversity and sink carbon naturally. Their loss reduces the ocean's capacity to sequester carbon, further accelerating global warming in a feedback loop leading to additional kelp loss.

Offshore Marine Permaculture offers a sustainable solution for expanding seaweed production. By regenerating nutrient availability through regenerative upwelling and deep cycling, Marine Permaculture platforms can foster the accelerated growth of seaweeds, and other marine flora, which in turn support biodiversity. This restoration of primary productivity helps rebuild marine ecosystems, enhances carbon sequestration, and contributes to food security for coastal communities.

Sir James Lovelock, in *Novacene*, reasoned that the greatest threat to life on Earth is overheating [18]. Life on Earth has played a pivotal role in regulating the planet's climate over geological timescales. Through processes such as photosynthesis and the biological carbon pump, marine organisms have been instrumental in drawing down carbon dioxide from the atmosphere and storing it in the deep ocean [12]. This natural regulation has kept the Earth's climate stable and habitable. Marine Permaculture leverages this intrinsic capacity of life to moderate climate. By restoring and enhancing marine ecosystems, it amplifies the natural processes that cool the planet. Seaweeds and other marine plants absorb carbon dioxide during photosynthesis and produce oxygen, contributing to the reduction of greenhouse gases in the atmosphere. The regenerative upwelling and deep cycling techniques used in Marine Permaculture mimic natural ocean processes, ensuring nutrient-rich waters reach the surface to support continuous primary production. This not only sustains marine life but also enhances the ocean's ability to act as a carbon sink, playing a crucial role in global climate regulation.

Over the past decade, the Climate Foundation Australia has leveraged philanthropic funding to advance Marine Permaculture platform technology to Technology Readiness Level 7. These platforms are submersible and flexible structures that provide substrate upon which seaweed is affixed. Marine Permaculture platforms can grow multiple species of seaweed on the same line to enhance nutrient utilisation as different species of seaweed often have varying nutrient requirements and uptake rates.

2 Regeneration of Deepwater Nutrient Supply

2.1 Deep Cycling

Deep cycling of seaweed platforms represents an innovative approach to addressing nutrient deprivation in marine ecosystems. Unlike regenerative upwelling, which brings deep water to the surface, deep cycling focuses on optimizing nutrient availability by lowering and raising the seaweed platform within the water column. This method is potentially much more energy-efficient [28].

While regenerative upwelling involves pumps and nutrient delivery systems that raise nutrient-rich water from the deep, deep cycling lowers the seaweed to the nutrients, preferably at night, raising them near the surface in the morning. Both of these provide nutrients from deep waters closer to the surface, ensuring a daily supply of essential nutrients to different layers of the ocean. By strategically providing nutrient supply to the seaweeds, deep cycling systems supply nutrients that regenerate primary production by restoring deep nutrient supply.

The energy efficiency of deep cycling can exceed that of regenerative upwelling by more than an order of magnitude [21]. This efficiency makes deep cycling a more sustainable option for large-scale deployment, especially when integrated with renewable energy sources such as solar, wind or wave power. The significant energy efficiency of deep cycling translates into both environmental and economic benefits. The reduced energy requirements make it possible to operate these systems using renewable energy sources, thereby minimizing the carbon footprint of the operation. This efficiency also contributes to the economic viability of large-scale implementations, as operational costs are significantly reduced when compared to more energy-intensive methods.

Deep cycling is particularly effective in promoting the growth of macroalgae, such as seaweeds and kelp. These organisms thrive in nutrient-rich environments and can grow rapidly when provided with adequate nutrients. By maintaining a steady supply of essential nutrients, deep cycling systems can support extensive macroalgae cultivation, which has numerous ecological, environmental, and economic benefits [21]. For example, macroalgae play a crucial role in carbon sequestration, absorbing carbon dioxide during photosynthesis and storing it in their biomass. When harvested sustainably, macroalgae can be used for various applications that include biofuels, bioplastics, biostimulants, and animal feed, providing renewable alternatives to fossil fuels and reducing greenhouse gas emissions. Additionally, the cultivation of macroalgae supports marine biodiversity by creating habitats and food sources for marine organisms.

The advantages of deep cycling extend beyond energy efficiency and macroalgae production. The system's versatility allows for deployment in various ocean environments, from coastal waters to open ocean settings. Its scalability makes it suitable for both small-scale experimental setups and potentially hectare-scale commercial operations. Furthermore, deep cycling minimizes disruption to natural ocean dynamics, as it works with existing currents and mixing patterns rather than introducing entirely new water movements. However, there are considerations to keep in mind when implementing deep cycling systems. While they provide excellent support for macroalgae growth, they may offer less direct fish habitat creation when compared to regenerative upwelling. This trade-off needs to be carefully weighed in areas where fish habitat restoration is a primary goal.

2.2 Regenerative Upwelling

While deep cycling has become the primary focus for Marine Permaculture due to its energy efficiency, regenerative upwelling remains a valuable tool in specific contexts, particularly for habitat restoration or regeneration offshore. Regenerative upwelling emulates natural upwelling processes, bringing nutrient-rich deep waters to the surface to invigorate primary production. The system typically involves submersible pumps or wave-driven devices anchored to floating platforms or buoys. These are strategically placed in nutrient-poor regions to maximize their impact. The pumps draw water from depths of 200 m or more, where nutrient concentrations are significantly higher. This nutrient-rich water is then released near the surface, creating a localized area of enhanced productivity. The influx of nutrients stimulates the growth of phytoplankton, which forms the base of the marine food web.

Regenerative upwelling has shown promise in habitat restoration efforts, especially for ecosystems such as kelp forests and coral reefs [29]. These habitats are heavily dependent on nutrient availability, and their decline is often attributed to nutrient scarcity [24], especially nitrogen in the Indo Pacific Ocean. In kelp forests, regenerative upwelling can create conditions conducive to the growth of kelp and associated marine life. By supplying the necessary nutrients, these systems promote the regeneration of kelp beds, which in turn provide shelter and food for a variety of marine species and act as carbon sinks. The restoration of kelp forests can have cascading ecological benefits, enhancing biodiversity and stabilizing coastal ecosystems [3]. For coral reefs facing severe bleaching due to warming waters, the influx of cooler, nutrient-rich water can help alleviate thermal stress and support the growth of symbiotic algae (which are crucial for coral health) and aid in the recovery of these vital ecosystems.

However, regenerative upwelling does have limitations that have led to the preference for deep cycling in many applications. The primary drawback is its higher energy requirements when compared to deep cycling. This increased energy demand can make large-scale implementation less economically viable and potentially less sustainable if not powered by renewable energy sources. Additionally, regenerative upwelling can potentially cause greater disruption to natural ocean dynamics compared to deep cycling. The introduction of large volumes of deep water to surface layers may have unforeseen consequences on local ecosystems if not carefully managed [16]. The implementation and maintenance of regenerative upwelling systems can also be more complex, requiring more robust infrastructure to handle the continuous pumping of water from great depths.

The carbon sinking potential of upwelling has been questioned by some researchers concerned that the deep carbon brought up in deep waters could offset the carbon being sunk through increased primary productivity. Global modeling [15] shows this concern is unfounded, at least in subtropical contexts, with upwelling more than doubling the effectiveness of the biological pump under high emissions scenarios, increasingly likely given present global trajectories continuing high emissions, suggesting that more than 4 gigatonnes of CO_2 per year can be fixed with upwelling under high emissions scenarios.

In sum, regenerative upwelling remains a valuable tool, particularly for targeted habitat restoration. The development of deep-cycling technology adds to this restoration of the pre-industrial biological pump with an energy-efficient approach for scaled mariculture operations. Selection among these methods depends on the project goals, local environmental conditions, and available resources for implementation and maintenance. By leveraging these techniques, ocean health and productivity can be regenerated, thereby ensuring a sustainable future for marine life and human communities alike.

3 Reflections on History of Offshore Marine Permaculture Platforms

Researchers and engineers have been working on creating specialized infrastructure for offshore seaweed cultivation since the Marine Biomass Program in the 1970s. This program was the first documented commercial-scale offshore seaweed cultivation programme. The 1/4-acre structure took the form of an inverted umbrella-like structure with a buoyed upwelling pipe in the centre of the structure to irrigate the seaweed with cold, nutrient-rich deep seawater. The six 'umbrella bones' were spread around the upwelling pipe, and the grow-out lines of seaweed were attached to the "umbrella bones" as shown in Fig. 1. The full life cycle of perennial kelp on the structure was about four years [4, 27]. However, the results of the experiments on the structures were extremely disappointing, with significant damage to infrastructure and extremely slow growth rates of the kelp. Efforts were made to strengthen the structure, but they all failed. It was realised later in 1981 that the marine environment offshore California, where the structures were tested, was considered too harsh for offshore seaweed cultivation at the time.

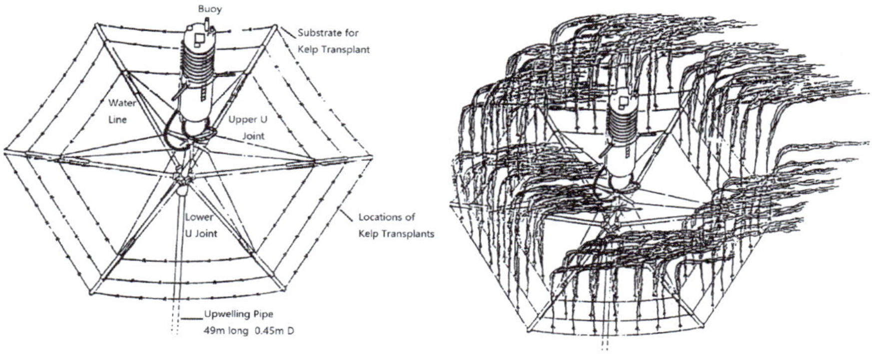

Fig. 1 Inverted umbrella structure in Marine Biomass Program [27]

In 2010, Seaweed Energy Solutions AS (SES) in Norway tested the Seaweed Carrier for commercial-scale kelp cultivation (see Fig. 2) [26]. The seaweed carrier is a single mooring system with a vertical cultivation net which is 6.5 m long and 5 m deep. The kelp grows on the submerged net. The design of the structure facilitates the harvesting of seaweeds. However, the structure has not been used for commercial seaweed cultivation due to high initial investments, operational costs and market prices of seaweed.

In the 2010s, a new kelp cultivation infrastructure called BioArchitecture Lab was trialled in Chile [4]. The cultivation system was designed like a square grid and composed of twenty-one 1-ha modules (100 m × 100 m) with 99 horizontal lines, each of 100 m (see Fig. 3). The best production yield for each module reached an average of 124 tonnes per hectare annually and the cost of operation is equivalent to that of wild, locally harvested kelp [4]. Although the system is the most commercially promising attempt, it has yet to be commercially operated [4]. Ocean Rainforest has

Fig. 2 Seaweed carrier [26]

Fig. 3 BioArchitecture Lab system and mature seaweed on one line (provided by Javier Infante, Patagonia Seaweeds)

conducted similar trials ending 2024 in the Santa Barbara Channel as part of the US DOE MARINER program, also at pre-commercial scale.

In 2012, Ecofys and Hortimare in The Netherlands proposed a submersible seaweed-growing farm as shown in Fig. 4 [9]. The entire structure is 120 m long and 10 m wide and is formed by a series of 'H' shaped frames arranged in two dimensions. Each 'H' shaped frame comprises a horizontal element connected to two vertical columns that act as SPAR buoys that have small holes in the pipework through which allow free passage of seawater. When the sea becomes harsh and the wave height increases, more seawater enters the buoys through the holes, causing the whole structure to sink. When the sea becomes calm, the excess seawater in the buoys flow out through the holes, causing the SPAR buoys to float. During deployment, a section of the structure detached, leading to the flooding of the H-frame's pipe. Consequently, the system submerged in January 2013 [4].

More recently, a team from the United States demonstrated a new seaweed farming system that can be thought of as a lattice mooring geometry as shown in Fig. 5 [20]. The mooring system allows for large, horizontally stable, continuous arrays of modular components. Each module has a size of 140 m by 70 m. The sea trials of the proposed system enhanced understanding of the structure's survivability and operability in offshore conditions. However, it was highlighted that better accessibility, enhanced tension control, and more precise anchor placement are needed [20].

4 Deep Cycling Architecture for Marine Permaculture

4.1 Design and Structure

The deep cycling architecture for Marine Permaculture represents a culmination of innovative engineering and ecological principles. These platforms are designed to be modular, scalable, and adaptable to various ocean environments, ensuring their

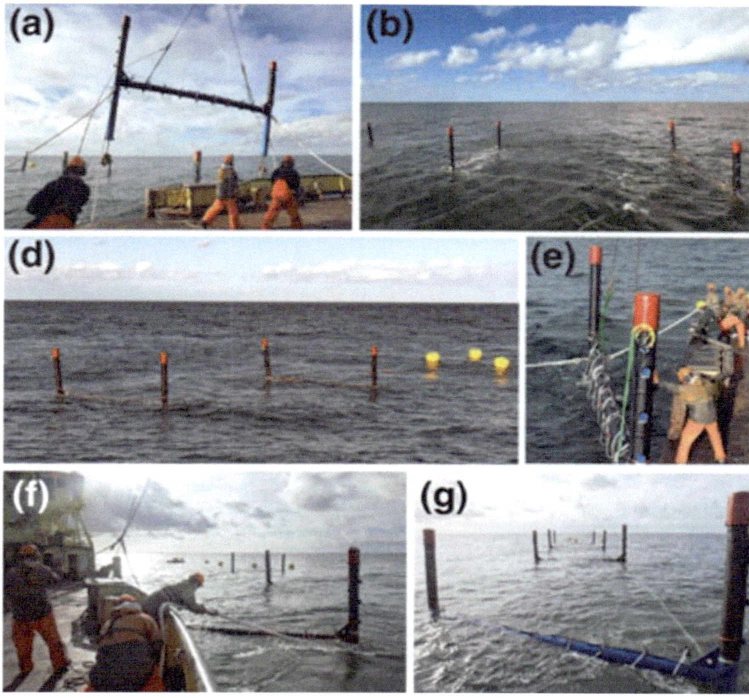

Fig. 4 **a–f** Installation of H-frame infrastructure offshore off Texel [9]

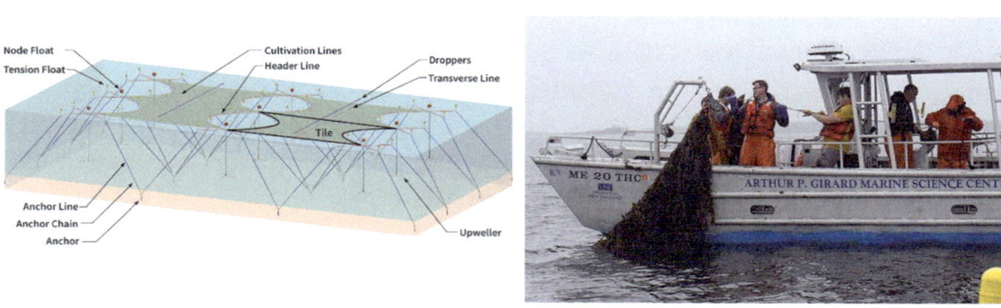

Fig. 5 structure drawing and harvest picture of kelp cultivation lattice mooring structure with 4 modular "tiles" [20]

effectiveness across different geographical and oceanographic conditions (see Fig. 6 as an example).

At the heart of the design is a floating structure, typically constructed using high-density polyethylene (HDPE) pipes. This material choice is crucial, as HDPE offers excellent durability, resistance to corrosion, and flexibility—all essential qualities for withstanding the harsh marine environment [10, 22]. The floating structure provides

Fig. 6 Concept of Marine Permaculture platforms

the necessary buoyancy and serves as the foundation upon which other components are mounted.

Attached to this main structure are the seaweed cultivation lines. These lines serve as the substrate for seaweed growth and are carefully designed to maximize growth potential while minimizing stress on the seaweed. The arrangement of these lines is optimized to ensure an even distribution of nutrients and sunlight for uniform growth across the platform.

The pumping and nutrient delivery system is a critical component of the deep cycling architecture. This system facilitates the movement of nutrient-rich water from deeper layers to the cultivation area. Unlike traditional upwelling systems, the deep cycling approach moves water more subtly, creating nutrient gradients that mimic natural oceanic processes. This method not only conserves energy but also minimizes disruption to the surrounding ecosystem.

The anchoring system is another crucial element, designed to keep the platform in place while allowing for some movements to accommodate ocean currents and waves. Depending on the specific deployment location and conditions, the anchoring system may use traditional mooring line and end plate anchors or incorporate more dynamic positioning technologies.

A key innovation in the deep cycling architecture is the use of ultra-high molecular weight polyethylene (UHMWPE) ropes. These ropes possess extraordinary strength and durability. They can withstand the constant stress and strain of ocean forces, resisting wear and tear far better than traditional materials. This durability translates into longer operational lifespans for the platforms, reducing maintenance needs and improving overall cost-effectiveness. The flexibility of UHMWPE ropes allows for easy installation and adjustment, especially for deep cycling operations.

4.2 Supporting Systems and Advanced Monitoring

Effective monitoring and control systems are the backbone of successful Marine Permaculture operations, especially when integrating SubTractor buoys into platform design. SubTractor buoys are new modular, redundant traction buoys that enable precise vertical movement of Marine Permaculture platforms, enhancing resilience and operational efficiency (see Sect. 6; Figs. 12, 13 and 14). These systems ensure that platforms function optimally, respond to changing environmental conditions, and maintain the health of cultivated seaweed.

A network of sensors is deployed across the platform and SubTractor buoys to continuously measure key environmental and operational variables. These include environmental parameters such as temperature, salinity, nutrient levels, pH, and dissolved oxygen, as well as structural parameters like tension in mooring lines, buoy performance, platform orientation, and traction mechanism status. Data collected by these sensors is transmitted to a central control unit, forming a real-time feedback loop that allows for dynamic responses to environmental changes.

The central control unit utilizes sophisticated algorithms to adjust platform depth based on nutrient levels and surface conditions, optimizing nutrient uptake and photosynthesis. The system can automatically raise or lower the platform using the SubTractor buoys, coordinating the traction mechanisms to enhance nutrient uptake across different platform areas. In anticipation of storms, the system can initiate protective measures, preemptively submerging buoys and adjusting configurations for maximum resilience.

Advanced technologies including machine learning enhance predictive capabilities, allowing for proactive maintenance and optimization of growth conditions. Predictive modeling based on rich data can help anticipate potential issues and guide interventions, further improving the efficiency and sustainability of Marine Permaculture operations.

Integration with the SubTractor buoys enhances supporting systems by providing distributed monitoring and redundancy. Each buoy serves as a monitoring node, increasing the spatial resolution of data collection. The redundancy ensures that sensor failures do not critically impact overall monitoring capabilities, and real-time diagnostics enable targeted maintenance actions, minimizing downtime.

4.3 Hurricane Resilience and Operational Efficiency

One of the most significant challenges facing offshore marine structures is their ability to withstand extreme weather events, particularly hurricanes. The ocean traction buoy architecture developed for Marine Permaculture platforms represents a groundbreaking approach to enhancing storm resilience while maintaining operational efficiency. SubTractor buoys play a crucial role in this aspect.

The ability of the traction buoys to adjust their positions dynamically within the tensegrity structure allows the platform to submerge to depths where wave energy is substantially reduced. As hurricane-force winds and waves approach, the system can lower the platform beneath the most destructive surface forces. The multi-anchored, multi-point mooring system enables the platform to maintain stability at depth, with the tensioned ropes and anchor weights ensuring that the structure remains secure despite strong currents and turbulence.

During Category 5 hurricanes and typhoons, such as Super Typhoon Rai in December 2021, a Marine Permaculture platform equipped with SubTractor buoys in the Philippines successfully weathered the storm. By submerging to just 5 m beneath the surface, the platform avoided significant damage, protecting both the infrastructure and the seaweed crops. In contrast, nearby shallow water seaweed cultivation suffered complete losses, highlighting the effectiveness of the SubTractor system. A few months later, we were able to provide neighbouring communities with seedlings from a Marine Permaculture® platform to regenerate their seaweed cultivation.

Maintaining stability at depth is critical for operational efficiency. Dynamic positioning systems equipped with automated control algorithms play a vital role in maintaining the platform's orientation and position. These systems use a combination of the platform's own buoyancy controls and the traction mechanisms of the SubTractor™ buoys to counteract currents and maintain the desired depth and alignment. Intermediate ballast weights passively provide the tension in mooring lines and connecting ropes needed to compensate for changes in buoyancy and movement, preventing slack that could lead to tangling or excessive strain.

The design of the platform itself incorporates protective housings and covers for sensitive equipment to ensure that vital components can continue to function even when submerged. These housings are designed not only to be waterproof but also to withstand the increased pressure at a few meters' depth when duck-diving larger waves.

SubTractors contribute to operational efficiency through reduced energy consumption. The efficient traction mechanisms lower the energy required for depth adjustments, utilizing buoyancy differences and mechanical advantage. The ability to adjust to environmental conditions allows for continuous operation in a wider range of scenarios, extending operational windows and improving crop yields by optimizing depth for nutrient access and sunlight exposure.

For SubTractor buoys to survive hurricane conditions, they need to be small and light with managed levels of buoyancy and need to survive periodic submergence ("duck diving") in steep and large waves. These are key to avoid coupling too much mechanical energy into the submerged seaweed platform. By submerging the platform a few meters at least, the seaweeds continue to get light but can survive the ravages of the storm. By limiting the size and buoyancy of each SubTractor buoy, system resilience can be ensured.

The integration of SubTractor buoys into Marine Permaculture platforms marks a significant evolution in offshore seaweed cultivation. By addressing previous challenges related to single points of failure and mechanical complexity, these modular redundant traction buoys enhance resilience, operational efficiency, and scalability.

The ability to raise and lower platforms with precision, combined with advanced monitoring and control systems, ensures that Marine Permaculture can adapt to changing conditions and withstand extreme weather events.

5 Climate Foundation Use of Deep Cycling Marine Permaculture Infrastructure

In recent development of offshore seaweed cultivation infrastructure, Climate Foundation trials in the Philippines in the 2020's have provided compelling evidence of the effectiveness of submersible Marine Permaculture platforms. These trials, conducted on platforms ranging from 100 to 1000 square meters, have demonstrated significant success in increasing nutrient levels, boosting seaweed growth rates, and enhancing overall marine ecosystem health. The 1000-square-meter platform is shown in Fig. 7 while Fig. 8 shows the plan and side views of the platform. Table 1 presents the material and geometrical properties of the various components of the platform and the mooring system.

The circular platform is constructed by two high-density polyethylene (HDPE) pipes connected by evenly spaced brackets. This material is commonly used for floating collars in open-net flexible fish pens [25]. By recycling HDPE pipes from

Fig. 7 Overhead drone photo of 1000 m^2 seaweed cultivation infrastructure with single-point mooring system tested in the Philippines in 2023. Photo courtesy of Eric Smith, Climate Foundation

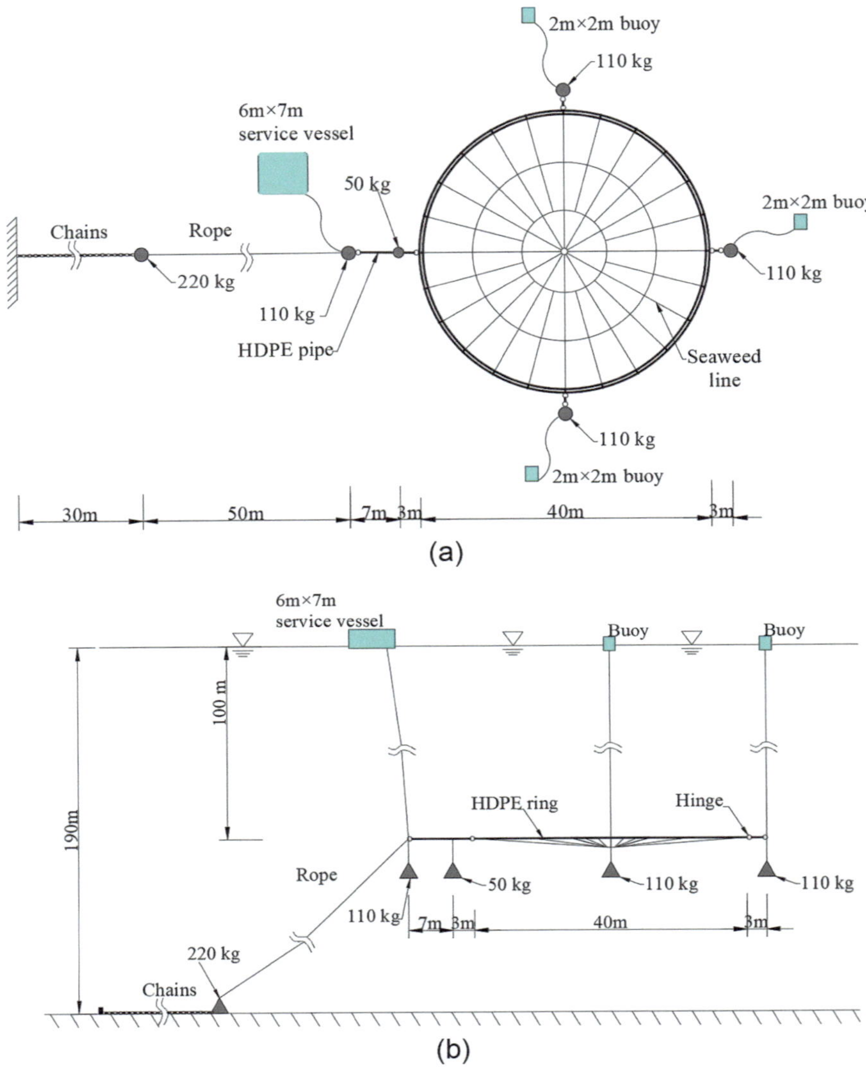

Fig. 8 Climate Foundation early design of a single-point submersible seaweed platform: **a** plan view, **b** side view in submerged state

fish farms and employing current fabrication methods, HDPE structures can be constructed for seaweed cultivation, improving waste management and reducing costs. HDPE pipes and connector systems are also being used in frame-typed fish pens, such as the Marine Donut [6] and the SeaFisher [30]. This HDPE circular ring structure supports seaweed grow-out lines and also act as sidewalks, allowing workers safe access from support vessels docked alongside.

Table 1 Design parameters of seaweed platform

Components	Description	Magnitude and unit
HDPE pipes, SDR17	Diameter of outer ring	40.5 m
	Diameter of inner ring	40 m
	Cross-section diameter of outer ring	0.225 m
	Cross-section diameter of inner ring	0.16 m
	Young's modulus	900 MPa
	Poisson's ratio	0.4
	Mass density	958 kg/m^3
Polypropylene ropes, anchor chains	Mooring rope diameter	0.04 m
	Seaweed line (or rope) diameter	0.02 m
	Rope Young's modulus	1500 MPa
	Rope mass density	910 kg/m^3
	Anchor chain diameter	0.032 m
	Chain Effective Young's modulus	54.4 GPa
	Anchor chain mass	16 kg/m
Cylindrical buoy	Buoy diameter	2 m
	Buoy height	2 m
Tube nets	Diameter of tube nets	7.5 cm
	Spacing between tube nets	1.2 m
	Submergence depth	0 to 100 m
	Mass density of seaweed	1030 kg/m^3

The test platform adopts a single-point mooring system consisting of a wire, rope and chain to keep it in place [5]. The abrasion resistant chain connects the wire rope to the end plate anchor. The single-point mooring system minimizes the number of mooring lines and anchors and minimizes the environmental loads on the platform structure and mooring forces [19]. Structures with this mooring system can be detached more easily, allowing them to be relocated efficiently.

The platform was utilized for cultivating red seaweeds such as *Gracilaria dura*, *Kappaphycus alvarez*ii, and *Eucheuma*. To protect the young seaweeds from strong environmental forces, they are placed inside tube nets, which are then attached to the seaweed grow-out lines. The initial seeded tube nets have a diameter of 75 mm, based on the standard tubenet arrangement for *Eucheumatoid* red seaweed cultivation. The spacing between the seaweed tube nets averages 0.5–1.2 m for adequate sunlight and nutrient supply.

The platform can be submerged to depths of up to 150 m at night to immerse the seaweed in cooler, nutrient-rich waters, but must be raised to the surface during the day for photosynthesis. The submergence is done by using a system of buoys, ropes and winches installed on the buoys. By submerging the platform, the platform's responses can be reduced significantly, increasing its survivability in storms. This is because the dynamic wave pressure is largest at the free surface, and decreases monotonically towards the seabed. For example, even for the 8-s maximum typical wave period and a water depth of 200 m, typical for most extreme site conditions for the Camotes Sea in the Philippines, the dynamic wave pressure decreases by 70% at 20 m, and decreases 95% at 50 m depth compared to surface wave pressure [11].

PV solar panels on the roof of the service vessel parked next to the seaweed platform (as shown in Fig. 7), each provide ~200 W of clean, renewable energy for operations and maintenance works.

Hydroelastic analyses of the platform floating on the surface and at a submerged depth of 100 m were performed to study the deformation and the stress distributions of the platform under wave action. The analyses were carried out by using the software AquaSim. In the AquaSim model, the HDPE pipes were modelled by beam elements and the mooring and grow-out lines were modelled by truss elements [23]. The surface vessels and the buoys were modelled by node-to-node spring elements.

Figure 9 shows the locations where the stresses and displacements are measured. Figure 10 shows the vertical displacement amplitude of the platform at different submergence depths for two wave conditions: (a) 2 m wave height, 4 s wave period, (b) 4 m wave height, 8 s wave period. Figure 11 presents the deformed shapes of the platform and von Mises stress distributions in floating and various submerged states in 2 m wave height, 4 s wave period. Furthermore, numerical results from the structures depicted in Figs. 13 and 14 indicate that the displacements and stresses in the platform are reduced significantly at the submerged state when compared to the floating state at the water surface.

The hydroelastic responses of the platform were observed to correspond to the trial platform responses in the Philippines. With the validation of the mathematical modelling, analysis and software, future optimisation of the platform and mooring configurations can be carried out.

One of the most striking outcomes from the trials in the Philippines has been the platforms' ability to support seaweed growth in areas where warming waters had

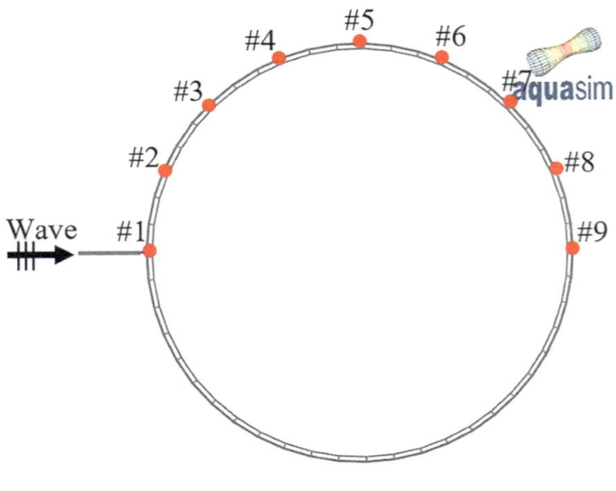

Fig. 9 Locations where stresses and displacements are computed

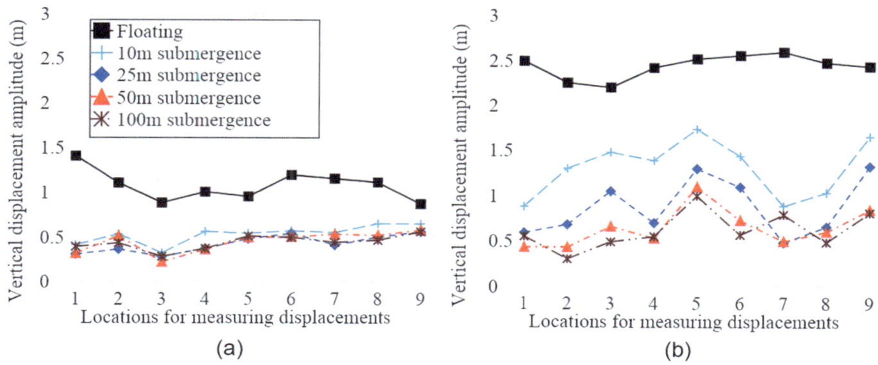

Fig. 10 Vertical displacement amplitudes of platform for different submergence depths: **a** 2 m wave height, 4 s wave period, **b** 4 m wave height, 8 s wave period

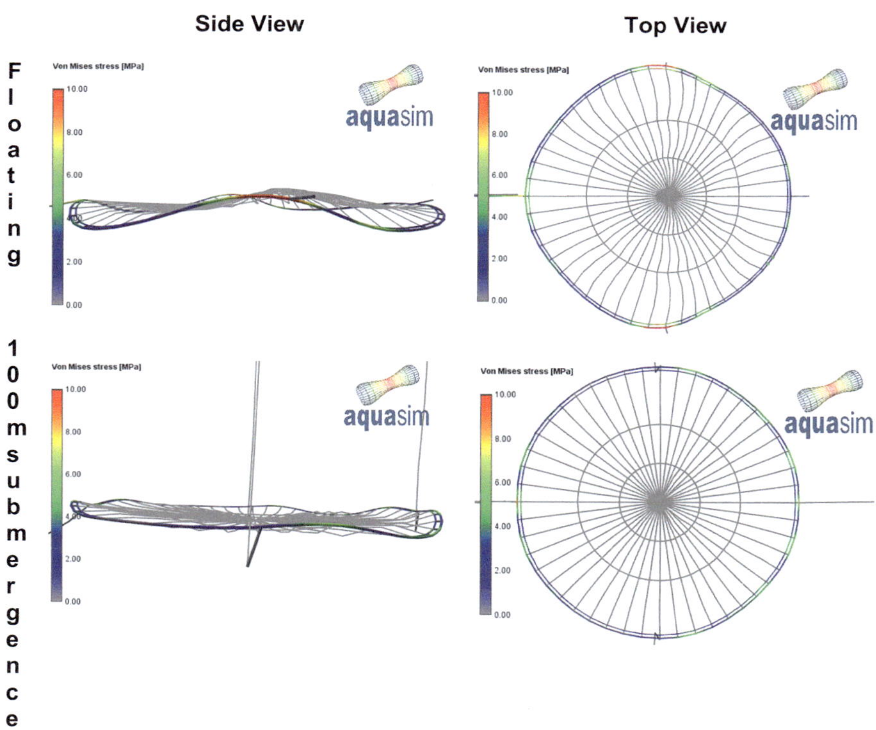

Fig. 11 Deformed shape of platform in floating and submerged conditions in 2 m wave height, 4 s wave period

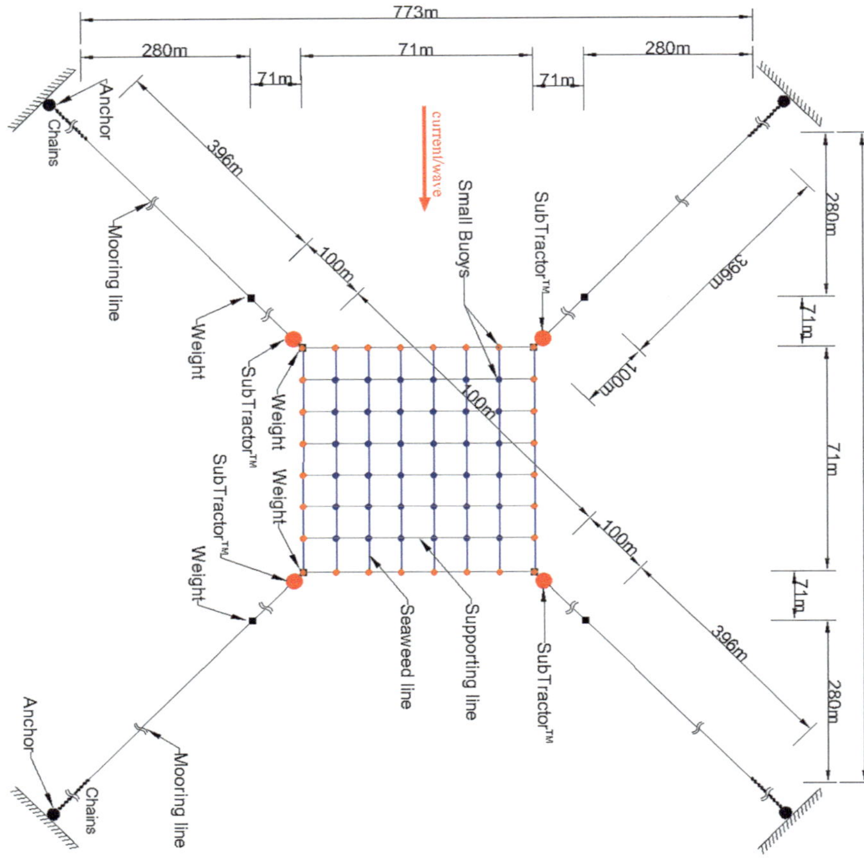

Fig. 12 Plan view of a hectare-sized Marine Permaculture platform with SubTractor buoys positioned around the perimeter, illustrating the multi-point mooring system

previously made cultivation challenging or impossible. The deep cycling technology effectively mitigated the impacts of elevated surface temperatures by providing a consistent supply of cooler, nutrient-rich water to the cultivation area.

Measurements taken during these trials showed seaweed growth rates on the Marine Permaculture platforms exceeding those in control areas by 100–300%. This dramatic increase in productivity not only validates the effectiveness of the deep cycling approach but also points to its potential for significantly boosting seaweed production in regions affected by climate change.

The platforms also demonstrated remarkable resilience in the face of extreme weather events. During the trial period, a Category 5 typhoon passed through the area. The ability of the platforms to submerge to a depth of 5 m allowed them to weather the storm with minimal damage, protecting the seaweed crop and the

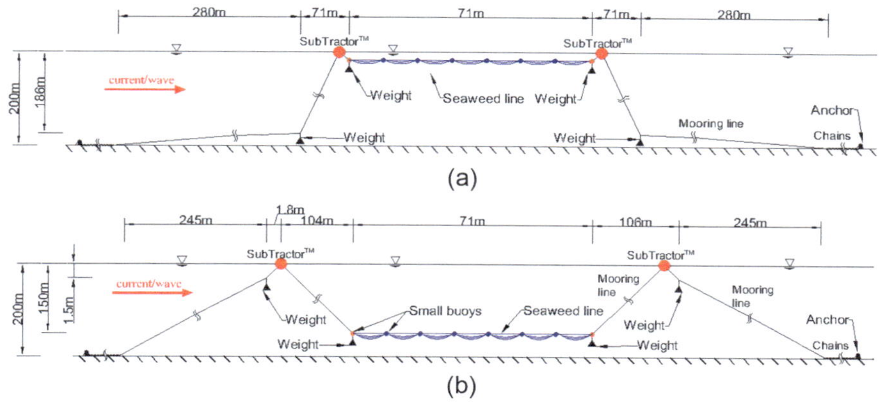

Fig. 13 Elevation view of a SubTractor buoy system, showing the traction mechanism, attachment points to the platform, and connections to anchor weights: **a** floating state, **b** submerged state

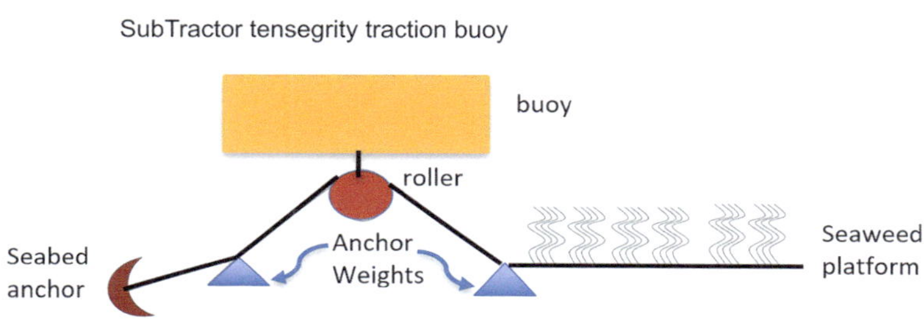

Fig. 14 The SubTractor buoy, similar to a cable ferry, utilizes ropes and a motor to raise and lower the platform in the water. Much in the way that a cable ferry is pulled back and forth across a river by a cable attached to a motor on the shore, the SubTractor buoys are pulled back and forth along the ropes by a motor inside the buoy, resulting in the raising and lowering of the seaweed platform

infrastructure itself. This resilience is a crucial factor in the viability of large-scale offshore seaweed cultivation in regions prone to tropical storms.

Beyond the direct benefits to seaweed cultivation, the trials revealed positive impacts on the surrounding marine ecosystem. Increased fish populations were observed around the platforms, suggesting they function as artificial reefs, providing habitat and food sources for various marine species. This unexpected benefit underscores the potential of Marine Permaculture to contribute to broader marine conservation efforts.

The trials also provided valuable insights into the operational aspects of Marine Permaculture. The modular design of the platforms proved to be advantageous, allowing for easy scaling from the initial 100 m^2 to 1000 m^2. This scalability is crucial for the future expansion of Marine Permaculture to hectare-sized implementations and beyond.

Engagement with local communities was another key aspect of the Philippine trials. The project team worked closely with local seaweed farmers and fishers, integrating their knowledge and addressing their concerns. This collaborative approach not only improved the design and operation of the platforms but also helped build local support for the project. The potential for job creation and economic diversification was well-received by community members, highlighting the social benefits that could accompany large-scale implementation of Marine Permaculture.

5.1 Harvesting and Processing

The Climate Foundation Marine Permaculture platform is designed for efficient and sustainable harvesting and processing of seaweed biomass. The red seaweed varieties cultivated on the platform, such as *Gracilaria dura*, *Kappaphycus alvarezi*, and *Eucheuma*, are typically harvested on a 45-day cycle to optimize yield and quality.

Harvesting is increasingly automated, reducing labour costs and ensuring consistent product quality. Specialized machinery can automatically harvest mature seaweed while seeding apical tips immediately back into the tube net structure, promoting continuous production [17]. The harvested seaweed is then transported to a nearby processing facility.

The Climate Foundation has developed a transportable, containerized seaweed biorefinery that enables affordable on-site processing, reducing transportation costs and preserving the freshness of the seaweed. This biorefinery utilizes a cold-press technology to separate the seaweed into liquid and solid fractions.

The liquid fraction, rich in bioactive compounds, is further processed into a biostimulant. This organic, cold-pressed biostimulant can be used in agriculture to enhance crop growth and yield. The solid fraction is dried and processed into various products, including food, feed, and fertilizer ingredients. This integrated biorefinery approach maximizes the value of the seaweed biomass and supports a circular economy model.

The distributed nature of these biorefineries allows for local processing and value addition, creating economic opportunities for coastal communities. By providing the infrastructure and technology for sustainable seaweed cultivation and processing, the Climate Foundation Marine Permaculture platform supports the development of resilient and regenerative blue economies.

5.2 Regulatory Frameworks

As with any new technology, there inevitably comes a time when governing bodies aim to set guidelines of regulation. With submersible platforms for seaweed cultivation comes a unique set of challenges. The largest being should these cultivation platforms be considered underwater structures or vessels. There are benefits and negatives to each mode of classification. With the key differences outlined in Table 2.

Should these systems be classified as vessels or structures? This key distinction will determine how regulators manage offshore seaweed mariculture systems in the future.

Environmental impacts are another important regulatory concern. The Climate Foundations Marine Permaculture systems have been shown to regenerate local biodiversity. Deep cycling has fewer environmental concerns than upwelling. In particular, upwelling has been suggested to potentially release carbon. The hypothesis here is that the deep water pumped to the surface also contains deep carbon—while extensive testing still needs to be done to validate any hypothesis on this subject, this claim lacks evidence. The evidence actually points in the opposite direction, the extra carbon brought up during upwelling is not more than the deep nutrients needed to stimulate significantly more carbon capture, and subsequent sinking by macroalgae and phytoplankton involved in the biological carbon pump. Some models even indicate that in a high carbon emissions scenario the solubility pump is equally effective as the biological pump, suggesting that any carbon brought up will certainly be returned to the deep [15]. However, further testing needs to be performed in certain local and regional settings as the microbial populations vary and require localized regimes and interventions. Writ large, regulatory agencies need to perform local experiments to determine the correct policy for their respective marine regions.

Table 2 Advantages and disadvantages of vessel classification and fixed permit structure classifications

Traits	Vessel advantages	Vessel disadvantages	Fixed permit structure advantages	Fixed permit structure disadvantages
Mobility	Free movement of the platform is not prohibited	Dynamic or other vessel positioning is required	Fixed location	Movement of the platform is not freely allowed
Licensing	Position does not need to be negotiated	X	X	Negotiated position with stakeholders, a lengthy process
Repairs and maintenance	System can be repaired both at sea and port	X	X	System must usually be repaired at sea
Operation	X	Requires more complex automation, guidance and/or control	Does not require steering	Requires other vessels to move the platform

6 Latest Developments and Future Directions

6.1 *Advancing Marine Permaculture Research and Technologies*

As Marine Permaculture continues to evolve and expand, several key areas of research and development will shape its future trajectory. Optimizing platform designs for different oceanic conditions, exploring new seaweed applications, and developing advanced monitoring and management systems will be critical for unlocking the full potential of this technology.

Continued research into the optimal design of Marine Permaculture platforms for various environments will be essential. While the basic principles of deep cycling and modular construction are broadly applicable, tailoring platform configurations to specific local conditions can maximize efficiency and productivity. This optimization may involve adjusting the size and spacing of cultivation arrays, fine-tuning the depth and frequency of cycling, or incorporating features to mitigate location-specific challenges such as biofouling or predation.

6.1.1 Marine SubTractor Buoys: Enhancing Platform Operation and Mechanics

The ability to raise and lower Marine Permaculture platforms is a critical feature of deep cycling architecture. Vertical mobility not only optimizes nutrient delivery and adapts to varying oceanographic conditions but also enhances the resilience of the platforms against extreme weather events. Traditional systems relied on buoyancy control or mechanical winches to adjust platform depth, but these methods often introduced single points of failure and operational inefficiencies.

The Climate Foundation has invented, designed and produced arrays of marine SubTractor buoys—modular, redundant traction buoys—which marks a significant advancement in platform operation and mechanics, addressing these challenges and bringing a new level of efficiency and reliability to Marine Permaculture systems.

The use of a multi-anchored, multi-point mooring system enables the formation of a tensegrity structure—a resilient and flexible balance of tension and compression elements—that allows the traction buoys to move in and out from a set of weights. By using the seafloor for horizontal compression and the buoyancy of the buoys for vertical tension in designs less than 500 m deep, this design enables the raising and lowering of the weighted rope array under tension as needed, providing precise control over the platform's vertical position. The traction buoys are connected to the platform and to anchor points via high-strength ropes (e.g. UHMWPE), forming a network that maintains structural integrity while allowing flexibility in movement.

The architectural design of SubTractor buoys centers on modularity, redundancy, and scalability. Each SubTractor buoy is equipped with a traction mechanism that permits it to move horizontally along the mooring line, adjusting the relative lengths of line from the anchor to the Subtractor buoy, and from the Subtractor buoy to the growth platform, allowing the length adjustments to raise and lower platform since the Subtractor buoy remains substantially on the surface. Each SubTractor buoy is powered entirely by onboard solar panels, generating up to 200W on sunny days and storing energy in 1kWh batteries. This renewable energy system enables continuous operation without external power sources. By increasing the number of smaller-sized buoys and integrating a multi-anchored, multi-point mooring traction structure, the system eliminates single points of failure and enhances overall resilience. This renewable energy system enables continuous operation without external power sources. By increasing the number of smaller-sized buoys and integrating a multi-anchored, multi-point mooring traction structure, the system eliminates single points of failure and enhances overall resilience.

In earlier designs, for example, the 1000 m^2 platform tested in 2023, the reliance on only two lifting winches meant that if one failed, immediate action was required to prevent the loss of seaweed productivity. The seaweed had to be raised while there was enough light left to enable the crop to photosynthesize; otherwise, the entire seaweed crop could be at risk. The updated design incorporates multiple SubTractor buoys—typically eight buoys with two spares for a hectare-sized array or four SubTractor buoys with one spare for a sub-hectare array. With this configuration, the failure of a single buoy affects only a small portion of the platform (12.5% or less), allowing most of the seaweed to continue its normal growth cycle.

Smaller buoys offer several advantages. Their reduced size allows them to "duck-dive" beneath strong waves and winds during hurricanes, minimizing the force exerted on the platform. By limiting their buoyancy, the amount of energy transferred from the buoy to the platform is decreased, reducing the risk of structural damage during extreme conditions. Additionally, smaller buoys are cheaper to manufacture, easier to handle, and more economical to ship and replace, enhancing the overall feasibility of large-scale deployment.

Figures 12 and 13 show the plan view and side views of the hectare-sized Marine Permaculture with SubTractor buoys deployed in Cebu waters in the Philippines.

The implementation of SubTractor buoys introduces several operational enhancements that improve the efficiency and resilience of Marine Permaculture platforms. The ability to raise and lower the platform is achieved through the coordinated action of the traction buoys within the tensegrity structure. By adjusting the tension in the ropes connected to the anchor weights, the platform can be submerged to access nutrient-rich deeper waters at night and raised to the surface during the day for photosynthesis (see Fig. 13). This method provides precise control over depth adjustments, enabling fine-tuned positioning without relying extensively on mechanical systems or high energy input. In fact, a key advantage of the SubTractor buoy modular architecture is its minimal energy requirements, aligning with the Climate Foundation's commitment to sustainable practices. Each buoy operates on less than 1 kWh of renewable energy per day, making the system highly energy-efficient and suitable for integration with renewable energy sources such as marine solar and wind power. Assuming an expected 8 SubTractor buoys per hectare and production of up to 8 tonnes of wet seaweed per hectare per week, that would imply renewable energy of approximately 7 KWh per wet metric tonne of seaweed (WMT), and total energy on the order of 10 kWh per WMT.

Figure 14 illustrates the traction mechanism around a SubTractor buoy, showing how vertical movement is achieved through tension adjustments in the ropes connected to anchor weights on both sides of the SubTractor buoy.

The multi-anchored, multi-point mooring system distributes mechanical loads evenly across the platform, reducing stress concentrations and extending the lifespan of structural components. The tensegrity structure maintains stability and integrity under various ocean conditions, allowing the traction buoys to adjust their positions dynamically in response to environmental forces. This flexibility is crucial during extreme weather events, such as hurricanes, where the ability to submerge and minimize exposure to surface conditions can prevent catastrophic damage.

Redundancy is a key feature of the SubTractor system. With multiple buoys in operation, the failure of a single unit does not compromise the entire platform. If one SubTractor buoy becomes inoperative, only a small fraction of the platform is affected, and the affected area remains limited to 12.5% for corner failures under low currents or potentially less for side-buoy failures. This redundancy allows for routine maintenance and hot-swap capability, where faulty buoys can be replaced swiftly without disrupting the entire system's functionality.

Logistics and maintenance are simplified due to the modular design and smaller size of the SubTractor buoys. Replacement buoys are easier to transport and deploy, reducing downtime and operational costs. Figure 15 shows the process of a hot swap, where a faulty SubTractor buoy is replaced with a spare without disrupting platform operations in several steps. First, the swap vessel takes up some slack in the tension line, then the new buoy is inserted, engaging the traction rope. Next, the faulty buoy is disengaged from the tension rope, followed by releasing the slack on both sides, and the faulty buoy is taken to shore. The modularity also allows for flexible configuration and scalability, enabling the system to be customized according to specific site conditions, cultivation practices, or expansion plans.

SubTractor buoys also enhance the monitoring and control capabilities of the platform. Each buoy is equipped with sensors and communication systems providing command, control, communications and telemetry, allowing for real-time monitoring of buoy performance, environmental conditions, structural integrity and system health. This distributed monitoring network increases the spatial resolution of data collection and enhances the system's ability to respond dynamically to changing conditions and report on any system errors they may need immediate intervention.

This system, coupled with the existing Fig. 14 traction system, allows for precise control of the platform's depth. The Fig. 14 system guides the horizontal movement of the SubTractor buoys, causing vertical displacement of the seaweed platform. The combined action of these two systems allows for efficient and controlled kelp farming operations.

Expanding the range of seaweed species cultivated on Marine Permaculture platforms is another promising avenue for future research. While initial trials have focused on a handful of well-known species, the incredible diversity of seaweeds presents a wealth of untapped potential. Identifying and domesticating new species

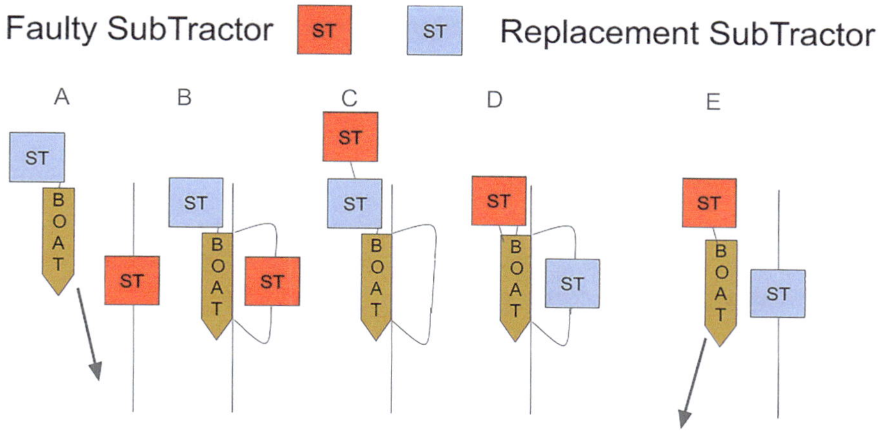

Fig. 15 Operational schematic depicting the process of replacing a faulty SubTractor buoy without affecting platform operations. **a** The BOAT vessel approaches the SubTractor for replacement. **b** BOAT accumulates slack rope in the tension line; **c** old SubTractor buoy is removed; **d** the replacement SubTractor buoy is inserted, engaging the traction rope. **e** The slack line is released from the BOAT, and the faulty buoy is taken to shore

with desirable traits such as rapid growth, high nutrient content, or unique chemical properties could open up new markets and applications. There are over 12,000 species of seaweed, many of which are known to be edible and often flavourful. Collaborative research with marine biologists, geneticists, and bioprospecting experts will be key to unlocking this potential.

Developing advanced monitoring and data analysis tools will be critical for optimizing the performance and environmental benefits of Marine Permaculture platforms. Integrating cutting-edge sensor technologies, remote sensing, and machine learning algorithms can provide unprecedented insights into the complex dynamics of these artificial ecosystems. Real-time monitoring of nutrient levels, seaweed growth rates, and biodiversity indicators can inform adaptive management strategies and help operators respond rapidly to changing conditions. Predictive modeling based on this rich data can also help anticipate potential issues and guide proactive interventions.

Exploring new applications for seaweed biomass is another exciting frontier for Marine Permaculture. While initial commercial efforts have focused on relatively straightforward products such as food, feed, and fertilizers, the potential uses of seaweed are vast and diverse. Seaweed-derived compounds are showing promise in fields ranging from pharmaceuticals and nutraceuticals to advanced materials and biofuels. Partnering with industry and academic researchers to develop and commercialize these innovative applications will be crucial for maximizing the economic and environmental benefits of Marine Permaculture.

As Marine Permaculture expands, integrating it with other sustainable aquaculture practices will become increasingly important. Developing synergistic multi-trophic aquaculture systems that combine seaweed cultivation with the rearing of fish, shellfish, or other marine organisms can enhance overall productivity and sustainability. These integrated systems can leverage the natural nutrient cycling and bioremediation capabilities of seaweeds to mitigate the environmental impacts of animal aquaculture while providing additional revenue streams.

6.2 Further Integrating Marine Permaculture with Offshore Energy Infrastructure

On a related front of ocean developments, co-location/integration of seaweed mariculture and offshore wind has received a lot of attention. This interest is motivated by the rapid development in offshore wind. The estimated total potential of offshore wind power is approximately 15 TW for water depths up to 200 m, and 5.5 TW for depths up to 50 m [33]. By the end of 2023, the total cumulative installed capacity of offshore wind has exceeded 75 GW, representing a 15% increase compared to one year earlier, and an over 1000% increase compared to a decade earlier [14].

Possible integrated infrastructure systems for seaweed farming and offshore wind turbines are presented in Fig. 16. In these systems, the offshore wind turbines also serve as part of the mooring system for circular seaweed platforms constructed from HDPE pipes. This approach may lower the costs for mooring systems because they can be shared for station keeping of wind turbines and seaweed cultivation platforms. However, as the wind turbines need to carry additional loading due to the seaweed cultivation infrastructure, some checks and modifications are needed to ensure the structural safety of wind turbines.

In order to avoid the need for modifications of the wind turbines which have been standardized and commercialized, seaweed cultivation infrastructure can be placed in between turbines, but with separate mooring systems as shown in Fig. 17. In a survey in Germany, most stakeholders supported in principle the idea of co-location of seaweed farms and offshore wind farms, but more research and sea trials are needed to gain better understanding of technological, biological, economic, and legal aspects of such a co-located farm [32].

Wave and wind energy can also provide potential energy sources for upwelling and even deep cycling-such work has been conducted over the past two decades by researchers such as Philip Kithil and others [1, 2]. Marine Permaculture provides a basis for seasteading communities to be deployed increasingly offshore as the maricultures themselves move further offshore. Storm-resilient architectures such as those proven with submersible seaweed platforms will be key to sustainability and longevity.

In conclusion, the future of offshore seaweed cultivation is brimming with opportunities for innovation, expansion, and positive impact. By continuing to push the boundaries of research and development, forging interdisciplinary partnerships, and

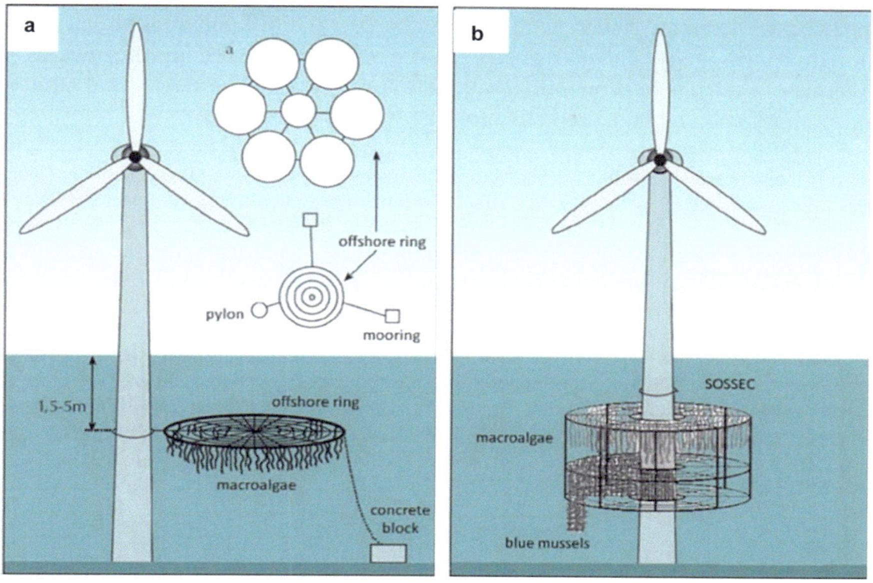

Fig. 16 Integrated aquaculture—wind turbine. **a** Ring system, **b** Integrated mussel-seaweed system [8]

Fig. 17 Co-located seaweed-mussel-wind farm [7]

engaging with diverse stakeholders, we can harness the full potential of this transformative technology. The path ahead is not without challenges, but the rewards—in terms of sustainable economic development, ecosystem restoration, and climate change mitigation and repaired climate—are well worth the effort.

Acknowledgements The authors acknowledge the financial support of the Blue Economy Cooperative Research Centre, established and supported under the Australian Government's Cooperative Research Centres Program, grant number CRC-20180101. The authors would like to thank Aquastructures AS for granting access to their novel software AquaSim (https://aquasim.no). The Climate Foundation development of Marine Permaculture and deep cycling has been supported from generous grants from the Woka Foundation, CO_2 Foundation and several other family offices. The Climate Foundation wishes to extend sincere gratitude for these contributions and for additional contributions from our supporters. The authors also wish to acknowledge the hard work of numerous contributors, including Steven Hamilton and the entire Marine Permaculture Philippines team in proving out the deepwater structures developed to date, including Jorito Rosal, Jayme Pepito, Regie Sanchez, Aldrin Agujar, Danly Lao, Manuel Rubi, Rene Barbon, John Mark Yangaw, John Jay Libre, Erwin Avila, Alejandro Conde, Cheryl Seno, Karizah Xyrah A. Seno, Joselito Ngujo, Ricky John Bandolon, Leonilo Bacalso.

References

1. Arvizu D, Bruckner T, Chum H, Edenhofer O, Estefen S, Faaij A, Fischedick M, Hansen G, Hiriart G, Hohmeyer O, Hollands KGT, Huckerby J, Kadner S, Killingtveit Å, Kumar A, Lewis A, Lucon O, Matschoss P, Maurice L, Mirza M, Mitchell C, Moomaw W, Moreira J, Nilsson LJ, Nyboer J, PichsMadruga R, Sathaye J, Sawin J, Schaeffer R, Schei T, Schlömer S, Seyboth K, Sims R, Sinden G, Sokona Y, von Stechow C, Steckel J, Verbruggen A, Wiser R, Yamba F, Zwickel T (2011) Technical summary. In: Edenhofer O, Pichs-Madruga R, Sokona Y, Seyboth K, Matschoss P, Kadner S, Zwickel T, Eickemeier P, Hansen G, Schlömer S, von Stechow C (eds) IPCC special report on renewable energy sources and climate change mitigation. Cambridge University Press, Cambridge, United Kingdom and New York, NY, USA
2. Arvizu et al (2011) Summary for policy makers: intergovernmental panel on climate change Special Report Renewable Energy Sources (SRREN). https://www.osti.gov/biblio/1050647. Accessed 9 June 2025
3. Babcock RC, Egli DP, Attwood CG (2023) Long-term marine protection enhances kelp forest ecosystem stability. Ecol Appl e2895
4. Bak UG, Gregersen Ó, Infante J (2020) Technical challenges for offshore cultivation of kelp species: Lessons learned and future directions. Bot Mar 4:341–353
5. Barltrop NDP (1988) Floating structures: a guide for design and analysis, vol 2. Oilfield Publications
6. Bluegreen (2023) Marine Donut. https://bluegreengroup.no/en/services/marine-donut. Accessed 15 Nov 23
7. Buck BH, Troell MF, Krause G, Angel DL, Grote B, Chopin T (2018) State of the art and challenges for offshore Integrated multi-trophic aquaculture (IMTA). Front Mar Sci 5:165
8. Buck BH, Langan R, Krause G, Pogoda B, Grote B, Wever L, Goseberg N, Schupp MF, Mochtak A, Czybulka D (2017a) The German case study: pioneer projects of aquaculture-wind farm multi-uses. In: Aquaculture perspective of multi-use sites in the open ocean: the untapped potential for marine resources in the Anthropocene. Springer

9. Buck BH, Nevejan N, Wille M, Chambers MD, Chopin T (2017b) Offshore and multi-use aquaculture with extractive species: seaweeds and bivalves. In: Aquaculture perspective of multi-use sites in the open ocean: the untapped potential for marine resources in the Anthropocene. Springer
10. Chu YI, Wang CM, Park JC, Lader PF (2020) Review of cage and containment tank designs for offshore fish farming. Aquaculture 519:734928
11. Dean RG, Dalrymple RA (1991) Water wave mechanics for engineers and scientists. World Scientific Publishing, Singapore
12. Elsler LG, Oostdijk M, Levin LA, Satterthwaite EV, Pinsky ML, Crespo GO, Wisz MS (2022) Protecting ocean carbon through biodiversity and climate governance. Frontiers in Marine Science 9:880424
13. FAO (2021) Seaweeds and microalgae: an overview for unlocking their potential in global aquaculture development
14. GWEC (2024) Global offshore wind report
15. Jürchott M, Oschlies A, Koeve W (2023) Artificial upwelling—a refined narrative. Geophys Res Lett 50(4):e2022GL101870
16. Levin LA, Le Bris N (2015) The deep ocean under climate change. Science 350(6262):766–768
17. Lewis N (2022) An Indian startup could revolutionize ocean farming with its 'sea combine harvester.' CNN Business. https://edition.cnn.com/2022/01/05/business/india-sea6-energy-seaweed-harvester-spc-intl/index.html
18. Lovelock J (2020) Novacene: the coming age of hyperintelligence. Penguin Books
19. Mobron E, Torgersen T, Zhu S, Riis J, Bye M (2020) Design of Havfarm 1. In: Proceedings of the Second World Conference on Floating Solutions, Rotterdam, The Netherlands
20. Moscicki Z, Swift MR, Dewhurst T, MacNicoll M, Chambers M, Tsukrov I, Fredriksson DW, Lynn P, Landon ME, Zotter B, MacAdam N (2024) Design, deployment, and operation of an experimental offshore seaweed cultivation structure. Aquacult Eng 105:102413
21. Navarrete IA, Kim DY, Wilcox C, Reed DC, Ginsburg DW, Dutton JM, Heidelberg J, Raut Y, Wilcox BH (2021) Effects of depth-cycling on nutrient uptake and biomass production in the giant kelp Macrocystis pyrifera. Renew Sustain Energy Rev 141:110747
22. Nguyen HP, Wang CM, von Herzen B, Huang C (2023) Hydroelastic responses of a submersible ring structure for offshore seaweed cultivation under wave action. J Mar Sci Eng 11:22238
23. Nguyen HP, Wang CM, von Herzen B, Huang C (2024) Hydroelastic responses of submersible seaweed cultivation platforms with single-point mooring systems. In: Proceedings of the ASME 2024, 43rd International Conference on Ocean, Offshore and Arctic Engineering, OMAE2024-124766. Singapore EXPO, Singapore
24. Polovina JJ, Howell EA, Abecassis M (2008) Ocean's least productive waters are expanding. Geophys Res Lett 35(3)
25. Shen Y, Greco M, Faltinsen OM, Nygaard I (2018) Numerical and experimental investigations on mooring loads of a marine fish farm in waves and current. J Fluids Struct 79:115–136
26. Solvang T, Bale ES, Broch OJ, Handå A, Alver MO (2021) Automation concepts for industrial-scale production of seaweed. Front Mar Sci 8:613093
27. Tompkins AN (1979) GRI marine biomass program annual report. Prepared for the Gas Research Institute
28. Tullberg RM, Nguyen HP, Wang CM (2022) Review of the status and developments in seaweed farming infrastructure. J Mar Sci Eng 10:1447
29. Upwelling (2024) Upwelling: coastal ecosystem cooling. https://www.upwelling.blue/. Accessed 31 Jul 24
30. Wang CM, Chu YI, Baumeister J, Zhang H, Qiao YP, Karampour H, Jeng DS, Savage L (2023) SeaFisher—a submersible high-density polyethylene offshore fish pen. J Mar Sci Eng 11:1795
31. Wernberg T, Bennett S, Babcock RC, De Bettignies T, Cure K, Depczynski M, Dufois F, Fromont J, Fulton CJ, Hovey RK, Harvey ES, Holmes TH, Kendrick GA, Radford B, Santana-Garcon J, Saunders BJ, Smale DA, Thomsen MS, Tuckett CA, Tuya F, Vanderklift MA, Wilson S (2016) Climate-driven regime shift of a temperate marine ecosystem. Science 353:169–172

32. Wever L, Krause G, Buck BH (2015) Lessons from stakeholder dialogues on marine aquaculture in offshore wind farms: perceived potentials, constraints and research gaps. Mar Policy 51:251–259
33. World Bank (2023) Global offshore wind technical potential. https://datacatalog.worldbank.org/search/dataset/0037787/Global-Offshore-Wind-Technical-Potential. Accessed 18 Jan 23

Dr. Brian von Herzen is the founder and Executive Director of the Climate Foundation, where he leads the development of Marine Permaculture technology to regenerate ocean ecosystems and mitigate climate change locally and regionally. He oversees community seaweed mariculture programs to create sustainable food, feed, and fertilizer value chains while supporting blue carbon sinks. Brian holds a degree in Physics from Princeton University (magna cum laude) and a Ph.D. in Planetary Science from the California Institute of Technology, where he was a Hertz Fellow. After two decades of work with companies like Intel, Disney, Xilinx and Microsoft, he founded the Climate Foundation to develop nature-based solutions to environmental challenges. His current work includes Marine Permaculture demonstration projects starting in the Philippines, advancing seaweed-based carbon sequestration and ecosystem regeneration.

Dr. Huu Phu Nguyen is a postdoctoral research fellow at School of Civil Engineering, The University of Queensland. His research interests are in the areas of ocean energy converters and large floating structures. He has developed numerical methods for analysing the interaction between ocean waves and several marine structures including large floating platforms, wave energy converters, and floating breakwaters. He currently works on the hydrodynamic analysis of floating structures with porous elements and developing platforms for seaweed mariculture.

Prof. Chien Ming Wang is a Professor of Structural Engineering at The University of Queensland. He is a Fellow of the Australian Academy of Technology and Engineering, a Member of the European Academy of Sciences and Arts, a Fellow of Academy of Engineering Singapore, a Fellow of the Institution of Structural Engineers, a Fellow of the Institution of Engineers Singapore and a Fellow of the Society of Floating Solutions Singapore. His research interests are in the areas of structural mechanics and floating structures and has published over 520 journal papers and 10 books on these subjects. He is the Leader of the Offshore Engineering Program of the Blue Economy Cooperative Research Centre that conducts research projects that combine seafood, renewable energy and offshore engineering, underpinned by a $329 million grant from the Australian Government and industry partners over a 10-year period. He is a founding member and Engineering Science Leader of the International Engineering Science Consortium, Vice President of the Society of Floating Solutions Singapore and the Emeritus Chairman of East Asia Pacific Conference on Structural Engineering and Construction steering committee. He has won many awards that include the 2019 Nishino Medal, 2019 JN Reddy medal, Singapore's Minister of National Development's R&D Award, IStructE Singapore Structural Awards, Keith Eaton Award, Lewis Kent Award, IES Prestigious Engineering Achievement Award, IES/IStructE Best Paper Award and the Grand Prize of the Next Generation Container Port Challenge. He was the consultant and advisor on many structural and floating projects that include the world's largest floating performance stage at Marina Bay, multi-purpose floating structures research project funded by the Land and Liveability National Innovation Committee and JTC Corporation, Maritime and Port Authority's project on uses of the underground space underneath container ports in Tuas, JTC project on floating hydrocarbon storage facility and HDB project on floating wetlands. He is currently conducting research on the next generation offshore fish pens and seaweed cultivation platforms.

Chenxuan Huang is a Ph.D. student in Civil Engineering at The University of Queensland. She holds a Master's degree in Civil Engineering from Nanyang Technological University and a Bachelor's degree in Civil Engineering from the China University of Mining and Technology. Her current research focuses on the design of offshore seaweed cultivation infrastructure, as well as the integration and colocation of seaweed mariculture with renewable energy farms. She aims to develop innovative approaches to utilizing marine resources, contributing to improved global food security and the reduction of atmospheric carbon emissions.

Perfecto Tubal is an aquaculture specialist with the Climate Foundation, bringing over 30 years of experience in aquaculture and seaweed cultivation. As a co-founder of the Climate Foundation's Marine Permaculture initiatives in the Philippines, Perfecto has been instrumental in advancing sustainable solutions for ocean and agricultural systems. He holds a degree in Aquaculture from Cebu Technological University and has held diverse roles throughout his career, including Prawn Farm Technician and District Technical Manager at Oversea Feeds Corp. Perfecto's expertise extends internationally, having managed seaweed mariculture and export operations in Zanzibar and fish cage projects in Zambia. Since joining the Climate Foundation in 2017, he has focused on developing biostimulants from seaweed and integrating aquaculture with agriculture to promote sustainable practices. His work exemplifies a commitment to innovation and environmental stewardship, contributing to the regeneration of marine ecosystems and the development of sustainable livelihoods.

Eric Smith is a passionate marine biologist with extensive experience in ecosystem regeneration across diverse locations, including India and the Philippines. Devoted to sustainable mariculture and Marine Permaculture, he empowers coastal communities to adapt to climate change. At the Climate Foundation, Eric drives the advancement of innovative seaweed mariculture techniques to regenerate healthy marine ecosystems and support sustainable livelihoods. His work exemplifies the integration of science and community-driven solutions in ocean conservation.

Dr. Joseph Rauch is a physicist specializing in biophysics and offshore aquaculture infrastructure. After earning his doctorate in Physics with a focus on biophysics from Brandeis University, he started with the Climate Foundation in engineering, software development, and project management. Over the past five years, he has developed offshore aquaculture systems in the Philippines to protect the industry from the effects of climate change. Joseph aims to expand these efforts globally to ensure food security, ecosystem survival, and carbon balance through innovative Marine Permaculture systems.

Sam Donohue is the EPC (Engineering, Procurement, and Construction) Manager at the Climate Foundation, bringing over 20 years of experience in mechanical engineering, maintenance, and sustainable technology. He has been instrumental in designing and constructing deep-sea ocean structures to support Marine Permaculture seaweed cultivation modules, helping to develop a scalable prototype recognized by the $1 M USD X Prize for sustainable ocean solutions. Before joining the Climate Foundation in 2020, Sam held leadership roles in service management and engineering at companies such as Outotec and AGC, where he honed his expertise in mechanical systems and project delivery. At the Climate Foundation, he focuses on advancing sustainable aquaculture and carbon sequestration through the development of submersible seaweed platforms. Sam's work exemplifies the integration of engineering and sustainability to create impactful solutions in ocean conservation and beyond.

Lucy Schlueter is a Climate Foundation Fellow with a background in chemistry and environmental science. She earned a Bachelor of Arts in Chemistry, graduating Magna Cum Laude from

Washington University in St. Louis, where she also minored in French and Dance. With experience in clinical research and science communication, Lucy contributes to the Climate Foundation's Marine Permaculture initiatives, integrating scientific research with practical applications to address climate change and regenerate healthy ecosystems.

Sergey Ingram, a Climate Fellow at the Climate Foundation, holds a Bachelor's degree in Biotechnology from the University of Manchester and is a Ph.D. candidate in Bioengineering at Utah State University. His research focuses on seaweed-based solutions to climate change, including developing hydrogen peroxide-containing organelles in seaweeds for innovative applications. Sergey has experience in laboratory biochemistry, grant writing, project management, and mentoring STEM students. He is passionate about advancing Marine Permaculture technologies to address climate change and regenerate healthy ecosystems, supporting sustainable development worldwide.

Theresa Theuretzbacher specializes in biomimicry-inspired engineering solutions that benefit both local ecosystems and economies. With hands-on experience addressing water, waste, and sanitation challenges in Ghana, Alaska, and DR Congo, she integrates environmental, social, and economic factors into her work. As part of the Climate Foundation, she contributes to the development of Marine Permaculture (MP) in the USA, Philippines, and Indonesia. Her current focus includes developing methodologies to measure seaweed carbon sequestration from MP platforms and advancing cascading seaweed value streams. Theresa holds a Master's in Water Management and Environmental Engineering from the University of Natural Resources and Life Sciences, Vienna.

Design and Fabrication of Floating Artificial Reefs: Enhancing Biodiversity and Food Production

Dan Luo and Valeriya Komyakova

Abstract The expansion of aquaculture and renewable energy infrastructure into offshore environments offers a unique opportunity to integrate ecological innovation and engineering expertise to address critical environmental and economic challenges. This chapter explores the development of floating artificial reefs (FARs) as a novel approach to enhance high-value food production, support marine biodiversity, and provide ecosystem services while mitigating the environmental impacts of offshore infrastructure. By examining the historical evolution and ecological benefits of artificial reefs, this chapter highlights the advancements in design, including state-of-the-art technologies such as modular construction and 3D printing, that enable FARs to function effectively as standalone habitats or co-located systems alongside offshore energy projects. Regulatory and governance challenges associated with deploying FARs are examined, emphasizing the need for multidisciplinary collaboration among engineers, ecologists, policymakers, and industry stakeholders. Case studies and research gaps identified by the Blue Economy CRC are provided to illustrate the potential for FARs to transform aquaculture and offshore development practices, and support sustainable marine industries. By combining ecological restoration and technological innovation, FARs present an opportunity to balance economic growth with environmental stewardship, paving the way for a more resilient and sustainable blue economy.

Keywords Artificial reefs · Offshore infrastructure · Floating structure · Blue economy

---

D. Luo (✉)
School of Architecture, Design and Planning, University of Queensland, Brisbane, Australia
e-mail: d.luo@uq.edu.au

V. Komyakova
Centre for Marine Socioecology, University of Tasmania, Hobart, TAS, Australia

Oceans Institute, The University of Western Australia, Crawley, WA, Australia

B. T. Wang et al. (eds.), *Large Floating Solutions*, Lecture Notes in Civil Engineering 636, https://doi.org/10.1007/978-981-96-5435-2_14

1 Introduction

In recent decades, the rapid expansion of offshore infrastructure and coastal development, driven by population growth and the increased exploitation of marine resources, has significantly impacted coastal ecosystems. This expansion has exacerbated conflicts over marine resource use and accelerated the degradation of critical coastal habitats [10, 21, 22, 66]. In response, a strategic shift has emerged to locate infrastructure further offshore, aiming to protect nearshore biodiversity hotspots and address the global decline in marine ecosystems. Concurrently, declining food security necessitates a renewed policy focus on sustainable food and energy production with minimal environmental impacts [60].

Australia's high-value aquaculture sector presents an unmet need for diversifying low-environmental-impact marine products suitable for offshore cultivation. Currently, the industry is heavily reliant on reef-based species, leaving significant opportunities for seaweed mariculture and other high-value marine products. To address this need, key components such as artificial habitats, support structures, and mooring systems are yet to be developed to align with industry-informed selections of marine products.

With the development of wind farms, particularly in the Bass Strait separating mainland Australia from Tasmania, there is considerable potential for co-locating multiple ecosystem services, including food production, biodiversity enhancement, and social benefits. However, limited stakeholder engagement has hindered a comprehensive understanding of these benefits and associated challenges. This underscores the pressing need to study floating artificial reefs (FARs) to provide insights into these issues and inform future planning.

The implementation of FARs also confronts a critical industry challenge: navigating the complex regulatory and permitting frameworks that govern offshore infrastructure. Classifying FARs as vessels, installations, or property holds significant implications for their deployment. By addressing these regulatory and ecological considerations, we can develop the next generation of industry, government, and research leaders in offshore developments, focusing on innovative approaches with global implications. This comprehensive understanding will facilitate the responsible integration of artificial habitats into offshore infrastructure, ensuring that such developments contribute positively to marine ecosystems and comply with environmental regulations.

In Australia, the Blue Economy CRC identified two key research gaps in 2022 to advance the potential of artificial floating benthic ecosystems in offshore development (Project ID 5.21.002). The first involves examining the design considerations of floating artificial habitats from both engineering and ecological perspectives. The second focuses on how domestic and international regulations must evolve to enable FARs to be deployed as standalone or co-located projects.

This chapter explores these critical research needs by examining the feasibility of standalone or co-located artificial habitats with offshore floating marine infrastructure to provide additional functions as aquaculture farms and marine habitats. Additionally, the chapter highlights the significant gap in the development of offshore infrastructure designed to enhance ecosystem services, minimize environmental impacts, and enable sustainable food and energy production. By addressing these gaps, FARs offer an innovative insight to supporting offshore industries while aligning with Australia's commitment to sustainable offshore development.

2 Background

Coral reefs provide unparalleled ecological, economic, and cultural benefits, contributing an estimated USD 375 billion annually in global socio-economic value. These ecosystems support marine biodiversity, fisheries, coastal protection, and tourism. However, in the recent decades, global coral reefs have seen significant declines with some regions reporting over 50% loss in coral cover due to rapid ocean warming and anthropogenic pressures [25, 28, 30]. Since 1950s, up to a 60% decline has been reported in the catches of coral reef-associated commercial fish taxa and 50% decline in coral reefs capacity to provide ecosystem services [17]. Incurring such losses already resulted in devastating implications for food security, biodiversity conservation, and coastal resilience, particularly for communities reliant on reef ecosystems [17].

To address these challenges, considerable efforts have been made since the 1950s to develop artificial reefs as a means to restore degraded habitats, enhance marine biodiversity, and support fisheries. Artificial reefs have demonstrated their potential to increase the biomass and density of marine organisms in targeted undersea environments [51], contributing to the improved productivity of algae, seaweed, lobster, shellfish, and fish [6, 13]. These structures also mitigate the loss of natural coral reefs by offering alternative habitats that help reverse coral cover declines and protect vulnerable coastlines from damaging wave forces [27].

While artificial reefs have shown promise, their success varies based on site-specific factors such as ecological compatibility, design complexity, and management strategies [3, 27]. The development of tailored approaches to artificial reef design and deployment has opened new avenues for addressing the challenges posed by climate change and coastal degradation. These efforts are further supported by technological advancements, such as modular and 3D-printed reef systems, which allow for greater adaptability to diverse marine conditions. Pioneering research has been conducted on the role of artificial reefs in improving marine biodiversity and supporting aquaculture initiatives, with researchers exploring the development and advancement of floating artificial reefs as innovative solutions for offshore environments [65]. Collectively, this precedent research offers an emerging understanding of the ecological and engineering facets of artificial reef systems and how these can be integrated by leveraging the buoyancy force of water with existing and emerging floating systems.

2.1 Artificial Reefs and Marine Habitats Enhancing Biodiversity

Artificial reefs (ARs) have garnered significant global attention as tools to enhance marine biodiversity and organism density. The European Artificial Reef Research Network (EARRN) defines an artificial reef as a man-made structure designed to mimic the characteristics of natural reefs, promoting ecological functions and supporting marine life [3]. However, the effectiveness of ARs varies depending on factors such as site selection, structural design, and management [3, 27]. Challenges such as destruction from storms, corrosion, sedimentation, and burial, as well as inadequate site selection and poor design have hindered some projects from achieving their objectives [3, 55].

Comprehensive understanding of factors that influence AR performance as a productive marine habitat is essential to overcoming these challenges and optimizing the performance. Studies of natural (e.g., [37, 56]) and artificial habitats (e.g., [24, 52]) provide critical baseline knowledge regarding habitat functionality and the factors influencing success, including size, materials, and structural complexity [7, 15, 62]. Several studies in natural and artificial habitats highlighted how reef complexity creates size-structured refuges, reducing prey mortality and enhancing fish recruitment and production [38, 57]. This knowledge is instrumental in designing artificial habitats that cater to specific ecological and commercial goals.

ARs are increasingly being utilized in aquaculture initiatives to cultivate species such as oysters, clams, and sea urchins, providing both ecological and economic benefits [12, 67]. For instance, concrete artificial habitats have been used in a commercial sea ranching initiative to cultivate abalone in Western Australia for at least a decade [49]. While not purposefully designed as marine aquaculture initiative, enhancement of the design of the wave power foundations deployed in Sweden resulted in the increased abundance of the commercially important brown crab (*Cancer pagurus*) [6]. Additionally, ARs have been successfully used to address habitat loss in degraded environments, offering a versatile solution for marine conservation [24, 27].

Emerging technologies, such as size-based habitat models and material innovations, enable the development of tailored AR designs to improve ecosystem services. The integration of these advancements into reef construction offers new opportunities to mitigate environmental degradation while supporting marine biodiversity and aquaculture expansion. By leveraging such innovations, ARs can serve as critical infrastructure in sustainable marine ecosystem management.

2.2 Historical Development of Offshore Artificial Reefs

Floating artificial habitats remain relatively rare, yet historical and scientific insights offer valuable guidance for their design and deployment. Research on sub-surface structures, such as oil rigs and semi-submersible platforms, has contributed significantly to the concept of repurposing industrial infrastructure into marine habitats. The Reef-to-Rigs program in the United States exemplifies this approach, converting decommissioned oil platforms into artificial reefs to support marine biodiversity. Although, some studies have demonstrated that offshore structures can be some of the most productive ecosystems [13], global meta-analysis challenged that assumptions, suggesting that oil and gas platforms and offshore energy installations may provide limited benefits to the production of biodiversity with purposefully designed and deployed artificial reefs being more effective to achieve conservation goals [41]. The same study, as well as several others [8, 43] however, hypothesized that the limited performance of offshore structures as marine habitats is likely linked to their design and placement—challenges that can be addressed during the planning and development stages. Addressing these issues should consequently enhance performance and help achieve desired environmental outcomes.

Historically, artificial reefs were predominantly bottom-founded, restricting their placement to shallow waters. However, advancements in construction technologies, such as 3D printing and modular designs, have enabled new design of floating artificial reefs for deeper offshore environments. These innovations address critical ecological challenges, offering scalable solutions to address negative impacts associated with inshore marine industries, mitigate habitat loss and promote biodiversity [20, 61]. Floating artificial reefs provide the added advantage of flexibility, allowing structures to adjust dynamically to environmental conditions, such as wave action and water temperature. In doing so, these innovations extend the available sites in which artificial reefs can be deployed, Floating artificial reefs are particularly valuable in heavily impacted coastal regions where pollution, sedimentation, and coastal hardening have degraded natural habitats. By positioning these reefs further away from the coastline and in the upper photic zone (depths $\leq$ 40 m), they avoid many stressors associated with nearshore areas while providing critical habitats for diverse marine species. Further, in the scoping study of BECRC (Komyakova et al. 2022), such habitats may enhance recruitment and support aquaculture initiatives, creating a dual benefit for ecological restoration and commercial productivity.

The growing demand for offshore renewable energy has also highlighted the potential for co-locating artificial reefs with infrastructure such as wind farms [40]. This integration can optimize the use of offshore spaces while enhancing ecosystem services and fostering sustainable resource management. For example, developments in the Bass Strait region could incorporate floating artificial reefs to support biodiversity while mitigating the environmental impacts of industrial activities [18].

3 Research Aims

To explore whether floating artificial reefs (FARs) can ensure sustainable offshore development, this chapter proposes that research in this field address the following questions:

1. How can we develop artificial habitat designs to enhance food production capabilities ("farm habitats") and ecosystem services ("eco-habitats")?
2. How can state-of-the-art engineering designs facilitate standalone floating artificial habitats for food production and co-located artificial habitats with offshore energy infrastructure?
3. What are the relevant regulatory frameworks, suggested regulatory reforms, permitting strategies, and economic and environmental viability models?

This chapter orients these research questions towards a critical market gap: how can floating artificial reefs address the provision of marine food for domestic food security and export markets. The development of FARs has the potential to create significant economic value, including businesses focused on supplying juvenile marine fauna and flora for offshore facilities and providing equipment for offshore infrastructure.

The following section addresses the ability of artificial reefs to enhance biodiversity and mitigate environmental impacts. If these benefits can be transposed to offshore sites through the use of floating structures, these results may offer benefits to urban development initiatives and broader ecological systems. By combining ecological insights with advanced engineering, FARs can serve as an innovative solution that supports sustainable aquaculture, fosters ecosystem services, and aligns with the goals of the blue economy (Komyakova et al. 2022).

4 Development of Artificial Habitat and Ecosystems with Floating Artificial Reefs

It has been suggested floating artificial reefs (FARs) have potential to enhance the biomass of marine species in environments where populations are constrained by the availability of suitable habitats. By providing additional resources, FARs can provide settlement habitats, reduce predation risks, increase food accessibility, and improve reproductive opportunities. Some studies have suggested that isolated habitats can support diverse and abundant communities by providing settlement areas in otherwise structurally simplistic environments, where hard substrate habitats are limited [33, 53]. They also allow propagules carried offshore by currents to establish new populations, which might have otherwise perished. However, in regions where population limitations are primarily influenced by recruitment availability, their effectiveness in boosting biomass may be constrained [47].

The community composition of associate fauna on artificial reefs often fluctuates, with some species dominating during particular seasons or following a successional trajectory (e.g. [35]). Over time, limited evidence has showcased that these systems have the potential to stabilize, reaching an equilibrium community structure (e.g., [35]). For targeted aquaculture applications, artificial reefs can be seeded with desired species, including clams, oysters, abalone, and sea urchins, enabling tailored ecological and commercial outcomes. Such tailored interventions can expedite ecosystem stabilization and optimize economic benefits for aquaculture.

Most conventional artificial reefs are seabed-based and deployed in shallow coastal environments, near the proximity of urban centres which exposes them to significant risks, including sedimentation, agricultural runoff, and a range of pollution types (e.g., chemical, noise, light) [10, 16] (Komyakova et al. 2022). For instance, fertilizer runoff near the Great Barrier Reef promotes algal blooms and increases the frequency of crown-of-thorns starfish outbreaks, causing severe damage to coral ecosystems. Similarly, sedimentation from upstream soil erosion decreases water clarity, disrupting seagrass habitats by limiting sunlight penetration [1]. These threats highlight the limitations of traditional seabed-anchored reefs.

In contrast, FARs, which are situated off-shore but within the upper photic zone (up to 40 m in depth), represent a sustainable alternative that avoids many of these coastal challenges. By leveraging offshore environments, FARs can minimize exposure to pollutants while maximizing ecological benefits. For example, the transformation of decommissioned offshore oil platforms into artificial reefs has gained global recognition. In the Gulf of Mexico, over 530 platforms were repurposed as artificial reefs between the 1980s and 2018, significantly enhancing fish biomass within a 300-m radius [7, 58].

Similar to the performance of artificial reef created by bottom wave foundations, the structural adaptability of FARs has the potential to enable the incorporation of customizable void spaces that attract coral reef fish and increase juvenile recruitment that worth future investigation. This targeted design approach not only enhances biodiversity but also creates specialized habitats for aquaculture species such as lobsters and seaweed. Additionally, co-locating multiple marine industries on FARs could unlock substantial economic opportunities while promoting ecological resilience [46, 61]. Emerging technologies, such as generative design and 3D printing, further enhance the flexibility and precision of FARs, allowing them to address site-specific environmental and commercial needs efficiently. Recent, rapid development of novel approaches (e.g. 3D printing, structure-from-motion (SfM) photogrammetry, AI agents) provide an opportunity to enhance the level of our understanding of structural drivers of biodiversity, and enhance our ability to mimic structural characteristics of natural habitats across environmental gradient. These advancements position FARs as integral tools for sustainable marine ecosystem development [19, 45, 50].

5 Engineering Solutions for Floating Artificial Reefs

The construction of artificial reefs has evolved significantly, employing a variety of structural elements. Historically, materials such as decommissioned ships, discarded train carriages, tanks, vehicular tires, and bridge rubble have been utilized as "materials of opportunity" [3]. Additionally, repetitive arrangements of concrete blocks or PVC pipes have been employed in early designs. However, modern approaches prioritize modular and highly specialized structures designed to enhance marine biodiversity and habitat suitability for targeted species [51, 62].

The structural configurations of artificial reefs vary based on project objectives, such as coral spawning, marine biodiversity enhancement, or habitat restoration. In fisheries, artificial reefs often serve multiple purposes: providing habitat for adult individuals, or supporting the recruitment and survival of juvenile fish [7, 9, 31]. These distinct goals influence the choice of materials and structural forms [7, 62].

Recent studies emphasize the importance of hydrodynamic characteristics in reef design. For instance, structures with low flow resistance provide suitable shelters for marine organisms, thereby improving habitat quality [68]. These findings underscore the need for responsive, highly customizable designs tailored to local environmental conditions and desired species. The advent of 3D printing technology has facilitated the creation of complex, purpose-built reef modules that meet these requirements [20].

The proposed floating artificial reef system comprises 3D-printed modules attached to floating pontoons submerged 2–25 m below the ocean surface. These modules can be rapidly fabricated to create new marine habitats customized for ecological or commercial purposes, such as oyster farming, seaweed cultivation, or coral restoration.

Generative design and topology optimization methods are critical for achieving efficient and adaptable reef structures [63]. These techniques enable the development of porous forms with high surface-area-to-volume ratios, which promote coral growth, attract juvenile fish, and reduce predation risks [29, 46]. This approach not only supports ecological outcomes but also allows for aesthetically appealing designs that attract divers and boost eco-tourism.

To ensure resilience under hydrodynamic forces such as waves and storms, computational fluid dynamics (CFD) simulations and bi-directional evolutionary structural optimization (BESO) methods are employed. These technologies enable the optimization of structural performance, ensuring the reef can withstand environmental stressors while maintaining ecological functionality [29, 30].

Traditionally, theoretical analyses and flume tests have been used to investigate the hydrodynamics surrounding ARs, examining flow fields and the forces exerted by these structures [32]. Recent advancements in technology have popularized the use of Particle Image Velocimetry (PIV) and CFD for analyzing the hydrodynamic effects of ARs. CFD, in particular, is favored for its cost efficiency and capability to simulate various water flow scenarios. With continuous advancements, CFD simulations have become indispensable in hydrodynamic analysis, enabling the quantitative characterization of flow fields and comprehensive evaluations of hydraulic characteristics around ARs.

To address limitations in current AR designs, an iterative process combining CFD analysis and topology optimization has been developed. This process involves computing surface pressures under flow conditions at each iteration, providing dynamic feedback to refine the AR design. The void ratio of the optimized structure is continuously assessed until it aligns with a predefined target based on a benchmark AR. BESO, a widely adopted topology optimization method, is central to this process. By iteratively removing inefficient material and adding necessary elements, BESO facilitates the creation of structures that meet predefined criteria derived from finite element analyses [29].

As a cutting-edge form-finding method in digital design [44, 69], topology optimization refines material layouts and explores innovative forms within contemporary design frameworks [4]. In the context of artificial reef design, BESO has gained prominence for its ability to generate diverse organic forms with superior structural performance. A hybrid generative design approach, integrating CFD and BESO techniques, has been employed to optimize reef designs by balancing environmental and structural considerations [20].

Biologically compatible materials are integral to 3D-printed reef construction. These materials mimic seawater pH levels and can include additives such as slow-releasing nutrients, glass-fiber polymers, or crustose coralline algae extracts to promote coral and algae growth [2, 11]. The submerged printing process offers several advantages, including reduced material deformation, increased construction speed, and the ability to fabricate geometrically complex structures at lower costs. Recent, unprecedent advances in modern technologies, such as development of the AI agents, provide a new opportunity to defining habitat complexity and evaluating features of success, allowing us to develop a holistic understanding of habitat structural characteristics in the marine environments and incorporate them into the design of nature-informed artificial habitats. These innovations offer the potential to make floating artificial reefs an effective, sustainable solution for biodiversity conservation and aquaculture expansion.

6 Challenges with the Regulatory Framework of Floating Artificial Reefs

The deployment of floating artificial reefs (FARs) as environmental offsets in the marine environment offers potential benefits but also faces significant regulatory and governance challenges. In Australia, such offsets could be implemented under mechanisms like the *Environmental Protection and Biodiversity Conservation Act 1999* (Cth). Environmental offset frameworks, while widely adopted, remain contentious due to ecological, legal, and political concerns. These mechanisms aim for a net neutral or positive environmental impact by allowing developers to undertake conservation efforts that compensate for environmental harm caused at another location [5, 48]. In the context of FARs, this could include their mandated inclusion as part of offshore development approvals to mitigate the effects of coastal hardening, where natural marine habitats are replaced with artificial structures.

Coastal hardening has long been recognized as a significant threat to natural ecosystems, particularly seagrass beds and coral reefs [21, 23, 26]. By incorporating

FARs as a requirement for development approvals, the complexities and financial burdens of habitat preservation and rehabilitation could be shifted to offshore developers. This approach could also foster innovative conservation solutions and reduce pressure on nearshore ecosystems [18]. In jurisdictions such as Australia, environmental offset policies further allow developers to outsource these responsibilities to the government or third-party organizations, potentially enabling government-led initiatives to deploy FARs as part of broader conservation strategies.

However, ocean governance in offshore areas is inherently complex, involving numerous regulatory stakeholders tasked with environmental protection and sustainable development. In Australia, the installation of a FAR would necessitate engagement with federal, state, and sometimes local government agencies. For example, deploying a FAR in Queensland may require Tidal Works development approval, a Marine Park permit, and compliance with federal legislation such as the *Sea Installations Act 1987* (Cth) and the *Environment Protection and Sea Dumping Act 1982* (Cth). Depending on the project's scale and location, a Commonwealth Sea Dumping Permit might also be required [34, 48]. These overlapping requirements illustrate the regulatory complexity that developers must navigate[64].

In scenarios where governments or third parties undertake offset activities, such as FAR deployment, regulatory safeguards must be established to mitigate risks of regulatory capture. This occurs when public entities tasked with environmental oversight become influenced by commercial interests, potentially compromising conservation goals [5]. To ensure accountability and transparency, rigorous monitoring and compliance frameworks are essential [42].

Moreover, offset mechanisms should be implemented as a last resort, only after all feasible measures to avoid or minimize environmental harm have been thoroughly explored and deemed impractical. This principle underscores the importance of prioritizing avoidance and mitigation strategies overcompensation. The effectiveness of FARs as offsets will also depend on their ecological performance and the long-term sustainability of the habitats they create [42].

Lastly, the role of FARs in addressing global conservation challenges will depend on international collaboration and the harmonization of regulatory frameworks across jurisdictions. This is particularly crucial for transboundary marine ecosystems, where coordinated governance can enhance conservation outcomes and reduce regulatory redundancies.

7 Conclusion

Floating artificial reefs (FARs) represent a transformative approach to addressing the intertwined challenges of marine biodiversity loss, sustainable aquaculture, and the environmental impacts of offshore development. By integrating ecological principles with cutting-edge engineering solutions, FARs offer a pathway to enhance marine ecosystems while supporting commercial activities such as aquaculture and renewable energy production. This chapter has explored the historical evolution of artificial reefs, the advancements in floating reef technologies, and the regulatory complexities surrounding their deployment.

The potential of FARs extends beyond their immediate ecological benefits. They offer a unique opportunity to co-locate multiple ecosystem services, from enhancing marine biodiversity and providing habitats for commercially valuable species to mitigating the environmental impacts of industrial offshore infrastructure. By adopting innovative designs such as 3D-printed modules, generative topology optimization, and biologically friendly materials, FARs can be tailored to specific ecological and economic needs while ensuring durability and resilience against environmental stressors.

However, the successful deployment of FARs requires overcoming significant regulatory and governance challenges. Harmonizing the interests of multiple stakeholders, ensuring compliance with jurisdiction-specific policies, and safeguarding against regulatory capture are critical to their widespread implementation. Furthermore, the use of FARs as environmental offsets should be approached cautiously, ensuring that these mechanisms are a last resort after all efforts to avoid or mitigate environmental harm have been exhausted.

As offshore industries expand, particularly in regions like Australia's Bass Strait, FARs can play a pivotal role in shaping sustainable blue economy (Komyakova et al. 2022) initiatives. They provide a blueprint for how ecological integrity and economic development can coexist, demonstrating the potential for human innovation to support, rather than exploit, marine environments. By fostering interdisciplinary collaboration and advancing research in engineering and ecological sciences, FARs can establish a new benchmark for sustainable marine development.

The deployment of FARs not only enhances ecological resilience but also positions countries like Australia as global leaders in innovative, ecosystem-focused offshore development. Future research and stakeholder engagement will be crucial to unlocking the full potential of FARs, ensuring their long-term success in addressing critical marine conservation and sustainability challenges.

References

1. Adams MP, Hovey RK, Hipsey MR, Bruce LC, Ghisalberti M, Lowe RJ, Gruber RK, Ruiz-Montoya L, Maxwell PS, Callaghan DP, Kendrick GA (2016) Feedback between sediment and light for seagrass: where is it important?. Limnol Oceanogr 61(6):1937–1955. https://doi.org/10.1002/lno.10319
2. Albalawi HI, Khan ZN, Valle-Pérez AU, Kahin KM, Hountondji M, Alwazani H, Schmidt-Roach S, Bilalis P, Aranda M, Duarte CM, Hauser CAE (2021) Sustainable and eco-friendly coral restoration through 3D printing and fabrication. ACS Sustain Chem Eng. https://doi.org/10.1021/acssuschemeng.1c04148
3. Baine M (2001) Artificial reefs: a review of their design, application, management, and performance. Ocean Coast Manag 44:241–259. https://doi.org/10.1016/S0964-5691(01)00048-5
4. Bao D (2022) Performance-driven digital design and robotic fabrication based on topology optimisation and multi-agent system. Thesis, RMIT

5. Bell J (2016) Implementing an outcomes-based approach to marine biodiversity offsets: lessons from the Great Barrier Reef. Australas J Environ Manage 23(3):314–329. https://doi.org/10.1080/14486563.2015.1081837
6. Bender A, Langhamer O, Sundberg J (2020) Colonisation of wave power foundations by mobile mega-and macrofauna–a 12 year study. Mar Environ Res 161:105053. https://doi.org/10.1016/j.marenvres.2020.105053
7. Blount C, Komyakova V, Barnes L, Smith ML, Zhang D, Reeds K, McPhee D, Taylor MD, Macbeth W, Needham E (2021) Using ecological evidence to refine approaches to deploying offshore artificial reefs for recreational fisheries. Bull Mar Sci 97(4):665–698. https://doi.org/10.5343/bms.2020.0059
8. Blount C, Komyakova V, Barnes L, Smith ML, Zhang D, Reeds K, McPhee D, Taylor MD, Macbeth W, Needham E (2021) Using ecological evidence to refine approaches to deploying offshore artificial reefs for recreational fisheries. Bull Mar Sci 97(4):665–698. https://doi.org/10.5343/bms.2020.0059
9. Bohnsack JA, Sutherland DL (1985) Artificial reef research: a review with recommendations for future priorities. Bull Mar Sci 37(1):11–39
10. Bugnot AB, Mayer-Pinto M, Airoldi L, Heery EC, Johnston EL, Critchley LP, Strain EMA, Morris RL, Loke LHL, Bishop MJ, Sheehan EV (2021) Current and projected global extent of marine built structures. Nat Sustain 4(1):33–41. https://doi.org/10.1038/s41893-020-00595-1
11. Buswell RA, Leal WR, Jones SZ et al (2018) 3D printing using concrete extrusion: a roadmap for research. Cem Concr Res 112:37–49. https://doi.org/10.1016/j.cemconres.2018.05.006
12. Chai Z, Huo Y, He Q, Huang X, Jiang X, He P (2014) Studies on breeding of Sargassum vachellianum on artificial reefs in Gouqi Island, China. Aquaculture 424:189–193. https://doi.org/10.1016/j.aquaculture.2013.12.038
13. Claisse JT, Pondella DJ, Love M, Zahn LA, Williams CM, Williams JP, Bull AS (2014) Oil platforms off California are among the most productive marine fish habitats globally. Proc Natl Acad Sci 111(43):15462–15467. https://doi.org/10.1073/pnas.1411477111
14. Clark S, Edwards AJ (1999) An evaluation of artificial reef structures as tools for marine habitat rehabilitation in the Maldives. Aquat Conserv Mar Freshwat Ecosyst 9(1):5–21. https://doi.org/10.1002/(SICI)1099-0755(199901/02)9:1%3c5::AID-AQC330%3e3.0.CO;2-U
15. Dodds KC, Schaefer N, Bishop MJ, Nakagawa S, Brooks PR, Knights AM, Strain EM (2022) Material type influences the abundance but not richness of colonising organisms on marine structures. J Environ Manage 307:114549. https://doi.org/10.1016/j.jenvman.2022.114549
16. Easton A, Komyakova V, Coughlin T (2024) Evaluating ecological risk in artificial habitat failure: a systematic review and risk assessment considering noise and light pollution in the marine environment. Environ Impact Assess Rev 107:107560. https://doi.org/10.1016/j.eiar.2024.107560
17. Eddy TD, Lam VW, Reygondeau G, Cisneros-Montemayor AM, Greer K, Palomares MLD, Bruno JF, Ota Y, Cheung WW (2021) Global decline in capacity of coral reefs to provide ecosystem services. One Earth 4(9):1278–1285. https://doi.org/10.1016/j.oneear.2021.08.016
18. Van Elden S, Meeuwig JJ, Hobbs RJ (2019) Offshore oil and gas platforms as novel ecosystems: a global perspective. Front Mar Sci 6:548. https://doi.org/10.3389/fmars.2019.00548
19. Evans AJ, Lawrence PJ, Natanzi AS, Moore PJ, Davies AJ, Crowe TP, McNally C, Thompson B, Dozier AE, Brooks PR (2021) Replicating natural topography on marine artificial structures–a novel approach to eco-engineering. Ecol Eng 160:106144. https://doi.org/10.1016/j.ecoleng.2020.106144
20. Feng Z, Gu P, Zheng M et al (2021) Environmental data-driven performance-based topological optimization for morphology evolution of artificial Taihu stone. In: Yuan PF, Chai H, Yan C (eds) Proceedings of the 2021 digital futures CDRF 2021. Springer. https://doi.org/10.1007/978-981-16-5983-6_11
21. Firth LB, Knights AM, Bridger D, Evans AJ, Mieszkowska N, Moore PJ, O'Connor NE, Sheehan EV, Thompson RC, Hawkins SJ (2016) Ocean sprawl: challenges and opportunities for biodiversity management in a changing world. Oceanogr Mar Biol Annu Rev 54:189–262. https://doi.org/10.1201/9781315368597-5

22. Firth LB et al (2024) Coastal greening of grey infrastructure: an update on the state-of-the-art. In: Proceedings of the institution of civil engineers: maritime engineering, vol 177(2), pp 35–67. https://doi.org/10.1680/jmaen.2023.003
23. Gittman RK, Scyphers SB, Smith CS, Neylan IP, Grabowski JH (2016) Ecological consequences of shoreline hardening: a meta-analysis. Bioscience 66(9):763–773. https://doi.org/10.1093/biosci/biw091
24. Hackradt CW, Felix-Hackradt FC, Garcia-Charton JA (2011) Influence of habitat structure on fish assemblage of an artificial reef in southern Brazil. Mar Environ Res 72:235–247. https://doi.org/10.1016/j.marenvres.2011.09.006
25. Harrison HB, Álvarez-Noriega M, Baird AH, Heron SF, Macdonald C, Hughes TP (2019) Back-to-back coral bleaching events on isolated atolls in the Coral Sea. Coral Reefs 38:713–719. https://doi.org/10.1007/s00338-018-01749-6
26. Heery EC, Bishop MJ, Critchley LP, Bugnot AB, Airoldi L, Mayer-Pinto M, Sheehan EV, Coleman RA, Loke LH, Johnston EL, Komyakova V (2017) Identifying the consequences of ocean sprawl for sedimentary habitats. J Exp Mar Biol Ecol 492:31–48. https://doi.org/10.1016/j.jembe.2017.01.020
27. Higgins E, Metaxas A, Scheibling RE (2022) A systematic review of artificial reefs as platforms for coral reef research and conservation. PLoS ONE 17(1):e0261964. https://doi.org/10.1371/journal.pone.0261964
28. Hoegh-Guldberg O, Kennedy EV, Beyer HL, McClennen C, Possingham HP (2018) Securing a long-term future for coral reefs. Trends Ecol Evol 33(12):936–944. https://doi.org/10.1016/j.tree.2018.09.006
29. Huang X, Xie YM (2010) Evolutionary topology optimization of continuum structures: methods and applications. Wiley. https://doi.org/10.1002/9780470689486
30. Hughes TP, Kerry JT, Álvarez-Noriega M, Álvarez-Romero JG, Anderson KD, Baird AH, Babcock RC, Beger M, Bellwood DR, Berkelmans R (2017) Global warming and recurrent mass bleaching of corals. Nature 543:373–377. https://doi.org/10.1038/nature21707
31. Ino T (1974) Historical review of artificial reef activities in Japan. In: Colunga L, Stone R (eds) Proceedings of the artificial reef conference. Texas A&M University, pp 21–23
32. Jiang Z, Liang Z, Zhu L, Guo Z, Tang Y (2020) Effect of hole diameter of rotary-shaped artificial reef on flow field. Ocean Eng 197:106917. https://doi.org/10.1016/j.oceaneng.2020.106917
33. Jones G, Barone G, Sambrook K, Bonin M (2020) Isolation promotes abundance and species richness of fishes recruiting to coral reef patches. Mar Biol 167(11):1–13. https://doi.org/10.1007/s00227-020-03772-0
34. Kate K, Inbar M (2012) Biodiversity offsets. In: Bayon R, Fox J, Carroll N (eds) Conservation and biodiversity banking: a guide to setting up and running biodiversity credit trading systems, Earthscan, London, UK, pp 188–204. https://doi.org/10.4324/9781849770842
35. Komyakova V, Chamberlain D, Jones GP, Swearer SE (2019) Assessing the performance of artificial reefs as substitute habitat for temperate reef fishes: Implications for reef design and placement. Sci Total Environ 668:139–152. https://doi.org/10.1016/j.scitotenv.2019.02.357
36. Komyakova V, Jaffrés JB, Strain EM, Cullen-Knox C, Fudge M, Langhamer O, Bender A, Yaakub SM, Wilson E, Allan BJ, Sella I (2022) Conceptualisation of multiple impacts interacting in the marine environment using marine infrastructure as an example. Sci Total Environ 830:154748. https://doi.org/10.1016/j.scitotenv.2022.154748
37. Komyakova V, Jones GP, Munday PL (2013) Strong effects of coral species on the diversity and structure of reef fish communities: a multi-scale analysis. PLoS ONE 8(8):e70924. https://doi.org/10.1371/journal.pone.0202206
38. Komyakova V, Chamberlain D, Swearer SE (2021) A multi-species assessment of artificial reefs as ecological traps. Ecol Eng 171:106394. https://doi.org/10.1016/j.scitotenv.2022.154748
39. Komyakova VW, B. Frusher, S. Mohajernasab, S. Shakourloo, A. Abdussamie, N. Haward M (2022) Identifying the potential of artificial floating benthic ecosystems to underpin offshore development. 5.21.002—final project report. Blue Economy Cooperative Research Centre, Hobart

40. Langhamer O (2012) Artificial reef effect in relation to offshore renewable energy conversion: state of the art. Sci World J 2012(1):386713. https://doi.org/10.1100/2012/386713
41. Lemasson AJ, Somerfield PJ, Schratzberger M, Thompson MS, Firth LB, Couce E, McNeill CL, Nunes J, Pascoe C, Watson SC, Knights AM (2024) A global meta-analysis of ecological effects from offshore marine artificial structures. Nature Sustainability 7(4):485–495. https://doi.org/10.1038/s41893-024-01311-z
42. Lima JS, Zalmon IR, Love M (2019) Overview and trends of ecological and socioeconomic research on artificial reefs. Mar Environ Res 145:81–96. https://doi.org/10.1016/j.marenvres.2019.01.010
43. Lima JS, Zalmon IR, Love M (2019) Overview and trends of ecological and socioeconomic research on artificial reefs. Mar Environ Res 145:81–96. https://doi.org/10.1016/j.marenvres.2019.01.010
44. Lin S, Chou N, Bao D, Zhang G, Xiong C, Fang J, Xie YM (2023) Design and fabrication of artificial brain coral: Evolution principle, turbulent hydrodynamics and matter interchange. Comput Struct 276:106955. https://doi.org/10.1016/j.compstruc.2022.106955
45. Lin S, Chou N, Zhao Y, Qin Y, Jiang H, Cui J, Li G, Xie YM (2022) Topological optimization of magnetic pulse welding coils with a connectivity-constrained particle swarm optimization algorithm. Mater Des 224:111337. https://doi.org/10.1016/j.advengsoft.2021.103065
46. Love MS, Schroeder DM, Lenarz WH (2005) Distribution of bocaccio (Sebastes paucispinis) and cowcod (Sebastes levis) around oil platforms and natural outcrops off California with implications for larval production. Bull Mar Sci 77(3):397–408. https://doi.org/10.1890/100112
47. Macredie PI, Fowler AM, Booth DJ (2011) Rigs-to-reefs: will the deep sea benefit from artificial habitat? Front Ecol Environ 9(8):455–461. https://doi.org/10.1890/100112
48. Madsen B, Carroll N, Kandy D (2011) State of biodiversity markets 2011: offset and compensation programs worldwide. Forest Trends
49. Melville-Smith R, Fotedar R, Pattiaratchi C, Adams B, Hart A (2017) Investigating critical biological issues for commercial greenlip abalone sea ranching in flinders bay. Western Australia, Fisheries Research and Development Corporation. Perth, Sept 2017. https://doi.org/10.4225/06/597abdc320ee7
50. Mirra G, Holland A, Roudavski S, Wijnands JS, Pugnale A (2022) An artificial intelligence agent that synthesises visual abstractions of natural forms to support the design of human-made habitat structures. Front Ecol Evol 10:806453. https://doi.org/10.3389/fevo.2022.806453
51. Paxton AB, Shertzer KW, Bacheler NM, Kellison GT, Riley KL, Taylor JC (2020) Meta-analysis reveals artificial reefs can be effective tools for fish community enhancement but are not one-size-fits-all. Front Mar Sci 7:282. https://doi.org/10.3389/fmars.2020.00282
52. Paxton AB, Taylor JC, Nowling LA et al (2019) Artificial reefs facilitate tropical fish range expansion. Commun Biol 10(1):1–10. https://doi.org/10.1038/s42003-019-0398-2
53. Polak O, Polak T, Shashar N (2014) Effects of habitat relocation on fish community structure-a case study. Environ Biol Fishes 97:1335–1345. https://doi.org/10.1007/s10641-014-0223-1
54. Polak O, Polak T, Shashar N (2014) Effects of habitat relocation on fish community structure-a case study. Environ Biol Fishes 97:1335–1345. https://doi.org/10.1007/s10641-014-0223-1
55. Reed DC, Schroeter SC, Huang D et al (2006) Quantitative assessment of different artificial reef designs in mitigating losses to kelp forest fishes. Bull Mar Sci 78(1):133–150
56. Rogers A, Blanchard JL, Mumby PJ (2014) Vulnerability of coral reef fisheries to a loss of structural complexity. Curr Biol 24(9):1000–1005. https://doi.org/10.1016/j.cub.2014.03.026
57. Rogers A, Blanchard JL, Newman SP, Dryden CS, Mumby PJ (2018) High refuge availability on coral reefs increases the vulnerability of reef-associated predators to overexploitation. Ecology 99(2):450–463. https://doi.org/10.1002/ecy.2103
58. Sammarco PW, Lirette A, Tung YF et al (2014) Coral communities on artificial reefs in the Gulf of Mexico: standing vs. toppled oil platforms. ICES J Mar Sci 71(2):417–426. https://doi.org/10.1093/icesjms/fst140
59. Schramski JR, Woodson CB, Steck G, Munn D, Brown JH (2019) Declining country-level food self-sufficiency suggests future food insecurities. BioPhys Econ Resource Q 4:1–9. https://doi.org/10.1007/s41247-019-0060-0

60. Schramski J, Woodson C, Brown J (2020) Energy use and the sustainability of intensifying food production. Nat Sustain 3:257–259. https://doi.org/10.1038/s41893-020-0503-z
61. Sherman RL, Gilliam DS, Spieler RE (2002) Artificial reef design: void space, complexity, and attractants. ICES J Mar Sci 59:S196–S200. https://doi.org/10.1006/jmsc.2001.1163
62. Vivier B, Dauvin JC, Navon M, Rusig AM, Mussio I, Orvain F, Boutouil M, Claquin P (2021) Marine artificial reefs, a meta-analysis of their design, objectives and effectiveness. Global Ecol Conserv 27:e01538. https://doi.org/10.1016/j.gecco.2021.e01538
63. Wang B, Bao DW, Ward S, Luo D (2022) Opportunities for further development of 3D-printed floating artificial reefs. J Aquacult Marine Biol 11(2):58–63. https://doi.org/10.15406/jamb.2022.11.00337
64. Wang B (2023) Regulatory reform and the potential use of floating artificial reefs under blue carbon and ocean carbon methodologies. In: International conference on ocean, offshore and arctic engineering, 2023-06-11–2023-06-16, Melbourne, Australia. https://eprints.qut.edu.au/240568/1/B01_OMAE2023_104421_Brydon_Wang.pdf
65. Wang B, Bao DW, Ward S, Luo D (2022) Opportunities for further development of 3D-printed floating artificial reefs. J Aquac Mar Biol 11(2):58–63. https://doi.org/10.15406/jamb.2022.11.00337
66. Williams BA, Watson JE, Beyer HL, Klein CJ, Montgomery J, Runting RK, Roberson LA, Halpern BS, Grantham HS, Kuempel CD, Frazier M (2022) Global rarity of intact coastal regions. Conserv Biol 36(4):e13874. https://doi.org/10.1111/cobi.13874
67. Xu Q, Zhang L, Zhang T, Zhang X, Yang H (2017) Functional groupings and food web of an artificial reef used for sea cucumber aquaculture in northern China. J Sea Res 119:1–7. https://doi.org/10.1016/j.seares.2016.10.005
68. Yaakob OB, Ahmed YM, Jalal MR et al (2016) Hydrodynamic design of new type of artificial reefs. Appl Mech Mater 819:406–419. https://doi.org/10.4028/www.scientific.net/AMM.819.406
69. Zhao B, Chen TY (2016) Application of topology optimisation to architectural design. Architect Cult 11:104–105. https://doi.org/10.1016/j.compstruct.2023.117603

Dan Luo is a Senior Lecturer in Architecture and Digital Design at the University of Queensland. She holds a Ph.D. in Architecture from Tsinghua University, a Master of Architecture from Columbia University, and a Master of Computer and Information Technology from the University of Pennsylvania. Dr. Luo's expertise in digital design to fabrication, data analysis, and applied robotic construction is strengthened by her interdisciplinary background in architecture and computer science. She has collaborated with industry partners across various fields, including natural resource management such as timber production and marine aquaculture, to develop state-of-the-art design and fabrication methods that enhance resource utilization while minimizing ecological impact. Her research integrates big data analytics, advanced fabrication technologies, and design-build practices to explore the future of intelligent design and automated construction.

Valeriya Komyakova is an environmental scientist with over 20 years of expertise in marine habitat functionality. She is committed to developing solutions for the sustainable coexistence of human activities and marine ecosystems, employing interdisciplinary approaches and cross-sector collaborations to drive innovation in conservation and resource management. Over the past decade, Dr. Komyakova's work has focused on designing ecologically friendly artificial reefs that enhance productivity while minimizing environmental impacts. She was the first to identify the formation of ecological traps caused by artificial structures and to propose mitigation strategies. As a co-lead of the Blue Economy Cooperative Research Centre (BECRC) study in 2022, she investigated the role of floating artificial reefs in recreational, commercial, and conservation applications. Her research also extends to sustainable fisheries management and mitigating the impacts of marine pollution. Dr. Komyakova's work has significant policy implications, particularly in the development of a risk assessment-based decision support tool for artificial reef implementation. Her expertise has contributed to policy development for the Great Barrier Reef Marine Park Authority and the State of the Environment report, among other initiatives.

Socio-legal Considerations to Address Social Acceptance and Legalities of Floating Structures

Designing for Buoyant Urbanism: Pedagogy, Architectural Paradigms and Social Acceptance of Floating Cities

Brydon T. Wang

Abstract Educational institutions have long been incubators for architectural innovation, producing groundbreaking theories and methodologies. By exposing students to the multifaceted challenges of buoyant urbanism, schools of architecture can nurture a new generation of thinkers who prioritise resilience and sustainability in their work. Buoyant urbanism, as a concept and design practice, represents a transformative response to the challenges of climate change, urbanisation and resource scarcity. Accordingly, teaching architecture students about floating structures not only equips them to tackle the emerging urban issues that arise from climate change but also fosters interdisciplinary innovation and societal acceptance of buoyant urbanism. This chapter outlines the approach and content of the architectural design studio that was run in the School of Architecture, University of Queensland in 2022. It argues that integrating buoyant urbanism into architectural education is essential for training new thinkers, advancing the field of research and enhancing public understanding and support for these solutions. The chapter sets out the six architectural paradigms that arise out of the historical development of floating structures and floating cities, and showcases studio participant work resulting in engagement with these paradigms.

Keywords Floating cities · Buoyant urbanism · Architectural pedagogy

1 The Pedagogical Imperative

As urban populations grow and climate change intensifies, architects are tasked with reimagining how humans inhabit increasingly vulnerable environments. Rising sea levels threaten low-lying coastal cities and Small Island Developing States (as discussed in Chap. "Buoyant Futures: Our Global Duty to Recognise Floating Cities

B. T. Wang (✉)
Centre for Policy Futures, The University of Queensland, St Lucia, QLD, Australia
e-mail: Brydon.Wang@uq.edu.au

KPMG Australia, Deal Advisory and Infrastructure, Brisbane, QLD, Australia

B. T. Wang et al. (eds.), *Large Floating Solutions*, Lecture Notes in Civil Engineering 636, https://doi.org/10.1007/978-981-96-5435-2_15

as 'Territory' in International Law"), demanding innovative and resilient solutions. Floating architecture has emerged as a viable response to these challenges, combining adaptive design, sustainability and technological innovation. Floating cities, their architecture and the technology required to harness the buoyancy force of water, remain an underexplored area within architectural research, offering vast potential for innovation in materials, construction techniques and urban planning. Despite their potential, floating cities face societal scepticism, often being perceived as impractical, elitist or environmentally disruptive [35]. Overcoming these misconceptions requires a concerted effort to educate both the public and the next generation of designers (architects and engineers) about the feasibility and benefits of buoyant urbanism.

Incorporating buoyant urbanism into architectural curricula enables students to develop creative solutions to these pressing issues. Educational institutions have long been incubators for architectural innovation, producing groundbreaking theories and methodologies. As architects confront the increasing vulnerabilities of urban environments, architectural education must evolve to equip students with the skills to design adaptable and sustainable built environments. By engaging with the complexities of buoyant urbanism, students develop a critical understanding of water-based settlements and the design strategies required to enhance their viability.

Integrating buoyant urbanism into academic design studios also encourages rigorous scholarly inquiry and fosters interdisciplinary collaboration between architects, engineers, environmental scientists and policy makers (as seen in the range of authors and backgrounds contributing to this edited collection). Advancing the academic discourse on floating structures creates opportunities for developing specialised subfields within architecture, such as amphibious urban design and adaptive planning. As architectural pedagogy evolves, it can position buoyant urbanism as a critical area of study, akin to sustainable design or parametric modelling. This academic foundation is essential for building a body of knowledge that informs both practice and policy.

This chapter outlines the approach and content of the architectural design studio that was run in the School of Architecture, University of Queensland (UQ) in 2022. The studio was run in parallel with an architectural studio on floating structures at RMIT, run by Dr. Nic Bao (see Chap. "Designing for Artificial Reefs: The Role of Architecture in Marine Ecosystem Restoration"). It argues that integrating buoyant urbanism into architectural education is essential for training new thinkers, advancing the field of research and enhancing public understanding and support for these solutions. The chapter articulates the six architectural paradigms that arise out of the historical development of floating structures and floating cities and features studio participant works that engage with these paradigms.

2 The Architecture Studio

The *Floating Cities: Buoyant Urbanism, Ecological Engineering and Marine Utopias* architectural design studio at the University of Queensland provided a platform for students to critically engage with floating structures as a response to urban and environmental challenges. The studio explored the historical evolution of buoyant urbanism and its contemporary applications, encouraging students to test speculative and practical design strategies. By situating floating architecture within a pedagogical framework, the studio fostered interdisciplinary collaboration and contributed to an emerging body of research on water-based urbanism. The studio was also designed to provide a space for participants to experiment with concepts like buoyant materials, ecological integration and modularity, contributing to the broader research field of buoyant urbanism. The studio outline is extracted below:

> This studio investigates the expansion of coastal cities onto its adjacent marine environments. In November 2021, Busan in South Korea agreed to host a floating city in collaboration with OCEANIX and UN-Habitat. Floating cities, like the one proposed, have been an idea we have returned to repeatedly over the past 70 years. This masterclass investigates the socio-political drivers behind these proposals: from historical floating settlements as fishing villages or settlements built in volcanic lakes for defence, to the explosion of ideas in the 50's through to 70's where architects like the Japanese *Metabolists* and Buckminster Fuller reconceptualised what the city could look like. These expansions of the city onto its adjacent body of water (whether sea or lake) fuelled fringe communities to think about how they could create their own marine utopias floating outside of the law. These ranged from luxury retreats for the wealthy to avoid taxes or indulge in speculative medical treatments to prolong their lives, to data centres that would sit outside data protection or information privacy laws, and to micronationalism (seasteading). But in the past ten years, we've returned to this idea again by positioning these floating settlements as a potential means of addressing rising sea levels and land scarcity (see Vincent Callebaut's *Lilypad* and the above-mentioned *Oceanix City*) and to provide homes for climate refugees. Through the University of Queensland Art Museum's exhibition *Oceanic Thinking*, students will be invited to think about the ocean and its adjacent terrestrial spaces through the eyes of artists from around the world. The ocean's complex cultural significance and its entangled relationship with human life will be central to our discussion.
>
> We will explore three dimensions to our designs: at the detailed level, we will consider ecological engineering techniques that can be embedded in offshore structures, such as crevices for marine life in our structures or cutouts in floating structures to allow light to penetrate into the marine environment below. At the street level, where we consider how to create circulation and spaces of refuge to enable urban occupants to enjoy an intimacy with the water, or how we harness NIMBYism to get people to emotionally safeguard their little patch of the ocean. And finally, at a more macro level, the massing and form of the floating structure in order for it to meet its function as a quarantine neighbourhood or campus or theme park, and still float. We will build these as models of pontoon or semi-submersible structures to test this on water. Finally, as a studio, your individual floating pieces of infrastructure will be assembled on water to present a masterplan of a hybrid floating city for South East Queensland.

The studio was timely given that it ran at the start of the *UN Decade of Ocean Science for Sustainable Development* (2021–2030) and followed on from the *2019 High-Level Round Table on Sustainable Floating Cities* (New York). But going back a little further, the studio followed in the wake of 'The Metropolis Issue', a visionary publication from the 1960s neofuturist architecture group *Archigram* that articulated an optimism of our wet urban future. In the issue, the seductive works of Kikutake's *Marine City*, Buckminster Fuller's *Underwater Island* and Warren Chalk's *Underwater City* project were put out into the public imagination, inspiring architectural design of ocean cities—a veritable smorgasboard of marine utopian proposals—over the next decade.

3 Taxonomy of Buoyant Urbanism: Six Design Paradigms

Buoyant urbanism is a dynamic field that draws upon historical practices, visionary proposals and contemporary innovations to address the challenges of living on water. The architectural history of floating cities can be broken down into six movements that are characterised by the motivation driving the actual design of the scheme. These movements provide a ready taxonomy allowing us to categorise these designs and provide a comprehensive framework for understanding the breadth and depth of buoyant urbanism and its diverse applications. Teaching students about these paradigms not only deepens their understanding of the design possibilities but also equips them to navigate the complexities of implementing such projects.

The studio began in the first three weeks with lectures that provided the historical context to the design of floating cities. This allowed studio participants to examine the utopian drivers or intellectual trends that characterise these design paradigms of floating cities and structures. Participants were then asked to begin their studio research on the drivers (anxieties, perhaps) that could potentially lead to a new wave of floating city proposals.

This chapter sets out these distinct paradigms chronologically and provides examples of student exploration of the social drivers that underpin designs falling within each category. These paradigms provide a useful taxonomy that continue to be relevant to contemporary floating projects, though there is necessary overlap between the categories—particularly as succeeding paradigms emerge out of the drivers and concerns of earlier paradigms. At the same time, given the complexity of infrastructure projects and the wide spectrum of proposed stakeholders and end-users, individual projects and their programmed spaces can span across categories (as seen in the featured projects below).

3.1 Traditional Floating Settlements

The concept of floating cities is not new. Historical precedents and visionary projects have laid the groundwork for contemporary buoyant urbanism. Floating settlements that exemplify how humans have adapted to life on water have supported communities for centuries, including the floating boat communities of Southeast Asia in Halong

Bay [32] and South Sulawesi [25], to modern floating suburbs in Vancouver and Amsterdam [35]. There are also floating landscapes, most notably the floating farms from Mesoamerica. These floating islands, or *Chinampas*, were created from weaving reeds, which were then built up with sediment from the water body, compost and mud [26].

These water-based settlements are part of a much broader global history of settlements involving the creation of habitable space from surrounding water. These settlements include the artificial islands of the Lau people of the Solomon Islands [10], the crannogs of Scotland that date back to the Neolithic times [23], *Nan Madol*—a temple complex in the Pacific [22], *Dejima*, created in 1635 in Nagasaki Bay [41], and the largest land reclamation project in the world—the Dutch *Flevopolder* [27].

However, historical and contemporary floating settlements tend to be an agglomeration of small floating structures, such as houseboats, floating houses, that are connected together by a variety of informal structures, or through water-based transport systems. As observed by Volz in Chap. "City Parts: Urban Form Design Guidelines for Floating Cities", these forms of floating settlements have maintained at this scale, and they are low-rise and oriented primarily on housing with informal planning arrangements (eg. Makoko in Lagos, Nigeria). Studio participant Edward Huang explored the idea of scaling up these floating settlements, focusing on how these urban forms related to the water at street-level and how residential density could be increased (Fig. 1). Huang's work leans on previous work by Koen Olthuis (co-author of Chap. "Brain Coral: Floating City in the Maldives") who designed the 'world's first floating apartment complex, The Citadel' [17], an apartment block with 60 units, car park and a floating road that would shift in height as it floats the rise and fall of floodwater.

The physical connections between the various proposals from the scheme were also investigated, with studio participants considering different ways of allowing different types of traffic to traverse between proposals (Fig. 2). These linkages then allowed for the agglomeration of the various proposals to form a floating city that was tested in an outdoor pool on campus.

Studio participants were also encouraged to explore the cultural connections to the water that led to the longevity of our history of building on water. The studio benefited from the guidance of Dr Catherine Keys, Senior Lecturer in Architectural Design, who provided additional lectures and consultations with participants to examine and challenge the idea of the sea as a form of tabula rasa (or '*Mare nullius*', see [24]) and to query how conceptions and previous cultural usages of the sea could be incorporated into the design concepts.

At the same time, given the difficulties of conducting a site visit and analysis in the middle of the sea off *Minjerribah* (the 'place of many mosquitos', Stradbroke Island), one of the studio sessions was conducted at the UQ Art Museum where participants were immersed in the exhibition '*Oceanic Thinking*' [5]. This exhibition was curated to present: 'new ways of understanding the ocean and the adjacent blue spaces of our planet. It [invited visitors] to consider how we may be able to think together with these liquid, vast, biodiverse and non-binary spaces to speculate on our collective future… Artworks [revealed] cutting edge perspectives and

Fig. 1 Huang's work explored the intimate spatial relationship between a floating settlement and its aquatic site, while examining how traditional floating housing forms could achieve greater residential densities

Fig. 1 (continued)

research, and [stimulated] discussions about race, the ongoing extractive colonial project, climate cruises, decolonisation' [34]. These discussions led studio participant, Annalise White, to observe that 'the sea played a significant integral part in the preservation and maintenance of these distinctive [local indigenous] values and traditions during European colonisation, forming a cultural moat that isolated these islands from the mainland'.

A number of studio participants explored how floating city proposals could sensitively engage with the cultural context of its site, to explore how traditional settlement forms can equip us with valuable insights into sustainable practices and community-oriented design that would co-exist with the aquatic landscape. In doing so, these studio participants suggested that architectural design needed to engage with the cultural practices and needs of the local community, the *Quandamooka* people. Studio participant Naomi MacKenzie drew inspiration from the shifting sandbanks that the community would inhabit and the various social and economic activities they would engage with (Fig. 3). She proposed that the external shell of the proposal would serve as an artificial reef for aquaculture (oyster production) (Fig. 4).

In the same vein, studio participant Annalise White was inspired by the eugarie shells that are found on *Minjerribah* that indicated gathering places (Fig. 5) and continue to be an important diet of the *Quandamooka* people.

The opened eugarie shells were used as a jumping off point for the planning and design of her project (Fig. 6), but importantly, inspired her to think about sand dunes and sheltering spaces, where she designed structural vessels to accommodate mangrove plants that could be connected to the floating structure and act as a floating breakwater and windbreaker (Fig. 7).

The enduring relevance of traditional floating cities lie in their ecological integration and adaptability. They provide insights into resource-efficient construction

Fig. 2 Studio participants explored floating walkways and floating marinas in deciding how to link up their proposals as a floating city. Chris Huang's work (bottom left) as a satellite installation and as part of the agglomerated whole

techniques, waste management, and community resilience, which are critical for modern buoyant urban development [39]. By studying these settlements, contemporary architects and planners can incorporate lessons from the past into designs that address present and future challenges. In particular, the emphasis on locally available materials and minimal environmental impact offers a roadmap for sustainable floating city development.

The subsequent five architectural paradigms concern the attempts to build new cities and capture contemporary innovations in buoyant urbanism. These designs take many forms, reflecting the diverse needs and aspirations of contemporary society to: first, answer the anxieties of overpopulation and resource pressures; second, escape to floating leisure colonies that serve as exclusive resorts or luxury enclaves; third, establish new city-states that self-govern; fourth, float infrastructure that would house sensitive facilities near to coastal cities but away from densely

Fig. 3 Naomi MacKenzie's proposal took inspiration from the shifting sand movements at Jumpinpin Point, Minjerribah (South Stradbroke Island) and conceived of her proposal as a space for the local Quandamooka people to live and farm oysters

populated areas; and fifth, provide floating housing and suburbs that respond to the emergent climate crisis and shelter communities.

3.2 *Marine Utopian Proposals Oriented on Population Anxieties*

Marine utopian proposals emerged from the 50s to the 70s largely in response to a global anxiety about overpopulation and the need to change the way our cities work.

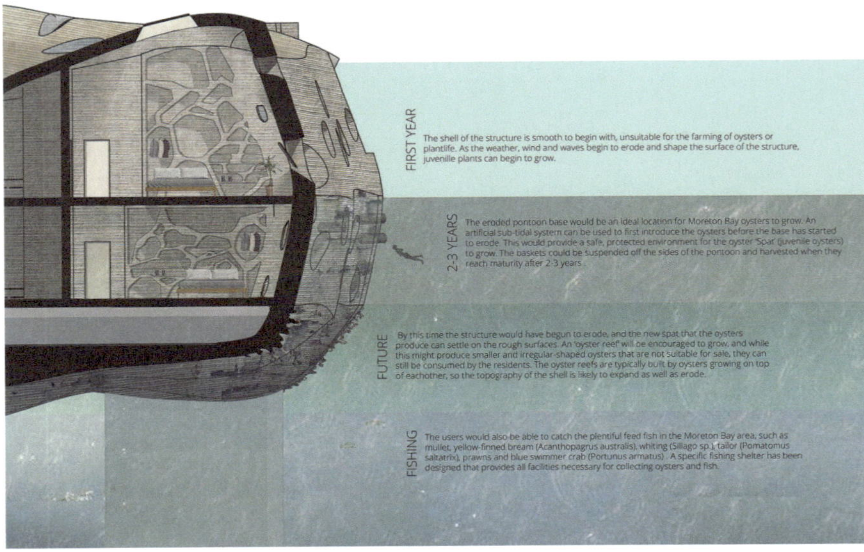

Fig. 4 Mackenzie conceptualised the outer shell of the floating structure as cladding that would gradually erode through wave and wind action, to then take on new life as an artificial reef for Moreton Bay oysters to grow

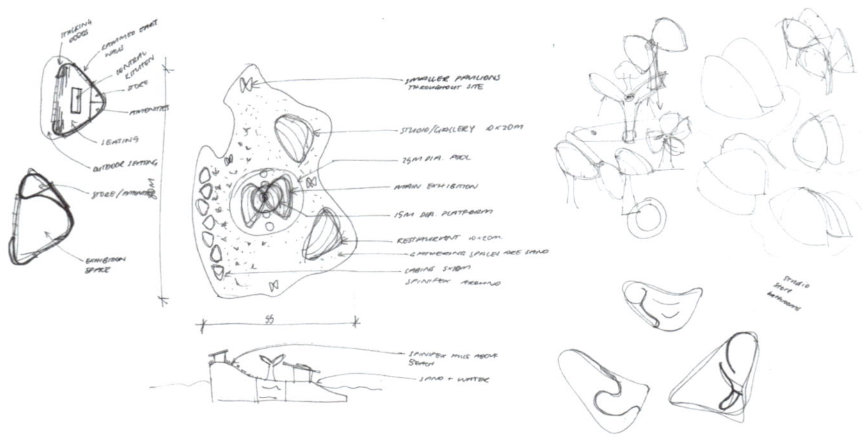

Fig. 5 Floating community and gathering spaces to support local cultural practices with spatial forms inspired by Eugarie shells

These proposals sought to address urban congestion and environmental degradation, leveraging our fascination with living with water to produce a plethora of designs for cities that inhabited the spaces on and under the surface of the sea. These included the Japanese *Metabolist* Kikutake's *Marine City for Tokyo Bay*. The Metabolists, a group of Japanese architects, proposed floating megastructures as part of their vision for a

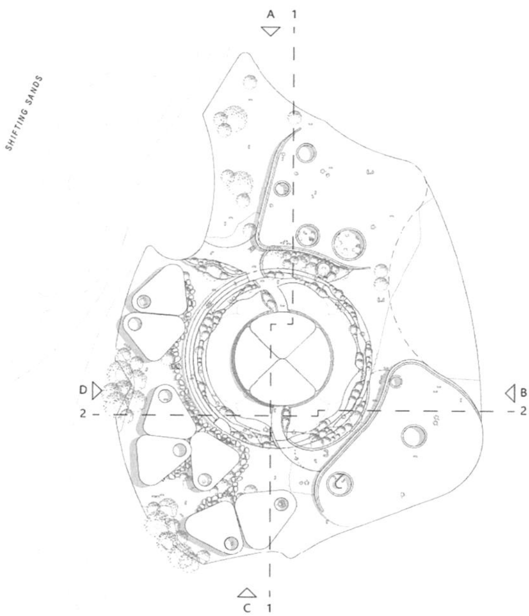

Fig. 6 Opened Eugarie shells inspired the overall planning and spatial forms of the project

flexible and scalable urban future. These projects were inspired by the idea of creating self-contained, expandable habitats that could evolve over time. Buckminster Fuller's *Triton City*, conceptualised in the 1960s, envisioned a floating tetrahedral structure in Tokyo Bay designed to house thousands of residents. Fuller later repackaged the idea for possible siting in the United States, underscoring the ability for such floating structures to be redeployed. This intellectual trend in the history of floating design led to the first UN Habitat conference (*Habitat I*) in Vancouver in 1976.

But the heady architectural experiments with ocean cities faded as utopianism was gradually erased from the practice of architecture at the end of the period. However, while many of these visionary marine utopian proposals were never realised, their conceptual contributions continue to influence the discourse on buoyant urbanism. They introduced key principles such as modular design, scalability, and adaptability, which remain integral to contemporary floating city projects. At the same time, while there persist valid critiques of utopian aspirations and the privileged few who are able to participate in shaping its ideologies and imagery in the architectural tradition, Hilde Heynen [14, pp. 382–3] offers a robust counterpoint as to why it is critical we continue to teach utopian thinking in architectural curricula:

> There is nevertheless much more to utopian thinking… One very important element is the critical capacity that is inherent to utopian thinking. As David Harvey remarks in Spaces of Hope, it is only by revitalizing the utopian tradition that we will be able to fuel a critical

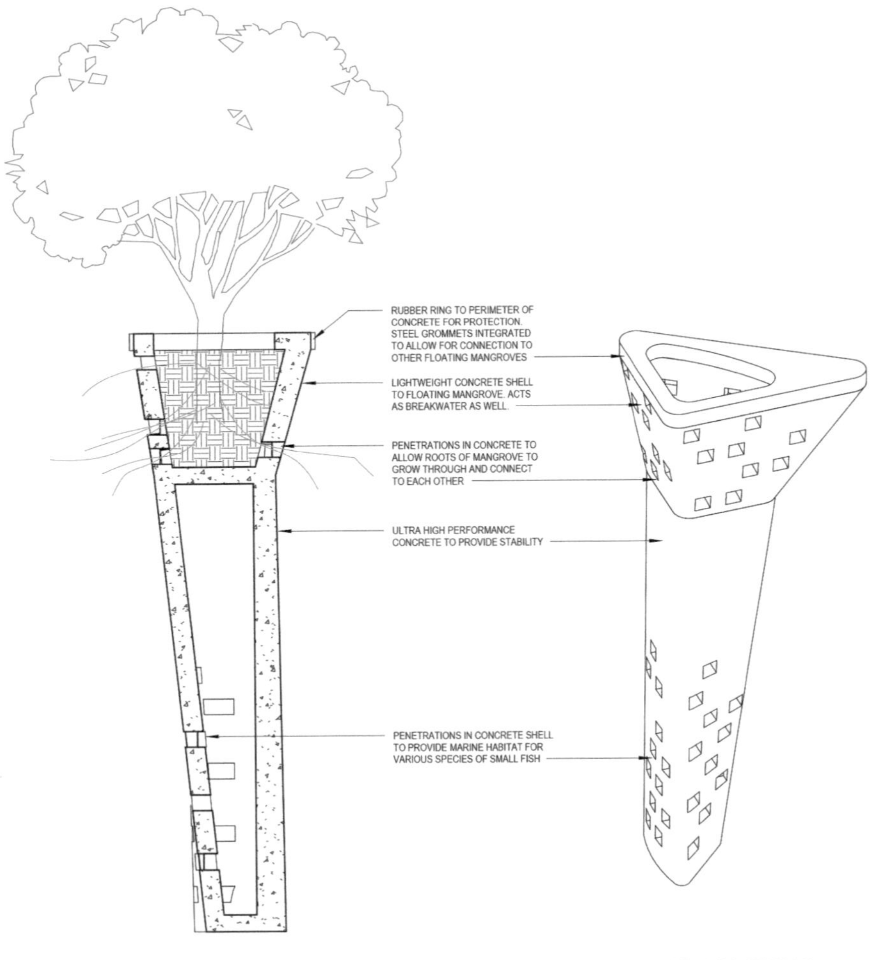

Fig. 7 Detail of a floating mangrove that form part of a floating breakwater, again planned out using the shape of the eugarie shell as inspiration

reflection that will help us to act as conscious architects of our fates rather than as helpless puppets of the institutions and imaginative worlds that we inhabit.

At the heart of this critical capacity is perhaps hope for new ways of living. The need to expand our cities to inhospitable territories and the marginal wet spaces of our cities lies in a simple promise: if we are bold enough to embrace a technological future that reaches over the surface of our adjacent water bodies, we are given the means to address many of the challenges that accompany city-making: overpopulation, the need for open spaces and access to sunlight, and to provide policy makers with options on how we can equitably share and alleviate pressures on diminishing land-locked resources.

Studio participants were introduced to over a half a century of floating city designs, with design commonalities identified. These proposals typically depict the marine utopian settlement from an altitude that allows views of the intervention in a highly formalised manner, and with readily deciphered geometric configurations (Fig. 8). In certain cases, the *absolute* quality of the geometry associated with the schemes can be perceived as symptomatic of a utopian driver that privileges image and the notion of order and perfection over the ahierarchical urban growth patterns of traditional urbanity. These images reflect a preoccupation with the scheme's fabricated origins and the intention to offer the scheme as an antidote to the chaos of their host cities. In fact, it is a rarity among marine utopian proposal for a scheme to not have an overall absolute geometric form that is immediately readable and iconographic.

From an architectural perspective, the development of marine utopian proposals has clear parallels with medieval ideal cities. The highly utopian iconography of star forts or the *trace italienne,* conceived in reaction to the walled-in chaos of medieval towns, find an architectural homologue with marine utopian proposals. Both are planned in strict adherence to rational and aesthetic requirements. In the case of medieval star forts, this rational requirement emerges from a military criteria, demanding absolute geometric configurations for ramparts and other fortifications in order for the city to be battle-ready [30]. Similarly, the prevalence of the hexagonal or pentagonal shape in marine utopian proposals is predicated on findings that this particular geometric shape is one of the more stable forms in water, able to withstand the various hydraulic forces acting on the floating structure. Thus, this geometric shape has evolved to become synonymous with 'perfection' on the water.

The overall geometric forms of these schemes are carried through in the architecture as well. Images of the floating city proposals found in the marketing brochures from the Floating Structures Association of Japan show startling similarities to the 1968 Triton City by Buckminster Fuller, Jacques Rougerie's 1976 *Archeoscope*, Odaiba's Tokyo Big Sight, and the Pyramid City TRY 2004 (Fig. 9). These designs

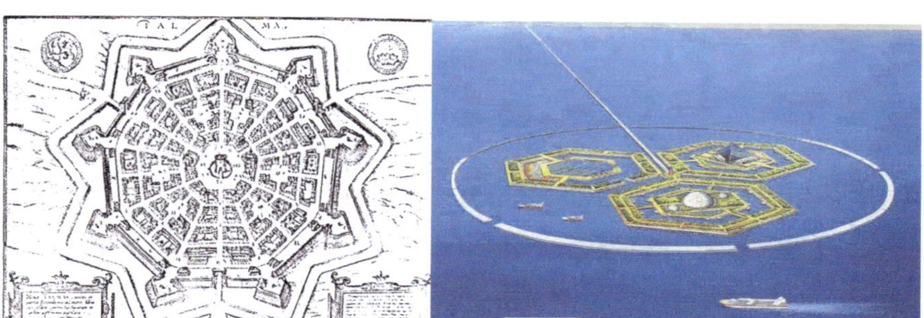

Fig. 8 The highly geometric forms of marine utopian proposals such as the one on the right from the Japanese Society of Steel Construction are often depicted at an altitude, reminiscent of the star forts of medieval ideal cities

were in turn influenced by the revolutionary illustrations of Frank R. Paul, such as the *City of the Future*, 1942, and the work of Paolo Soleri (Fig. 10).

Studio participant Nina Sappath's project explored the concept of pristine bubble architecture, drawing inspiration from Buckminster Fuller's geodesic domes and the *Amazon Spheres* in Seattle (Fig. 11). Her proposal envisioned a network of floating, self-sustaining arks designed to restore and protect fragile coastal landscapes that have been severely impacted by urban development. These transparent, domed structures, engineered to float on ocean currents, function as ecological sanctuaries, supporting marine biodiversity while simultaneously providing controlled environments for human habitation and research. Nina's work builds upon the historical lineage of marine utopian proposals, which have long imagined the sea as a site for regenerative architecture. In doing so, her project reinterpreted the classic utopian aspiration of oceanic settlements as floating refuges that actively heal degraded environments, offering a speculative yet pragmatic vision of buoyant ecological urbanism.

Fig. 9 Similar geometric forms shown in the work of: (top left) Jacques Rougerie Architecte's (1976) *Archeoscope*, (top right) the floating city proposal, 'Look at the Sea' by the Floating Structures Association of Japan, (bottom left) the Pyramid City TRY 2004 [30], and (bottom right) the Tokyo Big Sight in Odaiba (1996) reveal the influence of structural expressionism (high-tech architecture) in each of the schemes

Fig. 10 Science-fiction illustrations such as the *City of the Future*, 1942, by Frank R. Paul (left) and the utopian proposals of Paolo Soleri (right, [8]) bear significant influence on projects like the Tokyo Big Sight in Odaiba

A similar process of idealisation occurs with the Tokyo Big Sight that draws reference to the oil rig typology (the semi-submersible), a form that has been engineered for the ocean. As pointed out by Thomsen [30], these interventions reflect a design rationale in which 'utility and a practical approach to life are supplemented by precise ideas about the organisational form of [the utopian ideal], whose hierarchical structure is reflected even in the arrangement of the buildings'. These 'precise ideas' are utopian drivers that dictate the pristine geometric forms of these schemes, best summed in the fundamental considerations of leading Renaissance theorist, Leon Battista Alberti:

> The ideal of a city and its tasks, according to philosophers, may be seen in the fact that here the inhabitants lead a peaceful life, as free as possible of worries and care. I maintain, further, that a city should be conceived such that of all the disadvantages… none at all should be in evidence. And of all the things desirable for the necessities of life, none should lack [30].

However, as noted by Curtis [6, p. 65), 'utopian totalities – even pieces of them – do not mesh well with an urban grain that has taken centuries to evolve'. Accordingly, studio participants were pushed to consider how utopian architectural design can be solipsistic and static by turning away from responding to the site and the evolving needs of inhabitants.

In considering this point, the concept of the *Megastructure* is highly congruent to the understanding of many early marine utopian proposals. These projects possessed an 'object quality' which required each scheme to be read as a complete whole

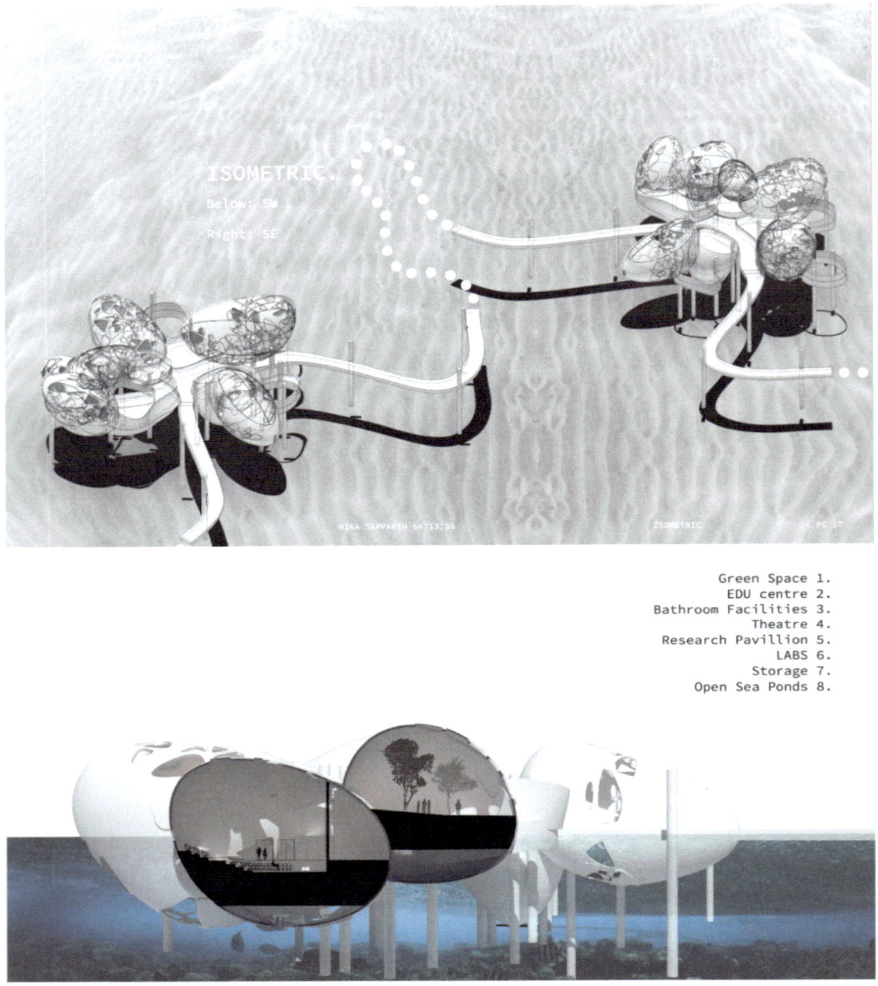

Fig. 11 Nina Sappath's proposal referenced water bugs and the bubble-architecture of technology campuses

Fig. 11 (continued)

with individual pieces subjugated under Beaux Art ideals of axial arrangement and symmetry. Ioan [15] criticises the megastructure, describing it as a 'mega/meta organism' that is fixated on the 'biological, internal functioning of the whole', privileging the structural grid, so apparent in the iconography of many utopian proposals, over the individual 'atomic' inhabitants distributed within.

In fact, Ioan posits that the megastructure is not a 'neutral grid' as proposed by Le Corbusier in his hospital project in Venice, stressing that the frame is 'dominant, permanent, fixed and structuring', and is constantly expressed in a monumental way, eliminating its neutrality. Thus, the frame cannot be considered the background against which the city projects its functioning, but 'the functioning mechanism turned into the very essence of the city environment' [15]. The structural grid, expressed through the urban elements of streets, pedestrian and vehicular traffic networks, city blocks, and roof planes, often becomes so integral to the architectural expression for many utopian proposals that it cripples the subject scheme from its ability to re-conceptualise its architectural identity. In its attempt at accommodating these new economic social units, it falls back on a generic urban congruence with traditional urban patterns that reduces the utopian driver to establish order and perfection to a mere methodical arrangement of these urban elements. This highly ordered form of planning is symptomatic of utopianism and is best summed by Le Corbusier:

> The modern city lives by the straight line, inevitably; for the construction of buildings, sewers and tunnels, highways and pavements. The circulation of traffic demands the straight line; it is the proper thing for the heart of the city. Geometry is the foundation... the material basis on which we build those symbols which represent to us perfection and the divine. [31, p. 139]

In studio, participants were asked to consider how the megastructure could be made less dominating, and instead be enlivened as a hive-like structure that is adapted to the changing needs of occupants. Studio participant Scott Kim's project on the *New Plug-In City* (Figs. 12 and 13) turns the limitations of the structural grid on its head by exploiting the original Metabolists' approach to megastructures—where repurposed shipping containers are spatially programmed for different functions that are then flexibly 'plugged into' the megastructure where and as needed. In doing so, Scott's project explores the grid as an organising principle to enable flexible programming of space to suit occupant needs.

The inherent attention to the dynamic needs of occupants and the resulting changes to the functional and spatial needs of a floating city recalls Christopher's [2] *A City is Not a Tree,* which criticised the undercurrent of structuralism inherent in such schemes as 'a compulsive desire for neatness' and proposed his notion of the 'spontaneous city'—a city that forms through autopoiesis. Alexander championed the structural model of the *semi-lattice* rather than a tree as it presented an ahierarchical network, allowing for 'an organic system capable of dynamic and complex growth'.

Despite having a similar pristine overall shape to many of the structuralist proposals, *Mirage City*, an artificial island proposed by Arata Isozuki in 1997, attempted to capture the spirit of morphogenesis through design interventions made by external participants via the internet. The process involved plotting out the distribution of churches in Venice onto the site to create nodal patterns, which were subsequently connected by the participants to form local street patterns. An amalgamation of these 'synapse-like networks' generated a resultant randomness that 'simulated the process through which urban sectors formed', imbibing an organic quality reflective of urban autopoietic agglomeration (Fig. 14).

Fig. 12 Scott Kim's *New Plug-in City* revisits the Metabolists' approach to the megastructure and its ability to plug in and plug out programmatic elements as and when needed

Fig. 13 Scott Kim's *New Plug-in City* comprises repurposed shipping containers refurbished to meet different functionality requirements, which are then plugged into the floating semi-submersible megastructure

A similar attempt at reflecting the morphogenesis of cities through synapse-like networks was examined in studio participant Jack Reboul's proposal, which sought to co-locate an artificial reef as part of a floating well-ness centre. The project sought to capture a floating structure that had a base that would continuously grow, allowing it to counterbalance and enable more and more structural pods to be built on (grown on) the existing superstructure. This was a different take to the traditional approach to the megastructure, one that was built on an organic floating foundation that extended

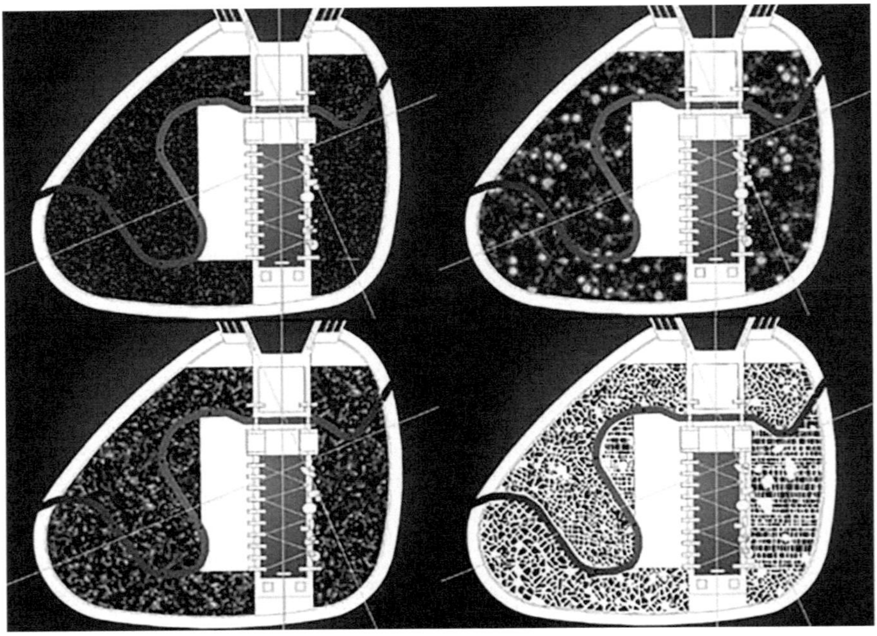

Fig. 14 Intervention by participants creates a simulation of autopoiesis in Arata Isozuki's *Mirage City* (1997)

the megastructure below the ocean surface and would progressively allow for more programmatic pods to be added onto the project (Figs. 15 and 16). In a way, it recalls Kisho Kurokawa's Nakagin Capsule Tower Building, except that pods can only be added once sufficient coral growth is achieved to stabilise the superstructure. A parallel architecture studio run out of RMIT explored this co-location of artificial reefs with adjacent floating structures in detail (see Chap. "Designing for Artificial Reefs: The Role of Architecture in Marine Ecosystem Restoration").

Isozuki's work on morphogenesis predated the algorithmic design tools available to us in recent times. If the generative force of utopianism is built on speculation, exploration and forecasting of potential design solutions, the advances in algorithmic, AI generated design—particularly the text-to-image diffusion models that are based on Generative Adversarial Networks (GAN) [7]—provides further opportunity for future-casting that may move marine utopian proposals beyond its top-down approach to designing floating cities. Likewise, Bao's Chap. "Designing for Artificial Reefs: The Role of Architecture in Marine Ecosystem Restoration" provides new parametric design tools to explore how we model these organic growth of cities and even non-human inhabitants of floating structures.

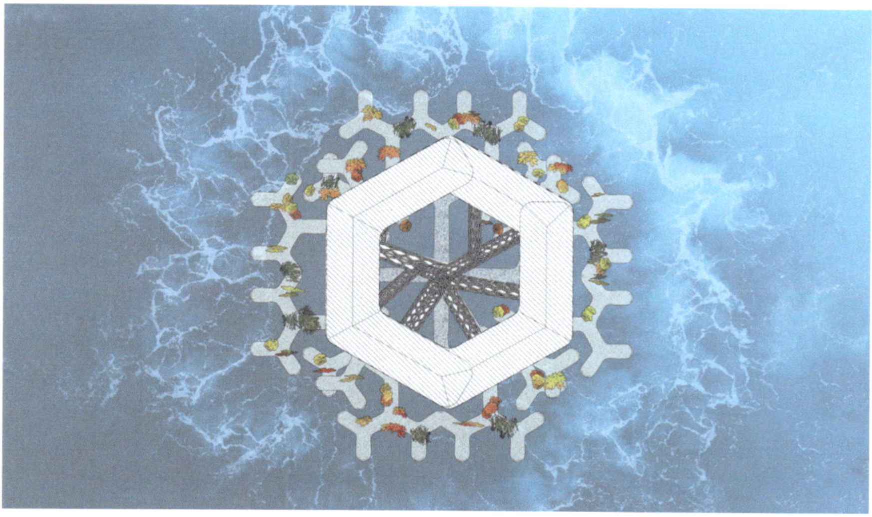

Fig. 15 Jack Reboul's floating proposal speculated that as the artificial reef foundation of the structure grew in size, it would increase the buoyancy force available for additional pods to be added to the superstructure

3.3 Floating Leisure Colonies

However, towards the end of preceding marine utopian paradigm, a pervasive cynicism emerged, leading to a significant hollowing out of funding for visionary marine utopian proposals. As financial support waned for these speculative projects, the architectural pursuit of marine cities shifted towards private corporations targeting affluent markets. These initiatives adopted the exclusive characteristics of gated communities, crafting proposals that engaged niche markets and fringe lifestyles [18].

Floating leisure colonies epitomise the commercial side of buoyant urbanism, catering to wealthy clientele through luxury resorts and exclusive residential enclaves. These developments prioritise aesthetics, exclusivity and high-end amenities, creating waterborne spaces that serve as symbols of wealth and innovation. While these projects are often criticised for their limited accessibility and environmental impact, they demonstrate the feasibility of creating liveable and highly seductive spaces on water.

Contemporary cruise ships can be seen as the modern counterparts to these historical leisure colonies. In fact, these have led to court cases where cruise ship operators have argued that these vessels are transient settlements on the sea and should be conceptualised as floating cities [36]. These vessels not only embody maritime mobility but also align with the concept of 'Ephemeral Cities'—a term used by Kotkin [19] to describe urban forms that evolve in response to new socio-economic paradigms, including Richard Florida's idea of the creative class. In these

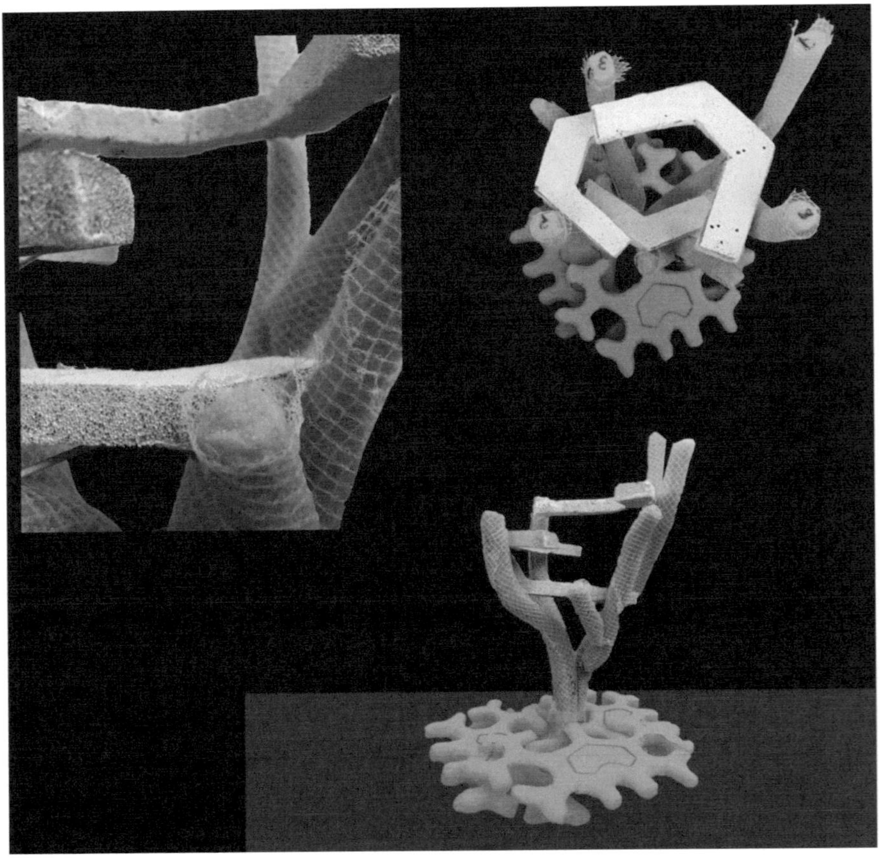

Fig. 16 The floating proposal for coral suggested that as the artificial reef grew in size, it would increase the buoyancy force available for additional pods to be added to the superstructure

emerging urban models, the emphasis shifts from traditional centres of commerce and political power to environments where curated experiences—embodied in galleries, boutique shops, vibrant street life, and tourism—take centre stage [21]. Other modern applications include floating sporting facilities and cultural centres.

Modern proposals have introduced floating sports and leisure facilities, notably the various floating Olympic infrastructures (proposed and realised) for the Olympic games in Paris and Rio, including the proposed floating stadium for Rio by Water-Studio (Fig. 17). The recent Parisian Olympic parade down the Seine showcased the integration of buoyant structures and leisure and entertainment. Likewise (and linking this back to cruise ships), the *Aaranui 5* served as the Tahitian Athlete Village as part of the Paris Olympics. Similarly, the Marina Bay Floating Platform in Singapore was used for the Youth Olympics Opening Ceremony and also as a football stadium.

Fig. 17 Proposed floating stadium by WaterStudio for the Rio Olympics Games

One of the initial exercises for the architecture studio was a rapid design task that encouraged participants to conceptualise a floating structure and appreciate its mobility—both physical and socio-political. The task required them to envision how functionality, leisure, and political aspirations might converge in buoyant design. Drawing inspiration from Rem Koolhaas' text on 'The Story of the Pool' from *Delirious New York*, participants produced A3 posters that responded to the interplay between mobility and form in floating leisure architectures. These schemes have a rich history as explored by Saharinen and Sirola in Chap. "Floating Leisure Infrastructure: Integrating the World's Largest Floating Infinity Pool in Lake Como". The exercise inspired studio participant Ryan Todd to develop a floating tennis centre for the 2032 Brisbane Olympics with associated health facilities (Fig. 18). Todd sought to capture the mobility inherent in a floating structure by building his design around the idea of the floating structure as circulation space, akin to a bridge that could be plugged into various parts of the city on its adjacent water space, extending its pathways and city functions over the water to connect disparate parts of the city (Fig. 19).

Another participant, Zoe Wells produced a floating performance centre (Fig. 20). This proposal integrated sound and water as phenomenological devices to explore marine cultures. The design was inspired by the interaction of sound waves with

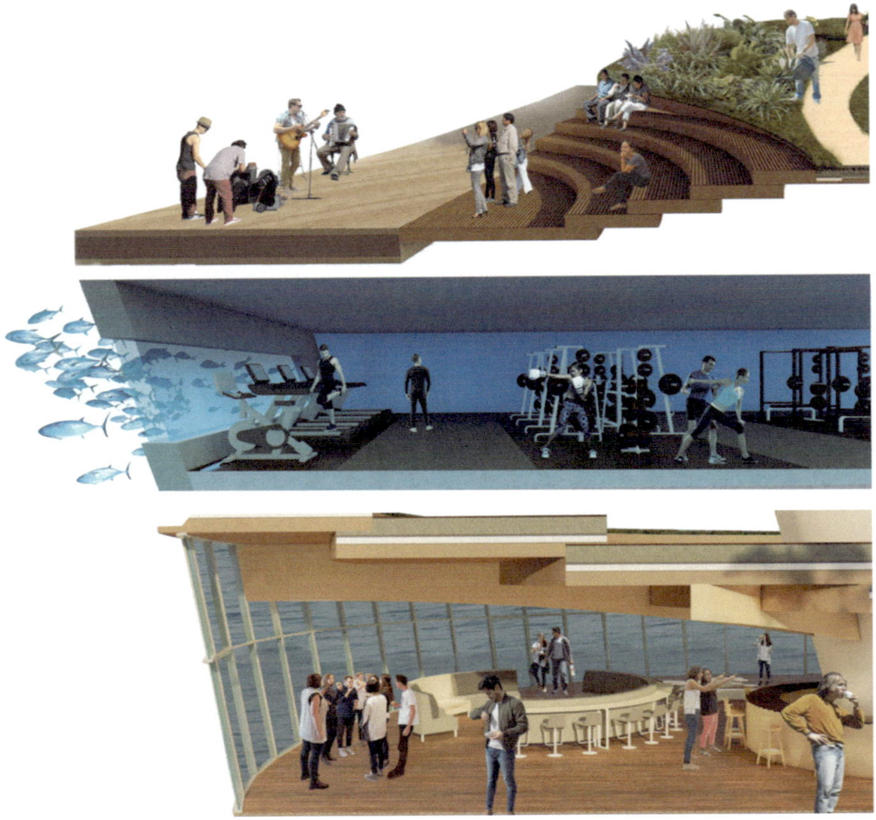

Fig. 18 The floating proposal for a multi-level health facility built around tennis infrastructure for the 2032 Brisbane Olympics

water, translating this concept into spatial dynamics of public vs. private, light vs. dark, and above vs. below, ultimately guiding visitors through a sensory journey culminating in an auditorium. The project drew on evolutionary psychology, emphasising the role of sound in human connection, memory and the formation of communities, positioning soundscapes as essential to spatial experience. By prioritising the acoustic dimension over visual representation, the proposal for an immersive, time-based engagement with place was oriented on evoking a profound sense of belonging.

Beyond their luxurious appeal, floating leisure colonies contribute significantly to the broader discourse on buoyant urbanism. They often incorporate advanced technologies including renewable energy systems, desalination facilities and innovative construction materials. For example, integrating solar panels and water purification systems into luxury floating villas not only enhances the aesthetic and functional

Fig. 19 The floating proposal prioritised circulation space to allow the structure to be plugged into various parts of the city to extend pathways and city functions over the water and connect disparate parts of the city

quality of these structures but also demonstrates the potential of sustainable technologies in maritime contexts. In this way, such sea colony projects serve as experimental laboratories, expanding the limits of architectural creativity and engineering to inform more inclusive and environmentally resilient urban applications.

3.4 *Sovereign Floating Territories*

The transition from floating leisure colonies to sovereign floating territories reflects a shift from exclusivity-driven developments to politically and economically motivated experiments in autonomy. While many early floating leisure colonies catered to wealthy individuals seeking luxury retreats that were an escape from the traditional urban structures of the time, these projects inadvertently laid the groundwork for more ambitious attempts at self-governing maritime settlements.

Accordingly, the intellectual trajectory of leisure colonies found an adjacent development in floating city proposals that sought to exploit regulatory grey zones or establish greater autonomy from terrestrial legal frameworks. The economic and legal advantages that made these floating resorts attractive—such as operating beyond conventional zoning laws, taxation regimes and regulatory oversight—drew the attention of libertarian and corporate actors interested in creating autonomous floating settlements. To facilitate such endeavours, industry and interest groups often asserted the potential legal sovereignty of such proposals, seeking to operate outside national

Fig. 20 Floating performance centre designed around the experience of sound on the surface of the water

jurisdictions, including as tax havens, medical ventures focused on anti-ageing and experimental treatments. These built on the tradition of micronationalism to construct autonomous enclaves in international waters.

Micronationalism, characterised by the creation of self-proclaimed sovereign entities, has long been intertwined with maritime autonomy. The ocean, as a space governed by international legal instruments such as the *United Nations Convention on the Law of the Sea* (UNCLOS), presents unique opportunities and constraints for such endeavours. UNCLOS delineates the limits of national jurisdiction, defining territorial seas, exclusive economic zones (EEZs) and the high seas (international waters), thereby structuring the possibilities for floating sovereign entities (see Chap. "Buoyant Futures: Our Global Duty to Recognise Floating Cities as 'Territory' in International Law"). While some micronations have successfully operated within the legal ambiguities of unclaimed maritime spaces, others have faced resistance from established states. The legal precarity of these projects underscores the tension between sovereignty and the regulatory reach of nation-states.

The shift towards sovereign floating territories was evident in projects like the *Freedom Ship,* a proposed floating city offering permanent residency in international waters [29]. In more recent times, marine cities arising from this intellectual trend have rebranded themselves as 'seasteads', with a variety of technological and legal designs seeking to establish self-sustaining, politically independent communities on water. Advocated by organisations such as *The Seasteading Institute*, these floating settlements are envisioned as platforms for social, political, and economic experimentation, offering a blank slate for testing alternative governance models and regulatory frameworks.

Some proposals have sought host nation partnerships, as exemplified by the ill-fated agreement between *The Seasteading Institute* and French Polynesia (Wong 2017) [28]. The proposed seastead was met with significant opposition, with concerns about sovereignty, territorial integrity, and the economic implications of ceding maritime space to private entities, leading to the project's collapse. Similarly, an attempt to establish a seastead within Thailand's EEZ resulted in legal action and threats of imprisonment, highlighting the geopolitical challenges of such undertakings [1].

This architectural paradigm highlights the complex intersection of legal frameworks, economic incentives and architectural experimentation in the development of buoyant urbanism. In particular, it exposed tensions between state authorities and floating developments. While luxury resorts and exclusive enclaves were intended to maintain contractual agreements with host nations, projects aiming for political autonomy, such as seasteads, have faced significant legal resistance. The collapse of the French Polynesia seasteading agreement and the legal challenges faced by floating micronations, illustrate the limitations of attempting to establish sovereignty at sea (the proposal later found its sea legs as the UN-endorsed *Oceanix City*). Thus, what began as floating leisure colonies designed to serve niche markets gradually evolved into speculative proposals for self-governing city-states and urban developments, engaging with broader questions of governance, sovereignty and self-sufficiency—even within the geographic and legal boundaries of host cities.

Beyond aspirations of sovereignty, floating city proposals have increasingly aligned with commercial and research-driven ideologies and agendas. Some initiatives have sought to create knowledge communities and innovation hubs that circumvent national regulatory regimes, particularly in the realms of data privacy and biomedical research. These same concerns would likely extend to emerging regulation on offshore applications of artificial intelligence systems (for an extended discussion on this, see Chap. "Navigating the Legal and Regulatory Frontiers of Urban AI, Autonomous Maritime Robots and Floating Cities"). By siting research facilities in international waters, proponents argue that these centres could push the boundaries of scientific progress unencumbered by domestic restrictions [35].

This concept has also permeated architectural experimentation. One such proposal, developed by studio participant Faris Asri, envisioned a marine research centre incorporating a coral nursery and public-facing aquatic facilities to foster environmental education, conservation and true academic freedom (Figs. 21 and 22). Another participant, Mardee, focused on autonomy through modular origami-inspired lightweight structures, which could be reconfigured by occupants to adapt to evolving spatial needs (Figs. 23 and 24). These speculative designs illustrate how buoyant urbanism intersects with both technological innovation and self-governance.

While autonomy is often positioned as the central appeal of these projects, popular culture frequently casts maritime settlements in a dystopian light. Science-fiction cinema has played a pivotal role in shaping public perceptions of floating urbanism. Films such as *Waterworld* portray the ocean as a hostile, lawless expanse, while disaster films like *The Day after Tomorrow* and *Deep Impact* and the Boxing Day Tsunami have etched the dangers of living on or near the sea firmly in the minds of the public, adding to the anxieties associated with maritime life. Additionally, films like *Logan's Run* and *Silent Running* presented a bleak picture of closed ecological life

Fig. 21 Floating marine research centre by Faris Asri

Fig. 22 The floating marine research centre by Faris Asri was designed around a floating coral nursery

support systems (CELSS) which have, in turn, influenced public attitudes towards self-sustaining floating habitats. These narratives contribute to the broader ambivalence surrounding floating cities, reinforcing concerns about isolation, governance and security.

Public scepticism is further compounded by comparisons to oil rigs, which are frequently cited as the closest real-world analogues to floating cities. The perception of oil platforms as industrial outposts fraught with environmental, security and logistical challenges informs public resistance to the idea of permanent oceanic settlements. These rigs are also characterised as inherently claustrophobic spaces, vulnerable to piracy, political activism and extreme weather events (Beal 2004). In fact, *The Principality of Sealand*, a self-proclaimed micronation based on an abandoned sea fort, provides a lesson on the governance challenges of maritime sovereignty. One of its principals was detained without food and water following a hostile takeover attempt. The recent television series *The Rig* has reintroduced these concerns into the public consciousness, depicting maritime platforms as sites of confinement and peril, once again raising questions of the viability of floating cities as secure, self-sufficient communities.

Despite these challenges, the concept of sovereign floating territories remains an important provocation in architectural, legal and geopolitical discourse. While such projects face significant logistical and legal hurdles, they raise important questions about the relationship between territory, sovereignty and innovation in urban design. As climate change accelerates the loss of habitable land, buoyant urbanism offers a potential avenue for future settlement, particularly in regions vulnerable to rising sea levels. Floating micronations challenge traditional notions of territoriality and statehood, prompting legal scholars and architects alike to reconsider the frameworks that govern space and sovereignty. Ultimately, the intersection of architecture,

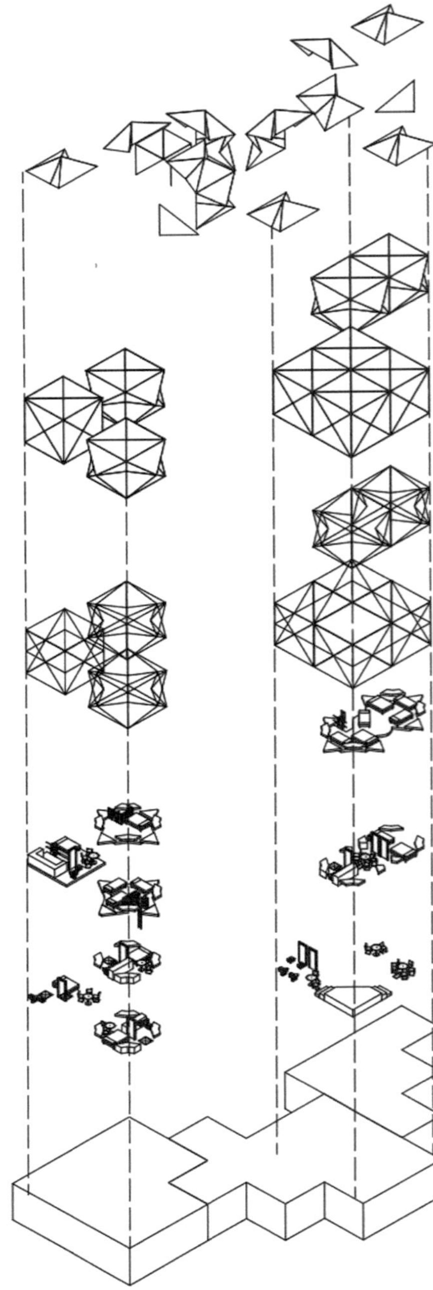

Fig. 23 Mardee's focus on autonomy and self-governance emerged as an origami-inspired super-structure that can be adapted by the community

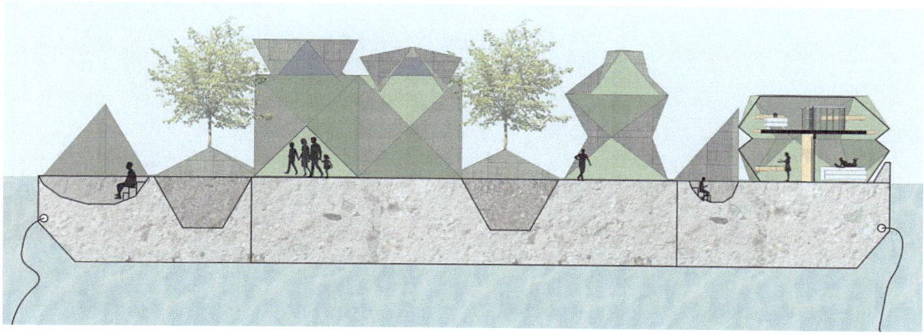

Fig. 24 Mardee's focus on autonomy and self-governance emerged as an origami-inspired superstructure that can be adapted to various functions by the community

law and political philosophy in these projects underscore the need for interdisciplinary collaboration in navigating the complexities of buoyant urbanism. A more fulsome discussion on the interaction of public international law,particularly the United Nations Convention on the Law of the Sea (UNCLOS), and these attempts at establishing sovereign micronations is discussed in Chap. "Buoyant Futures: Our global duty to recognise Floating Cities as 'Territory' in International Law".

3.5 *Floating Industrial Infrastructure*

Many marine utopian proposals, particularly those exploring undersea environments, were not 'not entirely altruistic' [18, p. 454]. Rather, they served as demonstrations of technological ambition, often aimed at securing defence contracts that required clear articulation of expertise in undersea systems [18]. The intersection of scientific and military research in this period led to significant advancements in subaquatic and floating technologies, with many architectural and engineering publications showcasing the resulting innovations. These developments highlight how offshore development and defence interests frequently overlap, driving investment into experimental projects that tested the feasibility of self-sustaining underwater habitats, submerged research stations and mobile maritime infrastructure. The resultant engineering outputs of the time then drove oil and gas exploration and production and led to greater innovation in large floating solutions (Fig. 25).

Over the past two decades, much of momentum in floating infrastructure projects has been driven by engineers, who have shifted the focus from speculative utopianism to large-scale industrial applications. As engineers transitioned into this space, filling the vacuum left behind by exiting architects, they have brought pragmatic, technical and large-scale industrial applications into focus. This shift resulted in projects such as floating oil storage facilities (Fig. 26), floating nuclear power plants, floating solar power platforms, floating emergency response centres [37], floating

Fig. 25 Innovations in large floating solutions, including the use of helical strakes in spars and semi-submersible floating structures to address vortex-induced vibrations that arise from ocean currents (above). A platform over a spar, a type of floating structure with a deep draft and long vertical hull (below) [13]

bridges (Chap. "Analysis and Design of Floating Bridges: Recent Advances and Challenges"), offshore wind farms (Chap. "Industrialization Challenges in Floating Wind") and fish pens (Chap. "Large Offshore Floating Fish Pens: Ocean Farm 1 and Havfarm 1").

Notably, Japan's *Mega-Float* project sought an alternative to land reclamation for Kansai International Airport, demonstrating how floating runways could offer a scalable and environmentally conscious alternative to land reclamation works [12]. Similarly, floating offshore wind farms have emerged as a promising avenue for renewable energy production, leveraging maritime space to generate electricity while avoiding the land-use conflicts associated with onshore wind installations. These projects illustrate how buoyant structures can host critical infrastructure that might otherwise be constrained by urban land shortages, the lack of social licence, or environmental concerns.

The increasing application of floating engineering solutions marks a departure from earlier leisure-oriented floating developments. The ability to relocate industrial and infrastructural elements to maritime zones has become a compelling driver of innovation, allowing cities to expand critical services without intensifying land-use pressures. While this book acknowledges the significant engineering achievements within this paradigm, studio participants were encouraged to push their design proposals in alternative directions and most participants focused on the other architectural paradigms. However, Milah Canon was a welcomed exception who was keen to explore the use of water in ceremony and the socio-political challenges of locating social infrastructure, such as cemeteries and funerary architecture (Figs. 27 and 28).

Fig. 26 Kamigoto floating oil storage base [33]

This continued the line of speculating thinking proposed by Bread Studios in 2017 for a floating cemetery as an alternative approach terrestrial graves and columbariums in land-scarce Hong Kong.

Fig. 27 In the same vein as locating socially-challenging infrastructure on the water, Milah Canon suggests new funerary architecture should be located on the sea

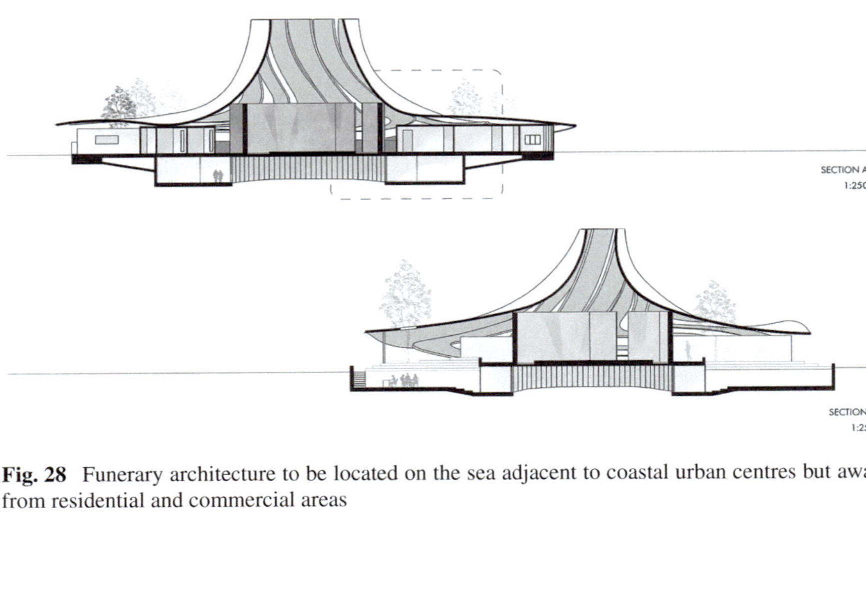

Fig. 28 Funerary architecture to be located on the sea adjacent to coastal urban centres but away from residential and commercial areas

Given the extensive body of work already dedicated to floating industrial infrastructure, such as those documented in this edited collection and the earlier *Large Floating Structures* [38], this chapter instead explores the remaining architectural paradigms to examine speculative and conceptual approaches that challenge conventional urbanism and reimagine waterborne architecture as a space for cultural, social, and political experimentation.

3.6 Climate-Resilient Urbanism

In a piece I wrote for The Conversation in 2019, I observed that the contemporary interest in floating cities as a response to the climate crisis gained significant attention in 2008 with Vincent Callebaut's *Lilypad* [35]. Callebaut's design was for a visionary floating ecopolis intended to house ecological refugees (Fig. 29). I included the project in the earlier *Large Floating Structures* as, in my view, the increasing encounters with the impacts of the climate crisis on our cities would inevitably lead us back down familiar pathways to examine if buoyant urbanism—floating homes, amphibious architecture, floating cities—could offer us a way to overcome the increasing costs associated with the unwanted incursion of water in our cities. This need not result in the creation of satellite communities in the ocean, but it would at minimum require a hybridisation of our cities where traditional structures sit beside permanently floating structures and amphibious structures that would rise with the incursion of water in our urban environments [39].

The catastrophic flooding in New Orleans in 2005 underscored the critical need for adaptive infrastructure in flood-prone cities. In response, Elizabeth English founded the *Buoyant Foundation Project* [9], focusing on developing amphibious foundation systems as a flood mitigation and climate change adaptation strategy. These systems allow structures to float during flooding events, reducing damage and enhancing resilience. Similar amphibious housing projects, utilising floating foundations have been increasingly investigated in various parts of the world.

In March 2022, as part of the *Asia Pacific Architecture Festival*, I was fortunate to be on a panel with flood historian Dr Margaret Cook, author of *River with a City Problem*. The urgency of exploring buoyant urbanism was keenly felt during our panel that coincidentally fell within the direct aftermath of the weeks following the 2022 floods in Brisbane, after the city experienced between 400 and 1100 mm of rainfall within a five-day period that lead to widespread inundation.

Studio participants were inspired by these developments and explored floating city proposals as a form of climate-resilient urbanism. Their designs emphasised ecological integration, modularity and accessibility, aiming to create habitats that not only protect occupants from harm arising from these climatic events, but are also environmentally sustainable and equitable. By envisioning how their projects would interact with marine ecosystems, they demonstrated the potential of floating cities

Fig. 29 Vincent Callebaut's Lilypad depicted as a proposal sited in Monaco [3]

to address both immediate environmental challenges and long-term sustainability goals.

Beyond environmental considerations, the studio also addressed social issues such as overpopulation and global migration. With increasing displacement of populations due to climate change and other factors, there is a pressing need to reconceptualise migrant settlements. Drawing inspiration from historical examples like the post-war *Silver Cities* in Australia, participants explored innovative designs to accommodate the globally dispossessed, particularly as inhabitants of low-lying islands are being relocated due to rising sea levels [4, 20]. These participants envisioned floating communities that offer dignity, safety, and opportunity for displaced individuals, emphasising the need for floating structures not only as extensions of the built environment but as essential components of future climate-adaptive urban strategies. Floating emergency centres and disaster relief hubs, for instance, have been proposed to enhance resilience in coastal regions vulnerable to extreme weather events. The modular nature of many floating designs also allows for flexible deployment and reconfiguration, ensuring adaptability to shifting economic and environmental demands.

Studio participant Jack Young's proposal for a self-sustaining high-rise community floating on the sea was particularly well-developed (Figs. 30 and 31). Young envisaged that these emergency accommodation facilities could be towed to areas requiring immediate support and space for displaced communities. This proposal can

Fig. 30 Jack Young's proposal for a self-sustaining high-rise community floating on the sea that could be towed to areas requiring immediate support and space for displaced communities

be conceptualised as the architectural expression of the discussions in Chaps. "Home Is Where the Heart Floats: Anchoring Legal Property Concepts for Environmental Migrants and The 'Bundle of Rights' of Property: A Close Examination of the Jural Relationship in Artificial Landscapes and Floating Structures as Land Replacements" of this edited collection where we explore solutions for climate-displaced communities.

The exploration of climate-resilient urbanism in these projects extends beyond immediate applications, serving as a catalyst for broader acceptance of buoyant

Fig. 31 Emergency housing that can be towed to coastal areas of need

urbanism as a necessary response to global crises. By aligning with international sustainability goals (such as the UN Sustainable Development Goals SDG 9 'Industry, Innovation and Infrastructure', SDG11 'Sustainable Cities and Communities' and SDG 14 'Life Below Water'), these designs contribute to efforts to combat climate change and promote equitable urban development. They exemplify the moral and practical potential of floating cities to shape a more sustainable and inclusive future.

4 Conclusion

Our historical relationship with water has profoundly influenced urban development. Settlements were originally established near water bodies to secure food sources, access essential materials and leverage natural defences. Today, this connection is evolving as water is increasingly viewed not just as a resource, but as a frontier for climate-adaptive urbanism. In response to rising sea levels and intensified flooding, architects and planners are reimagining buoyant and amphibious communities that are not only resilient to environmental challenges but also capable of redefining how cities engage with aquatic environments.

The six paradigms explored in this chapter illustrate the evolving role of buoyant urbanism in response to environmental, social and political pressures. While floating cities have long been theorised as speculative solutions, their increasing relevance in climate adaptation, urban expansion and resource management signals a shift from conceptual discourse to practical necessity. However, public acceptance remains crucial to their viability. Beyond technological feasibility, the success of floating cities will depend on their ability to integrate with existing socio-political frameworks and cultural contexts. This highlights the need for continued interdisciplinary collaboration, where architects, engineers, policymakers and local communities co-develop solutions that align with environmental sustainability and equitable urban development.

Architecture students play a pivotal role in advancing this discourse. Their engagement with buoyant urbanism during their education allows them to explore the social, environmental and political dimensions of floating structures, equipping them to advocate for their broader adoption. By integrating floating architecture into their design work, students contribute to a growing narrative that positions buoyant urbanism as both a viable and responsible urban solution, while gaining skills in interdisciplinary collaboration to refine and implement these complex concepts at scale.

References

1. ABC News (2019) Seasteading bitcoin couple charged with violating Thai sovereignty as navy boards floating home
2. Alexander C (1965) A city is not a tree. Architectural Forum 122(1):58–62
3. Callebaut V (2014) Lilypad: floating ecopolis for climatical refugees. In: Large floating structures: technological advances. Springer Singapore, Singapore, pp 303–327
4. Cañada G, Latourrette A (2025) We left pieces of our life behind': indigenous group flees drowning island. BBC
5. Chlanda J, Eshrāghi L, Rake P, Baker I, Flynn J, Wilson B (2023) Entangled and sustainable approaches to museum practices: a case study of the University of Queensland's Blue Assembly. Museum Int 75(1–2):96–107
6. Curtis WJ (1986) Le Corbusier: ideas and forms. Phaidon Press Limited, London
7. Deshmukh P, Ambulkar P, Sarjoshi P, Dabhade H, Shah SA (2024) Advancements in generative modeling: a comprehensive survey of GANs and diffusion models for text-to-image synthesis and manipulation. In: 2024 IEEE international students' conference on electrical, electronics and computer science (SCEECS). IEEE, pp 1–8
8. Eaton R (2002) Ideal cities: utopianism and the (un)Built environment. Thames & Hudson, London, p 223
9. English EC, Chen M, Zarins R, Patange P, Wiser JC (2021) Building resilience through flood risk reduction: the benefits of amphibious foundation retrofits to heritage structures. Int J Architectural Heritage 15(7):976–984
10. Fisher R, Hirshfeld J (2022) Why we are in "the age of artificial islands". BBC Future
11. Florida R (2002) The rise of the creative class: and how it's transforming work, leisure, community and everyday life. Basic Books, New York
12. Fujikubo M, Suzuki H (2014) Mega-float. In: Large floating structures: technological advances. Springer Singapore, Singapore, pp 197–219
13. Halkyard J (2014) Large spar drilling and production platforms for deep water oil and gas. In: Large floating structures: technological advances. Springer Singapore, Singapore, pp 221–260
14. Heynan H (2002) Engaging modernism. In: Henket H, Heynen H (eds) Back from Utopia; the challenge of the modern movement. 010 publishers, Rotterdam, pp 382–3
15. Ioan A (1999) Striated space versus refuge sites. In: Virtualia. International Centre for Contemporary Art, Bucharest
16. Intercommunication Centre (1997) Mirage City. http://www.ntticc.or.jp/Archive/1997/Utopia/Model/Internet/. Accessed June 2005
17. Jordana S (2009) The citadel: a floating apartment complex in Netherlands. Arch Daily
18. Kaji-O'Grady S, Raisbeck P (2005) Prototype cities in the sea. J Archit 10(4):443–461

19. Kotkin J (2005) The city: a global history. Princeton University Press, Princeton
20. Kottke J (2024) Hundreds of Indigenous families complete relocation off Gardi Sugdub due to rising sea levels. NBC News
21. Louv R (2004) San Diego, the 'Ephemeral City'. San Diego Union-Tribune
22. McCoy MD, Alderson HA, Hemi R, Cheng H, Edwards RL (2016) Earliest direct evidence of monument building at the archaeological site of Nan Madol (Pohnpei, Micronesia) identified using 230Th/U coral dating and geochemical sourcing of megalithic architectural stone. Quatern Res 86(3):295–303
23. Miller N (2021) Scotland's mysterious ancient artificial islands. BBC Travel
24. Mulrennan M, Scott C (2000) Mare Nullius: indigenous rights in saltwater environments. Dev Chang 31(3):681–708
25. Naing N, Santosa HR, Soemarno I (2011) Living on the floating houses for sustainable livelihoods at Lake Tempe, South Sulawesi. Environ Urbanization ASIA 2(1):93–108
26. Onofre SA (2005) The floating gardens in México Xochimilco, world heritage risk site. City Time 1(3):5
27. Raats PA (2015) Salinity management in the coastal region of the Netherlands: a historical perspective. Agric Water Manag 157:12–30
28. Ranganathan S (2019) Seasteads, land-grabs and international law. Leiden J Int Law 32(2):205–214
29. Steinberg PE, Nyman E, Caraccioli MJ (2012) Atlas swam: freedom, capital, and floating sovereignties in the seasteading vision. Antipode 44(4):1532–1550
30. Thomsen C (1994) Visionary architecture. Prestel-Verlag, New York, p 24
31. Todd I, Wheeler M (1978) The architect as visionary. In: Utopia. Harmony Books, New York
32. Trang NTT (2016) Architectural approaches to a sustainable community with floating housing units adapting to climate change and sea level rise in Vietnam. World academy of science, engineering and technology. Int J Civ Environ Struct Constr Architectural Eng 10(2):168–179
33. Ueda S (2014) Floating oil storage base. In: Large floating structures: technological advances. Springer Singapore, Singapore, pp 91–105
34. University of Queensland Art Museum (2022). Oceanic Thinking. 19 Feb–25 June. https://art-museum.uq.edu.au/oceanic-thinking
35. Wang B (2019) Floating cities: the future or a washed-up idea? The Conversation
36. Wang BT (2023) Development of the legal definition of the floating city: Judicial interpretation of structural characteristics of floating homes and developments. In: World conference on floating solutions. Springer Nature Singapore, Singapore, pp 139–157
37. Wang CM, Wang BT (2014) Great ideas float to the top. In: Large floating structures: technological advances. Springer Singapore, Singapore, pp 1–36
38. Wang CM, Wang BT (2015) Large floating structures. Ocean Eng Oceanogr 3
39. Wang CM, Wang BT (2020) Floating solutions for challenges facing humanity. In: ICSCEA 2019: proceedings of the international conference on sustainable civil engineering and architecture. Springer Singapore, pp 3–29
40. Wang JC (2017) Seasteading: tech leaders' plants for floating city trouble French Polynesia. The Guardian. 2 January
41. Willis DB (2007) Dejima: creolization and enclaves of difference in transnational Japan. In: Transcultural Japan. Routledge, pp 239–263

Dr. Brydon Wang is an expert on the trustworthy development and regulation of AI and Construction technology. He researches on trustworthy decision-making in infrastructure delivery and data governance, with a focus on how technology can help increase climate-resilience in cities. He is dually qualified in law and architecture and advises on multi-billion dollar infrastructure projects with KPMG. Brydon practised as a construction and technology lawyer with global top-tier law firm Allens Linklaters and has a previous career in architecture and contract administration on award-winning construction projects. He has over 15 years in the construction industry in both client-side (government) and contractor-side roles.

Brydon was recently lead editor of the book, *Automating Cities: Design, Construction, Operation and Future Impact* (Springer, 2021) and co-edited *Large Floating Structures* (Springer, 2015), a book exploring how we can embrace floating structures in our cities to respond to the increased incidences of floods and sea-level rises. He was featured on ABC Radio National's *Future Tense* discussing offshore architecture and marine urban sprawl, and on Seeker's popular documentary, 'How close are we to Living in the Ocean?' Brydon's expertise in technology and cities also extends to data-focused technologies such as digital twins, BIMs, smart construction contracts and the blockchain. He developed and taught the contract and technology law subjects at the Queensland University of Technology (data privacy, regulating the internet, EU AI Act), and taught architecture and law at the University of Queensland. His Ph.D. thesis on the 'Role of Trustworthiness in Automated Decision-making Systems and the Law' received the *QUT Faculty of Business and Law Executive Dean's Commendation for Outstanding Doctoral Thesis Award*.

Designing for Artificial Reefs: The Role of Architecture in Marine Ecosystem Restoration

Ding Wen Bao

Abstract This chapter examines the vital role of architectural design in addressing marine ecosystem degradation through the Master of Architecture Design Studio on Floating Artificial Reefs, called *Martine Utopia* at RMIT University. As climate change threatens coral reefs, architecture is redefined beyond spatial design to become an agent of ecological restoration and resilience. The studio challenges students to integrate computational methods, biomimicry, and environmental data-driven design to develop floating infrastructures and artificial reef structures that support marine biodiversity while fostering human engagement with fragile marine environments. Through analyses of student projects, including *Laminarosity, Undamaging, Heronia,* and *Mantaverse and Florid Utopia,* the chapter highlights innovative design approaches that synthesize ecological functionality with cultural, educational, and socio-economic considerations. This work underscores how architectural pedagogy fosters sustainable design practices and advances a paradigm where architecture actively contributes to environmental stewardship. By framing artificial reefs as both ecological interventions and cultural artefacts, the studio exemplifies how design can shape resilient futures, bridging human and non-human habitats in the face of climate change.

Keywords Artificial reefs · Buoyant urbanism · Architectural pedagogy · Computational design · Sustainability

1 Introduction

The escalating impacts of climate change pose an existential threat to coral reefs, with projections estimating that 70–90% could disappear within the next two decades (Fig. 1). As biodiversity hotspots, coral reefs support marine life while providing ecological, economic, and cultural benefits. However, warming oceans, acidification, and pollution demand urgent restoration efforts. Artificial reefs offer a proactive

D. W. Bao (✉)
School of Architecture and Urban Design, RMIT University, Melbourne, Australia
e-mail: nic.bao@rmit.edu.au

B. T. Wang et al. (eds.), *Large Floating Solutions*, Lecture Notes in Civil Engineering 636, https://doi.org/10.1007/978-981-96-5435-2_16

CNN World

Climate change could kill all of Earth's coral reefs by 2100, scientists warn

By Jessie Yeung, CNN

Updated 0728 GMT (1528 HKT) February 20, 2020

The race to save the Great Barrier Reef 03:49

(CNN) — Climate change could destroy almost all of Earth's coral reef habitats by 2100, according to new research.

About 70-90% of all existing coral reefs are expected to disappear in the next 20 years due to warming oceans, acidic water and pollution, said scientists from the University of Hawaii Manoa, who presented their findings Monday at an ocean sciences conference.

"By 2100, it's looking quite grim," said Renee Setter, one of the University of Hawaii Manoa researchers, in a press release.

Fig. 1 CNN world report in 2020

solution—not merely mimicking natural habitats but pioneering sustainable marine ecosystems. Master of Architecture Design Studio explores architecture's role in ecological restoration, positioning design as an active agent in marine conservation.

Architectural pedagogy, traditionally focused on terrestrial environments, must expand to address marine challenges. Designing artificial reefs presents a unique opportunity to bridge built and natural environments. Integrating architecture, computational methods, and marine science, the Master Studio developed dynamic reef structures that interact with hydrodynamics, sedimentation, and biodiversity.

Beyond their ecological function, artificial reefs serve as cultural and spatial artefacts. They challenge architects to reconsider materiality, form, and sustainability in new contexts. The studio explored reefs as cultural landscapes which symbolize identity, history, and environmental stewardship. Projects envisioned reefs as underwater public spaces for contemplation, education, and recreation, fostering deeper connections between people and marine ecosystems.

Artificial reefs also hold economic potential. Thoughtfully designed structures can support eco-tourism, diving, and aquaculture, benefiting coastal communities. The studio encouraged students to balance human access with environmental preservation, ensuring their designs enhanced both marine habitats and local economies.

This chapter introduces the Master of Architecture studio on Artificial Reefs at RMIT University School of Architecture and Urban Design, founded and led by Dr. Nic Bao in 2022. Through theoretical inquiry and experimental design, students developed ecologically responsive reef structures that integrate cultural and economic considerations.

Grounded in the alarming reality that "climate change could kill all of Earth's coral reefs by 2100," the studio embraced design as a tool for marine conservation. Inspired by the project from Floating Lab in San Francisco Bay (Fig. 2), students explored floating infrastructures and ecosystem-based approaches to create new coral habitats. Using parametric and bionomic design methods, they examined novel relationships between human and non-human habitation, addressing the urgent challenges of climate change. Their proposals envisioned naturalistic reef environments that support biodiversity, tourism, and economic sustainability, fostering a balanced coexistence between marine ecosystems and human engagement.

Fig. 2 Floating lab in San Francisco Bay by CCA in 2014

2 The Architecture Studio

The Master of Architecture Studio on Artificial Reefs at RMIT was designed to critically engage students with marine environments and explore how architectural design can mitigate the impacts of climate change. The studio's core objective was to redefine architecture as an active participant in ecological restoration, moving beyond aesthetics to embrace biodiversity support, socio-economic resilience, and climate adaptation.

The central aim of the studio was to develop an in-depth understanding of the complex interdependencies within marine ecosystems. Students explored how floating infrastructures, and underwater habitats can support both human activities and marine biodiversity. Artificial reefs were examined not only as ecological sanctuaries but also as cultural and economic assets that foster public engagement through tourism, recreation, and education. The studio challenged students to envision new architectural typologies that respond to the dynamic conditions of marine environments, expanding the discipline's scope beyond terrestrial design.

To address these challenges, the studio emphasized advanced design methodologies, particularly parametric and biomimetic approaches. These methods enabled students to integrate environmental data into the design process, developing structures that respond dynamically to site-specific hydrodynamic, climatic, and ecological conditions. Computational simulations allowed students to analyze water flow, wave dynamics, and structural performance under marine stresses, including typhoons and ocean acidification as shown in Fig. 3.

Students employed generative design techniques, swarm intelligence, and topological optimization to create structurally efficient and ecologically effective reef systems. By leveraging computational fluid dynamics (CFD) and environmental modeling, they refined reef morphologies to enhance water circulation, sediment control, and coral attachment. The studio also introduced additive manufacturing techniques, including 3D printing, to prototype reef structures optimized for marine biodiversity. Figure 4 shows the progress of generative design on reef designs.

Beyond technical innovation, the studio fostered a critical engagement with the socio-cultural dimensions of reef design. Students investigated how artificial reefs can serve as semi-natural monuments, reflecting environmental narratives and cultural identity. Projects considered how reefs could enhance eco-tourism, support coastal economies, and encourage community participation in marine conservation efforts.

The studio's outcomes balanced ecological, economic, and cultural considerations. Students developed proposals for floating infrastructures integrated with underwater artificial reefs, designed to sustain marine life while enhancing dive attractions and recreational experiences. These projects demonstrated a holistic approach, uniting technical precision with environmental responsibility and social engagement.

By the conclusion of the studio, students had not only advanced their skills in computational and generative design but also deepened their understanding of architecture's potential to address pressing environmental challenges. Their work

Fig. 3 Generative design and biomimetic reef visualization

underscored the role of design in fostering resilient, sustainable, and symbiotic relationships between built and natural ecosystems.

3 Discussion of Technology-Driven Design Approaches

The methodology underpinning the Master of Architecture Studio at RMIT School of Architecture and Urban Design is rooted in an interdisciplinary approach that integrates architectural design with environmental science, marine biology, and cutting-edge computational technologies. This particular studio, focused on architectural design for marine environments, draws extensively from Dr. Nic Bao's pioneering research in multi-agent algorithms, topological optimization and computational fluid

Fig. 4 Generative design on artificial reef modeling

dynamics (CFD). It provides students with both a theoretical foundation and practical tools to navigate the complexities of designing structures in marine ecosystems, ensuring that their proposals are not only innovative but also environmentally responsive as shown in Fig. 5.

At the heart of the studio's methodology is an environmental data-driven approach to design. This involves utilizing real-time ecological data—such as water temperature, salinity, current patterns, and biodiversity—to inform the design process. By

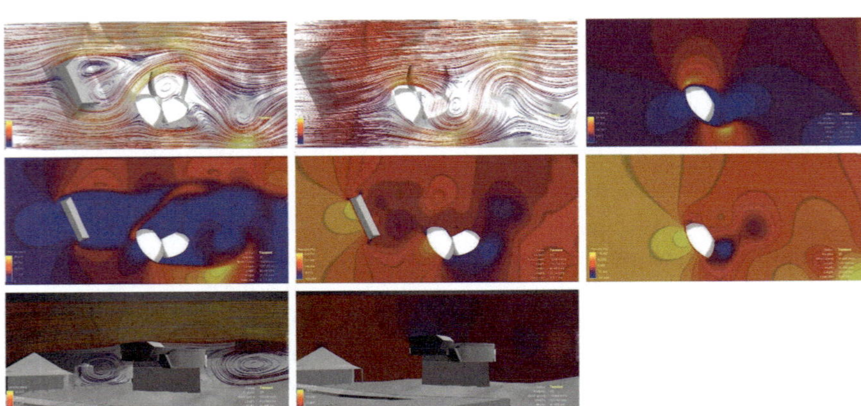

Fig. 5 RMIT research elective on CDF testing in architectural spaces by Paul Morgan and Nic Bao

grounding their work in this data, students develop design strategies that address the specific challenges posed by each site, promoting sustainability and resilience in artificial reef structures. The integration of ecological data ensures that the designs are not merely speculative but deeply connected to the real-world dynamics of marine environments.

Parametric design tools are employed to enable students to generate complex, adaptive forms that respond dynamically to fluctuating marine conditions. Biomimicry serves as a source of inspiration, drawing from natural reef structures to inform structural design principles. Computational workflows allow for real-time adjustments based on environmental feedback, optimizing both material usage and ecological performance. This integration of parametric tools and environmental data ensures that the design is both functional and responsive to the challenges of its marine context.

A central component of the studio's methodology is the use of Computational Fluid Dynamics (CFD). CFD simulations enable students to model the hydrodynamic behavior of their designs within real marine environments. Through these simulations, students can predict how water moves around and through their reef structures, assess the impacts of wave dynamics, and optimize designs for stability, durability, and ecological effectiveness. By visualizing fluid movement and understanding the forces at play, students can refine their designs to enhance water circulation, improve nutrient flow, and provide shelter for marine life. This approach ensures that each design element serves a dual purpose—functioning effectively as both a structural component and a habitat for marine species. Figure 6 shows the development and simulation of reef design and spaces.

Generative design techniques, which explore the morphological evolution of artificial reef forms, are also incorporated into the studio's approach. Drawing inspiration from natural processes such as coral growth, erosion, and sediment deposition, students use algorithms to generate complex geometries that mimic the adaptive behaviors of marine organisms. This design process is characterized by iterative experimentation, where form emerges through computational modeling rather than being predetermined. This methodology allows students to explore a range of possibilities, refining their designs to maximize both structural efficiency and ecological functionality. Figure 7 illustrates the development and form-finding process of reef design and the evolution of its morphology based on the generative design techniques.

Swarm intelligence and topology optimization are key computational methods integrated into the design process. These techniques draw natural processes to solve complex design challenges. Swarm intelligence, inspired by the collective behavior of

Fig. 6 CFD simulation on artificial reef spaces

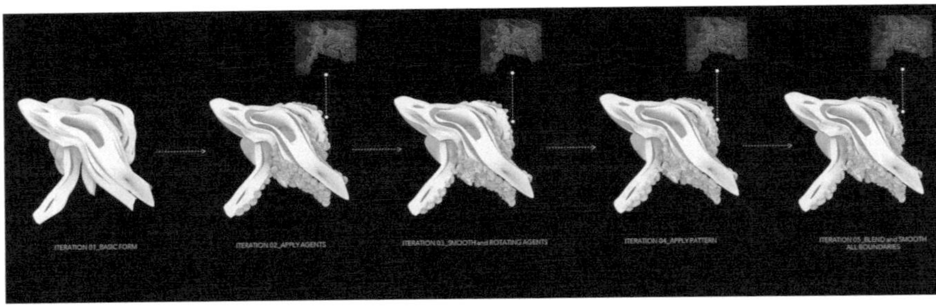

Fig. 7 Generative design on exploring the morphological evolution of reef forms

decentralized systems like ant colonies and schools of fish, allows for the generation of emergent solutions through local interactions. In the context of artificial reef design, swarm intelligence algorithms can simulate the behavior of marine organisms, such as coral polyps, to inform the arrangement and configuration of reef elements. This approach ensures that the design optimizes both the structural and ecological performance of the reef. As Dr. Nic Bao tested this integrated method on the Artificial Taihu Stone in 2021 as shown in Fig. 8, it has now been modified and reapplied to reef design.

Topology optimization, on the other hand, is a mathematical method that determines the optimal distribution of materials within a defined space, taking into account constraints such as structural integrity and performance. By iteratively redistributing or removing material, this method allows for the creation of lightweight, efficient structures that retain maximum strength. In combination with swarm intelligence, topology optimization ensures that the emergent designs are both structurally efficient and ecologically adaptive, offering a synergistic approach to addressing complex environmental and structural challenges in marine contexts.

The integration of swarm intelligence and topology optimization in architectural design is particularly relevant for creating sustainable and resilient structures in challenging environments as shown in Fig. 9. In the case of marine architecture, where hydrodynamic forces, erosion, and environmental degradation pose constant threats, these techniques offer innovative solutions that enhance the stability, durability, and ecological function of artificial reefs. By simulating the behavior of marine organisms

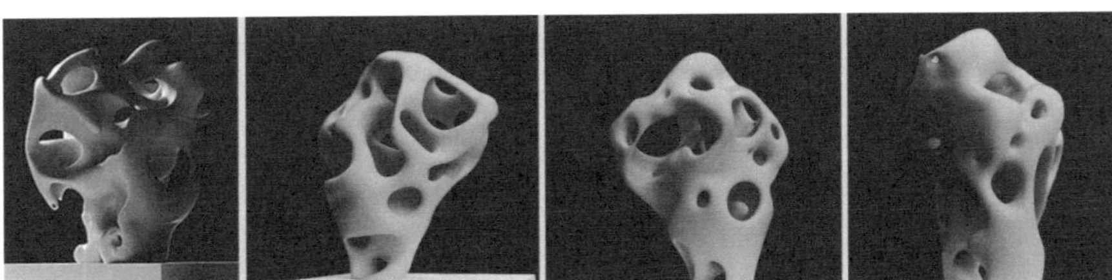

Fig. 8 Topological optimisation for morphology evolution of artificial Taihu Stone led by in 2021

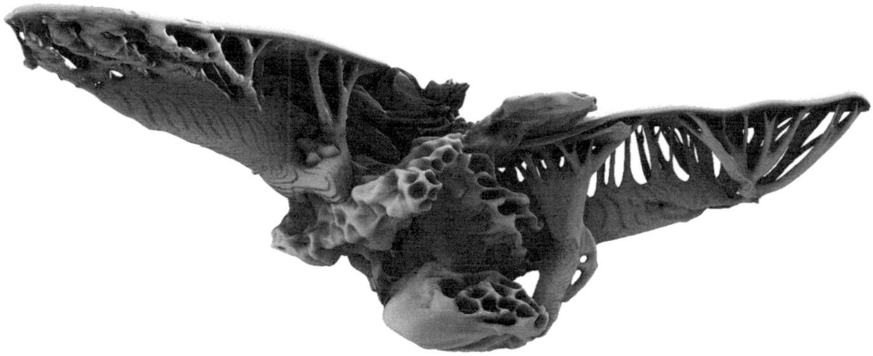

Fig. 9 Floating artificial reef design generated through the integration of swarm intelligence and topology optimization

and refining material distribution, architects can design reefs that are not only resilient to environmental forces but also conducive to marine life.

One of the key strengths of these computational methods is their ability to provide real-time feedback, enabling an iterative design process that adapts to new data and insights. In artificial reef projects, as new environmental data becomes available, such as changes in water flow patterns, coral growth rates, or species interactions, the design can be adjusted and optimized. This adaptability makes these methods particularly valuable for large-scale environmental projects, such as coral reef restoration, where the ability to respond to changing conditions is crucial for long-term success.

In addition to these computational methods, additive manufacturing, particularly 3D printing, plays a crucial role in the prototyping and fabrication of artificial reef structures. Students experiment with a variety of materials and fabrication techniques to create physical models that replicate the structural and textural conditions necessary for coral attachment and growth as shown in Fig. 10. This hands-on approach bridges the gap between digital models and real-world outcomes, providing valuable insights into the practicalities of implementing their designs in marine environments. The use of bio-compatible materials and innovative fabrication techniques opens new possibilities for creating sustainable, locally tailored solutions for marine environments.

Additive manufacturing, particularly 3D printing, offers a transformative approach to the fabrication of architectural and environmental structures. Unlike traditional manufacturing methods, which rely on subtracting material from a solid block, 3D printing allows for the layer-by-layer deposition of material based on digital models. This method not only enhances precision but also enables the production of intricate geometries that would be difficult or impossible to achieve through conventional methods. In the context of artificial reefs, 3D printing allows for the creation of custom modules optimized for marine biodiversity and coral attachment, ensuring that each piece is structurally sound while minimizing material waste.

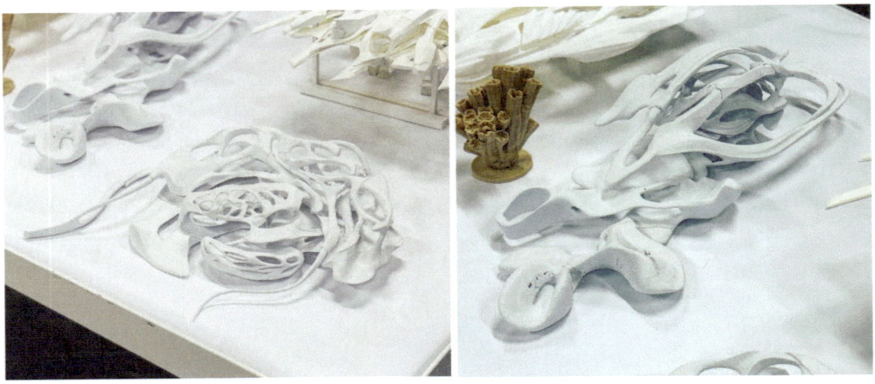

Fig. 10 Additive manufacturing/3D printing tested on intricate reef structures

One of the key advantages of 3D printing is its ability to optimize material usage. Traditional construction methods often lead to significant material waste, but with additive manufacturing, material is deposited only where it is needed, significantly reducing waste. Additionally, 3D printing offers flexibility to work with a wide range of materials, including biodegradable substances, which are crucial for developing environmentally friendly reef structures. As this technology evolves, it has the potential to revolutionize the way we design and construct structures, particularly in challenging environments like marine ecosystems.

The studio's methodological framework encourages an iterative, research-based design process, where students are guided to move fluidly between analysis, simulation, design development, and prototyping. This dynamic process fosters not only technical proficiency but also critical thinking, as students are constantly refining their projects in response to new insights and data. By engaging with both the scientific and cultural dimensions of marine environments, students develop a holistic understanding of how architecture can contribute to the restoration and resilience of coral reef ecosystems. This approach promotes sustainable, adaptive, and ecologically mindful design practices that will shape the future of marine architecture.

4 Case Studies from the Studio

The studio has produced a diverse array of student projects that offer compelling architectural responses to the challenge of marine ecosystem restoration. These projects not only demonstrate innovative design solutions but also reflect a nuanced understanding of the complex environmental, social, and cultural contexts that define marine architecture. This section delves into selected student works, providing architectural critiques that celebrate the strengths of each project while identifying potential areas for further refinement and development.

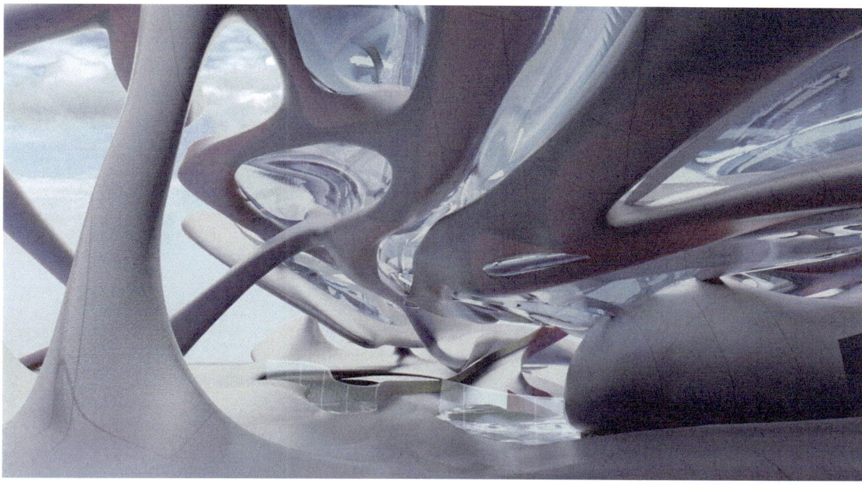

Fig. 11 Internal renderings of the Laminarosity project

4.1 Laminarosity *by Carol Hu and Dorothy Zhang*

In *Laminarosity*, Carol and Dorothy explore the concept of porosity and interstitial space through a floating infrastructure inspired by laminar coral formations as shown in Fig. 11. Their design biomimics the layered, plate-like structures of laminar corals, translating these biological forms into habitable spaces floating on the sea. The central core of their floating structure, surrounded by peripheral open spaces, functions as both a public park and an artificial reef, integrating seating areas within porous façades and underwater structures that facilitate coral growth. The project's strength lies in its seamless integration of ecological and architectural forms, with a computational design approach that effectively leverages biomimicry.

However, while the design demonstrates a sophisticated understanding of coral morphology, the architectural expression could benefit from a more critical exploration of how human activities interact with these ecological systems. Greater attention to material sustainability and long-term resilience in harsh marine conditions would further enhance the project's viability.

4.2 Undamaging *by Jingyan Qin and Jiacan Tan*

Figure 12 presents Qin and Tan's *Undamaging* project, which proposes a coral-themed floating public park designed to educate visitors about coral ecosystems while supporting reef restoration. Inspired by the skeletal structure of bleached corals and Taihu stones, the design incorporates porous architectural forms that promote marine

Fig. 12 The underwater part of undamaging project

biodiversity. The project effectively conveys a strong narrative that links architectural design with ecological processes, particularly in how the porous structures enhance light penetration and water flow, both essential for coral health.

Architecturally, the project presents an elegant synthesis of form and function, with fluid transitions between above-water and underwater spaces. However, the project could be strengthened by a deeper investigation into the dynamic forces acting on floating structures, such as wave energy and storm surges. Addressing these factors through structural simulations and resilient material strategies would add robustness to the design.

4.3 Heronia *by Joseph Luo and Harlan Guo*

In *Heronia*, Joseph and Harlan reimagine the floating platform as a hybrid coral research center and educational theme park, as shown in Fig. 13. The design features a fragmented, iceberg-like form, symbolizing the fragility of marine ecosystems under climate change. The project's innovation lies in its creation of an immersive journey for visitors, from marine museums to underwater observation decks, effectively raising awareness about coral bleaching and climate issues. Computational fluid dynamics inform the platform's design, optimizing it to enhance water currents that support coral growth beneath.

The project's strength is its strong educational focus and integration of interactive spaces, which bridge scientific research with public engagement. Nevertheless, the architectural language, while visually striking, could be more cohesive in its structural

Fig. 13 Floating platform of Heronia project

logic. A more nuanced exploration of materiality and modular construction methods would enhance both the aesthetic and functional dimensions of the design.

4.4 Mantaverse *by Bill Zhang and Cynthia Chen*

Bill and Cynthia's *Mantaverse* envisions a self-sufficient floating city that addresses urban population pressures and climate-induced sea-level rise. The project explores the potential of ocean cities as long-term human habitats, integrating renewable energy systems, waste recycling, and sustainable living environments. The architectural concept draws from science fiction narratives and existing floating city prototypes, proposing large-scale, adaptable infrastructures capable of supporting diverse marine and human life. Figure 14 depicts the floating biomimetic platform, while Figs. 15 and 16 show the underwater artificial reef with coral growth.

This ambitious project is commendable for its systems-thinking approach, considering not just architectural form but the broader socio-economic and environmental systems in which it operates. However, the project's scale raises questions about feasibility, particularly concerning resource management, governance, and environmental impact. A more detailed exploration of modular design principles and phased development strategies could provide a clearer pathway from conceptual vision to practical implementation.

Fig. 14 Floating part of *Mantaverse* project

Fig. 15 Underwater part of *Mantaverse* project

4.5 Florid Utopia *by Renling Zhang and Zejia Huang*

Florid Utopia designed by Renling and Zejia is a visionary project that integrates ecological restoration with floating urbanism. Situated adjacent to Heron Island, this large-scale floating theme park features an underwater artificial coral reef structure designed to support marine biodiversity while fostering public engagement and local

Fig. 16 Underwater artificial reef with corals

economic development. Beyond serving as a tourist destination, the project aims to create a resilient habitat that attracts and sustains diverse marine life, contributing to environmental conservation efforts. Figure 17 shows the floating artificial reef with architectural natural spaces, while Fig. 18 presents the underwater section of the artificial reef interacting with marine life.

They investigated the challenges affecting marine ecosystems and explored biomimetic design strategies to develop floating infrastructures that respond to these issues. The project reflects a critical understanding of how climate change influences the evolution of non-human habitats, prompting innovative approaches to designing symbiotic environments that balance ecological and economic interests.

These strategies ensured that the artificial reef contributes positively to marine life reproduction while maintaining structural resilience in dynamic oceanic conditions. With the ongoing expansion of urban populations, major coastal cities are increasingly exploring offshore developments as a sustainable alternative to traditional land-based expansion. Floating structures offer a means to reduce environmental impact by alleviating pressure on terrestrial ecosystems such as green belts and arable land. This project exemplifies the potential of innovative floating architecture in enabling sustainable urban transitions and advancing ocean-based resource exploration. The large-scale implementation of floating infrastructures not only addresses urbanization challenges but also underscores their role in shaping future coastal and marine developments.

Fig. 17 Floating artificial reef with natural spaces

Fig. 18 Underwater porous reef forms

4.5.1 Reflection

Collectively, the projects reveal the transformative potential of architectural design in tackling marine ecological challenges. The designs demonstrate how architecture can transcend its traditional role as a built environment to become a dynamic agent of environmental stewardship, cultural expression, and socio-economic development. Each project illustrates a commitment to both enhancing marine biodiversity and fostering a greater public understanding of environmental conservation through innovative and thoughtful architectural interventions.

The incorporation of biomimicry, parametric design, and computational simulations in these projects reflects a sophisticated engagement with both natural systems and technological innovations as shown in Fig. 19. The use of biomimetic principles, inspired by the forms and structures of coral reefs and marine organisms, ensures that the designs are not only visually resonant but also ecologically effective. These approaches allow students to explore the inherent complexities of marine environments, ensuring that their designs can foster the resilience and regeneration of underwater ecosystems.

Similarly, parametric design and computational simulations offer a deepened understanding of the environmental conditions that influence the performance of artificial reefs. The ability to test and optimize designs before they are built in physical space empowers students to make informed decisions about how their structures can function most effectively within their natural context. By simulating water flow, sediment transport, and biological growth, the designs engage with the dynamic, ever-changing conditions of the marine environment.

However, as designs push the boundaries of creativity and conceptual exploration, future iterations could benefit from a deeper integration of material research, structural engineering principles, and long-term environmental impact assessments as reflected in Fig. 20. While many of the projects demonstrate impressive formal innovation, there is room to further develop the material choices used in these designs, ensuring they are not only ecologically responsible but also durable and resilient in the harsh marine environment. Investigating sustainable, bio-based materials or

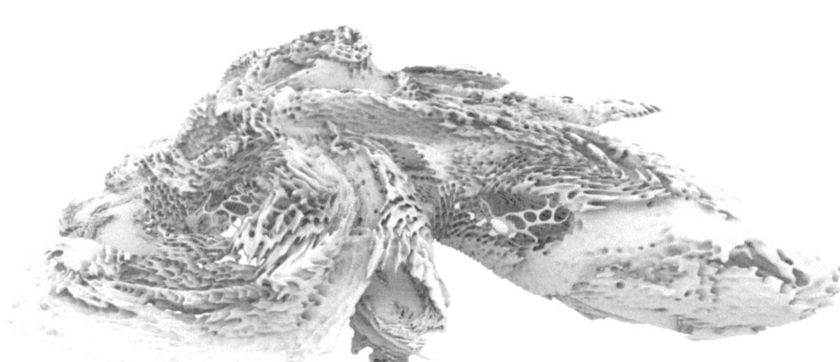

Fig. 19 Biomimetic design of above-water artificial reef space

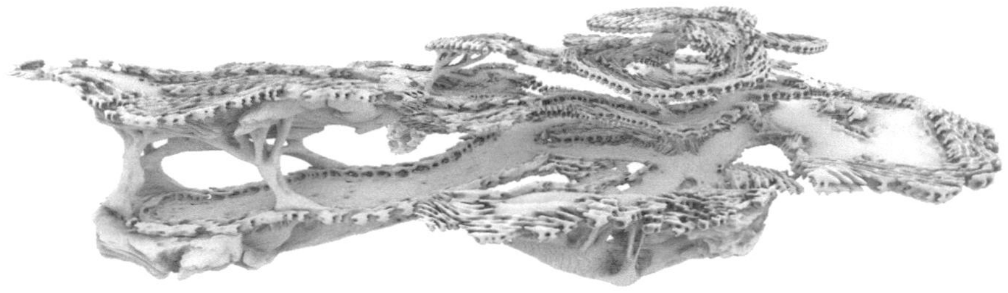

Fig. 20 Design of underwater artificial reef space with structural performance

exploring advanced composite structures that mimic the self-healing properties of natural reefs could elevate the ecological integrity of the designs.

In parallel, reinforcing structural engineering principles would further ensure that these designs are both feasible and robust under real-world conditions. While the conceptual and aesthetic aspects of the projects are well-executed, a deeper understanding of structural performance—particularly in relation to wave forces, marine corrosion, and long-term stability—would add critical technical depth to the designs.

Most importantly, a long-term environmental impact assessment would allow for a more comprehensive understanding of how these artificial reef structures interact with their surrounding ecosystems over time. This includes evaluating potential risks, such as the disruption of existing marine habitats or the unintended consequences of introducing new materials into the ocean. Long-term monitoring and adaptive design strategies could ensure that these projects not only survive but thrive in the evolving context of marine environments.

By strengthening these aspects, the projects would not only enhance their technical feasibility but also reinforce architecture's potential as a mediator between human aspirations and ecological imperatives. Architecture's role in responding to pressing environmental challenges is not simply about creating beautiful forms, but about ensuring that these forms are both functional and responsible stewards of the environment. This deeper integration of material science, engineering, and environmental impact would enable architecture to evolve into a more sustainable and resilient practice that serves both human and ecological needs, fostering a balanced relationship between the built environment and the natural world.

5 Conclusion

The Master of Architecture studio on Artificial Reefs at RMIT exemplifies how architectural education can respond to pressing global environmental crises through innovative, interdisciplinary approaches. As climate change accelerates the degradation of coral reefs, architecture transcends its traditional role as a discipline concerned solely with the built environment, emerging as an active agent in ecological restoration and resilience. The projects presented in this chapter demonstrate how design can be leveraged to create adaptive, sustainable structures that not only support marine biodiversity but also foster meaningful human interaction with fragile ecosystems.

Through the integration of advanced computational techniques, biomimicry, and environmental data-driven design, students have explored novel methodologies that push the boundaries of traditional architectural practice. Their work reflects a nuanced understanding of the complex interrelations between form, function, and ecological systems, offering solutions that are both aesthetically compelling and environmentally responsive. The incorporation of floating infrastructures, porous structures, and dynamic forms reveals an architectural language that is sensitive to the unique challenges of marine environments. This approach directly addresses critical issues such as climate adaptation, habitat restoration, and socio-economic development.

Moreover, these projects highlight the transformative potential of architecture as a cultural and educational tool. By embedding spaces for public engagement, research, and ecological awareness, the designs foster a deeper connection between communities and the marine world. They challenge conventional perceptions of architecture's role, positioning it as a mediator between human aspirations and environmental stewardship. In this way, architecture becomes more than a built artifact, becoming a bridge between human culture and the natural environment, fostering both awareness and action. Recently, we also explored more advanced design methods to integrate artificial intelligence with brainstorming for conceptual designs of coral reef flood resilience, as shown in Fig. 21, to examine new tools for pedagogy. It is much efficient and more creative.

In conclusion, the studio underscores the importance of architectural pedagogy embracing ecological challenges as opportunities for design innovation. As future architects engage with the complexities of climate change and marine conservation, their work will not only shape resilient infrastructures but will also inspire a more harmonious coexistence between humanity and the natural world. The studio's projects demonstrate the power of architecture to influence environmental consciousness and catalyze positive change, proving that design can be a force for both ecological resilience and societal progress.

Fig. 21 AI-assisted flood-resilient architecture inspired by Coral Reefs

References

1. Bao DW (2022) Performance-driven digital design and robotic fabrication based on topology optimisation and multi-agent system. RMIT University
2. Bao DW, Yan X (2023) Data-driven performance-based generative design and digital fabrication for industry 4.0: precedent work, current progress, and future prospects. Archit Des Ind 4.0: Theory Pract 943–961
3. Yan X, Bao DW, Xiong Y, Snooks R, Xie YM (2023) Structural topology optimization based on a multi-agent model. Eng Struct 296:116978
4. Luo D, Wang B, Bao DW, Ward S (2022) Opportunities for further development of 3D-printed floating artificial reefs. J Aquac Marine Biol 11(2):58–63
5. Wang BT and Wang CM (2021) Automating cities: design, construction, operation, and future impact
6. Kaji-o'grady S, Raisbeck P (2007) Prototype cities in the sea. J Archit 10(4):443–461
7. Wang CM, Wang BT (2006) Colonization of the ocean and VLFS technology. In: Very large floating structures, pp 15–34
8. Wang CM, Wang BT (2015) Great ideas float to the top. In: Large floating structures, pp 1–36
9. Lin S, Chou N, Bao DW, Zhang GB, Xiong CW, Fang J, Xie YM. (2022) Design and fabrication of artificial brain coral: evolution principle, turbulent hydrodynamics and matter interchange. Mater Des
10. Feng Z, Gu P, Zheng M, Yan X, Bao DW (2021) Environmental data-driven performance-based topological optimisation for morphology evolution of artificial Taihu Stone. In: Proceeding of the 2021 DigitalFUTURES, pp 117–128
11. Lin S, Bao DW, Xiong CW, Fang J, An HW, Sun Z, Xie YM and Zhou SW (2021) Human-made corals for marine habitats: design optimization and additive manufacturing. Adv Eng Softw 1–12
12. Guneri GD. Peter cook beyond ArchiGram: towards a critical utopianism

Ding Wen 'Nic' Bao is a Senior Lecturer in Architecture and Architecture Technology Course Coordinator, at RMIT University's School of Architecture and Urban Design, where he leads FormX Research Lab. His expertise spans architectural technology, computational design, robotic fabrication, and intelligent construction. Nic is a senior member of RMIT Centre for Innovative Structures and Materials and a member of the RMIT Post-Carbon Centre. With over 60 publications in high-impact journals, book chapters, and international conference proceedings, he has also led several research projects funded by Australian Research Council and industry partners. His work has been exhibited globally, including at Melbourne Design Week, IASS Expo, and Venice Biennale. Nic is a registered architect in Australia and US and a RIBA Chartered Architect in the UK. He is a director at B.W Architects (BWA) and a partner at Ameba. He also serves on the Australian Institute of Architects (AIA) committee, is the Secretary of the Association for Computer-Aided Architectural Design Research in Asia (CAADRIA) and is a core member of the organizing committee for DigitalFUTURES, IASS 2023 Symposium, and LSAA 2024 Conference. Additionally, he is an editorial board member of two high-impact journals: *Sustainable Structures* and *Architectural Intelligence*. Nic has received numerous prestigious awards, including 2024 DigitalFUTURES Project Award, 2023 A' Design Award, 2023 DigitalFUTURES Young Award, 2023 RMIT Award for Research Excellence (ECR), the third prize in the IASS Design & Competition of Innovative Lightweight Structures (2023), 2022 IAI Design Award, 2021 Grand Prix Design Paris Gold Award, 2021 MUSE Design Award, 2021 RMIT Prize for Research Impact (Design), and 2021 Young CAADRIA Award.

City Parts: Urban Form Design Guidelines for Floating Cities

Kirsty Volz

Abstract Urban form design guidelines provide frameworks that ensure sustainability and equal access to quality urban life in cities. This chapter discusses the application of urban form design guideline principles to floating cities. Drawing from existing literature on urban design governance and case studies from floating cities in Vietnam, the Philippines and the Netherlands, this chapter proposes a framework for analysing the proposed design for Oceanix in South Korea. Using the concept of City Parts, devised by Oswald Ungers and Rem Koolhaas, the analysis reveals a discussion about the challenge in balancing adequate public open space to ensure amenable cities that foster social cohesion, while also preserving affordable housing that is accessible to everyone in floating cities.

Keywords Urban design · Density · Floating cities · Design governance · Design guidelines

1 Introduction

In the literature on urban development of floating cities, there are two contrasting stories about the people live in floating communities. The first depicts floating villages designed by architects, built by private entities that promote leisure via a unique, idyllic lifestyle on water. These exist in developed countries and are often financially out of reach for much of the general population [1]. In this scenario, living in floating urban settlements is an attractive choice, made by individual residents to enhance their quality of life, and engage with water-based recreation in their daily existence. In contrast to this, the second tells the story of floating communities, especially in Southeast Asia, that exist out of necessity, where living on the water allows access to resources and needed income for residents. The physical condition of these settlements is often poor, with simple construction, little to no infrastructure

K. Volz (✉)
School of Architecture and Built Environment, Queensland University of Technology, Brisbane, Australia
e-mail: kirsty.volz@qut.edu.au

B. T. Wang et al. (eds.), *Large Floating Solutions*, Lecture Notes in Civil Engineering 636, https://doi.org/10.1007/978-981-96-5435-2_17

or shared amenity, and neglected maintenance of buildings and infrastructure that negatively impact residents' lives both socially and economically. These settlements may even cause damage to the physical environment and ecology that they occupy [2]. These two opposing circumstances for floating communities both diminish the sustainability of floating urban development in different ways. The inequity generated by both scenarios creates barriers to the accessibility and therefore long-term occupation of floating cities.

One cause for this inequity is a lack of governance structures that support and articulate the delivery and maintenance of high-quality, accessible floating cities in terms of procurement, design and construction. This lack of governance stems from a void in judicial consideration or legislation reform to support the sustainable development of floating cities [3]. In Nguyen's [2] research on floating settlements in Vietnam, she asserts that the lack of governance for development and operation of developments is the most prominent cause for underprivileged quality of life experienced by residents. While there are many aspects of governance that relate to floating cities, this research focuses on design governance and the need for urban form design guidelines to improve accessibility and sustainability of floating cities. Design governance for urban form extends beyond the look and feel of a settlement, it improves people's lives in cities. A lack of design governance contributes to poorer outcomes for residents in floating villages in terms of socio-cultural, economic and equal access to services and opportunities.

A significant theme from this research on urban form design guidelines for floating cities is the importance of well-designed, fit for purpose shared open space for communities. In fact, the existence of public space helps to support governance structures and democracy in cities, and as such, the development of these spaces is mutually beneficial to citizens as well as governing bodies [4]. This chapter starts by defining urban form design guidelines and then synthesises this definition with the existing literature on urban design and development of floating cities. From this synthesis, a framework for urban form design guidelines in floating cities is developed. There were two main themes that emerged from the research, the first is "clustering" or city parts and the second was public open space. To further explore these two themes, this chapter tests the framework for urban form design guidelines in floating cities through an analysis of the early, prototype design proposals for a case study floating city, Oceanix in South Korea. The analysis reveals the complexity in striking the right balance in density of "clusters" or smaller neighbourhoods within a floating city and providing the right amount of public open space. Too much shared open space and amenity will drive up housing prices, making living on water unaffordable and inaccessible to large sections of the community. It also has the potential to negatively impact on water ecology by covering too much surface area on the water as discussed in Chap. "Exploring Design of Floating Urban Developments: Integrating Density, Engineering and Environmental Considerations" of this edited collection. However, the alternative of providing too little open space risks poor living environments, which reduces urban form quality and disincentivises long term inhabitants.

2 Urban Form Design Guidelines for Floating Cities

Often people will only interact with urban form design guidelines when engaging in development assessments with planning departments of municipal governments. All building types have relative urban form design guidelines to address, from small individual houses, shopping centres and large apartment complexes. These guidelines are often measured by physical built outcomes, such as building height or setbacks from public and shared boundaries, but what underpins this guidance is the desire for local governments to shepherd cohesive, equitable cities that promote good social and economic outcomes for citizens [5]. As a simple example, the illustration in Fig. 1 depicts an increasingly common street scape in the Australian city of Brisbane. A city that is impacted by flooding and where the predominant approach to urban fabric is to make space for water. Architecturally, these strategies take the form of houses raised on columns above a predicted flood level, which are then tethered to street frontages by access and services that connect individual houses to the city. Devices such as stairs, driveways, sewerage, stormwater drainage and electricity act as anchors to terrestrial structures that appear to float above the street and yet possess no such buoyancy in their design.

While there is obvious differentiation between the height of House B and the other three houses shown on the streetscape illustration, there is also a social and economic story behind the lack of cohesiveness in this street scape. It could be that House B is located at the lowest point in the street, or an area with drainage issues and therefore must be raised. This apparent visual difference between dwellings in the streetscape can conjure assumptions about the quality and liveability about the site that the house is located on. Alternatively, the entire street is impacted by flooding or overland flow of water and the owners of House B are the only residents who can afford to raise their house above the flood water. Both scenarios reflect an inequity in housing on the same street [6]. The role of urban form design guidelines is to provide structures for the fabric of cities, specific to their contexts—be it on land or water—to ensure the quality and equity of built environments [7].

Fig. 1 Illustration of a Brisbane street scape where drainage issues or seasonal flooding has impacted dwellings altering suburban visual identity and uniformity in urban built form. Illustration by author

2.1 *Defining Urban Form Design Guidelines*

Urban form design guidelines aim to influence cohesive, functional, safe, sustainable, accessible and visually amenable urban environments [8]. Universally informed by research and on the ground best practices, urban form design guidelines cover multiple aspects of city and suburban built and natural environments [9]. From the location of green infrastructure and open space to car accommodation for individual dwellings. They also cover transportation and accessible movement throughout the city. Many aim to foster social cohesion and neighbourliness, addressing the need for safety and defensible space, passive surveillance or "eyes on the street." At the very core, these guidelines cover built form through control over setbacks, building heights, density of buildings and provision of quality, shared open space and amenity [10].

As design guidelines they provide more detailed information about the quality of built outcomes. Generally, though, they start with, or refer to, a desire to promote a unified urban structure including the layout of streets, blocks and shared public space and amenity [11]. This overarching level of organisation ensures that cities are efficiently organised in a manner that promotes connectivity and ease of movement to conduct business and day-to-day activities. Importantly, they maintain the urban fabric, and this involves guidance on building form and massing-height, setback from other buildings, and density—to ensure that new buildings fit in well with their specific contexts [10]. This guidance can extend to the architectural design, including material selection and colour to improve scenic amenity and aesthetic appeal of neighbourhoods. While these might seem trivial, an identifiable visual strategy for cities that encourages character definition for a place can enhance people's connection and sense of belonging to a city [12].

Regarding the social side of considerations for the city, urban form guidelines contend with the public realm, promoting high quality spaces for people to socialise, gather and freely interact outside of their homes. Public realm spaces include parks, plazas, streets, footpaths, the location of public seating and lighting for safety and accessibility, and importantly, planting for street trees [13]. The way individual buildings interact with the public realm is also a consideration, especially for safety and security. This means providing guidelines for external lighting on buildings, sightlines and passive surveillance and the importance of activated street frontages and the promotion of eyes on the street [14]. The provision of shared open public space has to be carefully managed. Too little open space minimises the quality of urban life in cities impacting on residents' experience, which flows on to how they engage with economic activity, especially recreational and hospitality businesses. Too much open public space can cause unnecessary competition for amenity, which can jeopardise the sustainability of businesses and in some cases drive up housing prices to an unaffordable range [15]. Striking the right balance in the ratio between open green space and built form is a significant consideration for newly created floating cities.

Economically, these guidelines are integral to the sustainable development of cities. From climate resilience strategies for designing in response to natural hazards, down to detailed assessments for energy efficiency to reduce the impact on city energy infrastructure [16]. They also outline methods for reducing urban heat island effect through green infrastructure and the preservation of natural resources. Other important economic factors include ease of city transportation in a variety of accessible modes including both public and private transport, walking, cycling and e-commuting [17]. Urban form design guidelines also provide critical oversight for the right mix of developments within a neighbourhood across residential, commercial, hospitality, civic and health care. Regulating the types of buildings that a precinct can support enhances community vibrancy while also ensuring sustainable economic growth [12].

From this review of literature on urban design guidelines for land based cities, nine key themes were derived to define a framework to synthesise existing literature on urban form design guidelines and the urban development for floating cities. The categories are (1) Open space, (2) Density, (3) Space use planning (mix of residential, commercial, civic, health, education, etc.), (4) Economic activity, (5) Social cohesion, (6) Visual amenity, (7) Shared amenity (lighting, footpaths etc.), (8) Green infrastructure, (9) Transportation systems. In table 2, these categories will be synthesised with the frameworks and design principles that emerged from individual studies on urban design governance. The aim of this process to define general overarching urban form design guidelines for floating cities.

2.2 *Frameworks and Urban Design Principles for Floating Cities*

Existing studies on urban design and development for floating cities focus on specific case studies in different international contexts. One contribution from this study is the synthesis of these studies to produce an overarching urban design governance framework that can be adapted for specific contexts. This review draws on three recent studies on urban design and development for floating citied in Vietnam [2], Netherlands [1] and Philippines [18]. The results of this review are presented in Table 1. There are several common overlapping themes that emerged from the literature across very different contexts. There were also divergent themes relating to specific historic, economic, cultural and social settings.

The studies in Vietnam and Philippines highlighted the cultural ties to floating villages, where generations of families have occupied houses floating on various river and oceanic settings. These houses connect people to their work and livelihood and deep community bonds that have developed along culturally specific living traditions over time [18]. They are vibrant hubs that support living, working, socialisation, and food production. Vietnamese communities established floating villages in the Mekong Delta in the eighteenth century [2]. However, there are negative associations attributed to living in these villages due to the poor condition of housing and lack

Table 1 Frameworks and urban/architectural design principles from international studies

Study Author/s, year and country context	Framework for urban form	Urban design Principles
Ham, [18] Philippines	Building elements	
	Infrastructure elements	
	Public space	
	Sanitation units	
Thomas et al., 2023 Netherlands	Urban	Mobility
	Affordable	Financial feasibility
	Attractive	Spatial quality
	Embedded part of the city-on-land	Energy and installations
	Ecological	Ecology
	Technically feasible with today's know-how	Construction
Nguyen, 2022 Vietnam	Socio-cultural	Liveability
	Environmental	Adaptability
	Economic	Affordability Self-sufficiency
	Technology	Efficiency

Table 2 Framework for urban form design guidelines

Urban form guideline category	Framework Criteria	Floating city urban design principle
Open space	Public space [18]	Spatial quality [1]
Density	Urban [1]	
Space use planning (mix of residential, commercial, civic, health, education, etc.)	Building elements (Ham 2023) Technically feasible with today's know-how [1]	Construction [1] Adaptability [2]
Economic activity	Economic [2] Technology [2]	Financial feasibility [1] Efficiency [2]
Social cohesion	Affordable [1] Socio-cultural [2]	Liveability [2] Affordability [2]
Visual amenity	Attractive [1]	Spatial quality [1]
Shared amenity (lighting, footpaths etc.)	Sanitation units (Ham 2023) Infrastructure elements (Ham 2023) Embedded part of the City-on-land [1]	Energy and installations [1] Self-sufficiency [2]
Green/blue infrastructure	Environmental [2]	Ecology [1]
Transportation systems	Infrastructure elements (Ham 2023)	Mobility [1]

of sanitation. Nguyen [2] found that there was a gradual migration of people from floating villages to land based housing.

Vietnam's established floating villages provide interesting case studies for the development and morphology of floating villages. Having existed through changing environmental conditions on land and water, they demonstrate strategies for adaptation and resilience in floating communities. They are also valuable examples of providing housing diversity with housing solutions that address individual needs. The co-existence of this diversity from houses half on stilts and half on the ground; stilt houses; boat houses; and floating houses all tie together to create vibrant neighbourhoods on water [2].

Diversity, and in fact hybridity, is a key part to successful, liveable cities. Urban form design guidelines often require diversity in building types and planning for spatial activation as a means of offering citizens a sense of interest both visually and experientially. Diversity is also an approach to city resilience in terms of long term financial sustainability and the ability to adapt to changing external factors such as the climate, economy, and population or demographic fluctuations. A distinctive feature of floating cities is the capacity to offer diversity through hybrid methods for inhabiting land, water or both simultaneously. This hybridity means that floating cities are uniquely capable of adapting to fluctuations caused by both predictable environmental changes as well as weather events and conditions that are challenging to forecast. Hybridity also means that floating cities can adapt to changing economic conditions and cultural preferences for living on land or water.

While Vietnamese floating villages have been successful in promoting hybridity in their development over time, a lack of design governance has resulted in uncontrolled and rapid growth. The resulting lack of infrastructure and amenity has delivered poor living environments for communities. Efforts to improve the living conditions in these villages needs to incorporate the protection of people's livelihoods, especially fishing, their social networks and the physical environment at the same time [2]. In her study, Nguyen [2] provides an overarching framework with three pillars to achieve these outcomes: "(1) Social and Cultural Considerations: reserving and promoting local culture, improving social facilities and social activities, establishing management regulations, (2) Economic Considerations: livelihood development and poverty reduction and cost-effective floating houses, and (3) Environmental Considerations: environmental protection, prudent use of natural resources climate change and rising sea levels."

Further developing this framework to actionable outcomes Nguyen [2] provides five urban design principles for floating villages: "(1) Adaptation: including adaptation with local conditions (environment, socio-culture, economy, architecture), and adaptation with climate change and social change,(2) Self-sufficiency: including abilities to secure sufficient food, sufficient water, sufficient energy and sufficient waste treatment; (3) Liveability: including the provision of comfortable living conditions, enhance of education, social activities and socio welfare that contribute to community, and protection of human health; (4) Efficiency: including effective protection of environment, prudent use of nature resources/renewable energies, and increasing profitability and productivity of floating buildings; (5) Affordability:

including affordable construction, operation, and maintenance of floating houses". These principles are summarised and presented in Table 1, alongside the findings from the studies in the Netherlands and the Philippines.

Several mutual themes emerged from the findings in Thomas et al [1] research on urban design for floating villages in the Netherlands and Nguyen [2] research in Vietnam. These included the importance of housing diversity with a broad range of housing types, including homes that are accessible by foot as well as homes that are only accessible by boat [1]. The provision of diversity of housing extends to procurement methods, suggesting that the housing mix should include social housing to ensure financial stability of floating cities. On the other hand, the research in the Netherlands found that, in contrast to the studies in Vietnam and Philippines, the affordability of floating houses is a significant concern. There was also guidance on housing diversity in terms of density with a desire for a mix of low-rise apartments and detached dwellings [1]. The need for housing diversity will be further discussed in relation to the Oceanix case study analysis.

A shared theme emerging the literature was the consideration of water ecology and impact of developments on the environment. The broader framework generated by Thomas et al [1] was similar to Nguyen's [2]. For the Netherlands it was derived through broad categories including: "Urbanity (density, function mix), Inclusivity (affordability, accessibility), Ecology and feasibility (financial, constructive)." [1]. Both studies in Vietnam and Netherlands discussed the benefits of self-sufficiency in terms of energy systems, especially systems that eradicated the need for cables or other infrastructure attached to the land or that anchor to the river or seabed.

Where the study in the Netherlands differed from the other two was in the explicit discussion about the attractiveness of floating cities, the provision of a variety of transport services and that floating villages should be planned as an extension of land-based cities [1]. This last theme was briefly shared in the study of the Philippines, with the suggestion that floating villages be designed to continue to rely on amenities and infrastructure, such as health precincts, public open space and major transportation, on adjacent land-based cities to make them more liveable. That floating cities should be designed as hybrid cities, part on land and part on water with shared public space and amenity as the conduit or "node" between the two settlements. It would provide pleasant urban experience for land-based residents to interact with water side activities and connect people in floating villages with the larger city and necessary amenities [18]. They should be designed as extensions of existing cities to save on doubling up on existing resources [1]. This approach would mean that urban form design guidelines could be extrapolated from existing municipal governments and that the structure for design governance would be relatively established.

While studies implicitly set out a desire for attractiveness, the study on the Netherlands made this an explicit requirement, arguing that high quality attractive environments possess the ability to bring communities together and enhance social interaction [1]. Interestingly, this study was the only one to suggest the potential of designing with water to enhance and generate delightful experiences and beautiful environments in floating cities. Attractiveness extended to transportation, where designing precincts in a way that did not require cars would contribute to improving the overall look and feel of places [1]. Cities located on water provides an opportunity to engage with a wonderful array of transportation options, including large and small watercraft. There should be transportation planning systems on the water that provide safe means to utilise individual transport like canoes or kayaks or even swimming (the walking and bicycling of the sea or river). Where there is an adequate density in population, shared public transport such as ferries and well-designed ferry stops should be incorporated. This is one aspect of the Oceanix case study appears to have managed in an exemplary manner. There is also an opportunity to foster shared transportation modes as part of floating city developments [1] The potential for a variety of transportation modes should be supported by hubs that provide safe storage and organise different modes of transportation.

A common theme across all three studies was the lack of shared open space and how this inhibited the liveability and sustainability of floating villages. In the Netherlands, current floating villages are predominately detached dwellings with limited diversity in type and no dedicated provision of shared open space or amenity. They therefore lack any sense of identity, spatial quality and cohesion [1]. The need for shared open space and the "clustering" dwellings was a key finding from Ham's [18] study of the Philippines. Based on community participatory methods, the study found that it was vital for floating villages to be designed both at the neighbourhood and individual house level [18]. A discussion about "Clusters" and "City Parts" is in the next section of this chapter is central to the urban design strategy for Oceanix.

Consideration for thermal comfort in the planning of floating villages from climate and air movement was only briefly mentioned [18]. Human comfort in terms of thermal (temperature), olfactory (smell), visual (glare from the water), and acoustic (sound), are essential concerns in the successful design of cities. Further, these conditions for human comfort are supported when the local environment and ecology are respected as part of the design. This is an area of research that requires much greater exploration and analysis in the future. The study on the Philippines provided a simple, but essential framework for urban design of floating villages [18]. The principles and frameworks for each study are outlined in Table 1.

2.3 *Establishing a Framework for Urban Form Design Guidelines for Floating Cities*

To bring the three studies together and synthesise them with the framework for urban form design guidelines framework set out earlier in this chapter, Table 2 establishes a broad framework for urban design and the generation of quality urban spaces in floating cities and villages. This framework will be further developed through the analysis of the Oceanix case study in this chapter.

3 Clusters and City Parts for Floating Cities

In the study on community participatory design for floating villages in the Philippines, the concept of "clustering" was introduced [18]. This term refers to a need for designing floating villages at a neighbourhood scale, where groups of houses (a neighbourhood) are arranged around shared open space. These open spaces provide shared areas for private gatherings and social events as well as public organised events, such as sport or theatre and music. Importantly, they provide for a variety of activities between residents. In the Philippines, these clusters are intentioned for low density, detached dwellings, with groupings of 2–4 dwellings shared between familial groups as mostly private family space [18]. These shared public spaces could be relatively small for low densities of housing and only require a single entry point that private residences expand out onto, as shown in the left-hand diagram in Fig. 2 [18]. Where densities for dwellings are higher, such as low to medium rise apartment housing, larger shared space is required [19]. This is shown in the diagram in Fig. 3. Additionally, higher density buildings require multiple entry points for buildings users and services, with open space on all sides of the building, as shown in the diagram to the right in Fig. 2 [1].

While the concept of clustering is useful in low density developments, a more complex and comprehensive framework is required for higher density and large-scale projects like Oceanix. An extension on the concept of clustering is a theory that emerged in a speculative plan for the redevelopment of post war Germany in the 1970s by Oswald Ungers and Rem Koolhaas called "city parts". Although an urban concept generated for land-based systems, "city parts" sees the city as a series of archipelagos, or islands, as a network of microcities within a larger city. This more complex system for understanding clustering in floating cities is very useful in an analysis of Oceanix.

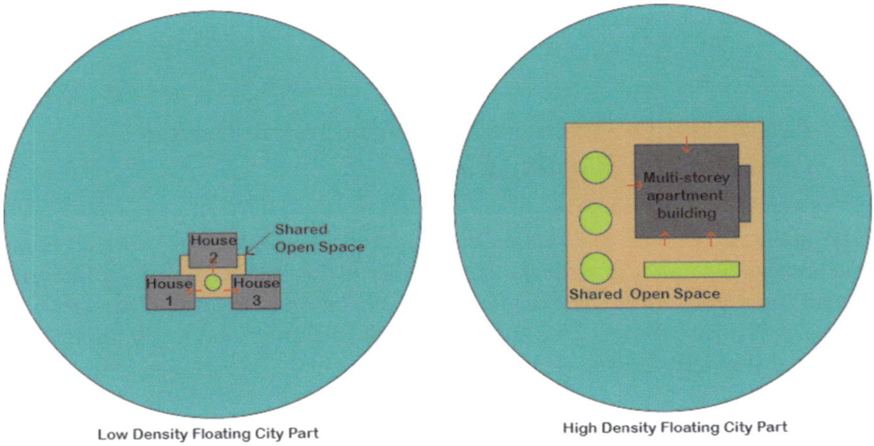

Fig. 2 Clustering open space requirements relative to density

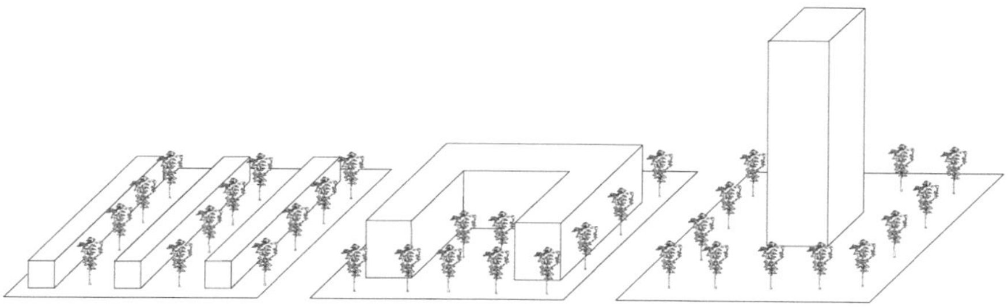

Fig. 3 Density to open space diagram drawn from Lehmen [19] study. Diagram by Author

3.1 Clusters, City Parts and Floating Cities

While studying architecture at the Architectural Association in London, Rem Koolhaas undertook a field trip to West Berlin in 1971. At the time Berlin was a city still healing from the wounds of war and suffering a decline in population. On his return from the trip, Koolhaas began to question some of the teaching at the AA. He found from his field work in Berlin that there were definite limitations of what architecture could achieve alone within the city. The following year, Koolhaas moved to New York to study, and later work with, Ungers at Cornell University. In 1977 they returned to Berlin to run a studio where they developed the manifesto *The city within the city—Berlin: A green archipelago* [20].

The concept of the green archipelago can be described as a 'system of fragments', which was developed by Ungers as negotiating the need for community and connection while maintaining the individual within the city. Ungers' critique of the modern city was that the excessive focus on the unified whole of urban environments had eroded any sense of individual self, rendering contemporary cities as anonymous. The city within the city provides a framework for a 'pluralistic urban concept' which counters modernist urban design theories that define a unified whole city. While there are several concepts presented within the manifesto, the two that relate most to floating cities are "City Parts" and "Green Archipelagos" [21]. City Parts refers to a breaking down of the urban fabric into smaller legible pieces. Each 'part' functions as its own autonomous city—a micro city within a larger collection of cities—an archipelago of cities [22].

Decentralizing the city in this manner assists with several important social and economic factors. Firstly, it creates flexibility for each part to be expanded or contracted as situations change and populations fluctuate. Applied as a climate resiliency strategy, it could provide places for populations to be distributed to in times of climate emergency, while key infrastructure is repaired or re-built. Unlike the rigidity of Ungers' land-based City Parts, the inherent mobility of floating microcities would facilitate the capacity for an interchangeable organisation of separated parts. Each microcity contains its own amenities and is accompanied by infrastructure that can support its own urban identity through smaller urban units to foster a stronger sense of community. Additionally, each micro-city should have responsive, localised governance dedicated to that part of the whole. It is probably this last aspect of the governance structure that is conceptualised in City Parts that is missing from current floating city proposals such as Oceanix; a governance structure that encourages diversity, where each section can develop its own unique functions and characteristics.

Proposals like Oceanix are master planned as archipelagos but designed to be read as a functional whole with minimal discernible characteristic between each island. While the buildings take on varied forms, there is a distinct repetition in density with almost all islands hosting buildings between 4 and 5 storeys (20–25 m in height)—a density outcome that likely arises from yield calculations to support the financial feasibility for the project. Again, while there is variation in the appearance of built form, each island boasts similar scale and there is repetition in formal expression in terms of fenestrations, flat roofs and so on [23]. This is where appropriate oversight through urban form design guidelines in negotiation with individual development proposals on islands could address the gap between individualism for each island—city within a city—and a cohesive whole for an overarching precinct development.

The other concept developed by Ungers was Green Archipelagos, a concept easily applied to floating cities by replacing "green" with "blue". It involves shrinking the footprint of each microcity to its most compact version, freeing space for surrounding natural spaces, such as landscape or water. Each concentrated island would then be separated by space for agriculture or aquaculture and more-than-human centres for wildlife only interrupted by minimal connecting public spaces and infrastructure where necessary [20]. Taking advantage of the mobile capacity of each floating city part would allow for a multitude of configurations for clustering microcities that could be altered as necessary to meet infrastructure, repair, maintenance or even spectacular event gatherings. The blue spaces between micro cities would act to enhance life on each city island in terms of quality of open space and connection to the natural environment. It would promote sustainability, improve air quality and support biodiversity.

Of most significance to the urban development of floating cities, the intervening spaces between islands leave space for adaptive urban structures, expansion and contraction, and this is especially valuable for cities promoting sustainable degrowth and in the case of population fluctuations, especially declining populations [24]. Ungers' concepts of Green Archipelagos and City Parts offer a strategic framework for rethinking urban development in the face of changing demographic and environmental conditions. They emphasize the importance of adaptability, sustainability, and community in urban planning [20]. The adaptability and flexibility of city parts is especially relevant to tidal fluctuations, littoral flows or the design of amphibious architecture, where buildings will vary in their operation from land-based structures to floating structures. The ability to generate variable conditions of urban fabric will be especially critical for these buildings.

3.2 City Parts, Density and Open Space

The extra amount of public open space required for higher density buildings, as illustrated in Fig. 3 [19], means seeing each building as its own island, its own "city part". Rather than reading floating villages as individual floating dwellings, proposals like Oceanix, force a reading of each building as its own independent city within a larger city–as though each island is a cruise ship, relatively capable of independent existence for a period of time, but reliant on other islands or port stops to sustain long term existence. The very master plan for Oceanix provides a fascinating illustration for Ungers' and Koolhas' concept for City Parts. And this is a condition that largely results from an increase in density and the required open space to support medium to high density buildings (see for example Fig. 2).

4 Analysing City Parts, Density and Urban Form Design for the Oceanix City Proposal

Oceanix is located offshore from South Korean city, Busan. It is a prototype proposal for a resilient and sustainable floating community. This prototype has been developed in collaboration with UN-Habitat and the Busan Metropolitan City and it aims to demonstrate how floating infrastructure can help coastal cities expand sustainably while adapting to rising sea levels [23]. The current proposal consists of three prototyped platforms that are dedicated to singular functions, including lodging (accommodation), research (workplace) and living (recreation). Each building type is around the same massing and density, standing at 4–5 storeys (20–25 m) [23]. There are slight variations in the footprint for each island and its built form occupation as outlined in Table 3. The project aims to positively contribute to the ecological environment that it occupies. Oceanix is designed to address immediate urban needs but also hopes to be seen as a precedent for future developments that prioritise sustainability and resilience [25].

Initial work for the project commenced in July, 2021. Oceanix is projected to cover an area of approximately 750,000 m^2 and consist of six hexagonal floating islands, each of which would serve public purposes such as healthcare, education, shopping, and cultural centres, with a cumulative capacity of approximately 10,000 residents. Each hexagonal island is designed to accommodate 300 residents over 2 ha., reaching a density of 150 people per hectare. As a benchmark, the current density for the neighbouring city of Busan is 45 people per hectare, whereas the density of the capital city, Seoul, is 155 people per hectare. For comparison, a medium density city like Barcelona has 156 people per hectare and a high density city like Manhattan has 283 people per hectare [26]. A cluster of six islands, as diagrammed in Fig. 4 constitutes a Village, accommodating 1650 people over 12.2 ha., delivering a density of 135 people per hectare. A city is then a further clustering of six villages, accommodating 10,000 people over 75 ha. and a density of 133 people per hectare.

Table 3 Open space to built form percentage and green space in the proposed master plan for oceanix

Village Island (as labelled in Fig. 4)	Percentage of Island occupied by built form (%)	Proposed percentage of open green space (%)
Village Neighbourhood Island A	46	≥ 30
Village Neighbourhood Island B	50	≥ 30
Village Neighbourhood Island C	58	> 20
Village Neighbourhood Island D	46	≥ 30
Village Neighbourhood Island E	47	≥ 30
Village Neighbourhood Island F	40	> 30

The masterplan essentially folds out from neighbourhood, to village, to city. Or in Koolhaas and Ungers' terms, a city, within a city, within a city.

The larger city master plan incorporates separated smaller islands for shared amenity and there are many good design approaches that align with the urban form

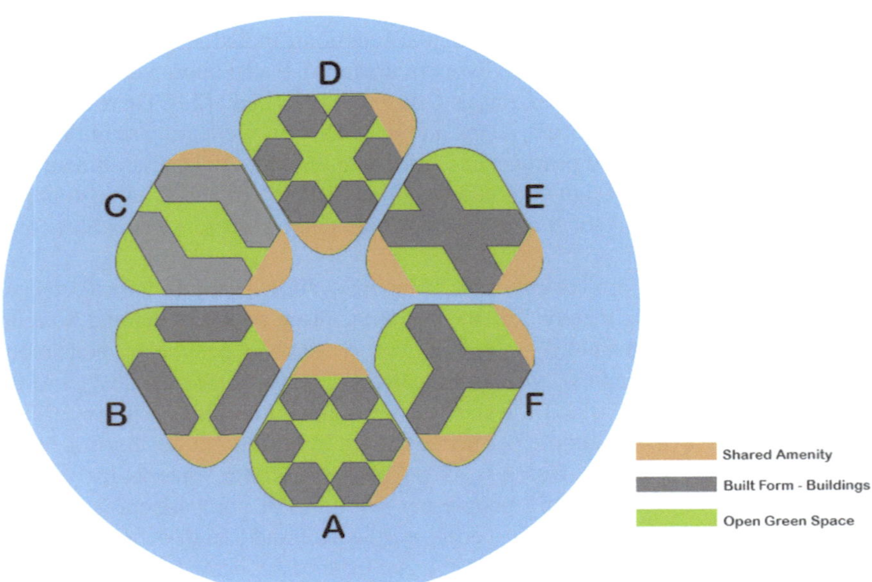

Fig. 4 Open space and open green space to built form ratio diagram Oceanix Village cluster

design framework established at the start of this chapter. These include the incorporation of clear mobility strategies for public and other private transport. A variety of ways to move between villages and neighbourhoods. Each island provides a generous amount of shared open space and especially shared green space. This has obviously been a driving concept behind the master plan. However, this high level of amenity risks driving up housing costs and make living in Oceanix unattainable to the general population [27], reflecting the findings from studies in the Netherlands that showed floating cities were generally unaffordable for most [1]. The percentages shown in Table 3 are based on calculations drawn from the diagram in Fig. 4. This diagram was generated from the Village plan shared on the World Landscape Architect [23] website. Of note, the general guidelines for the provisions of open spaces in medium density cities is between 15 and 20% [28]. Two of the neighbourhood villages in Oceanix provide more than 40% of open spaces and the other four provide more than 50%.

As stated earlier, there is a lack of diversity in built form, however, it is understood that perhaps densities were driven by a minimum yield calculation for the project to be financially viable. In their study on Netherlands Thomas et al. [1] identified the lack of data on yield calculations for floating cities as a significant barrier to their successful proliferation. This data is particularly crucial in terms of delivering accessible and affordable housing for everyone who needs to live in a floating city, and is a key area that requires further research.

Further, a significant challenge lies in balancing access to adequate shared open space for medium and high-density developments without significant, negative impact on the ecology of the water. The more surface area a floating village consumes, the greater the risk that it may inflict ecological damage [1]. While the need for large open space is driven by Ocenix's proposed autonomy from land cities [23], perhaps, if better synergies were shared between Oceanix and Busan, the shared amenities on land may reduce the need for such large areas of open space for each neighbourhood. At the same time, land-based residents could benefit from the water based amenities of Oceanix as well. A hybrid approach between land and water was a strategy shared by Ham [18] and Thomas [1] in their studies on the Netherlands and the Philippines, as discussed earlier in this chapter.

5 Conclusion

The importance of properly planned open urban space to bring communities together is integral to the success of floating cities—especially for communities that share a unique living situation, such as co-existing in a floating village or city. The biggest challenge going forward for floating cities will be to develop planning, governance and development systems that support quality, while also providing affordable access to these developments. The literature points out that there is a clear disparity between floating villages in developing countries that are poorly serviced with adequate infrastructure and amenity and those in developed countries, which are designed in a way to be out of reach for the general population. This needs to be addressed because without such consideration, future developments risk dividing social classes across countries and cultures in floating cities. Good design governance and the development of urban form frameworks and design guidelines can establish methods and measures that address this gap and promote the sustainable development of floating cities.

References

1. Thomas L, de Haan E, Gerards L, de Haan G, Brouwer J (2023) Designing floating urban realities: on the urban design of floating high-density environments in the dutch context. In: World conference on floating solutions. Springer Nature Singapore, Singapore, pp 33–51
2. Nguyen TTT (2022) Architectural design guideline for sustainable floating houses and floating settlements in Vietnam. In: WCFS2020: proceedings of the second world conference on floating solutions, Rotterdam. Springer Singapore, pp 461–474
3. Wang BT (2023) Development of the legal definition of the floating city: Judicial interpretation of structural characteristics of floating homes and developments. In: World conference on floating solutions. Springer Nature Singapore, Singapore, pp 139–157
4. Yogesh S (2022) Are floating cities the urban future? In: Chias P, Hernandaz S (eds) Eco-architecture IX: harmonisation between architecture and Nature. WIT Press, UK, pp 210–215
5. Holden S, Volz K (2023) Design governance: Leveraging the value of architects. Archit Aust 112(2):54–55
6. Volz K (2018) Natural disasters, undesign and the absent interior. In: Undesign. Routledge, pp 175–185
7. Volz K, Healy E (2023) Keeping it simple: public messaging and climatic design guidance for the built environment. In: ASA 2023 proceedings: sustainability and health: the nexus of carbon neutral architecture and well-being-56th international conference of the architectural science association 2023. Architectural Science Association (ANZAScA), pp 462–475
8. Brogden L, Volz K (2023) Developing climatic design guidelines for Australia's growing regional centres: proposing an interdisciplinary, place-based methodology. In: Place-based sustainability: research and design extending pathways for ecological stewardship, pp 108–129
9. Steiner FR, Butler K (2012) Planning and urban design standards. Wiley
10. Talen E (2012) City rules: how regulations affect urban form. Island Press
11. Jankovič Grobelšek L (2015) Public spaces and private spaces open to the public: spatial planning and development using urban design guidelines. Open Urban Stud Demography J 1(1)
12. Carmona M (2021) Public places urban spaces: the dimensions of urban design. Routledge
13. Ferdous F (2013) Examining the relationship between key visual characteristics of urban plazas and aesthetic response. SAGE Open 3(2):2158244013485581
14. Thompson CW (2002) Urban open space in the 21st century. Landsc Urban Plan 60(2):59–72
15. Wu J, Wang M, Li W, Peng J, Huang L (2015) Impact of urban green space on residential housing prices: case study in Shenzhen. J Urban Plann Dev 141(4):05014023
16. Parolek DG, Parolek K, Crawford PC (2008) Form based codes: a guide for planners, urban designers, municipalities, and developers. Wiley
17. Boarnet MG, Crane R (2001) Travel by design: the influence of urban form on travel. Oxford University Press, USA
18. Ham P (2024) From floating prototype to floating community: designing resilient floating communities by using participatory approaches. In: World conference on floating solutions. Springer Nature Singapore, Singapore, pp 67–80

19. Lehmann S (2016) Sustainable urbanism: towards a framework for quality and optimal density? Future Cities Environ 2:1–13
20. Shannon K (2014) The city in the city. Berlin: A Green Archipelago–Oswald Mathias Ungers and Rem Koolhaas with Peter Riemann, Hans Kollhoff and Arthur Ovaska. A critical edition by Florian Hertweck and Sébastien Marot
21. Aureli PV (2008) Toward the archipelago. Log 11:91–120
22. Walker J (2015) Islands-in-the-City: Berlin's urban fragments. J Archit 20(4):699–717
23. World Landscape Architect (2024) Oceanix. https://worldlandscapearchitect.com/big-and-partners-unveils-oceanix-city/?v=8bcc25c96aa5
24. Aureli PV (2011) The possibility of an absolute architecture. MIT press
25. Yang H, Zhao S, Kim C (2022) Analysis of floating city design solutions in the context of carbon neutrality-focus on Busan Oceanix City. Energy Rep 8:153–162
26. Statistica (2024) Korean population per square kilometre. https://www.statista.com/statistics/1112322/south-korea-population-density-by-province/
27. Irwin EG (2002) The effects of open space on residential property values. Land Econ 78(4):465–480
28. Ibrahim PH, Dali MM, Yusoff SYM (2013) Implementation of open space: the need for uniform policy. J Sustain Dev 6(7):16

Kirsty Volz Kirsty Volz is an architect and academic. Her research focuses on design governance and design-led policies for the built environment. Kirsty is especially interested in how design-led policies impact on housing quality and its relationship to human health and well-being, as well as more than human factors to enhance sustainable environments.

Navigating the Legal and Regulatory Frontiers of Urban AI, Autonomous Maritime Robots and Floating Cities

Brydon T. Wang, Bernadette Hyland-Wood, and Michael Guihot

Abstract Artificial Intelligence (AI) systems, such as environmental and geospatial AI systems, have significant potential for a range of scenarios involving offshore deployment. These AI systems integrate satellite imagery, sensor data and environmental models to design floating cities, inform planning, model offshore renewable energy, monitor aquaculture facilities, and manage marine and offshore ecosystems. *Urban AI* refers to the integration of AI technologies to plan, design and implement urban environmental strategies to enhance efficiency, sustainability and governance. These Urban AI systems may oversee and use autonomous maritime robots to manage a range of offshore infrastructure requirements. For example, they may be used to track fish movements and optimise energy and water use in offshore fish pens. In turn, these maritime robots fall under the umbrella of AI systems that autonomously make decisions and take actions on their own based on algorithms and learned experiences, which are broadly referred to as *Agentic AI*. Unlike traditional AI, which typically respond to predetermined outputs, Agentic AI can evaluate situations, set goals, report on outcomes and adapt its behaviour to achieve those goals. Often, Agentic AI operate in dynamic environments, learning from interactions and improving their performance over time. Relevant to practitioners and scholars of AI systems are the ethics, control, legal and regulatory implications of granting agency in decision-making processes in offshore settings, particularly in settings with growing geopolitical contestation. Maritime laws such as the *United Nations Convention on the Law of the Sea* (UNCLOS) are not designed to address the unique issues presented by AI-driven technologies specific to ocean users. This

B. T. Wang (✉)
Centre for Policy Futures, The University of Queensland, St Lucia, QLD, Australia
e-mail: Brydon.Wang@uq.edu.au

KPMG Australia, Deal Advisory & Infrastructure, Brisbane, QLD, Australia

B. Hyland-Wood
Centre for Data Science, Queensland University of Technology, Brisbane, QLD, Australia
e-mail: b.hylandwood@qut.edu.au

M. Guihot
School of Law, Queensland University of Technology, Brisbane, QLD, Australia
e-mail: michael.guihot@qut.edu.au

B. T. Wang et al. (eds.), *Large Floating Solutions*, Lecture Notes in Civil Engineering 636, https://doi.org/10.1007/978-981-96-5435-2_18

chapter assesses the following challenges: the use of AI systems to plan, design and implement city management strategies and maritime robots deployed *autonomously* to monitor, evaluate and report on tasks. These present previously uncontemplated pressures on international maritime laws. We review scenarios for the use of Urban AI and Agentic AI, and discuss regulatory ambiguities, data ethics and privacy issues. Our review of recent AI regulation in the European Union and United States show that offshore development activities and maritime robots present novel considerations that have yet to be contemplated and urgently require coordinated governance and regulatory oversight.

Keywords AI regulation · Agentic AI · Urban AI · Automated decision-making · Autonomous systems · Drones · Maritime robots · Offshore development

1 Urban AI and Autonomous Maritime Robotics in Floating Cities

Growing challenges posed by climate change, rising sea levels and the rapid deployment of AI systems require cooperative transnational solutions. This chapter examines the potential of Urban AI and Agentic AI integrated in autonomous maritime robotics to overcome surveying, planning, design, logistics and operational challenges in floating cities and offshore environments. We show that the international legal frameworks for monitoring and responding to our oceans have not kept pace with the increasing datafication of our seas and the automation of infrastructure within them. For example, these legal frameworks have yet to address how the datafication and automation of the maritime transport industry has seen an exponential advance of cyberattacks that is growing both in numbers and sophistication [1]. During a period of increasing geopolitical contestation [13], the maritime sphere is facing concerns described as ‘cyber-fragility’ that requires coordinated and clearer regulatory responses. By examining how AI-driven strategies can be incorporated into the development and governance of floating cities and its associated energy and food production infrastructure, this chapter distils the barriers and facilitators in international law and emerging AI regulation that will have relevance in the wider context of the datafication of vessels and ports.

On one hand, Urban AI offers transformative solutions to planning and delivery of urban services in cities [6, 8, 50] that can be extended to their maritime counterparts. In the planning phase, Urban AI systems can be used to model climate impacts and optimise the placement and design of floating structures, while ensuring resilience and sustainability (see Chap. “Exploring Design of Floating Urban Developments: Integrating Density, Engineering and Environmental Considerations”). For instance, machine learning algorithms now analyse vast datasets to predict environmental changes, enabling proactive adaptation strategies [7, 33]. Additionally, these Urban

AI applications can facilitate the efficient allocation of resources within floating cities, enhancing their self-sufficiency and reducing their ecological footprints.

At the same time, the risks associated with AI-driven solutions also extend to other scenarios of AI integration, particularly with the use of Agentic AI to operate autonomous maritime robots. Robots for construction, operations and maintenance of terrestrial infrastructure are extant [42]. These robots can respond to site conditions and autonomously shift and re-design the construction process 'with design changes automatically calculated and "seamlessly" programmed back into construction processes, 'in which conventional human involvement is merely supplementary' [23, p. 98].

The use of maritime robots and Agentic AI in design, construction, operation and maintenance of infrastructure to the maritime domain has led to a vast field of maritime robotic systems. Rødseth and Vagia [32] observe that the different fields in which research into 'autonomy' arises has led to definitional challenges around the word 'autonomy' and the need to distinguish autonomous applications from 'automated' systems. They argue that the common argument tying autonomy of an agent to that of being 'independent of external control' is 'meaningless' given that there is 'absolutely no point in having an autonomous system that does not follow orders or mission plans' [32, p. 012003–04]. Other developments of these systems in the maritime sphere include Autonomous Underwater Vehicles (AUVs) that are used to collect a variety of data including biological, physical, chemical metrics from marine environments [10], sensor networks, remotely controlled unmanned surface vehicles (USVs) and autonomous surface vehicles (ASVs). Our chapter refers to recent developments in autonomous applications of Robotics and AI (RAI) in the maritime sphere as *'autonomous maritime robots'*. These autonomous robots emerge from the preceding wave of automation technologies [31] and are currently being developed and deployed to offshore sites to enhance operational efficiency, safety, and sustainability of offshore wind energy [2, 26] and offshore fish pens [14, 22].

Urban AI in floating city contexts and autonomous maritime robots are interlinked. For example, an Urban AI system would need to respond to and plan around the movement of autonomous marine vessels or draw from sensor data fed to it by autonomous maritime robots. Likewise, autonomous maritime robots and the Agentic AI systems that drive them also raise similar challenges to Urban AI that require careful legal and regulatory consideration. For example, there is a growing gap between AI development and regulatory frameworks, particularly at the international level [16]. Maritime laws such as the *United Nations Convention on the Law of the Sea* (UNCLOS) are not designed to address the unique issues presented by AI-driven technologies specific to ocean users.

This chapter addresses how existing regulatory frameworks must evolve to accommodate these advancements, including an urgent call for new international standards for liability, governance, and environmental protection in the context of offshore AI systems. As autonomous systems become more integral to automated decision-making systems deployed in the planning context, and in offshore operations, questions regarding algorithmic accountability, transparency and human oversight are raised. We begin with an exploration of the convergence of AI, autonomous maritime robots and offshore urban development. We undertake a discussion on AI systems and maritime robotics in offshore development, including the use of such technologies in the planning and design of floating cities. The chapter then turns to the deployment of Agentic AI and autonomous maritime robots in the context of the infrastructure needs of floating cities, focusing on offshore wind energy and fish farming. Focus is given to how AI systems, including maritime robots and Agentic AI approaches have the potential to enable sustainable practices, reduce risk associated with human error and optimise operations. In doing so, we highlight several aspects of the complex legal, environmental, and ethical concerns to be navigated. Following this, the chapter assesses the regulatory landscape, exploring the common legal considerations posed by the deployment of RAI systems in maritime and offshore environments. We then focus on the need for ethical guidelines to ensure that Urban AI operates within human rights-based frameworks, and addresses unintended and intentional algorithmic bias to reinforce legal legitimacy in offshore governance settings. We conclude by advocating for proactive legal frameworks that support the sustainable and responsible integration of AI and maritime robotics in offshore development. Through this analysis, we set the stage for a deeper understanding of the legal and technological challenges that arise at the intersection of AI, robotics, and offshore industries, while offering insights and recommendations for floating futures.

2 Urban AI: AI-Driven City-Making for Floating Cities

Urban AI encompasses applications ranging from predictive analytics in urban planning to real-time management of public services and environmental monitoring [6, 24]. Urban AI extends beyond traditional smart city initiatives by incorporating autonomous decision-making systems that dynamically adapt to urban challenges and overcome gaps in information, facilitating a shift from automation to autonomy in city governance [8]. This approach has been applied to address complex urban issues such as traffic congestion, resource distribution, and environmental sustainability [49].

Floating cities, conceptualised as self-sustaining urban settlements on water, are gaining attention as a response to rising sea levels, urban overpopulation and the need for climate resilient infrastructure. These developments are envisioned as modular, scalable and environmentally adaptive, incorporating renewable energy sources [46], AI-managed waste recycling and autonomous maritime transport systems. Proposals, such as *Oceanix City* [39], that aim to establish floating urban settlements could integrate RAI-driven resource management and environmental monitoring systems to ensure sustainability. In doing so, floating cities can leverage existing AI systems to optimise spatial planning, dynamically adjust urban services to respond to climatic and population requirements and ensure efficient governance through data-driven decision-making [11, 19]. In fact, given floating structures are mobile, the city can benefit from AI systems that could dynamically manage and shift these structures to reconfigure the city to meet the changing needs of its inhabitants.

However, the design of any city is challenging given they 'epitomize complex systems par excellence... full of contestations, conflicts, and contingencies that are not easily captured, steered and predicted' [3], p.1; see discussions in Chaps. "Exploring Design of Floating Urban Developments: Integrating Density, Engineering and Environmental Considerations", "Brain Coral: Floating City in the Maldives, "Designing for Buoyant Urbanism: Pedagogy, Architectural Paradigms and Social Acceptance of Floating Cities" to "City Parts: Urban Form Design Guidelines for Floating Cities"). These competing demands are heightened by the unpredictability of climate change, pandemics, capital flows and shifts in demographics, requiring data-driven technologies that enable urban computing and intelligence [3]. The integration of AI into buoyant urbanism represents a critical evolution in urban computational planning, where data is employed to manage and optimise the planning and design of floating cities. Urban AI can be utilised at multiple levels, from predictive analytics in urban design to real-time monitoring of marine environments [25]. For example, AI-powered simulations allow urban planners to assess the impact of climate change on floating infrastructure, optimising structural resilience against extreme weather events. Similarly, AI-driven digital twins enable continuous adaptation by modelling interactions between built structures and existing environmental conditions in real time [2, 44]. These digital twins could extend real-time modelling to floating cities to consider maritime conditions and other ecosystem needs.

AI-enhanced governance mechanisms also play a pivotal role in floating cities, enabling decentralised and participatory urban management. AI-assisted policy-making can help authorities regulate resource distribution and automate public service provision [49]. This is because AI systems have the potential to bridge fragmented data sources often generated by sensors, location-based digital services and social media content to offer a more nuanced understanding of urban data subjects [43]. Additionally, autonomous systems in maritime transport, security surveillance, and environmental conservation further underscore AI's role in maintaining operational efficiency within floating urban ecosystems.

3 Agentic AI and Autonomous Maritime Robots

Deployments of maritime robotics and AI in offshore environments are growing and offer significant advantages in terms of efficiency, safety, and sustainability. Whether in offshore wind energy, fish farming, or deep-sea exploration, these technologies are improving operational practices, reducing risks to human workers and enabling industries to operate in increasingly challenging offshore environments [2]. As RAI continue to evolve, their integration into offshore infrastructure will likely play an even greater role in shaping the future of maritime industries.

Agentic AI systems are designed to operate autonomously with a degree of self-direction, problem-solving capabilities and decision-making independence. These AI systems exhibit agency; that is, they are able to perceive their environment, plan actions, adapt to changing conditions and execute tasks without continuous human oversight and intervention. They are also able to undertake multi-agent coordination to collaborate with other AI agents or systems.

Agentic AI technologies are being integrated with maritime robots in offshore environments to rapidly transform various sectors, including offshore wind energy and fish farming. Offshore environments are inherently hazardous due to extreme weather conditions, deep waters and complex logistics. Robotic systems that are equipped with advanced sensors and AI algorithms enable these systems to perform detailed inspections of offshore structures, monitoring their condition, often in challenging environmental conditions [32]. These technologies can reduce the need for frequent human intervention and enable remote operations. For instance, autonomous vessels and drones can transport equipment, inspect infrastructure and monitor conditions at offshore sites without the need for costly and complex logistics [32].

RAI systems can also enhance operational flexibility by allowing operators to make real-time decisions about resource allocation based on the data collected by autonomous systems [29]. For example, RAI systems can autonomously detect cracks, corrosion or structural degradation in subsea infrastructure such as pipelines, subsea cables and wind turbines. These inspections, which are traditionally carried out by human divers or vessels, can now be conducted by robots with high precision and speed, significantly reducing the time and cost of offshore maintenance [20]. RAI systems can also analyse the data collected during inspections to predict when repairs or replacements will be needed, allowing for predictive maintenance and more effective resource allocation. This real time maintenance reduces risks by identifying potential issues before they develop into costly or catastrophic failures, thus enhancing safety, operational efficiency and sustainability. Thus, by deploying autonomous robots, operators can reduce the exposure of human workers to these dangerous environments, minimising the risks associated with offshore operations and improving safety standards. We examine two examples of such deployments.

3.1 Deployment of Autonomous Maritime Robots in Offshore Wind Farms

Remotely operated vehicles (ROVs) are especially useful for inspecting remote or difficult-to-access offshore installations where human presence is either dangerous or impractical. They are widely used in subsea inspections, particularly for inspecting the floating platforms, foundations, mooring lines, anchors and subsea cables that connect offshore wind turbines to the grid without the need for divers or manned vessels [20]. These ROVs are equipped with high-definition cameras, sonar systems, thermal sensors, ultrasonic imaging sensors, LIDAR and other sensors, allowing them to capture detailed visual and environmental data from beneath the water's surface [20]. Their ability to operate in deep water, challenging weather conditions, strong currents, and with little down time, makes them an invaluable and tireless tool for maintaining offshore wind farms.

AI systems are increasingly oriented on enhancing the capabilities of ROVs by automating the identification of potential issues in the data they collect, enabling faster decision-making and reducing the need for human intervention. In contrast, while AUVs, like ROVs, are used for subsea inspections, they differ in that they are fully autonomous. The integration of Agentic AI into AUVs allows these vehicles to make real-time decisions based on sensor data, optimising their route and operational tasks. For example, AUVs can autonomously adjust their course to avoid obstacles, change depth to assess specific infrastructure components and autonomously optimise battery usage to maximise the duration of operations. This autonomy reduces the need for continuous human supervision, decreases operational costs and increases the operational flexibility of offshore wind farm maintenance.

AUVs are typically used for tasks such as mapping the seabed, inspecting offshore structures, and conducting environmental monitoring. Equipped with sonar, cameras, and other specialised sensors, AUVs can autonomously navigate the subsea environment and collect data that is then analysed by other AI systems [51]. These vehicles are highly effective in conducting deep-water surveys and inspections that would otherwise require complex, high-cost operations. In this way, new sites for floating wind infrastructure (and by extension, floating cities) could potentially be identified without impacting fish movement and in a manner that reduces site-specific requirements that may limit the design approach to the floating infrastructure asset.

Unmanned Aerial Systems (UAS), commonly known as drone systems, are likewise increasingly being used to inspect the topside components of offshore wind turbines, including the blades, nacelles, and towers. UAS are equipped with high-resolution cameras and advanced sensors, including near infrared (NIR), hyperspectral and thermal sensors that enable them to conduct detailed visual inspections of turbine blades for signs of damage or wear [34]. By equipping UAS with infrared cameras, these systems are able to detect issues such as overheating or mechanical stress that may not be visible through regular visual inspection. AI technology is then used to process the large amounts of data captured by UAS during inspections. This allows for rapid analysis of turbine conditions, where any issues can be automatically identified and maintenance tasks prioritised.

When integrated with other RAI systems, UAS can also be used for site surveys and mapping, helping offshore wind farm operators better understand the geographical and environmental conditions of the offshore site [48]. At the same time, their ability to access difficult-to-reach turbine components quickly and safely makes them a critical tool for offshore wind farm operators, reducing the need for costly and dangerous manual inspections. In stepping in to replace the need for a human performing a hazardous task, these autonomous maritime robots improve the efficiency of inspections and are instrumental in enhancing risk mitigation.

3.2 Aquacultural Applications

In the aquaculture industry, RAI systems are transforming fish farming by improving monitoring capabilities, optimising feeding practices and maintaining the structural integrity of fish pens [22, 45]. Autonomous vessels and underwater robots are increasingly deployed to monitor water quality, track fish health, automate feeding processes and clean the overall structure. These systems are designed to collect data from sensors placed within fish pens, including temperature, pH levels, oxygen content and ammonia concentration, which are critical indicators of water quality and fish well-being. Agentic AI systems can analyse this data in real time, allowing for immediate adjustments to the environmental conditions in these remote offshore pens. For example, if the water temperature rises above optimal levels or oxygen levels drop, the AI system can adjust the ventilation or initiate water pumping systems to restore ideal conditions, all while minimising human intervention [30].

Robotic systems are also deployed for feeding the fish in a more efficient and controlled manner. Autonomous feeders ensure that fish are fed at appropriate intervals and with the right amount of food, reducing waste and preventing overfeeding, which can lead to environmental pollution [45]. Further, RAI systems help in the early detection of disease by continuously monitoring the fish for behavioural changes that may indicate illness. This allows for swift intervention, such as isolating affected fish or adjusting conditions to prevent further spread of diseases [45].

Likewise, the ability to assess environmental factors is crucial for the sustainability of aquaculture. RAI systems are also employed to evaluate the environmental impact of fish farming operations by analysing the effect of operations on local ecosystems, water quality and biodiversity. This helps ensure that fish farms comply with environmental regulations and mitigate their ecological footprint. As the aquaculture industry grows, AI will play an increasingly important role in optimising production while maintaining environmental responsibility.

Accordingly, AI-enabled robotic systems offer numerous advantages that are transforming offshore operations, making them more autonomously responsive and efficient. These benefits are particularly evident in areas such as subsea exploration, infrastructure inspection, marine ecosystem monitoring and fish farming. By leveraging advanced AI algorithms and sensor technology, these systems enhance the precision and granularity of data, scalability, and sustainability of offshore operations, while also reducing the reliance on human intervention in dangerous and remote environments.

4 Legal Considerations of Deploying Urban AI and Autonomous Maritime Robots

4.1 Common Legal Considerations

Both scenarios of Urban AI and autonomous maritime robots in RAI systems raise common legal considerations. First, the integration of AI in maritime operations presents complex legal challenges, particularly concerning liability, jurisdiction and regulatory adaptation. Floating cities and their associated infrastructure (such as, offshore wind farms and fish pens) can operate in international waters in which no single nation-state has full sovereignty. This creates ambiguity regarding which legal system governs AI-related disputes, including issues of liability, privacy and algorithmic decision-making. Further, there is a lack of a universally accepted legal framework for AI in international waters, which complicates enforcement mechanisms and raises questions about regulatory accountability.

Key among these legal issues is AI liability in maritime environments. For example, when RAI-operated offshore systems are deployed, who bears the liability in the event of system malfunctions, cybersecurity breaches or environmental damage? RAI systems may fail or make incorrect decisions. In cases of malfunction, determining the cause and extent of the issue, and assigning responsibility and liability can be difficult, especially when the Urban AI or autonomous maritime robotic system involves multiple interconnected parts and operates autonomously and without direct human intervention. For example, if an autonomous vessel conducting inspections of offshore infrastructure collides with a wind turbine due to a failure in the navigation system, it may be unclear whether the fault lies with the design of the RAI system, the infrastructure or whether it may be due to external factors, such as weather conditions. In many cases, the liability may not rest with the owner or operator of the technology but with the developers or manufacturers of the AI systems that control the robots, which can lead to significant legal disputes over accountability. Establishing clear legal frameworks for liability, including defining the roles of AI developers, operators and manufacturers, is essential to mitigate the risk of legal uncertainty and disputes [15, Chap. 7]. We discuss this in further detail in Sect. 4.3.

Cybersecurity threats also pose significant risks to the operation of AI-driven offshore systems. Boghossian et al [4] reported on similar concerns about confidentiality, integrity and availability in embedded and connected technologies in precision agriculture. Autonomous systems are highly dependent on software and communication networks, making them vulnerable to hacking, malware and other cyberattacks. A successful cyberattack on an autonomous offshore vessel, for instance, could lead to significant infrastructure damage, operational disruption, environmental pollution, or in extreme cases, the loss of life. The legal implications of cybersecurity breaches in the context of AI-operated systems are complex, as they involve both the technology providers and operators. Ensuring robust cybersecurity measures and establishing clear liability for breaches is critical in protecting these systems from external threats [17].

The use of autonomous systems in offshore operations also raises concerns about environmental harm. Urban AI systems and autonomous maritime robots, such as those used in fish farming, have the potential to damage marine ecosystems if not properly managed. For example, in fish farming, autonomous maritime robots must operate within stringent Commonwealth, State and Territory environmental regulations aimed at minimising the ecological impact of aquaculture operations. Autonomous maritime robots are also used to monitor the health of fish populations and ensure that the farming process does not lead to overfishing or contamination. Likewise, Urban AI systems that fail to manage waste systems correctly could contribute to environmental degradation. Determining liability for environmental damage caused by these systems is difficult because traditional environmental laws do not account for autonomous technologies. As mentioned above, where these AI systems are deployed in international waters or offshore environments with a number of overlapping regulatory requirements, questions of standing and which laws and standards apply become increasingly complex. Accordingly, international legal frameworks must evolve to address these issues and ensure that AI-operated offshore systems comply with environmental protection standards.

To address these governance challenges, international cooperation is required to develop new legal frameworks that govern the use of autonomous systems in offshore environments. AI-managed floating infrastructure, including autonomous vessels, smart energy grids and automated security systems, will require robust liability frameworks and payment mechanisms to determine accountability and quantum in the event of system failures, disruption and delay, accidents or cyberattacks; and how risk is allocated between parties and calibrated to account for external factors such as force majeure. The EU AI Act has introduced risk-based regulations, categorising AI applications based on their potential harm, but its applicability to floating installations remains unclear. AI in floating cities may require a hybrid governance model, incorporating aspects of international maritime law [38], AI-specific regulations and self-regulatory mechanisms. Similarly, Karim et al. [21] argue that there needs to be greater consideration by nation states on how international law will apply in the maritime cyber security context.

Another major concern is data sovereignty and cross-border data governance. Floating cities will likely rely on AI-driven data analytics for public services, such as environmental monitoring and predictive maintenance. However, determining who owns, processes and regulates this data presents a legal challenge. In traditional urban settings, data protection laws such as the EU *General Data Protection Regulation* (GDPR) set standards for data privacy and security. Floating cities, on the other hand, may fall outside the territorial scope of these laws (although the GDPR does have extraterritorial effect), necessitating the creation of new international agreements on AI-driven data management or careful calibration of national laws to consider if extraterritorial requirements that capture offshore deployments in international waters are desirable. Given the complexity of regulating autonomous systems across multiple jurisdictions, there is a need for a unified approach to ensure that AI technologies are used safely, ethically and in a trustworthy manner in offshore operations.

At the same time, governance structures must be flexible enough to accommodate rapidly advancing technologies. This requires international cooperation to harmonise the regulation of autonomous systems, ensuring that AI-operated offshore systems adhere to common operational and safety standards and environmental regulations, while appreciating the complexity of overlapping (and, at times, conflicting) laws. Given the global nature of the maritime industry, coordination between international bodies, such as the International Maritime Organization (IMO), national governments and private sector stakeholders, is essential to create cohesive governance frameworks for RAI systems. Karim et al. [21] observe that the successes of the IMO in developing and evolving the regulatory regimes on the safety and security of international shipping and on issues relevant to a wide spectrum of ocean users, makes it a strong forum to develop legal solutions for data regulation. It is clear that the IMO and other international bodies must play a key role in establishing globally recognised standards for AI and robotics in offshore industries. These regulations should cover a range of issues, including safety standards, liability, environmental impact and data protection, and where agreed to could then be adapted and applied to Urban AI and Agentic AI systems in the maritime domain.

4.2 Legal and Ethical Considerations for Urban AI in Floating Cities

The transition from traditional urban planning to AI-driven buoyant urbanism presents several challenges, including regulatory ambiguities, ethical concerns and (as mentioned above) data privacy issues. Given the extraterritorial nature of floating cities that can be deployed in a variety of contexts outside territorial waters, jurisdictional complexities arise in defining legal responsibilities for AI-managed governance [29]. The global AI regulatory landscape, including frameworks such as the EU AI Act and the OECD AI Principles, will play a crucial role in shaping the governance of AI-integrated urban development, including the development of floating cities. As AI systems increasingly mediates urban decision-making, ensuring transparency, accountability and alignment with human-centred design principles will be essential to fostering sustainable and equitable marine urbanism [18, 43].

One of the challenges raised by Urban AI is the potential for algorithmic bias in AI-driven urban planning and public service allocation. AI models are trained on historical data, which can reflect and reinforce existing socio-economic inequalities [47]. For instance, biased datasets may result in unfair distribution of resources, favouring affluent residents over marginalised communities. In floating cities, which are envisioned as international urban hubs, ensuring equitable AI-driven governance requires transparency in AI decision-making processes [49].

Another ethical concern is with privacy and surveillance in AI-managed environments. Floating cities are expected to rely heavily on real-time surveillance through data collection ('dataveillance') for infrastructure maintenance, security and environmental monitoring. But other aspects of AI-driven surveillance systems, such as facial recognition and predictive policing, have been criticised for disproportionately targeting certain demographic groups and infringing on civil liberties. Exacerbating this concern is that unlike traditional cities, floating urban environments may operate under complex jurisdictional arrangements and even contractual arrangements between developer and inhabitants, making it difficult to enforce robust privacy protections and data governance frameworks [29]. Addressing these issues requires adherence to international AI ethics guidelines, such as the OECD AI Principles, which advocate for fairness, accountability and transparency in AI deployment, or the adoption of robust contractual frameworks between individual inhabitants of these floating cities, and with the public or private developer. We note, however, that in the context of the latter contractual approach, it has been shown that individualised contractual relations providing consent override user autonomy and are a poor control for abuses of privacy [36].

Additionally, the rise of AI-mediated decision-making in urban governance introduces concerns regarding human oversight and democratic participation. Floating cities may employ AI to manage public utilities, regulate environmental policies and even automate aspects of governance. However, over-reliance on AI in decision-making can undermine public engagement and reduce opportunities for citizen-led governance [8]. Ensuring that AI operates as a decision-support tool rather than a decision-maker, keeping community-in-the-loop, is crucial for maintaining democratic legitimacy and public trust in floating city governance structures.

The ethical and social risks associated with AI in floating cities necessitate proactive regulatory frameworks that emphasise fairness, transparency, consensus-building and human oversight. Establishing AI ethics committees, implementing bias audits and incorporating human-in-the-loop decision-making processes will be critical to mitigating these challenges and ensuring that AI serves as a tool for social good rather than a mechanism for systemic inequities.

4.3 Legal Considerations for Autonomous Maritime Robots

One of the most complex legal issues related to the use of AI and robotics in offshore operations is determining liability in the event of system malfunctions, accidents, or failures. In traditional offshore operations, human operators are held accountable for any mishaps or damages that occur. However, with autonomous systems, liability becomes more ambiguous. If an AI-powered system malfunctions or causes damage, such as a robot failing to detect a fault in an offshore wind turbine or a drone crashing into infrastructure, it may be unclear who is responsible for the incident.

The absence of clear legal precedent exacerbates the issue. Unlike human-operated vessels, AI systems make autonomous decisions based on algorithms and machine learning processes. The challenge lies in determining whether errors in AI decision-making arise from software defects, insufficient training data or operator oversight [15]. The development of regulatory frameworks must address these uncertainties by clarifying liability allocation among stakeholders involved in AI-driven maritime operations.

Existing legal frameworks governing maritime activities, such as IMO regulations and national maritime laws, generally place liability on vessel owners and operators. However, the introduction of AI in autonomous maritime systems complicates the allocation of legal responsibility when these systems malfunction. Traditional liability principles, including strict liability and negligence, assume a human actor who exercises control over decision-making processes. In AI-driven operations, it is unclear whether liability should be attributed to the vessel owner, the software developer or the manufacturer of the AI system.

This question of liability raises concerns for manufacturers, operators and AI developers. Establishing clear legal frameworks for accountability is crucial to minimise the risks associated with autonomous systems and to ensure that victims of accidents or damage are appropriately compensated. A comprehensive liability framework must be developed to address these concerns, as current maritime laws are not equipped to handle issues arising from AI-driven systems.

At the international level, the IMO has begun discussions on revising key maritime conventions, such as the *Convention on the Safety of Life at Sea* (SOLAS) and the *International Convention on Standards of Training, Certification and Watchkeeping for Seafarers* (STCW), to accommodate AI-operated vessels. These discussions aim to establish global standards for AI safety, training, and operational compliance. The *Maritime Safety Committee* (MSC) of the IMO has been developing a regulatory framework for Maritime Autonomous Surface Ships (MASS), with a non-mandatory MASS Code due to be finalised and adopted in 2026, and a mandatory Code to come into force by 2032 (MSC 109th session, December 2024).

At the national level, autonomous maritime robots may traverse areas governed by different legal regimes, requiring a clear understanding of which national and international laws apply. The ambiguity regarding jurisdiction and sovereignty over floating assets, such as autonomous vessels or offshore structures, complicates the enforcement of legal responsibilities and standards [26]. For example, autonomous maritime robots can operate across multiple maritime zones, including territorial seas, exclusive economic zones (EEZs) and the high seas. Current maritime laws, such as UNCLOS, were not designed to regulate or govern AI-driven systems, leaving a significant gap in the legal framework. While UNCLOS provides an established structure for maritime activities, it fails to account for the complexities introduced by the autonomy of AI and the use of robotics in offshore environments. Accordingly, where, for example, an autonomous maritime robot traverses different regulatory environments without human intervention and there is an accident, the spectre of jurisdictional ambiguity arises, raising questions about which national laws would apply. As a result, there is an urgent need to develop supplementary regulatory frameworks that can effectively govern the use of AI-operated systems in offshore contexts.

The diversity of national regulatory approaches further complicates enforcement. This lack of clear liability frameworks necessitates regulatory intervention to define responsibility among vessel owners, software developers and AI manufacturers. The EU AI Act, while primarily focused on broader AI applications, includes provisions that could impact autonomous maritime operations by imposing risk-based regulatory requirements. We note, however, that not all nations may adopt risk-based regulatory frameworks [35], which would compound the challenges and complexities in enforcement. As a result of these considerations, some national agencies, such as the

UK Maritime and Coastguard Agency, have begun exploring AI-specific maritime regulations [12]. However, critical questions remain, particularly when considering the complexities of movement of autonomous maritime robots. For example, can a coastal State or port State under UNCLOS exercise enforcement rights over any autonomous maritime robots or MASS exercising rights of passage in waters under their national jurisdiction to prevent pollution, or safeguard the safety of life at sea?

Consequently, the harmonisation of these regulations at an international level remains a significant legal challenge. Without coordinated governance, AI-driven maritime operations risk facing conflicting regulatory requirements depending on the jurisdiction in which they operate. Ongoing initiatives, including the EU AI Act and IMO regulatory discussions around MASS and other autonomous systems, are critical in shaping the future legal landscape of AI-driven maritime operations.

5 Conclusion

The operational benefits of Urban AI and autonomous maritime robots in offshore environments are undeniable. Reducing human intervention in hazardous environments, improving efficiency and minimising costs are some of the key advantages offered. However, there are significant challenges that need to be addressed. In particular, the evolving legal and regulatory landscape requires the development of new frameworks that address liability, governance and compliance issues, ensuring the safe, ethical and trustworthy deployment of these technologies in offshore industries and environments.

As such, the integration of AI-driven solutions and autonomous systems necessitates a reassessment of existing legal and regulatory frameworks to ensure the safe and ethical deployment of AI and autonomous systems, particularly the need for better standards, governance arrangements and laws. Proponents of these AI systems should be aware of, and address, the complexity of managing and integrating these technologies across various offshore sites, which requires coordination between AI developers, offshore operators, ports, other stakeholders, agencies and regulatory bodies.

While existing maritime laws provide foundational governance, they require significant modifications to address AI-specific risks. As such, the governance of AI-operated offshore systems remains a critical legal issue. As AI technology continues to evolve, proactive legal frameworks must be established to ensure safe, efficient and legally compliant deployment in offshore environments. To address these challenges, floating cities must establish legal clarity through AI governance frameworks that balance regulatory flexibility with accountability. This could involve international AI accords that define liability rules for AI-operated urban environments; cross-jurisdictional agreements to regulate AI-driven data management; and hybrid regulatory models that integrate elements of maritime law, urban governance and AI ethics. Ultimately, the success of AI in floating cities and other floating assets will depend on the development of robust, adaptable legal structures that ensure safety, transparency and accountability while fostering innovation in buoyant urbanism.

References

1. Akpan F, Bendiab G, Shiaeles S, Karamperidis S, Michaloliakos M (2022) Cybersecurity challenges in the maritime sector. Network 2(1):123–138
2. Ambarita EE, Karlsen A, Osen O, Hasan A (2023) Towards fully autonomous floating offshore wind farm operation and maintenance. Energy Rep 9:103–108
3. Bibri SE (2021) Data-driven smart sustainable cities of the future: Urban computing and intelligence for strategic, short-term, and joined-up planning. Comput Urban Sci 1(1):8
4. Boghossian A, Linskey S, Brown A, Mutschler P, Ulicny B, Barrett L, Bethel G, Matsen M, Strang T, Wagner Ramsdell K, Koehler S (2018) Threats to precision agriculture, report to the office of the director of national intelligence and the department of homeland security. https://www.hsdl.org/c/view?docid=826417
5. Bolbot V, Kulkarni K, Brunou P, Banda OV, Musharraf M (2022) Developments and research directions in maritime cybersecurity: a systematic literature review and bibliometric analysis. Int J Crit Infrastruct Prot 39:100571
6. Caprotti F, Cugurullo F, Cook M, Karvonen A, Marvin S, McGuirk P (2024) Why does urban artificial intelligence (AI) matter for urban studies? Developing research directions in urban AI research. Urban Geogr 45(5):883
7. Cheong SM, Sankaran K, Bastani H (2022) Artificial intelligence for climate change adaptation. Wiley Interdisc Rev Data Mining Knowl Discovery 12(5):e1459
8. Cugurullo F (2020) Urban artificial intelligence: from automation to autonomy in the smart city. Front Sustain Cities 2:38
9. Department of Agriculture, Fisheries and Forestry 2016 'National Aquaculture Statement—DAFF'. https://www.agriculture.gov.au/agriculture-land/fisheries/aquaculture/national-aquaculture-strategy/national-aquaculture-statement
10. Eiler JH, Grothues TM, Dobarro JA, Masuda MM (2013) Comparing autonomous underwater vehicle (AUV) and vessel-based tracking performance for locating acoustically tagged fish. Mar Fish Rev 75:27–42
11. Engin Z, van Dijk J, Lan T, Longley PA, Treleaven P, Batty M, Penn A (2020) Data-driven urban management: mapping the landscape. J Urban Manage 9(2):140–150
12. Fenton AJ, Chapsos I (2023) Ships without crews: IMO and UK responses to cybersecurity, technology, law and regulation of maritime autonomous surface ships (MASS). Front Comput Sci 5:1151188
13. Friedman G (2020) The storm before the calm: America's discord, the coming crisis of the 2020s, and the triumph beyond. Random House
14. Fu J, Liu D, He Y, Cheng F (2024) Autonomous net inspection and cleaning in sea-based fish farms: a review. Comput Electron Agric 227:109609
15. Guihot M, Bennett Moses L (2020) Artificial intelligence, robots and the law LexisNexis
16. Guihot M, Huggins A (2025) Regulation: mapping evolving AI regulatory responses. In: Arvidsson M (ed) Routledge handbook on AI, law and society
17. Hajkowicz SA, Karimi S, Wark T, Chen C, Evans M, Rens N, Dawson D, Charlton A, Brennan T, Moffatt C, Srikumar S, Tong KJ (2019) Artificial intelligence: Solving problems, growing the economy and improving our quality of life. CSIRO Data61, Australia
18. Hyland-Wood B, Snoswell A, Reddy S, Ouyang, Perrin D, Fielt E, Price A, Mengersen K (2024) Response to proposals paper on introducing mandatory guardrails for AI in high-risk settings. Submission to the Australian Department of Industry, Science and Resources. Analysis and Policy Observatory
19. Kamrowska-Załuska D (2021) Impact of AI-based tools and urban big data analytics on the design and planning of cities. Land 10(11):1209
20. Khalid O, Hao G, Desmond C, Macdonald H, McAuliffe FD, Dooly G, Hu W (2022) Applications of robotics in floating offshore wind farm operations and maintenance: literature review and trends. Wind Energy 25(11):1880–1899

21. Karim MS, Haataja S, McKenzie S, Guihot M (2023) Maritime cybersecurity and the Australian cyber security strategy. Submission on the 2023–2030 Australian Cyber Security Strategy Discussion Paper
22. Livanos G, Zervakis M, Chalkiadakis V, Moirogiorgou K, Giakos G, Papandroulakis N (2018) Intelligent navigation and control of a prototype autonomous underwater vehicle for automated inspection of aquaculture net pen cages. In: 2018 IEEE international conference on imaging systems and techniques (IST). IEEE, pp 1–6
23. Luo D, Yu L (2021) From factory to site—designing for industrial robots used in on-site construction. In: Automating cities: design, construction, operation and future impact, pp 87–109
24. Luusua A, Ylipulli J, Foth M, Aurigi A (2023) Urban AI: Understanding the emerging role of artificial intelligence in smart cities. AI Soc 38(3):1039–1044
25. Major PY, Li G, Zhang H, Hildre HP (2021) Real-time digital twin of research vessel for remote monitoring. In: Proceedings of 35th European council for modelling and simulation
26. Mitchell D, Blanche J, Zaki O, Roe J, Kong L, Harper S, Robu V, Lim T, Flynn D (2021) Symbiotic system of systems design for safe and resilient autonomous robotics in offshore wind farms. IEEE Access 9:141421–141452
27. Nash JZ, Bond J, Case M, McCarthy I, Mowat R, Pierce I, Teahan W (2021) Tracking the fine scale movements of fish using autonomous maritime robotics: a systematic state of the art review. Ocean Eng 229:108650
28. Nezhad MM, Neshat M, Sylaios G, Garcia DA (2024) Marine energy digitalization digital twin's approaches. Renew Sustain Energy Rev 191:114065
29. Novelli C, Hacker P, Morley J, Trondal J, Floridi L (2024) A robust governance for the AI Act: AI office, AI board, scientific panel, and national authorities. AI Board, Scientific Panel, and National Authorities
30. Ohrem SJ, Kelasidi E, Bloecher N (2020) Analysis of a novel autonomous underwater robot for biofouling prevention and inspection in fish farms. In: 2020 28th Mediterranean conference on control and automation (MED). IEEE, pp 1002–1008
31. Pribyl ST, Weigel AM (2018) Autonomous vessels: how an emerging disruptive technology is poised to impact the maritime industry much sooner than anticipated. RAIL 1:17
32. Rødseth ØJ, Vagia M (2020) A taxonomy for autonomy in industrial autonomous mobile robots including autonomous merchant ships. In: IOP conference series: materials science and engineering, vol 929, no 1. IOP Publishing, p 012003
33. Sahil K, Mehta P, Bhardwaj SK, Dhaliwal LK (2023) Development of mitigation strategies for the climate change using artificial intelligence to attain sustainability. In: Visualization techniques for climate change with machine learning and artificial intelligence. Elsevier, pp 421–448
34. Shafiee M, Zhou Z, Mei L, Dinmohammadi F, Karama J, Flynn D (2021) Unmanned aerial drones for inspection of offshore wind turbines: a mission-critical failure analysis. Robotics 10(1):26
35. Snoswell A, Hyland-Wood B, Mengersen K, Orth T, Cook B, Wickramanayake B (2023) Draft G7 guiding principles for organizations developing advanced AI systems. Analysis and Policy Observatory
36. Solove D, Hartzog W (2024) Kafka in the age of AI and the futility of privacy as control. Boston Univ Law Rev 104:1021
37. Spirling A (2023) Why open-source generative AI models are an ethical way forward for science. Nature 616(7957):413
38. Tanaka Y (2023) The international law of the sea. Cambridge University Press
39. UN Habitat (2022) UN-Habitat and partners unveil OCEANIX Busan, the world's first prototype floating city
40. Wang B (2019) Floating cities: the future or a washed-up idea? The Conversation
41. Wang B (2021) The seductive smart city and the benevolent role of transparency. Interac Des Architect (s) 48:100–121

42. Wang BT (2021b) The machine metropolis: introduction to the automated city. In: Automating cities: design, construction, operation and future impact, pp 1–21
43. Wang BT (2023) An updated model of trust and trustworthiness for the use of digital technologies and artificial intelligence in city making. In: Proceedings of the 6th media architecture biennale conference, pp 69–80
44. Wang BT, Burdon M (2021) Automating trustworthiness in digital twins. In: Automating cities: design, construction, operation and future impact, pp 345–365
45. Wang C, Li Z, Wang T, Xu X, Zhang X, Li D (2021) Intelligent fish farm—the future of aquaculture. Aquac Int 1–31
46. Wang CM, Wang BT (2020) Floating solutions for challenges facing humanity. In: ICSCEA 2019: proceedings of the international conference on sustainable civil engineering and architecture. Springer Singapore, pp 3–29
47. Weatherall K et al (2023) Safe and responsible AI in Australia discussion paper: ADM+S submission. Analysis & Policy Observatory
48. Wu YK, Lee CY, Chen CR, Hsu KW, Tseng HT (2013) Optimization of the wind turbine layout and transmission system planning for a large-scale offshore windfarm by AI technology. IEEE Trans Ind Appl 50(3):2071–2080
49. Yigitcanlar T, Desouza KC, Butler L, Roozkhosh F (2020) Contributions and risks of artificial intelligence (AI) in building smarter cities: Insights from a systematic review of the literature. Energies 13(6):1473
50. Zhang J, Hua XS, Huang J, Shen X, Chen J, Zhou Q, Fu Z, Zhao Y (2019) City brain: practice of large-scale artificial intelligence in the real world. IET Smart Cities 1(1):28–37
51. Zhao L, Bai Y (2024) Unlocking the ocean 6G: a review of path-planning techniques for maritime data harvesting assisted by autonomous marine vehicles. J Marine Sci Eng 12(1):126

Dr. Brydon T Wang is an expert on the trustworthy development and regulation of AI and Construction technology. He researches on trustworthy decision-making in infrastructure delivery and data governance, with a focus on how technology can help increase climate-resilience in cities. He is dually qualified in law and architecture and advises on multi-billion dollar infrastructure projects with KPMG. Brydon practised as a construction and technology lawyer with global top-tier law firm Allens Linklaters and has a previous career in architecture and contract administration on award-winning construction projects. He has over 15 years in the construction industry in both client-side (government) and contractor-side roles.

Brydon was recently lead editor of the book, *Automating Cities: Design, Construction, Operation and Future Impact* (Springer, 2021) and co-edited *Large Floating Structures* (Springer, 2015), a book exploring how we can embrace floating structures in our cities to respond to the increased incidences of floods and sea-level rises. He was featured on ABC Radio National's *Future Tense* discussing offshore architecture and marine urban sprawl, and on Seeker's popular documentary, 'How close are we to Living in the Ocean?' Brydon's expertise in technology and cities also extends to data-focused technologies such as digital twins, BIMs, smart construction contracts and the blockchain. He developed and taught the contract and technology law subjects at the Queensland University of Technology (data privacy, regulating the internet, EU AI Act). His PhD thesis on the 'Role of Trustworthiness in Automated Decision-making Systems and the Law' received the *QUT Faculty of Business and Law Executive Dean's Commendation for Outstanding Doctoral Thesis Award.*

Dr. Bernadette Hyland-Wood is a Research Fellow at the Centre for Data Science, Queensland University of Technology, specializing in responsible data science and deploying generative AI technologies. Her research explores the adoption of rapidly emerging AI platforms by regulators and industry, focusing on data ethics and governance. She has led international initiatives on data exchange standards to support digital research and evidence-based policymaking. Dr. Hyland-Wood holds a Ph.D. from The University of Queensland, where her dissertation, *Bridging the*

Open Data and Public Policy Gap, examined barriers and facilitators to government data sharing. A seasoned technology entrepreneur, she has founded multiple startups, including two that were acquired—one by a U.S. Fortune 100 company. Her work bridges academia, industry, and policy, emphasizing applied research and research translation to ensure responsible AI and data-driven decision-making.

Dr. Michael Guihot is Associate Professor at the Queensland University of Technology. His work examines the nexus between emerging technologies and legal frameworks, focusing on the governance and regulation of artificial intelligence (AI), as well as examining the broader implications of technological advancements on power structures and legal institutions. His research explores power relations created by AI and the influence on both private and public governance, as well as the transformative impact of new technologies on the legal landscape. He is the co-author of the text "Artificial Intelligence Robots and the Law" a second edition of which will be published in 2025.

From Land to Water: Policy Considerations Supporting Why Floating Cities Can Help Combat Climate Change

Lucy M. Cradduck

Abstract The chapter presents matters that should be considered in government land planning processes and policies targeted to accommodating growing populations; and that are relevant to seeking to overcome issues arising from the adverse impacts of climate change. Climate change has had, and will continue to have, significant impacts upon coastal and riverine areas in particular. This is more obviously seen in adverse weather events, such as flooding and erosion, and the issues arising due to the continued development, and redevelopment, of land in these 'at risk' areas. In some regions, climate change impacts have seen land degradation in rural and regional areas; and yearly flooding of highly populated regions. In other areas climate change has resulted in the permanent loss of arable and development-appropriate land, which in turn adversely impacts communities located in those areas. Examples of where land planning has failed will be considered and, in doing so, the chapter will identify why migration to floating cities can be a core component of a broader policy and legal solution to address the impacts of climate change related risks.

Keywords Climate change · Land planning · Coastal · Government responsibility

1 Introduction

One of the most obvious adverse impacts of climate change is the impact to real property (land), which is being damaged, destroyed, or disappeared by a variety of events including sea-level rise [1, 2]. As discussed by other authors, any area close to the coast, or surrounded by the sea, is at a real and tangible risk from the adverse actions of the sea [3, 4]. This includes cities [5, 6] and islands [7, 8]. However, despite the real risks from the sea, coastal populations are increasing and are projected to continue to increase [4]. Some individuals choose to live in coastal areas by choice, others do so because they have no choice. Irrespective of the 'why', all those living

L. M. Cradduck (✉)
QUT Law School, Queensland University of Technology, Gardens Point Campus, Brisbane, QLD, Australia
e-mail: l.cradduck@qut.edu.au

B. T. Wang et al. (eds.), *Large Floating Solutions*, Lecture Notes in Civil Engineering 636, https://doi.org/10.1007/978-981-96-5435-2_19

and working in these at-risk areas are themselves at-risk. Climate change events also put the flora and fauna in those at-risk areas at-risk [9]. Land planning therefore needs to better consider and support individuals, businesses, and our environment.

While the politicians, activists, and scientists debate the issues, protest, and try to find a solution for climate change, real communities continue to feel its impacts, with many now finding that where they live leaves them at an even greater risk due to the high cost of insurance caused by these events [10]. There is therefore an ever-pressing need for issues of climate change to be appropriately addressed in land planning policies to protect the whole of a community [6, 11]. There is also a need to ensure that this planning occurs in a transparent and inclusive manner [9].

Past planning decisions have a significant impact on our everyday lives today. Many of the issues currently being faced have been caused, or are being exacerbated, by poor planning decisions made decades, or in some instances centuries, ago. This can be seen, for example, in the flooding issues now facing those living in areas adjacent to the Rhine River which, due to its being straightened in the mid-1800s to create more arable land and to facilitate the easier passage of steam vessels, now causes those areas to suffer the adverse effects of water that rushes too quickly downstream after heavy rains (see discussion in [12], pp. 104–110). The impacts of poor planning decisions are also reflected in the developments permitted on flood plains [13–15]. Future land planning must deal with overcoming the shortfalls of the past, the present concerns raised by climate change, and an uncertain future. Some risks may be able to be avoided or mitigated against [4], however other risks, particularly some climate change related risks, will not be able to be easily avoided [16].

This chapter commences the policy and socio-legal considerations in this section of the edited collection by exploring the rationale for why floating cities can aid in the fight against the adverse impacts of climate change. It does this by identifying but some of the land planning issues facing coastal communities, and why, as an alternative of relocation to higher ground or costly barricades, the use of in situ floating cities can assist with maintaining existing connections and communities. Significantly, the use of land planning policies to embrace floating cities can maximise existing infrastructure, maintain population bases, and give those populations, if needed, the time for a more considered and orderly relocation to higher ground.

The chapter begins by considering what communities are, and how these can be impacted by an adverse climate change event, before identifying related land planning considerations. It then considers various solutions available and, in doing so, explains the role that floating cities play in combatting climate change.

2 Communities

A city can be a community [17, 18], albeit a very large physical community made up of many smaller communities, for examples inner-city neigbourhoods, arrondissements, or zones [19]. Physical communities also are based in the countryside, in the hills, and near the coast. Such physical communities have grown up over time as people move, settle, and then move again. Any physical community is also a grouping of other communities where people live, work, and play and have contact and interactions with others within various interest, social, or cultural communities [18–22].

A 'community' therefore has various elements including psychological elements, which can include: a recognition of belonging; a recognition of connection; and a means by which many identify 'who' they are in the world [23]. These other forms of community may be centred in a physical or virtual place, or it may not have a set place. A common factor is connection with place and or others. In this regard, technologies have forever changed how our communities are created and how individuals engage, as well as enabling our engagements [24, 25].

Even in times of common risk, such as the recent pandemic, the desire of members to continue to be part of their community/ies remains, albeit that these communities were transferred to the online environment as erecting boundaries between ourselves and others became essential [22, 26]. However so caused, a loss of community can have significant negative social impacts for both the individual and their community [19]. Where the relocation is forced [1] or due to destruction of a community [27] the impacts can be even more so. Communities have a need for connection and a place in which to connect or be based. Planning for continuity of community will be essential for individual and group survival, and the solution is not (just) digital or to 'run for the hills'.

Therefore, while many communities are now focussed on-line [22, 28], everyday life is not purely digital. Even participants who seek to function primarily on-line continue to have a physical presence on land [29]. How physical communities operate is not flexible. In comparison, those who engage in on-line communities have a level of flexibility as to how they can and do operate, and which rules of operation they will engage with in what they do [30]. Noting however, that where the community is located in an online virtual world, the community member is at the mercy of the world's creator [31].[1] However, still, the same level of flexibility as applies to online communities does not apply to physical communities.

Land use, generally, is *"fixed because there is a finite amount of land available"* [32, p. 199];[2] and, in the context of necessary approvals, which are regulated by the governments of the country in which the land is located, is not flexible. The reality of land's physicalness therefore impacts what can be done, and where it can be done, in the physical world. Importantly, if you need to leave where you live and work, due to some external risk such as a climate change related event [5, 11], you may lose one or more of your communities.

---

[1] Noting of course that, as Ramshaw identifies, the player needs permission of the developer to be in the virtual world in the first place, and that a virtual world can be destroyed (closed) without any input from a player: *"Those in a virtual world enjoy the limited permission of the developer to enjoy the world per an End User Licence Agreement. The virtual world typically remains the intellectual property of the developer and so if a developer chooses to close the virtual world the user has little recourse to resist."* [31, p. 3].

[2] Noting that land can be reclaimed from the sea, this does not aid as a solution to climate change due to the fact of sea level rises. Such land is the first to be at risk from such events.

3 Land Planning as a Concept

Land planning, as an arm of governments, is one consequence of the industrial revolution. 'Land use planning', also referred to as 'town and country planning', developed as a response to the resulting massive urban development across Europe, and in the various colonies of the European nations and of England. As seen in 'younger' countries, land planning typically is devolved to state or local governments, sometimes with, sometimes without, an overarching federal policy/ies or law/s [9, 11]. Land planning regulates where we work, play, build our schools and communities, and live. It enables us to live in various places within the boundaries of the planning authorities' controls, and sometimes beyond it.

However, if you like your home but do not like where you live, unless your home is a caravan or other easily portable abode, you cannot just move your home to somewhere else. Equally, in this context, a community is place-based and, unless that place is truly virtual, relocating an entire community to another place is not an easy task. If there is a safety risk or significant adverse amenity issue with your home or your community, depending on what the risk or issue is, if it cannot be resolved then you will need to relocate, whether you want to or not [8]. The impact forced relocation can have cannot be underestimated: the more established the person or community, the less likely the person or community is to want to leave (See discussion of the failed relocation of Byron Bay in [8] at p. 43).

Issues of industrialisation and urbanisation have been considered in earlier land use planning literature by authors such as Howard [33], Sulman [34], and MacPherson [35]. It has also been considered more recently by others such as Turner and Neale [36], who reflected on the vexed relationship globally between post-colonial settler law and surviving Indigenous law.[3] Land planning is complicated where regulation and oversight of development involves different layers of government. In the context of the complexities of climate change management, this requires support of municipal and higher-level planners at all levels of government, and ensuring those persons are appropriately informed and skilled [37].

The last few decades have seen increasing populations, which has resulted in an increase in developments and a corresponding increase in development applications, and continuing development in at-risk areas. In Australian coastal areas, for example, permission has been and continues to be given for developments on low-lying land; and for the raising and under-building of old high-set homes. This, combined with the desire to live near services, and the finite nature of land, has resulted in planning approvals for developments in known flood-risk areas [11]. Unfortunately, Australia's position is consistent with that of many other countries. In the US, for example, while *"approximately 7% of all land … is designated as having a significant flood risk"* [38], referring to [39] development still continues in those areas. As Henninghausen and Suter [38] noted, this may *"explain some of the high costs associated with flooding"*.

---

[3] Acknowledged but beyond the scope of this chapter to consider.

Seeking to rely on local knowledge of issues is no longer sufficient. Ongoing variable weather patterns, as described in Chap. 1, have added to the problem. While in the last few years there have been significant rain events, before that there were several decades of drought, non-typical low rain summers, plus an influx of those seeking a sea- or tree-change. Added to this is the death, relocation, or retirement of those professionals and other individuals with true long-term local knowledge of an area. What this effectively means is that local knowledge of past bad weather conditions had dissipated with the sunshine. Ensuring appropriate planning for land use in the future therefore is imperative [40].

4 Land Planning in Practice

The impacts of the adverse impacts from climate change events for coastal areas are receiving specific attention, in particular by the United Nations [5]. Unfortunately, effective implementation of international policies is reliant upon domestic governments and are impacted by their policy development processes.

The process of any policy development, that is from the adoption of a government policy to the enactment of an enforceable law, is a cyclical process in which government policy influences and is influenced by the market and the law [41]. After a law is made, it is 'tested' by the market (as seen in court decisions). This 'testing' can result in changes being made to the law, which then can have a flow-on impact on government policies. If the (current) government then does not like the law's impact, it will change the law, and the cycle continues. The process is also adversely affected by disruptions. As depicted in Fig. 1, in the context of land use policy development, as was posited by Cradduck et al. [42], these disruptions are: **political control**, perhaps most evidenced at a time of a change of government, **developer imperatives**, primarily to make a profit from the land, and **climate change**, the impacts of which are yet to be fully realised.

Arguably, the current methods of land planning, and the granting of development approvals, leads to issues. One issue is that of consistency. As schemes are subject to change, at any one time, even in the same street, there may be developments occurring to which different planning regimes apply. Another issue is that government policies and law are most usually reactive. For example, current planning policies, which can take years to develop or amend, have been developed in isolation from current understandings of the future impacts of external issues such as climate change. This is, as we are seeing on the news daily, very problematic [9, 43].

The risks arising from living in any at-risk area, for example at the coast, often can be outweighed by the amenity. However, while the risk from sea level arise can reduce the amenity value of coastal land [4], this is not reflected in planning policies as the development of land planning policies appears disconnected from their impact on land value. While valuers are required to give due consideration to planning schemes and conditions imposed on development applications, planning schemes do not have a similar requirement imposed upon their operations or considerations. This is seen

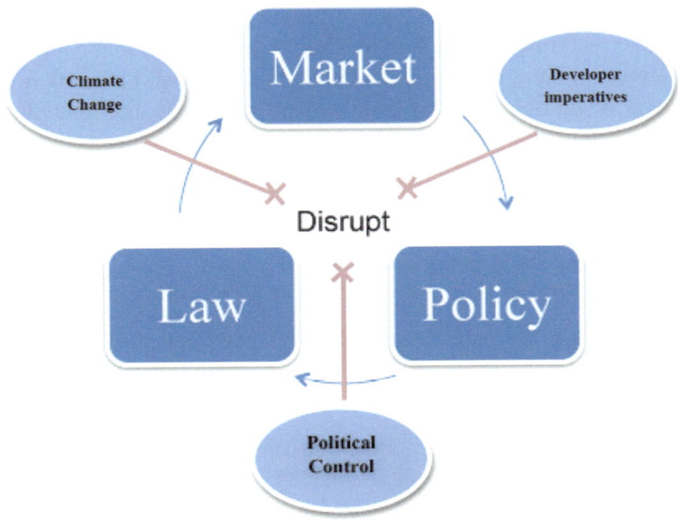

Fig. 1 Cradduck et al. [42], as adapted from Kee et al. [41] p. 175

internationally, for example, in the United States. In the US, programs such as that of the National Flood Insurance Program, arguably, have had negative impacts on land values. This scheme enables insurance to be provided in at-risk areas (usually by a government agency, i.e. the Federal Emergency Management Agency, or by other government subsidy), when the market would not otherwise provide such cover [38]. The ability to obtain insurance in these areas, supports planning policies that permit ongoing developments, and use of existing facilities, in an otherwise at-risk area. It also enables these properties to be valued such that financiers are willing to offer finance to prospective buyers to buy land or businesses in these areas despite the clear risks from climate change events. In the face of land policies and financiers that enable buyers to buy in such at-risk areas, the uninformed, inexperienced buyer may misunderstand the true nature of the underlying risk.

The consideration of such conduct that appears to be missing from the current discourse, is whether (why?) planning policies should continue to permit such behaviour.

5 Climate Change Impacts

In late August 2005 Hurricane Katrina caused extreme devastation across the United States [13]. Areas impacted included its cities due both to winds and storm surges, with New Orleans being the notable example of the devastation caused. The lack of preparedness of the city [13], as well as the government response subsequently, as well as the failures of citizens to leave when directed, have all been subject to criticism [14]. In part the actions of citizens may have been a misguided reliance on the levees that ultimately gave way. In any event, the resulting inundation of water caused immense loss: of property, of life, and to community [44].

The adverse impacts of climate change are seen all too readily on the current evening news. In inland areas across the world this is seen on nightly news stories showing homes disappearing due to landslips; and towns flooding as rivers break their banks following unseasonal deluges. Worse impacts still are being felt in areas close to the sea. As Goodall [45] observed *"sea-level rise is one of the central facts of our time, as real as gravity. It will reshape our world in ways that most of us can only dimly image"* [45, p. 8]. As Dalkiran et al. [5] more recently observed "[e]*specially in coastal cities with high population density, the rise in sea level is a cause for concern"* [5, p. 2612]. Beaches are disappearing [27], homes are falling into the sea as cliff faces are being eroded, and island nations such as Tuvalu are disappearing literally before our eyes as the seas rise [7].

Regrettably, the adoption (or not) of effective policies is impeded by the relevant government's *ability* (which includes both capacity and specific legislated planning processes) to implement policies to engage with the issues arising [46, 47]. However, even where that ability exists, as Cradduck and Warren-Myers [11] observed in the context of the New South Wales Central Coast Council, the ability to create an effective policy to regulate development in at-risk areas does not necessarily lead to effective implementation.

Development in some at-risk communities continues, with businesses and people being put at risk despite those risks being clearly known to government [11]. In other communities, managing for risk has led to increasing development in adjoining 'safe' areas, leading to overdevelopment in those areas. What is needed is a solution that supports communities, while protecting citizens, and which understands vulnerabilities to climate change [16]. However, as Goodall [45] observed *"nobody wants to spend money to build a more resilient city because nobody owns the risk"* (p. 103). What then is the solution for coastal communities?

6 Current Solutions

"The aim of global climate change adaptation policies is to make cities resilient against the impacts of climate change" [5, p. 2612]. To this end, many countries and communities have investigated and or implemented a variety of solutions that are more commonly focussed on *"reducing risks through hard engineering of coasts, designing strategies to manage and live with risk, and relocating people and infrastructure away from hazardous places"* [1]. In other areas the more natural planting of vegetation on or near dunes, to strengthen those dunes as a natural barrier against the impacts of the sea, have been implemented [48].

Although *"the specific direct and indirect impacts at the local scale are less well understood*" [49, p. 2] the 'hard engineering' solutions include: building barricades between the sea and buildings, which include 'dikes and levees' [4, p. 2], 'giant sea walls' [16, p. 2], and retreat/relocations. Unfortunately, some solutions create they own problems. As Lopez [9] considers, sea walls, can adversely impacts turtles (see discussion at p. 68).

Even more unfortunately for those relying on those solutions, they can fail. As was seen in New Orleans [13, 14], and in Lismore, Australia [15], barricades and levees work well, until they do not. Despite the protections in place, New Orleans suffered both physical damage and significant loss of life when those barriers failed and the city flooded [13]. As Landy [14] observed New Orleans' *"misery resulted from the floods that took place when several of the levees protecting the city broke."* [14, p. 156]. A similar impact was seen in Lismore when the levees were too short to hold back the rising water levels, and water rushed into the centre of the CBD causing havoc and damage.

Banning future development in at-risk areas, as Cradduck and Warren-Myers [11] previously identified, and as the Australian Standing Committee recommended after the Inquiry into insurers' responses to 2022 major floods claims ([50], See: Recommendations 71, 72, and 83) will prevent others being put at risk. In some countries, for example France, current government focus has in fact seen the implementation of prohibitions on future development in at-risk areas [4]. In other countries, the focus has been on relocation, which means that those at-risk populations must be located elsewhere. However, the wholesale movement of populations to other areas causes other problems, both for those who are forced to move, and for those already in the areas those being relocated move into as services are put under strain [5]. Wholesale movement also fails to fully appreciate the communities that were built in the at-risk areas are not purely physical. Consequently, as Chap. 22 considers, in some at-risk areas those communities have rejected relocation as a solution.

Current solutions can come at significant financial [7], emotional, and societal costs. Equally, the costs of relocation can be high, both financially and in the destruction of community/ies [1], and relocation places those relocated to areas under strains that the area may not be able to maintain. Further, a prohibition on development does not address the needs of growing populations. Communities need alternative solutions to mitigate or migrate. It is this gap that floating cities can fill [51].

7 Floating Cities as a Solution

The need to think more broadly for solutions to climate change is reinforced by various international documents. This includes the United Nations *Agenda 21,* which notes *inter alia* the importance of the appropriate *"selection of environment and development policies in the decision-making process"* [52]. This also is reinforced by the call to action of the Sustainable Development Goals, particularly that of Goal 11 with its focus on "sustainable cities and communities" [53]. The adverse risks from climate change events must be considered in land planning [10, p. 27]. However, there needs to be a greater social focus in land planning policies: one that is designed to keep communities together; and keep communities connected to their history. Therefore, future planning must also include consideration of social impacts of proposed solutions [1].

Dal Cin et al. [3] suggested, albeit in the context of the future design of public spaces on the waterfront, that *"the water body must be integrated into the design"* (p. 21 of 25). In the context of this chapter's focus, and in a world now living with the reality of climate change, it is submitted that it is now essential that *any* future planning in *any* coastal or riverine area must ensure that all water bodies in that area are integrated into the design of its future planning policies, and that all proposed developments are resilient to the incursion of water. As such, as part of the related planning policies and laws, the placement of floating cities in an at-risk area's water bodies for use for homes and businesses should be enabled.

Floating cities are a clear solution to the adverse impacts of climate change in coastal regions [5]. As Cradduck et al. [54] identifies, floating cities can *"provide habitable space in areas where land is scarce or at risk of disappearing"* and have the *"promise of creating dynamic, adaptive communities that can respond to environmental changes"* (p. 1). Floating structures also can be adapted and used to provide their own energy [55–57], and can be constructed to enable farming and other food production activities [56, 58]. The mooring of such structures may be achieved by a variety of means [58] depending upon where, and how close to any coastal shelf, it is proposed to be located.

Examples of *proposed* floating cities are Oceanix City [59], Maldives Floating City [5] and Buckminster Fuller's Triton City (see discussion in [54]). However, while there are examples throughout the world of actual floating houses (i.e., in Sweden and Amsterdam, see discussion in [37]), floating buildings (i.e., as used for conferences, see discussion of "Floating Pavilion Rotterdam", see discussion in [59], and smaller floating communities) (i.e., in Cambodia's Tonle Sap Lake, see discussion in Keskinen [60]), at the date of writing this chapter, no floating cities have been built to the author's knowledge.

Imagine if cities such as Lismore or New Orleans had been relocated to floating platforms rather than hiding behind levees. Instead of water building up and then rushing through the storm damaged levees causing devastation to property and life, water levels would have risen more slowly and with those rising water levels, and with minimal impacts to either people or properties, the floating platforms on which the cities were based also would rise.

Floating cities can offer a variety of flexible solutions facilitated by their flexible construction [51]. They can have mid-term use, as a place to relocate communities while individuals determine their future; a longer-term use, as a refuge while communities establish or reinforce their physical mitigation measures; and a perpetual one for a relocated, or 'new', community. While there are legal obstacles to overcome [51, 54], the planning path seems clear—floating cities are now options that we must consider in our land use planning, not mere ideas for discussion or for works of fiction.

8 Concluding Remarks

Irrespective of where you live, proper planning, that is planning that now manages for climate change impacts will be essential to future land use [4]. Absent proper land planning and support many residents will continue to be adversely impacted by such events. Poor planning decisions will affect those living in the areas directly impacted by the sea, who will be forced to move, and the broader community. Poor planning decisions also will impact current residents living in the areas that the former coastal residents move to, as those population bases grow and the services provided become stretched [5].

Floating cities are not without their critics and criticism. One notable criticism is that, similarly to the artificial floating breakwaters deployed to assist in reducing coastal erosion by lowering wave activity (see discussion in [61]), floating cities do not in fact protect the land-based structures from the effects of sea level rise [37, p. 2282]. However, this author suggests that criticism merely serves to reinforce an uncomfortable truth: the sea is rising, and unless it is possible to relocate what is based on the land then, yes, those structures and places will be lost [9]. If that land is an island, then all could be lost. As Lincke and Hinkel [7] noted "*where* [to locate] *the population from islands that cannot be protected and are eventually abandoned is a key issue in the climate change debate*" (p. 10). This is an uncomfortable truth that the Pacific and other islands are already facing, and planning for, and which, yes, means relocation. A floating city on the other hand can be placed above the sinking city [59] and, as Chap. 22 considers, can maintain a connection to the place beneath rather than losing that connection altogether as happens in any relocation.

Floating cities therefore can do more than provide spaces to work and live. By enabling a continuing connection to a place, albeit one that may now be under water, the floating city can enable a continuing sense of community. Floating structures also can enable the creation of completely new communities, including those designed to move around the world. Whatever the impetus for a floating city—to maintain connection to what was, or to build something new—the very nature of these structures means that we need a new way of thinking about 'land' planning and use. There is no easy solution for many coastal communities to the problems caused by climate change. In the next decade, and beyond, it is only logical that governments work to facilitate the construction of floating cities to solve problems of flooding and rising sea levels, and to create space on the water bodies adjacent to climate risk safer cities for urban expansion.

Moving forward therefore, not only will we need to move to floating cities, but our laws will need to change to facilitate our uses. Matters requiring international consideration and protection will include treaties and laws to address the mooring rights and obligations of such structures [58], to ensure the proper protection of the rights and interests created in those floating cities [54], and, most importantly, to protect the rights of the citizens that inhabit them.

References

1. Abu M, Heath S, Adger W, Codjoe S, Butler C, Quinn T (2024) Social consequences of planned relocation in response to sea level rise: impacts on anxiety, well-being, and perceived safety. Sci Rep 14(3461):1–9. https://doi.org/10.1038/s41598-024-53277-9
2. Warren-Myers G, Cradduck L (2023) Tackling the wicked challenge of climate change risks to property: are Australian valuers prepared? J Property Investment Finan 41(4):429–452
3. Dal Cin F, Hooimeijer F, Matos Silva M (2021) Planning the urban waterfront transformation, from infrastructures to public space design in a sea-level rise scenario: the European Union prize for contemporary architecture case. Water 13(2):218. https://doi.org/10.3390/w13020218
4. Tierolf L, Haer T, Athanasiou P, Luijendijk A, Botzen W, Aerts J (2024) Costal adaptation and migration dynamics under future shoreline changes. Sci Total Environ 917:170239. https://doi.org/10.1016/j.scitotenv.2024.170239
5. Dalkiran C, Yörük E, Ülger M (2023) The effects of climate change on coastal areas and coastal property and floating cities as a solution proposal. Kent Akademisi Dergisi 16(4):2611–2632
6. Fiorentino S, Sielker F, Tomaney J (2024) Coastal towns as left-behind places: economy, enviroment and planning. Camb J Reg Econ Soc 17:103–116
7. Lincke D, Hinkel J (2021) Coastal migration due to 21st century sea-level rise. Earth's Future 9:e2020EF001965. https://doi.org/10.1029/2020EF001965
8. Torabi E, Dedelorkut-Howes A (2021) When it's time to let go: re-imagining coastal urban living in the face of rising seas. In: Baumeister J, Bertone E, Burton P (eds) SeaCities. Cities research series. Springer, Singapore, pp 1–58. https://doi.org/10.1007/978-981-15-8748-1_3
9. Lopez J (2024) The climate is changing and so must we: the need to prioritize: at-risk communities and ecosystems. Wash Univ J Law Policy 74:44–87
10. Climate Council of Australia Limited (Climate Council) (2022)
11. Cradduck L, Warren-Myers G (2022) Development in a state of climate change: an Australian case study of government response. J Property Investment Finan 40(4):362–380
12. Crump T (2007) A brief history of the age of steam: the power that drove the industrial revolution. Robinson, London
13. Davis T (2006) A failure of initiative final report of the select bipartisan committee to investigate the preparation for and response to hurricane katrina (U.S. G.P.O.) 109th Congress, Report, pp 109–377. https://books.google.com.au/books?id=8vebAAAAMAAJ&ots=vwiMf41EPo&dq=hurricane%20katrina&lr&pg=PR2#v=onepage&q=hurricane%20katrina&f=false
14. Landy M (2008) A failure of initiative: final report of the select bipartisan committee to investigate the preparation for and response to Hurricane Katrina (House of representatives, February 15, 2006 www.gpoaccess.gov/katrinareport/mainreport.pdf) The federal response to Hurricane Katrina lessons learned (White house staff, February 2006 www.whitehouse.gov/reports/katrina-lessons-learned/). Publius 38(1):152–165. https://doi.org/10.1093/publius/pjm024
15. Mortimer A, Egbelakin T, Sher W (2024) Drivers, services gaps and improving disaster management for displaced people: a case study of prolonged displacement following the 2022 floods in Lismore, Australia. In: Handbook of flood risk management and community action. Routledge, pp 66–84
16. Purnomo A, Kurniawan T, Farandy A, Apriliani T, Imron N, Sajise A (2024) Revisiting the climate change adaptation strategy of Jakarta's coastal communities. Ocean Coast Manag 253
17. Prezza M, Schruijer S (2001) The modern city as a community. J Commun Appl Soc Psychol 11:401–406
18. Scheffer J (2001) Keeping the city alive. J Commun Appl Soc Psychol 11:491–494
19. Abramovitz M, Albrecht J (2013) The community loss index: a new social indicator. Soc Serv Rev 87(4):677–724
20. Bailey T (2007) Continuity of community. Can Fam Physician 53(1):183
21. Brennan M, Phillips R (2018) Culture, community, and development: a critical interrelationship. In: Brennan M, Phillips R (eds) Culture, community, and development. Taylor & Francis Group, pp 3–14

22. Lynn T, Rosati P, Conway E, Curran D, Fox G, O'Gorman C (2022a) Digital technologies and civil society. In: Digital towns. Palgrave Macmillan, Cham, pp 91–108. https://doi.org/10.1007/978-3-030-91247-5_5
23. Colombo M, Mosso C, De Piccoli N (2001) Sense of community and participation in urban contexts. J Commun Appl Soc Psychol 11:457–464
24. Cradduck L (2011) The future of the internet economy: addressing challenges facing the implementation of the Australian national broadband network. Professional Doctorate thesis, Queensland University of Technology
25. Lynn T, Rosati P, Conway E, Curran D, Fox G, O'Gorman C (2022b) Defining, rationalising and measuring digital towns. In: Digital towns. Palgrave Macmillan, Cham, pp 1–21. https://doi.org/10.1007/978-3-030-91247-5_1
26. Iossifova D (2020) Reading borders in the everyday: bordering-as-practice. In: Scott JW (ed) A research agenda for border studies. Edward Elgar, pp 89–105
27. Phillips C, Murphy C, Bresnihan P (2022) The impacts of and responses to place loss in a coastal community in Ireland. Local Environ 27(7):879–896. https://doi.org/10.1080/13549839.2022.2078294
28. Baym N (1998) The emergence of on-line community. In: Jones S (ed) Cybersociety 2.0: revisiting computer-mediated community and technology. Sage Publications, pp 35–68
29. Marcoccia M (2023) The physical-digital interface: what does ICI ("here") mean in a written online discussion? In: Parini A, Yus F (eds) The discursive construction of place in the digital age. Taylor & Francis Group. ProQuest Ebook Central. https://ebookcentral.proquest.com/lib/qut/detail.action?docID=7248452. Created from QUT on 2024–09–09 04:33:42
30. Priharsari D, Abedin B (2021) What facilitates and constrains value co-creation in online communities: a sociomateriality perspective. Inf Manage 58(6):103433
31. Ramshaw A (2020) 'World of warcraft is my home from home': an argument for the protection of virtual worlds. J Law, Technol Trust 1(1):1–23. ISSN 2634-4343
32. Wan ZJ (2023) Evaluating the theory and implementation of land value taxation. Int J Soc Sci Humanity 13(3):198–202
33. Howard E (1902) Garden cities of to-morrow. Faber and Faber, London, 1965
34. Sulman J (1919) Town planning: a sketch in outline. Government Printer, Sydney
35. MacPherson CB (1962) The political theory of possessive individualism: Hobbes to Locke. Oxford University Press, Oxford
36. Turner S, Neale T (2016) First law and the force of water: law, water, entitlement. In: Neale T, Turner S (eds) Other people's country: law, water and entitlement in settler colonial sites. Routledge, Abingdon, pp 111–121
37. Storbjörk S, Hjerpe M (2022) Stuck in experimentation: exploring practical experiences and challenges of using floating housing to climate-proof waterfront urban development in Sweden. J Hous Build Environ 37:2263–2284. https://doi.org/10.1007/s10901-022-09942-4
38. Henninghausen H, Suter JF (2020) Flood risk perception in the housing market and the impact of a major flood event. Land Econ 96(3):366–383
39. Holway JM, Burby RJ (1990) The effects of floodplain development controls on residential land values. Land Econ 66(3):59–271
40. Mitsova D, Esnard A (2012) Holding back the sea: an overview of shore zone planning and management. J Plan Lit 27(4):446–459
41. Kee K, Cradduck L, Blodgett B, Olwan R (2011) Cyberinfrastructure inside out: definition and influences shaping its emergence, development, and implementation in the early 21st century. In: Araya D, Breindl Y, Houghton T (eds) Nexus: new intersections in internet research. Peter Lang, New York, pp 157–189
42. Cradduck L, Rayner K, Sheehan J, Armitage L (2023) An Australian relationship between land values and land policies. Presentation at Pacific rim real estate society conference 2023, 15–17 January 2023. Sydney, Australia
43. Norman B (2023) Urban planning for climate change. Routledge, Abingdon.
44. Powers M (2024) The aftermath of Hurrican Katrian: stories of loss, resilience, and returning home. Am J Qual Res 8(1):89–106

45. Goodall J (2017) The water will come: rising seas, sinking cities, and the remaking of the civilized world. Littlebrown and Company. NY, NY
46. Harvey N, Clarke B (2019) 21st century reform in Ausralian coastal policy and legislation. Mar Policy 103:27–32. https://doi.org/10.1016/j.marpol.2019.02.016
47. Rogers N, Adams V, Byrne J (2024) Agenda-setting and policy leadership for municipal climate change adaption. Environ Sci Policy 161(103869):1–11. https://doi.org/10.1016/j.envsci.2024.103869
48. Feagin R, Furman M, Salgado K, Martinez M, Innocenti R, Eubanks K, Figlus J, Huff T, Sigren J, Silva R (2019) The role of beach and sand dune vegetation in mediating wave run up erosion. Estuar Coast Shelf Sci 219:97–106
49. Elrick-Barr C, Smith T, Thomsen D (2024) Is 'hope' helpful or a hindrance? Impolications for coastal governnance. Ocean Coast Manag 248(204):106953. https://doi.org/10.1016/j.ocecoaman.2023.106953
50. House of Representatives (House of Reps) (2024) Flood failure to future fairness: report on the inquiry into insurers responses to 2022 major flood claims, Standing Committee on Economics, October 2024
51. Wang B (2024) Development of the legal definition of the floating city: judicial interpretation of structural characteristics of floating homes and developments. In: Ikoma T, Tabeta S, Lim SH, Wang CM (eds) Proceedings of the third world conference on floating solutions: WCFS2023; 28–29 August, Tokyo, Japan. Springer, Singapore, pp 139–157
52. United Nations (1992) Agenda 21, conference on environment and development. https://sustainabledevelopment.un.org/content/documents/Agenda21.pdf. Accessed 12 Nov 2024
53. Sustainable Development Goals (2015)
54. Cradduck L, Wang B, Weinert K (2024) Underpinning the value of buoyant Urbanism: returning Torrens to the seas. In: Proceedings of the ASME 2024 43rd international conference on ocean, offshore and arctic engineering (OMAE): volume 9: Philip Liu honoring symposium on water wave mechanics and hydrodynamics; blue economy symposium. American Society of Mechanical Engineers (ASME), United States of America. OMAE 2024–123759, pp 1–9
55. Nguyen D, Sølvberg S, Aas-Hansen T (2024) Digital twin-driven optimal design of connector stiffness for floating multi-modular structures—an application for floating solar energy harvest. In: Proceedings of the ASME 2024 43rd international conference on ocean, offshore and arctic engineering (OMAE): volume 9: Philip Liu honoring symposium on water wave mechanics and hydrodynamics; blue economy symposium. American Society of Mechanical Engineers (ASME), United States of America. OMAE 2024–125883, pp 1–10
56. Wan l, Tay Z, Shi W (2024) Numerical modelling and analysis of a new combined concept of floating wind turbine and fish farm structure. In: Proceedings of the ASME 2024 43rd international conference on ocean, offshore and arctic engineering (OMAE): volume 9: Philip Liu honoring symposium on water wave mechanics and hydrodynamics; blue economy symposium. American society of mechanical engineers (ASME), United States of America. OMAE 2024–126926, pp 1–8
57. Zhang S, Zheng X, Zhang X, Lei Y, Liu Y, Jiang Y, Qian K, Li X (2024) Experimental study on the dynamic responses of a novel 10 MW floating offshore wind turbine integrated with a fish cage. In: Proceedings of the ASME 2024 43rd international conference on ocean, offshore and arctic engineering (OMAE): volume 9: Philip Liu honoring symposium on water wave mechanics and hydrodynamics; blue economy symposium. American Society of Mechanical Engineers (ASME), United States of America. OMAE 2024–126028, pp 1–10.
58. Wang C, Tay Z (2011) Very large floating structures: applications, research and development. Procedia Eng 14(2011):62–72
59. Yogesh S (2022) Are floating cities the urban future? In: Chias P, Hernandez S (eds) Eco-arcitecture IX. WIT Press, Southampton, UK, pp 215–224
60. Keskinen M (2006) The lake with floating villages: socio-economic analysis of the Tonle Sap Lake. Water Resour Dev 22(3):463–480
61. Risheharan K, Baumeister J, Adame M (2025) Are floating mangrove breakwaters effective for wave attenuation? A literature review. Ecol Eng 211(107436):1–13. https://doi.org/10.1016/j.ecoleng.2024.107436

Lucy M. Cradduck Associate Professor Lucy Cradduck is an academic in the QUT Law School, a former Assistant Member to the Valuations Board of Queensland, and an Independent Chair for Queensland statutory site valuation objections conferences > $5 million. Lucy's interest in property law, practice, and policy began during her time in private practice. She worked for various firms that acted for private clients in a range of property and commercial matters. As well, her interest and experience was enhanced through the different perspective she gained in various in-government legal roles. Her current research interests include real property law, tenure issues, and property and technology related issues. This culminates in her focus on the impacts of extreme events for land, as well as the impact of those events for landowners, occupiers, businesses, and property professionals. In addition to her academic writings, she regularly contributes submissions to government reviews and, in her spare time, she coaches students in mooting (mock court) competitions.

Home is Where the Heart Floats: Anchoring Legal Property Concepts for Environmental Migrants

Kim D. Weinert, Apoorva Abeysundera, Gauravi Baral, Liluz Maria Biscocho, Rose Foster, Zoe Hennessy, Kit Leach, Shiina Oiwa, Sophie Oudyn, Amber Sheridan, and Leanna Zhang

Abstract Climate change-induced displacement is a growing global challenge. As sea levels rise and coastal areas become increasingly vulnerable, innovative solutions like floating structures have emerged as potential responses. This chapter explores the socio-legal implications of floating structures, focusing on their role in preserving and redefining the concept of 'home' for displaced communities. While traditional legal perspectives often prioritise property rights and regulatory frameworks, this chapter argues for a more holistic approach that considers the cultural, social, and psychological perceptions of home. By examining the diverse experiences of communities in the Solomon Islands and the American state of Florida, this chapter demonstrates how floating structures can provide not only physical shelter but also a sense of belonging and continuity. To ensure the successful implementation of floating structures, policymakers, legal professionals, and engineers must collaborate to develop comprehensive frameworks that balance individual property rights with communal needs. By prioritising the well-being of displaced populations and fostering sustainable development, we can realise the potential of floating structures to create resilient and equitable communities in the face of climate change.

Keywords Floating structures · Home · Environmental migrants · Property rights · Solomon Islands · Florida

1 Introduction

Millions of people globally are either at risk of displacement or have been displaced owing to the rapidly changing climate. One critical cause of displacement under climate change is rising sea levels. This phenomenon risks partial or total damage to

K. D. Weinert (✉) · A. Abeysundera · G. Baral · L. M. Biscocho · R. Foster · Z. Hennessy · K. Leach · S. Oiwa · S. Oudyn · A. Sheridan · L. Zhang
T.C Beirne School of Law, University of Queensland, Brisbane, Australia
e-mail: k.weinert@uq.edu.au

B. T. Wang et al. (eds.), *Large Floating Solutions*, Lecture Notes in Civil Engineering 636, https://doi.org/10.1007/978-981-96-5435-2_20

coastal areas, with potentially irreversible environmental, economic, and humanitarian damages [53]. This presents a complex challenge for both the international community and individual governments. As such, decisions on public policy, domestic laws, and international conventions and decrees in this area must navigate the complicated relationship between human activities and the natural world. Encompassed within this challenge is a complex and dynamic relationship of loss. This is because the loss of non-living things (for example, physical structures) is deeply connected to the loss of living things (for example, humans).

Generally, loss can manifest on three levels: individual, relational, and collective [47]. In this chapter, we consider that there is scope to recognise a proprietary claim between living and non-living things in a way that connects the individual and the community. While people continue to negotiate the rights and content of the dominion of non-living things, it is asserted in this chapter that the concept of 'home' actually encompasses an enduring connection between living and non-living things. But when the physical non-living aspect of a home is destroyed and lost forever by a changed natural environment, where is and what now becomes of home?

People who live in coastal areas who are then forced to move as land becomes uninhabitable, their sense and relationship of home becomes fragile. Article 11(1) of the *International Covenant on Economic, Social and Cultural Rights* [26] recognises individuals having the right to an adequate standard of living for themselves, which includes adequate food, clothing and housing, and to the continuous improvement of living conditions. The breadth of Article 11(1), by including housing as a right to an adequate standard of living, captures the proprietary relationship between non-living physical infrastructure and its tenure with the living individual and community. As such, it can be asserted that living and non-living things blend the individual and the community through the concept of home.

Climate change and rising sea levels threaten the positive and negative rights under Article 11(1), particularly in the face of potential forcible displacement of individuals, communities and climate-vulnerable nations. Prima facie, the forceable moving of people from their homes raises the question of how State parties to the *Convention Relating to the Status of Refugees* [9] meet their principal obligation under international law to protect vulnerable individuals. Currently, the concept of 'environmental refugees' is not formally recognised under international law. This chapter adopts the International Organisation of Migration (IOM)'s 2007 working definition of 'environmental migrants' that encompasses:

> persons or groups of persons who, predominantly for reasons of sudden or progressive change in the environment that adversely affects their lives or living conditions, are obliged to leave their habitual homes, or choose to do so, either temporarily or permanently, and who move either within their country or abroad" [27].

Environmental migration draws attention to loss driven by climate change. This includes loss of physical structures, space, land and home. One emerging area where this environmentally-linked physical loss has been addressed is through advancements in engineering techniques and materials to develop floating structures. These structures provide for the spatial distribution of housing stock that touches on tangible and intangible notions of property for individuals and the community. Our chapter examines how the concept of property could endure between individuals and their community through floating structures that serve as homes to those facing a climate diaspora. Across this chapter's three parts it brings into focus the concept of home across specific contexts, such as the distinct needs of each country and its individuals, family and community when considering the legality of floating structures.

The first part of this chapter will explore the loss of land and non-living things, and the potential role of floating structures to address this loss. The second part will discuss the concept of home concerning floating structures, as something more than mere physical property. Specifically, this discussion will focus on the socio-legal concepts of home and floating structures. The third part of the chapter will present contrasting case studies in the Solomon Islands and Florida, where established floating structures are home to individuals and communities. These case studies provide clear examples of the relationship between the non-living floating structure and the home as a living resource for human sustainability for environmental migrants. The key distinction between the presented case studies of the Solomon Islands and Florida is how each different societal group views property through a collective communal lens or as an individual right. This should be a consideration when developing both the law and engineering behind floating structures, as it is fundamental that both reflect a community's critical needs.

2 Replacing Loss with Floating Structures

The global sea level average has risen 21–24 cm since 1880 [32]. In 2023, the annual sea level average set a new record high of 101.4 mm above the levels in 1993 [32]. On one hand, terrestrial environments provide the foundation for human well-being and livelihood through primary productivity, food supply and other ecosystem services. However, these terrestrial environments are also integral in establishing identities of individuals and societies. Land provides numerous intangible benefits to humans, including a sense of belonging and cognitive and cultural enrichment. Thus, where land is lost due to climate change impacts, individuals and societies likewise suffer significant loss.

The United Nations [52] has identified sustainable floating homes as a potential solution to the millions of people who have lost their homes due to climate change. Engineers, architects and urban planners have been exploring and developing the viability of floating settlements as a potential solution to rising sea levels and housing shortages to replace the loss and sustain life. Notable engineering examples where the designs of floating structures prioritise and provide long-term sustainability include proposals that feature recycling rainwater and the construction of climate resilient infrastructure [37]. Yet these climate resilient solutions are visualised reminders of loss–the loss of land and homes that are replaced by substitute structures positioned on bodies of water.

The central division within the law of property is between real property and personal property (tangible), and moral rights (intangible). Beyond this simple dichotomy, the meaning of property remains elusive. Indeed, Gummow J of the Australian High Court case of *Yanner v Eaton* [64] 166 ALR 258 emphasised this:

> as a striking example of the inherent ambiguity and looseness in legal terminology. The risk of confusion is increased when, without further definition, statutory or constitutional rights and liabilities are so expressed as to turn upon the existence of 'property'. The content of the term then becomes a question of statutory or constitutional interpretation. [85]

Justice Gummow offers a methodological solution that simplifies the problems of understanding property to a constitutional and statutory interpretation to overcome the inherent definitional legal challenges of understanding property. This methodology assists in understanding a floating structure's legal status. However, Gummow J's suggested methodology has limitations with novel categories of non-living things, such as floating structures and cities, as there is an absence of a statute in common law jurisdictions that defines and allocates property rights and interests in these structures.

Whether floating cities can be considered real property remains ambiguous, as explored in Weinert's chapter of this edited collection (Chap. 21). Nonetheless, the tensions arising under general law question whether the floating structure itself is a chattel that has been sufficiently annexed to land, or whether a floating structure could perhaps acquire legal status by virtually occupying the seabed below it or by serving as an extension to the pre-existing city [19], (also [54]). Otherwise, a floating structure remains bogged down in the circular debate that a floating structure is to be viewed as personal property akin to vessels [56]. Despite these distinctions, there is currently a significant absence of statutory governance relating to the ownership of floating structures and its regulation. However, it is critical to note that Florida has legislation that views floating structures as separate from vessels. This is discussed further in Part III of the chapter.

Despite the absence of a transparent statutory regime inherent in the familiar problems with defining and understanding property, the law must balance conflicting interests, rights and values. The problem with defining or understanding property to incommensurables induces the law into resolving the conflict based on the normative values that advance the individual's ownership interests. The obvious gaps in this area of the law, which disregard minority values, interests, and rights, raise problems for combating new emerging issues caused by rising sea levels. In order to better resolve these issues, the law must adapt to and consider all sections of society more equally.

3 Home

Home is a highly contested concept [21]. Its meaning is influenced by multifaceted political and personal factors, making a universal definition impossible [34]. The varying nature of the concept of home is best understood as reflexive, 'self-conscious' and 'self-aware' [48]. There are four general understandings of home: a physical structure, a territory, an identity, and a cultural and social sphere [20]. These four main taxonomies of home reveal the relational aspect of how the non-living physical structure, is constitutively bound to personhood. It is this tightly bound relationship that can show how the concept of home transcends strict doctrinal legality of ownership, and it is marked by psychological and social needs. The unique status of home is recognised in law by offering special protections for its defence in tort and criminal law and supremacy rights over other interests [4].

However, when the land and physical structure are lost, and the home is displaced, what then becomes property law's main policy objective of protecting excludability rights (the right of a property-holder to exclude the world)? As extreme climate change events become more frequent, the immediate response to the humanitarian crises is establishing temporary physical shelter, whether the shelter provided for environmental migrants in the form of an evacuation centre or a camp of humanitarian assistance. Excludability rights are now hollow when the 'world at large' (*Powell v Macfarlane* (1977) per Slade J) cannot be excluded. Replacing loss now requires new forms of housing such as floating structures. The expanding importance of floating structures not only provides the physical requirements of shelter, but through the physical property has the potential to intrinsically connect the emotional, psychological, and cultural needs that make a space feel like home. For floating structures to be practical homes in the psycho-social lens, they must provide security, permanence, and the ability for individuals and communities to recreate their identities and social networks, ensuring that the psychological and social foundations of home are not lost in the process of displacement [51].

The definition of home is varied; some legal frameworks view home primarily as an individual right, whilst others emphasise its broader social significance (This will be addressed further in Part III). Property is often interwoven with cultural and societal norms [44]. The concept can range from the western absolute individual homeownership as a private and inviolable right [33] or even virtue [16], to community living where homes are not just private assets, but spaces that contribute to communal identity and well-being. Regardless of jurisdiction, one common thread is the recognition of home as more than just physical property. This reflects the deep societal recognition that a home holds a unique position in individual lives, offering stability, privacy, and security [46]. Sprankling [49] argues that the right to property is increasingly recognised as a global issue, with international law beginning to accommodate the needs of displaced individuals and regulating non-traditional forms of housing property in the global common areas including the high seas. This is particularly relevant in discussions about environmental migrants and floating homes, where traditional land-based property rights may no longer apply. In addition to property rights, it will be proposed that there are a '*bundle of rights*' essential to identify a '*home*', which include security and permanence, identity and autonomy for cultural flexibility, collective cohesion and social networks, and sufficient infrastructure for self-sufficiency.

3.1 Home as Territory, Permanence and Cultural Continuity

Home can also be understood as something that has 'permanence and continuity' [12]. According to Carole Després, the view is that home is "a temporal process that can only be experienced along time" and through "memories … indicating a connection with past experiences" [12]. It is arguable that Després is suggesting that a home is not only a physical structure but a mental construct. Therefore, even though the physical element of the home or environment may be destroyed, the concept of home could remain through connections and memories formed in the physical location. The importance of Després' observation is that despite displacement and the loss of physicality, it does not result in an instant loss of home.

When environmental migrants are settled to different areas, out of the reach of rising sea levels, there is a real risk that time will reshape and even eliminate memories of home. One solution is to build floating structures close to where home has been lost. Where it is sound for floating structures to be placed in the exact location, environmental migrants have the chance to retain some connection with their home. Floating homes must be designed with more than mere survival in mind: these structures must foster a sense of place attachment [45], which refers to the emotional bonds individuals form with a physical location through their experiences and interactions. For environmental migrants, floating homes must offer stability and permanence, providing not just shelter but also a space where individuals can rebuild their lives and maintain cultural practices and memories, and moreover, where communities are allowed to thrive.

Floating structures cannot always be safely placed in the exact location where environmental migrants' homes on the land once laid. The archipelago of Fiji has been required to confront the difficult task of relocating environmental migrants, communities and homes. In Fiji, the territory of Vunisavisavi is traditionally guarded or controlled by the 'Tui Cakau' (a chief) [38]. If Fiji were to be completely submerged in the future, simply purchasing land in other countries (like Kiribati) and moving Fijians to other countries would not be a viable nor permanent option, primarily due to the extended sense of connection and home that Fijians have with the physical territory in Fiji [40]. It has been suggested by some Fijian villagers that maintaining traditional values and faith through this relocation process are vital, of which the village chiefs play an essential role in [8].

In an attempt to address situations such as this, in 2018, the UN adopted the Global Compact on Safe, Orderly, and Regular Migration which outlined that governments should work to protect environmental migrants and devise planned relocation and visa options if they are unable to return home [22, 23]. The protection of environmental migrants is a vital issue that must be addressed as mass displacement results in a number of catastrophic effects on both the environment itself and the communities that they move to. It is well known that when mass displacement occurs and there are no satisfactory and safe homes for environmental migrants to move to, they must set up camps, which results in deforestation, overexploitation of scarce natural resources, increased urbanisation, and the consumption of contaminated food and water.

The concept of home also entails physical spatiality, expanding the understanding of home to be territorial: a 'self/other boundary mechanism that involves personalisation or marking of a place or object, and communication that a person or a group owns it' [2] and Carole [12]. Home as a territory provides people with a sense of control over what happens to their territory and provides physical and psychological security from the dangers of the world outside [12]. Floating homes would more than likely fit this territorial definition of home, but they could possibly also be recognised as territory under international law.

How the floating structure is understood primarily impacts ownership of floating homes, the legality of floating cities, and 'defined territory' for the purposes of being recognised within international law as a state [54] (*Montevideo Convention*). For example, it has been suggested that mortgages and insurance may especially be impacted if the floating structure were defined as a vessel (being a chattel) because insurers may give owners of floating homes insurance for a vessel and not a home [54]. This debate over the legality of floating structures has been addressed in various courts and jurisdictions, such as in the United States [58, 60, 62]. Ultimately, floating structures can be understood as either personal or real property on a case-by-case basis (See Chap. 21). Whether the floating structure is personal or real property will determine what rights and liabilities are available to owners of floating homes—for example, if defined as real property, could a floating home owner acquire indefeasible title? This issue regarding territory is also relevant to statehood according to the *Montevideo Convention*, since one of the requirements to establish a state is having 'defined territory' (*Montevideo Convention*). Since there is some suggestion that sunken states could lose their statehood [28], if a state were to completely sink, but its population were relocated to floating homes above their sunken state, could the floating homes or floating cities be 'defined territory' for the purposes of being recognised as a state if the homes were classified as 'real' property? That question is yet to be answered by international courts and tribunals as discussed further in Chap. 22.

3.2 *The Emotional Aspect and Home as an Identity: Psychological and Social Needs for a Home to Be a Home*

The concept of home carries significant psychological and social significance, which cannot be reflected in a legal definition of home. As described above, a home has two facets: a shelter and a psychological space where individuals experience security, identity, and emotional wellbeing [45]. Legal scholar Margaret Radin's [39] personhood theory supports this view, arguing that individuals form personal connections to their homes, embedding them in their sense of self and identity. The home becomes a psychological anchor, providing continuity, stability, and a sense of rootedness that is essential to mental health.

This idea is evident in the emotional trauma that occurs when people are forcibly displaced from their homes. Displacement disrupts not only the physical structure in which people live but also their social ties, sense of identity, and emotional security [7]. Homes serve as a base for individuals' interactions with their broader social networks—be it family, friends, or the community at large [30]. These psychological ties to the home are so profound that dislocation can result in significant negative psychological outcomes, including feelings of rootlessness, instability, and loss of identity [39]. Property laws that allow for flexibility in housing (such as the ability to modify or personalise floating structures) can help mitigate the psychological impact of displacement [30].

Additionally, the home is often where individuals feel the greatest degree of control over their environment as it allows individuals to create a secure and predictable environment. Rykwert [42] notes Sir Edward Coke, an early seventeenth century English judge as first revering home in the law: "The house of everyman is to him as his castle and fortress, as well as his defence against injury and violence, as for his repose". This has developed into the doctrine of *home as a castle* where one's home is their fortress against external threats, reflecting the deeply ingrained need for safety and autonomy. The security provided by a home is not only physical (in terms of shelter from external dangers) but also psychological, offering a sense of control, stability, and personal freedom [10].

Radin's [39] argument extends further by suggesting that the law ought to protect this deep connection to home. She contends that when people are so attached to their homes that these spaces become part of their personal identity, the law should prioritise the protection of this attachment over competing interests, such as purely economic consideration. This argument is particularly relevant for housing policies and property laws, which must recognize that homes are more than just commodities—they are central to individuals' emotional and psychological wellbeing.

3.3 Floating Structures as Floating Homes

As seen with the displacement of many Fijian populations, a solution is needed that lessens the security trauma experienced by those relocated on land. When compared with displacement to other countries, floating structures located in or close to migrants' original home could mitigate this trauma. Environmental migrants require suitable housing solutions that provide adequate resources, security, safety, and the ability to become self-sufficient, with an ideal connection to their cultural ties. To ascertain the suitability of floating structures as a solution for displaced populations, the concept of home must be interpreted in this context. For displaced populations, *home* must address their physical, psychological and social needs, such as a place of permanence that offers safety, shelter, and the opportunity for self-sufficiency—jobs to support their families and the ability to create social networks [18] Floating structures pose a theoretical solution to the mass number of environmental migrants who have been displaced due to catastrophic natural disasters. Provided that said floating structure not only provides housing but also sufficient infrastructure to facilitate a functional economy and jobs for populations who are placed there, this definition of *home* maybe met. With housing and urban infrastructure, displaced populations could become self-sufficient which subsequently fosters the creation of social networks within their designated structure [25].

The notion of sustainable floating homes and cities has been argued by the UN as a possible solution to the immense number of people who have been pushed out of their homes due to climate change. Floating homes have been utilised by a number of nations for many years as floating semi-detached homes that are often seen in coastal areas as they are connected to land [31]. These floating homes are often singular dwellings that are connected to local sewer and utility lines which like traditional land-based dwellings, cannot be easily moved. Floating homes like these (or floating cities) could be a solution to the growing risks associated with rising sea levels for sinking islands. Entire cities of environmental migrants could seek solace at one time without many of the risks that accompany large-scale migration. These floating cities would include urban infrastructure as well as houses, so that they may effectively function as regular cities. To effectively house displaced populations, floating homes and floating cities should have the following qualities: (1) security and permanence, (2) identity and autonomy for cultural flexibility, (3) collective cohesion and social networks, and (4) sufficient infrastructure for self-sufficiency.

The following section will discuss how the idea of home and property is currently dealt with in different jurisdictions.

4 Case Studies: Solomon Islands and The American State of Florida

4.1 Solomon Islands

Unsurprisingly, perceptions and understandings of property ownership are diverse, owing to different cultures, beliefs, lifestyles and demands of the people. This section focuses on the Solomon Islands in considering a legal system that departs from the Eurocentric Blackstonian theory of property.

4.1.1 Customary Tenure and Capital Developments

Many postcolonial jurisdictions of the global south heavily rely on customary land tenure or long-term possession, concepts rooted in a communal idea of property ownership. Elements of family, community and group living are also embedded in their understanding of property which are not legally required to attain freehold rights under state law. Communal living is the hallmark characteristic of customary land tenure, with advocacy of exclusive possession carrying connotations of individualised greed. The perception of land ownership as an individual right is completely alien to the customary principles of land use and ownership. Parliaments are yet to find a solution to reconciling government policies and legislation that is able to convert land from a cultural or social element to a form of property that can be owned [17].

Although former colonies of the United Kingdom (UK) have inherited the common law system, many Pacific nations, including the Solomon Islands, have granted constitutional recognition of customary law. The *Constitution* (Constitution of the Solomon Islands Sch 3, para 2(1)(c)) expressly states that custom shall prevail over common law; however, this is often ignored, and in practice UK legislation and case law is more likely to prevail for the commitment of acknowledging both legal systems. Currently, there are challenges surrounding land tenure in the Solomon Islands, generally described as the Melanesian 'development dilemma' [6]. These problems arise from contradictions between tribal and colonial value systems throughout the history of governmental development policy. This has raised issues of legal plurality, where western colonial legal values contradict customary legal values [11]. Land in the pacific region is fundamental to tradition, culture and spirituality which cannot be simply substituted with monetary value. Case law and new land tenure legislation based on western ideas of property create serious consequences on Solomon Islands' society. This may include capitalist ideals which cause resource exploitation, environmental impacts and commercialisation of sacred lands. There has been local frustration against the Land and Titles Act which has been amended multiple times, as it achieves nothing to solve most significant problem of balancing the Solomon Islanders' desire to retain the "ancestral trust" with the ability to use the land for commercial purposes and exploit its resources in a way that would benefit local communities [11].

Traditional houses in the Solomon Islands, particularly Western Solomon Islands and islands such as Nusa Baruka and Niu Manra, were built on stilts in the water [24]. Generally, these floating structures and villages were built between the sea and mangroves. People would travel between these floating structures using canoes, which in cases of natural disasters, impeded the ability to efficiently escape.

In the case of *Allardyce Lumber Company Limited v Laore*, the High Court accepted the customary claim of ownership over the foreshore but rejected customary ownership of seabed and reefs [57]. This decision was heavily influenced by the English common law interpretation of the statutory definition of land—'land covered by water' in section 2 of the *Land and Titles Act* Ch 133 (Solomon Islands) [71] to not include seabed and reefs [29]. However, Ward CJ recognised the possibility of some customary rights existing over the sea with such rights superseding and replacing the common law position. The supplantation of common law by customary law is supported by the hierarchical status of customary law in the Solomon Islands [29]. This position is enhanced given the significant statutory recognition of customary law in the Solomon Islands under section 75 of the *Constitution of Solomon Islands* [67], which recognises that "Parliament shall make provisions for the application of laws, including customary law...". Further, Paragraph 3(1) of Schedule 3 of the *Constitution of Solomon Islands* [66] also outlines that "...customary law shall have effect as part of the law of the Solomon Islands". Hierarchically, Philip Kanairara ranks customary law higher than common law and equity, and below the *Constitution* and other legislative material. Accordingly, the restriction of ownership of seabeds and reefs that (by extension) limits the ability of floating homes to maintain collective cohesion and social networks, directly contradicts the custom of communal sharing of land, which would extend to the right of access to reefs, lagoons, and fishing grounds [55].

Contrasting from *Laore*, the case of *Combined Fera Group and Three Others v Attorney General* adopted a liberal approach to defining 'land' as including the seabed below it [59]. Previously, common law in the Solomon Islands had excluded land permanently covered by the sea to be customary land. However, the reformulation of the land definition in the *Land and Titles (Amendment) Act* [70] (Solomon Islands) permitted land below the low-water mark to be registrable as public land owned by the Commissioner of Lands. Justice Palmer of the High Court, adopted the view that land which is capable of public ownership could have no other possible conception than to also be land capable of customary ownership [15]. Customary ownership and native title of land is granted where traditional use and occupation of the land is capable of being proved prior to 1 January 1969 (when the current *Land and Titles Act* Ch 133 came into force). The test applied to determine whether 'land' constituted customary native title was whether the Appellants had sufficiently demonstrated rights of ownership, use or occupation [15].

The common law of the Solomon Islands permits individual land ownership along with the use and occupation by others. Therefore, the crux of customary land recognition of land and sea, appears to lie in the use and occupation of land. This provides some form of physical security and permanence, as mere occupation of the floating home reduces risk of forcible removal. There are limits, however, as interest in land does not necessarily amount to land ownership. Moreover, permitting the ownership of the seabed below a floating home would address the issues of cultural independence which currently exist in the laws of the Solomon Islands, with multiple communities within the South Pacific Islands adopting their own belief systems. For example, the Trobriand Islanders from Papua New Guinea refuse to build floating structures without certainty of the land (sea) below as the local superstitions which govern them profess a sorcerer may light a fire below their home [43]. Therefore, in order to provide certainty to homeowners and broader practical application of floating homes in a manner which is sensitive to the unique cultural practices and superstitions of each community across the Pacific Islands, a more liberal interpretation of 'land', such as that in *Combined Fera Group and Three Others v Attorney General,* should be adopted.

4.2 Florida

In contrast to the Solomon Islands, American exclusive possession and individual ownership of property is perceived synonymously with freedom rather than (as it would be framed in more collectivist societies) an individual greed [20]. Senior US District Court Judge for the Southern District of New York, Judge Rakoff, once noted that '[t]he house, particularly the owner-occupied house, served as a powerful symbol of order, continuity, physical safety, and a sense of place or belonging' for the American people [20]. This statement followed a study which identified that homeownership held deeper ontological meanings for the American people, extending beyond mere means of financial investment [20]. The broader significance of home within property carried meanings encompassing family, personal identity, status, and the home as a place of permanence, security, and refuge from the outside world [12, 20]. Ownership provides the greatest control, security, and status in a home.

4.2.1 American Concepts of Home and the Emphasis on Exclusivity

The American homeownership dream stems from Blackstone's conception of property as an absolute right in which individuals possess 'despotic dominion' over land 'in total exclusion of the right of any other individual in the universe' [5]. It is important to note that contemporary scholars have cautioned against discussing Blackstone's theory, advocating for a revised perspective on American property rights [50]. Despite this criticism, it is contended that there is merit in discussing Blackstone's framework on account of its particular influence in shaping the American legal understanding of property ownership and, by extension, homeownership as an absolute right [5].

Consider the 'bundle of sticks' metaphor. Adopted by the Supreme Court of the United States as the starting point for property jurisprudence, it hyper-focuses on individual rights. While the concept purports to encompass a collection of interests pertaining to individual, community and public rights [14], the very illustration of a 'bundle of sticks' enables the suggestion that each category of interest, comprising "ownership", can be entirely isolated from anything else [50]. In American property jurisprudence, in particular, the rights of individuals are often privileged over the rights of the surrounding community, public rules and regulations, and the rights of other individuals [50]. For example, Florida's unlimited homestead exemptions—which prohibit the foreclosure of the home by creditors—prioritise shelter for the debtor and protect their private rights at the expense of the creditor interests [4]. In fact, the very interpretation of property is enshrined in the American Constitution as private and inviolable [23]. Public and community interests are often portrayed as restrictions on individual rights as opposed to a constituting part of the property regime itself [50]. Thus, there is a tendency to equate the term property with "private property". Owners of single-family homes or condominiums may view themselves as private owners of their properties. In the context of homeownership, property theory subsumes a family as a single 'individual' or entity regarding the regulation of their property rights. The idea of the traditional family comprising part of the American dream is reduced into a fictional individual to perpetuate the notion of 'individual ownership' [41].

4.2.2 Legal Practicality of Floating Homes in Florida

Focusing on Florida specifically, houses that are illegally built on stilts for fishing, hunting and other recreational purposes are common to the Apalachicola River [36]. Also known as river camps, these houses attach to the riverbed and bank, and some are also connected to electricity. However, many of these river camps are abandoned following natural disasters [36]. Although conservation groups attempt to tag abandoned vessels and fishing camps on the river, the lack of registered title for these structures means owners are generally not held liable for hazards, waterway damage or the removal of abandoned waste [35]. The affordability of these camps in comparison to land-based dwellings have made them attractive to locals and slowly increased their popularity over time [35]. However, Floridian property developers may be more likely to consider building floating structures for commercial or real estate purposes before residential solutions due to the profitable incentive. Property developers in Florida have listed variations of dockside floating homes, however, they may only be designed for luxury lifestyles.[1] For example, Floridian architect company Arkup, founded by Arnaud Luguet and Nicolas Derouin, have proposed a floating yacht with retractable motorised spuds designed to semi-permanently attach itself to the seabed in shallow waters [3]. Although the homes are innovative and offer a housing alternative, they are not designed to be affordable to those who have been displaced by natural disasters.

There are a number of legislative instruments throughout Florida counties that deal with the definition and legality of floating structures. For example, under section § 6.2(d)(1) of the Palm Beach County Code of Ordinances, it is unlawful for any person to anchor, moor, tie off, or otherwise affix a floating structure upon or to the waters of the County [72]. Floating structure is defined under subsection (c) to mean a floating entity typically associated with a structure or improvement, for example residences, places of business, office, hotel, restaurant, storage facilities or mining platform [72]. Subsection (e) sets out exceptions, including floating structures in private mooring, for construction purposes, expressly permitted or at established marinas when there is compliance with local applicable laws [72]. The Supreme Court has expressly clarified that floating structures should not be considered boats or vessels [62]. Boats and vessels, while they can be registered and are subject to tax exemptions under the Florida State Constitution, are considered personal property. Movement does not

[1] Penning-Rowsell E (2020) Floating architecture in the landscape: climate change adaption ideas, opportunities and challenges, Journal of Landscape Research 45. https://doi.org/10.1080/01426397.2019.1694881.

preclude the entity from being a floating structure, nor does registration of the structure. However, in *Lozman*, the Supreme Court considered whether it was problematic that marinas offering services to immoveable floating structures would not receive the security of maritime liens under admiralty law similarly awarded to creditors of vessels, which allows for automatic seizure upon default [62]. The Court concluded that where a floating structure is precluded from being a vessel due to lack of self-propulsion and difficulty of towing, it would pose no risk to creditors.[2] In Florida, the ability of the structure to navigate waters will affect the securities that will attach to the property, and as such, the movement of the floating structure should be a legal concern in construction. Controlling movement will provide greater protection for communities that assist floating homes, including docking space landlords.

Although Floridian legislation has attempted to define floating structures, it does not go as far as to differentiate it from a houseboat or "live-aboard vessels" analogous to mobile homes [61]. The attempt at a statute definition for "live-aboard vessels", which was explicitly for the purpose of exclusion, failed to distinguish from boats defined in the constitution [61]. By way of example, under the Florida Statutes section § 320.015, a mobile home may only be classified as real property when the homeowner also owns the land to which the mobile home is permanently affixed to [69]. This is because the mobile home ceases to be mobile and becomes part of the real property [65]. Under section § 253.03(1) of the Florida Statutes, the Board of Trustees of the Internal Improvement Trust Fund of the State of Florida has registered title over waterways and substantially all submerged lands as state-owned land [68]. Unlike mobile homes, it may be dependent on the discretion of the State for owners of permanently affixed floating homes to have a legal interest over the seabed which they occupy. A practical example of this issue was argued in Florida's District Court of Appeal, where an application for the lease of dock space where a houseboat was permanently affixed was denied, on the basis that the regulations of floating structures did not give a superior right over the Trustees' "proprietary authority as lawful owners of sovereignty submerged land" [63]. As discussed above, exclusivity is an essential right to Blackstone's property law as derived from the legal title. In Florida, it is presently unlikely floating structures will be classified as real property if the owners have no exclusive possession over the waters upon which the structures reside.

4.3 A Framework of Interests for Floating Homes for Environmental Migrants

The 'bundle of rights' essential to home will vary across contexts and for different populations. To unify the view on property rights for displaced populations on floating

---

[2] Yankowski K (2013) Whatever Floats the 'Reasonable Observer's' Boat: An Examination of Lozman v. City of Riviera Beach, Florida and the Supreme Court's Ruling that Floating Homes Are Not Vessels. University of Miami Legal Studies Research Paper. http://dx.doi.org/10.2139/ssrn.2264866.

structures, a reflexive framework of 'home interests' grounded in the human dimension is most suited [13]. In many cultures, such as the Solomon Islands case study explored in this chapter, home is not a private, individual possession but a shared space tied to communal practices and social structures. In comparison, American concepts of home place emphasis on exclusivity and security in title, which is an important consideration for legislating the physical design of floating structures.

The Solomon Islands and the U.S. differ significantly in their integration of home in property ownership. This is evident in their reflection of different cultural values and legal frameworks. While ownership is central to identity in both contexts, land in the Solomon Islands is perceived as a cultural asset that embodies traditions and facilitates customary practice. Thus, its ownership serves as a communal right. In contrast, homeownership in the U.S. is integral to personal security in providing control and freedom for the individual. The ability to exclude all others is held as key.

Both the Solomon Islands and Florida are prone to natural disasters, but they experience different impacts. Rising sea levels and land submersion pose a significant risk to the Solomon Islands, while Floridan homes are regularly impacted by hurricanes and flooding. Faced with the possibility of losing their way of life, the issue of floating homes in the Solomon Islands is tied to the challenge of balancing cultural preservation with the need for economic growth in local communities. Comparatively, in Florida affordable floating homes, such as river camps, exist supporting a small number of environmental migrants. However, such development has begun to shift away from affordable housing in favour of luxury developments which are driven by market incentives. As a result, future floating home projects in Florida are likely to become more commercialized, prioritizing profitability over the needs of displaced communities.

While both regions find difficulty with the legality of rights over submerged land, for the Solomon Islands, the issues in law are primarily about recognition of customary ownership and the legal treatment of land and sea as interconnected entities. Whereas, with Florida, legal issues revolve around whether floating homes should be classified as real property and the complications of claiming ownership over water or seabeds, especially with state-owned submerged lands. Despite their differences, however, the manifestation of home in property law for both the Solomon Islands and America embodies values central to those societies. Legislation for floating structures should, therefore, support a wide range of ownership models for floating homes, allowing environmental migrants to maintain their cultural traditions and social cohesion even in the face of environmental displacement.

In the context of climate change and increasing displacement, the idea of *home* must be reimagined, particularly for those who may no longer have access to traditional land-based homes. Home for displaced populations varies in contexts, geography and culture, but their importance is universal [1]. As this chapter has discussed and illustrated, floating homes could be a potential solution to this growing issue.

5 Conclusion

This chapter departed from the traditional legal discussion surrounding the proprietary status of floating structures. It instead presented a different focus towards the socio-legal understandings of 'home' and how this affects the existence of floating structures in different communities. While individuals may retain proprietary interests in floating structures, the crucial conclusion from this chapter is that the extent of these interests must be determined with reference to the specific countries' legal systems and perceptions of what 'home' means for them.

While this chapter presented a perhaps unconventional perspective on this topic, it is an especially important discussion because the consequences of climate change present many challenges for environmental migrants worldwide. Viewing floating structures through this lens provides this community with a holistic solution that would take into consideration their cultures, practices, traditions and beliefs. As such, this chapter proposes that both legal professionals and engineers should adopt this focus with floating structures in order to secure a successful and holistic implementation of floating structures in the world.

References

1. Akesson B, Basso AR (2022) From bureaucracy to bullets: extreme domicide and the right to home. In: Genocide, political violence, human rights. Rutgers University Press, pp 3–35
2. Altman I (1975) The environment and social behavior. Brooks/Cole
3. Arkup (2024) Orders for Arkup 50 are heating up. August 26. https://arkup.com/orders-for-our-latest-product-arkup-50-is-heating-up/
4. Barros DB (2006) Home as legal concept. Santa Clara Law Rev 46(2):255–306
5. Burns RP (1985) Blackstone's theory of the absolute rights of property. Univ Cincinnati Law Rev 54(1):67–86
6. Burt B (1994) Land in Kwara'ae and development in Solomon Islands. Oceania 64(4):317–335
7. Carroll B, Morbey H, Balogh R, Araoz G (2009) Flooded homes, broken bonds, the meaning of home, psychological processes and their impact on psychological health in a disaster. Health & Place 15(2):540–547. https://doi.org/10.1016/j.healthplace.2008.08.009
8. Charan D, Kaur M, Singh P (2017) Customary land and climate change induced relocation: a case study of Vunidogoloa village, Vanua Levu, Fiji. Limits to climate change adaptation. Springer, pp 345–358. https://doi.org/10.1007/978-3-319-64599-5_19
9. Convention Relating to the Status of Refugees (1951) https://www.ohchr.org/en/instruments-mechanisms/instruments/convention-relating-status-refugees
10. Conway H, Stannard J (2015) Property and emotions. Emot Rev 8(1):38–41. https://doi.org/10.1177/1754073915601225
11. Corrin J (2011) Customary land in Solomon Islands: a victim of legal pluralism. In: Angelo A, Sage Y-L (ed) Droit Foncier Et Gouvernance Judiciaire Dans Le Pacifique: land law and governance in the South Pacific. University of Queensland TC Beirne School of Law Research Paper No. 12–11, pp 223–227
12. Després C (1991) The meaning of home: literature review and directions for future research and theoretical development. J Architectural Plann Res 8(Summer):96–115
13. Du Plessis J, Leckie S (2006) Property rights and the need for more inclusive concepts, laws, policies, and practices. Realizing property rights.

14. Duncan ML (2002) Reconceiving the bundle of sticks: land as a community-based resource. Envtl L 32:773
15. Farran S (2001) South Pacific land law: some regional challenges, cases and developments. Victoria Univ Wellington Law Rev 32(4):953
16. Fischel WA (2002) The homevoter hypothesis: how home values influence local government taxation, school finance, and land-use policies. Harvard University Press, Cambridge, MA
17. Fitzpatrick D, Compton C, Foukona J (2019) Property and the state or the folly of Torrens: a comparative perspective. Univ New South Wales Law J 42(3):953–980
18. Flatau P, Smith J, Carson G, Miller J, Burvill A, Brand R (2015) The housing and homelessness journeys of refugees in Australia, AHURI final report No. 256. Australian Housing and Urban Research Institute Limited, Melbourne, pp 55–57
19. Flikkema MMB, Lin F, van der Plank PPJ, Koning J, Waals O (2021) Legal issues for artificial floating Islands. Front Mar Sci 8:619462. https://doi.org/10.3389/fmars.2021.619462
20. Fox L (2002) The meaning of home: a chimerical concept or a legal challenge? J Law Soc 29(4):588–601. https://www.jstor.org/stable/4150519
21. Gallie WB (1956) Essentially contested concepts. Proc Aristot Soc 56(1):167–180. https://doi.org/10.1093/aristotelian/56.1.167
22. Global Compact for Safe, Orderly and Regular Migration (2018) Global Compact for Migration
23. Goodman LS, Mayer C (2018) Homeownership and the American Dream. J Econ Perspect 32(1):31–58
24. Hagen K, Petterson MG, Humphreys D, Clark N (2021) Why disaster subcultures matter: a tale of two communities: how and why the 2007 Western Solomon Islands tsunami disaster led to different outcomes for two Ghizo communities. Geosciences 11(9):387
25. Hulse K, Stone W (2006) Housing, housing assistance and social cohesion
26. International Covenant on Economic, Social and Cultural Rights (1966). https://www.ohchr.org/en/instruments-mechanisms/instruments/international-covenant-economic-social-and-cultural-rights
27. International Organisation of Migration (2007) Discussion note: migration and the environment. https://environmentalmigration.iom.int/sites/g/files/tmzbdl1411/files/MC_INF_288.pdf
28. Juvelier B (2017) When the levee breaks: climate change, rising seas, and the loss of island nation statehood. Denver J Int Law Policy 46(1):21–44
29. Kanairara P (2017) An analysis of the recognition of customary law and common law and equity in the legal system of Solomon islands. Solomon Islands law reform commission. http://www.lawreform.gov.sb/publications/other-reports/conference-papers/16-underlying-law-conference/file.html
30. Kim K, Smets P (2020) Home experiences and homemaking practices of single Syrian refugees in an innovative housing project in Amsterdam. Curr Sociol 68(5):607–627. https://doi.org/10.1177/0011392120927744
31. Lin FY, Spijkers O, Van der Plank P (2022) Legal framework for sustainable floating city development: a case study of the Netherlands. In: WCFS2020: proceedings of the second world conference on floating solutions, Rotterdam. Springer Singapore, pp 433–460
32. Lindsay R (2023) Climate change: global sea level. https://www.climate.gov/news-features/understanding-climate/climate-change-global-sea-level
33. McCavana C (2020) Property is a social right: a comparison of American, European and Australian governmental acquisition jurisprudence and property theory. W Aust Student Law Rev 4:1–4
34. Meers J (2021) Home as an essentially contested concept and why this matters. In: Housing studies, vol 38, no. 4. Routledge, pp 599–608. https://doi.org/10.1080/02673037.2021.1893281
35. Mossa J, Chen Y (2018) Floating fish camps on the Apalachicola River, Florida: increases and implications. Florida Geogr 49. https://ojs.test.flvc.org/flgeog/article/view/105572
36. Mossa J (2020) Controversial camps in conservation lands: Apalachicola river's floating structures, Florida. Southeast Geogr 60(2):93–96. https://www.jstor.org/stable/26912602

37. Nakajima T, Yamashita Y (2023) August. Sustainable water city—a case study in Singapore. In: World conference on floating solutions. Springer Nature Singapore, Singapore, pp 19–32
38. Nand M, Clissold R, McNamara K, Yee M (2024) Guiding our responses to climate change by what people value: insights from Fiji. Geoforum 155:1–13. https://doi.org/10.1016/j.geoforum.2024.104080
39. Radin MJ (1982) Property and personhood. Stanford Law Rev 34(5):957. https://doi.org/10.2307/1228541
40. Ristroph B (2024) Using community knowledge to inform planned relocation in the Pacific Islands. Suffolk Trans Law Rev 47(1):1–28
41. Rovit EH (1960) The American concept of home. Am Sch 29(4):521–530
42. Rykwert J (1991) House and home. Soc Res 58(1):51–62
43. Sainim B (2007) Traditional architecture in the Pacific
44. Salkin PE, Gross D (2012) International comparative property rights: a cross-cultural discipline comes of age. Brigham-Kanner Property Rights Conf J 1:41
45. Scannell L, Gifford R (2017) The experienced psychological benefits of place attachment. J Environ Psychol 51:256–269. https://doi.org/10.1016/j.jenvp.2017.04.001
46. Schultz Kleine S, Menzel Baker S (2004) An integrative review of material possession attachment an integrative review of material possession attachment. Acad Market Sci Rev
47. Sedikides C, Gaertner L, Luke MA, O'Mara EM, Gebauer JE (2013) Chapter five—a three-tier hierarchy of self-potency: individual self, relational self, collective self. Adv Exp Soc Psychol 48:235–295. https://doi.org/10.1016/B978-0-12-407188-9.00005-3
48. Spicer MW (2018) What do we mean by democracy? Reflections on an essentially contested concept and its relationship to politics and public administration. Adm Soc 51(5):730–740. https://doi.org/10.1177/0095399718786881
49. Sprankling JG (2013) The global right to property. Columbia J Trans Law 52:464
50. Turčan KA (2021) U.S. property law: a revised view. William and Mary Environ Law Policy Rev 45(2):319–376
51. Tyrer S (2020) Home in Australia: meaning, values and law?. Univ New South Wales Law J 43(1):340–383. https://doi.org/10.53637/ggos1001
52. United Nations (2019) Sustainable floating cities can offer solutions to climate change threats facing urban areas, deputy secretary-general tells first high-level meeting. https://press.un.org/en/2019/dsgsm1269.doc.htm
53. United Nations (2023) Climate change-induced sea-level rise direct threat to millions around world, secretary-general tells security council. https://press.un.org/en/2023/sc15199.doc.htm
54. Wang BT (2023) Development of the legal definition of the floating city: judicial interpretation of structural characteristics of floating homes and developments. In: World conference on floating solutions, pp 139–157. Springer Nature Singapore, Singapore
55. Ward RG (1997) Changing forms of communal tenure. Gov Common Property Pacific Reg 19:29–30
56. Worthington S (1996) Proprietary interests in commercial transactions. Oxford University Press

Case Law

57. Allardyce Lumber Company Limited v Laore (1989) High court Solomon Islands civil case no 64/1989
58. Board of Commissioners of Orleans Levee District v M/V Belle of Orleans 535 F 3d 1299 (2008)
59. Combined Fera Group and Three Others v Attorney General (1997) High court Solomon Islands land appeal case No 4/1993
60. De La Rosa v St Charles Gaming Co 474 F 3d 185 (2006)
61. *Department of Revenue v. Fla. Boaters Ass'n, Inc.*, 409 So. 2d 17 (1981)

62. *Lozman v. City of Riviera Beach*, 568 U.S. 115 (2013)
63. *Marrone v. City of Key W.*, 814 So. 2d 478 (2002) LEXIS 3926
64. Yanner v Eaton (1999) HCA 53
65. Zapo v Gilreath, 779 So. 2d 651

Legislation

66. Constitution of Solomon Islands 1978 (rev. 2018) sch 3 para 3(1)
67. Constitution of Solomon Islands 1978 (rev. 2018) sec 75
68. Fla. Sta. § 253.03(1) (2024)
69. Fla. Sta. § 320.015 (2024)
70. Land and Titles (Amendment) Act 1964 (Solomon Islands)
71. Land and Titles Act Ch 133 (Solomon Islands) sec 2
72. Palm Beach County Florida Code of Ordinances Chapter 6-1 sec. 6.2

Kim D. Weinert is a legal academic and lawyer. Kim researches the intersection of law, society, and culture. She has a particular focus on charity law, not-for-profit governance, and property law. Kim's not-for-profit research has been cited by the Queensland Supreme Court and the Western Australian parliament. Kim was the recipient of the Australian Postgraduate Award (APA) Scholarship at Griffith University, where she completed her doctoral thesis on speech legality in Australian culture. Kim's research interest in law and culture in Australia and the South Pacific has been published in international journals. The exploration of cultural legalities in Kim's research provides a unique perspective on the impact doctrinal law has on minority groups. Through this lens, Kim can identify and address numerous shortcomings in the law which paves the way for legal and governance reform. Exploring the traditional boundaries of doctrinal law and its theoretical foundations in relation to emerging and new societal challenges, her research unites several strands of legal inquiry linked to social justice and equality. Her charity law research has been cited by the Queensland Supreme Court, the Western Australian Parliament, and eminent legal charity scholars. The impact of Kim's research has produced an impact on how charity law is discussed and applied in Australia.

The 'Bundle of Rights' of Property: A Close Examination of the Jural Relationship in Artificial Landscapes and Floating Structures as Land Replacements

Kim D. Weinert

Abstract This essay will critically examine the nature of property rights in artificial landscapes and floating structures within a system of private property to consider if they can be treated as 'real property' (land). The legal concept of private property is primarily tied to an individual's right to 'exclusivity', or the sole right of access to use, occupy, and exploit a terrestrial piece of land. However, this grounding of the legal concept of private property ownership in the individualised right of exclusivity, that is, whether ownership should only be understood by this one defining right and duty, has been subject to ongoing examination and debate. This chapter analyses the bundle of rights that attach to the legal concept of real property in order to understand how the law can translate the existing legal concept of the right of exclusivity to the resources of artificial landscapes and floating structures. It examines the idea of rights and the concept of ownership in property through the lens of contract and torts law to articulate the legal concept of exclusivity (and other property rights) with greater clarity and how these concepts need to be adapted when the context of property shifts from land, to the water. This is a critical question that needs to be addressed in order for floating structures to have utility as land. This chapter proposes that some rights need to be linked for artificial landscapes, within a context of land scarcity and water, for the right of exclusivity to make sense to property-holders.

Keywords Property theory · Bundle of rights · *In rem* rights · Artificial landscapes · Floating structures

K. D. Weinert (✉)
Centre for Public, International and Comparative Law, Brisbane, Australia
e-mail: k.weinert@uq.edu.au

Visiting Fellow 2025, State Library of New South Wales, Sydney, Australia

T.C Beirne School of Law, University of Queensland, St Lucia, QLD, Australia

B. T. Wang et al. (eds.), *Large Floating Solutions*, Lecture Notes in Civil Engineering 636, https://doi.org/10.1007/978-981-96-5435-2_21

1 Introduction

Intense and frequent environmental events worldwide have caused severe property damage and displaced too many communities. These defining environmental moments coincide with unprecedented pressures on sustainable governance and management of land. In response to these deepening issues, innovative solutions such as artificial islands and floating structures have gained credible traction as the panacea to the ongoing humanitarian issue caused by climate change. Artificial islands can be conceptualised as bottom-founded construction of land through land reclamation techniques, which may likely be treated the same as real property (land), this chapter uses the term 'artificial island' to describe any floating object that is use to displace space occupied by water, that is then treated as a simulacrum of land.

Despite the increasing volume of research and public interest in the viability of floating structures and artificial islands as potential replacement for terrestrial environments, this shift from the land to adjacent water bodies requires a conclusive property framework regarding these engineered spaces that remains underdeveloped. Accordingly, discrepancies in the law become known when the courts are called upon to address numerous challenges arising from the innovative technological developments of artificial islands and floating structures that are currently treated as anomalies in property law. While there might be a natural tendency to eliminate any perceived issues by categorising these engineered spaces as 'real property' (land), such thinking overlooks context and the strains of theoretical tensions in a system of private property. Unlike land, the utility of artificial islands and floating structures, which are attached to or located within a coastal zone of a territorial interior water body, manifests a closer dialectical relationship between the natural resource of water.

Water is an unpredictable natural resource of the global commons (that is, something that is owned by everyone as a shared natural resource of the Earth). As described in Chap. 1, disastrous climate events have caused extreme terrestrial water scarcity on every continent. At the same time, rising sea levels and floods have submerged many other countries. These parallel catastrophes manifest how the same natural resource causes harm and loss in different contexts. It is a bitter irony that a property-holder can lose their land to a resource that cannot generally, in itself, support clear individual ownership. Indeed, efforts to replace private property have primarily involved the occupation and exploitation of water to create manufactured floating objects that are then treated as a simulacrum of actual land. As such, the proposed solutions of artificial landscapes and floating structures can be perceived as a historical continuum of converting the 'common lands [are] into private property' [9, p. 8]. But while a Lockean understanding of property would find that engineered land is far from being a common natural thing, [14] the basic legal question that arises is simply: what private property rights will be afforded in or upon these structures? The answer to this question is not always straightforward and depends on which respective jurisdiction the answer is being presented.

Nonetheless, generally, there are three possible answers to the question. First, the obvious answer that reflects a liberal and neo-liberal paradigm would give absolute individualised private ownership rights in the resource. Second, an Aristotelian way of thinking about property ownership for the communal use of resources means that a community as a whole determines how the common property is managed. Third, by combining private ownership rights and communal use of resources vested in the state to promote social prosperity (utilitarian), these structures would be to maximise the total satisfaction that would be obtained from the resource. These answers manifest a proliferation of different kinds of property rights across various jurisdictions. Confident defenders of their respective answers will do so according to property jurisprudence manifested in national constitutions and legal systems.

Indeed, all property rights arguments have their problems, and this chapter does not undertake a granular critique of each of these arguments nor advocates or decides which answer is the most correct. Rather, this chapter is grounded in the assertion that if artificial islands and floating structures are to be built as a necessary replacement for land, then it is only inevitable that the rights-based vacuum will be called upon to sustain a system of private property.

Before progressing any further, it must be made clear that all sustainable efforts aimed at preserving, restoring, rehabilitating, and, moreover, preventing land and waters from the consequences of past and ongoing institutional neglect should always be the paramount strategy and aim; this chapter must not be read or understood as a separate concern to taking all necessary actions of avoiding and limiting the destruction of the natural world and cultural life. Instead, this chapter should be understood from the perspective of an anti-diasporic experience. As CM and BT Wang correctly point out, territorial waters provide the space to bolster a colonisation agenda of resource exploitation [28]. If artificial islands and floating structures are to be land replacements, the engineered structures and spaces will be analysed and viewed through the characteristics and identity of real property. The Global North's state power to control and exploit these natural resources is inextricably bound to this homogeneous understanding of land.

In this legal context, the state (that is, the government of the day) can easily utilise artificial islands and floating structures to facilitate its power under a thin veneer of neoliberal policies concerned with identity and community. The rights vacuum in property legal theory will neatly serve this agenda by respecting a system of private property rules which strives to find and protect individualised property rights. But to some extent, artificial islands and floating structures represent something different to real property, and since property is a shifting concept, the central question of this chapter is how the law should understand the *in rem* rights (that is, the right or an action over property that can be enforced against the rest of the world, or the right of exclusion) in artificial islands and floating structures through the bundle of rights theory.

The predominate bundle of rights theory of property has always been a useful starting point in conceptualising a resource as property owing mainly to its coherence. This chapter's focus on understanding the *in rem* rights in these engineered spaces might not speak of transforming property law, nor work with sticks from the bundle in a wholesale manner. Instead, this chapter proposes that specific sticks in the bundle can be fused for *in rem* rights to have meaning and applicability in an artificial island. This assertion will predominantly take a liberal standpoint grounded in a Legal Realist understanding of property to answer the chapter's question.

Not dissimilar to many other legal property scholars, this author's natural scepticism towards the Legal Realist's understanding of property and ownership as a bundle of sticks treated as atomised rights ignores and obscures the fascinating nuances in and around the concept of property. Why does this chapter adopt a reductionist, atomised bundle of rights view of property in the context of artificial islands and floating structures? Well, to put it simply, as land becomes increasingly scarce, the predominated reliance on engineered landscapes and spaces in and on the water will perpetuate modernity's liberal ethos of the individual's right to exclude. If liberal ideals are to remain central to upholding a private property system that includes artificial landscapes and floating structures, the impulse to account for these things as land will follow the simple bundle of rights theory.

This chapter continues as follows: Sect. 2 outlines the main legal principles of understanding the concept of property through a small offering of theoretical categories of property. This part will offer a broad overview of how the law theoretically developed the concept of property through the metaphor of a bundle of sticks from Hohfeld's jural relationship theory. From here, Sect. 3 explores how the artificial landscape is a feature of the global commons; unlike land, specific sticks from the bundle need to be fused for these structures to have utility. By teasing out the malleability of the bundle of sticks, it aims to enrich the abstract nature of *in rem* rights.

Section 4 seeks to examine and fill the gap that lies in the scant judicial attention paid to propertising and classifying floating structures as land or personal property (chattel). This section presents two case studies that demonstrate the application of property law in propertising floating structures. The respective case study reveals that the approaches used in common law and civil jurisdictions when categorising floating structures as property differ. Apart from highlighting the difference that emerges from national domestic property law in two distinct legal systems, this section illustrates that if floating structures and artificial island scapes are to be the panacea to a changed natural environment, parcels of land still need to be preserved for the proper timing and functionality of these engineered spaces.

Finally, a conclusion is presented that identifies future research and discusses the legality and regulation of propertising these novel substitutes of land in the global commons.

2 Understanding Property as a Bundle of Rights/Sticks

Property is an elusive concept that has evaded a universal definition. Despite well-heeded judicial, parliamentary and scholarly efforts across numerous legal systems to precisely understand property, it is a moot point [1]. Nevertheless, property is a distinct and explicit idea; and there is much legal debate concerning how the law can best understand what property is [6, p. 1]. Despite these complexities, the better view of lawyers and the courts when defining property is drawn from its legal dimensions. The law provides three definitional approaches: exclusionist, the bundle of rights and the social function [1]. However, owing to space (and the impossible task of covering the vibrant social, political and moral aspects of property theory) it is only the first two approaches that this section will address.

The exclusionist definition of property concerns the factors of dominion, exclusion and external things [4]. Blackstone refers to the factor of dominion, which is the control a person (whether real or juristic) has over a thing. Exclusion refers to the legal enforceability to prevent others from using or enjoying a thing. Lastly, the external thing refers to and includes corporeal objects (for example, land and personal property or chattels, and as the law has expanded the categories of property to include incorporeal things such as intellectual property). A key characteristic of the thing here is that the subject matter is definable, stable, and identifiable by third parties, and the law will attach certain rights to the thing, such as enforceability and alienability (capable of being consumed by a third party) [18]. Property understood through these three factors is mainly reflected in Western-centric jurisdictions. Exclusionists strongly believe that property can only be defined and understood through control over a thing, giving a person the right to exclude [16].

The second and popular approach to understanding property extends the exclusionist approach. It is known as the bundle of rights—often referred to by lawyers as the ‘bundle of sticks or rights.’ The metaphorical bundle of sticks picture is claimed to have its roots in the jurisprudential school of Legal Realism [22]. However, the picture’s beginnings are traced to the jural relations theory of Wesley Hohfeld [11]. Hohfeld broke down the concept of property into discernible components paired with a legal right or privilege (liberty). Whether the right or privilege is considered one of substance or not, it will correspond to a duty that legally compels an individual to do or refrain from some action (Ibid). For instance, the relationship between individuals stems from the baseline of land ownership whereby A, the fee simple landowner, has rights to exclude B and C, who have no corresponding rights to A’s land. Further, A’s rights and privileges allow A to use the land (such as building a house, growing crops, etc.), and B and C have a duty not to interfere with A’s liberty of use and enjoyment. The jural relationship that Hohfeld conveys is less about the thing or resource and how property emphasises the interconnectivity of rights, privileges and duties.

Understanding property as discernible pieces allowed the Legal Realists to expand upon Hohfeld’s theory with ‘nominalist impulses’ [20, 23]. Legal realism, as a complex body of American intellectual thought, has two dominant stages: (1) deconstruction and (2) reconstruction [7]. According to David Frazer, deconstructing involves challenging the hegemonic vision of the law, arguing that having a proper understanding of judicial decisions consists of chipping away at the veneer of deductive reasoning, legal rules, and concepts that are presented as being neutral. According to the Legal Realists, once the veneer of mechanical reasoning is peeled off, judges’ decisions are indeed based on social policy and personal or political biases and constructed from hunches [3]. American legal realists argue against legal formalism (black letter law), and their approach is grounded in Frazer's second stage of reconstruction.

Reconstruction is an analytical tool for social purposes, and if those purposes are not being served well, the law should be changed [3]. Reforming the law involves legal standards that purportedly reflect people's expectations and customs rather than subjecting people to legal technicalities emanating from doctrinal indeterminacy [3]. Exhibiting such hostility towards the untidiness of doctrinal traditions, the Legal Realists see themselves as 'systematists, but in a functional sense' [26]. Indeed, the conventional wisdom of the Realists' bundle of rights metaphor of property by taking Hohfeld's discrete components and packaging of a series of sticks as interconnected and divisible rights seeks to limit the common law's vagueness around understanding property. The benefit of this bundle of rights is that it brings a list of qualities that uses scientific language to locate and scope whether an individual owns a resource.

In 1960, A M. Horné set about to list the incidences (sticks) to work out if a person is an owner of the property. The inherent wisdom in Horné's list (at least for law students) does offer a simple functional and analytical framework that explains property ownership. The list is made up of the following:

1. 'The Right to possess—to have exclusive physical control of a thing;
2. The Right to use—refers to the owner's enjoyment and use of the thing;
3. The Right to manage—a right to decide how and by whom a thing shall be used. This right depends, legally, on a cluster of powers, chiefly powers of licencing acts that would otherwise be unlawful and powers of contracting;
4. The right to income—refers to the benefits derived from foregoing personal use of a thing and allowing others to use it;
5. The right to capital—involves the power to alienate the thing and the liberty to consume, waste, or destroy the whole or part of it;
6. The right to security—legally, this is, in effect, an immunity from expropriation;
7. The right of transmissibility—the power to devise or bequeath the thing;
8. The absence of term—the indeterminate length of one's ownership rights;
9. The prohibition of harmful use—the duty to refrain from using the thing in certain ways harmful to others;
10. Liability to execution—refers to having the thing taken away for repayment of a debt;
11. Residuary character—the existence of rules governing the reversion of lapsed ownership rights' [12].

How the law makes sense of these incidences has to do with the nature of a stick's function in response to a unique set of facts [8]. In affirming this approach, the courts find that the more sticks a person will hold, the greater 'property interest' a person will have in the thing [32].

Consequently, the case of *Knapp* demonstrates the functional simplicity inherent in the bundle of rights. However, acquiring as much of this bundle as possible oversimplifies what is legal property ownership. For instance, the economic sticks (i.e., to manage, to capital, etc.) are not legally recognisable rights [19]. Although these sticks are liberties providing an individual choice, it is to the point that James Penner makes that if the bundle of rights are to be an accurate reflection of property ownership, there remains a need for most of the so-called rights to be vindicated by another legal genius. Whether it is tort, contract or equity that is called upon to regulate and/or legitimise property ownership, for Legal Realists, the notion of property can be reduced to one distinctive character: the right of exclusion (*in rem* right).

The functional and imaginative ease that the 'right to exclude' brings to property's messy and paradoxical nature brings a sense of social order that the courts are content to impose [31]. As Thomas W Merrill asserts, 'give someone the right to exclude others from valued resources…and you give them property.' [17] Deny someone the exclusion right, and they do not have property [16]. There is a neatness to this account of property ownership, which makes the individual the central concern rather than the material thing itself. Manoeuvring an individual into a strong pro-vested rights position of power armed against the rest of the world evokes a David and Goliath battle narrative. Armed with a stick to keep the rest of the world from physically interfering with the owner's land appears futile. Yet, it is not. Although the Legal Realists have constructed the right that befalls upon the individual, Simon Douglas and Ben McFarlane argue that this right, under a Hohfeldian claim-right tradition, binds the rest of the world to a duty not to interfere with an individual's ownership [15].

Holding the many (the world) to a rights-based duty that secures the property of individuals against derogation is an interesting characterisation of property ownership. Property ownership can be configured either by the individual themselves making the excluding or by the collective restraining their interference. The main point to grasp here is that the individual has rights that have strong legal power. The law considers some sticks in the bundle more valuable than others in different circumstances, but the right to exclude remains the definitive characteristic of property ownership.

The attraction of understanding property as a bundle of sticks postulates rights and limits the use of property rights, but it is not without its defects [1]. General discontent concerning the bundle of rights theory is that it presents a simplified view of property ownership, often overlooking the complex and dynamic nature of property rights. One significant criticism is its failure to adequately address the social and economic context within which property exists. The theory tends to focus on individual rights and entitlements, neglecting the broader societal implications of property ownership. This can lead to a neglect of issues such as equitable distribution of resources, environmental sustainability, and the impact of property on marginalised communities. Additionally, the bundle of rights concept can be overly rigid, struggling to accommodate the evolving nature of property rights in the modern era. The traditional bundle of rights picture is often challenged when new forms of property emerge, such as intellectual property. Regardless of these challenges, new forms of property or things remain subject to the rigorous bundle of rights analysis. There is little doubt that artificial islands and floating structures will be treated as property with legal exclusionary rights as a newer representation of resources.

3 Artificial Islands a Special Category of Property?

As advancements in constructing artificial islands and floating structures mature, the potential for these structures to alleviate land based pressures on resources and a population becomes a tangible prospect. The solutions these structures have already provided include creating new habitable solutions or offshore renewable energy sources.

However, the legal landscape must be more prepared alongside this technological progress. A robust legal framework will be required to govern the ownership, uses, exploitation, and environmental impact of these maritime structures. A myriad of legal issues will need to be addressed and resolved if these structures are to achieve their potential. One of the critical issues for a legal framework to succeed is jurisdictional matter over structures in open waters, as explored in Chap. 22. While international legal frameworks are important, this chapter will explore how individual countries approach the legal aspect of artificial islands and floating structures, highlighting the difference that emerges from national domestic property law.

3.1 *Artificial Islands*

There is little doubt that creating artificial islands was the first iteration of resolving the problems of overpopulation and displaced populations by expanding the mainland. Since the 1970s, numerous countries have expanded their lands by exploiting the technology within their territories [29]. Successfully and well-known inhabited artificial islands such as the Solomon Islands' Lau Lagoon, which contains approximately 60 artificial islands built on the reef off the northeast coast of Malaita Island, and the United Arab Emirates city of Dubai have developed and managed the largest artificial islands illustrate the viability of this structure. However, outside of these celebrated examples, artificial islands occupy a grey area in international law.

Artificial islands cannot claim territorial sea or constitute an exclusive economic zone and should not disrupt international navigation under the United Nations Convention on the Law of the Seas 1982 (UNCLOS) (Article 2). However, the UNCLOS does not address ownership, resource rights or environmental concerns surrounding these structures. Except Article 123 of UNCLOS requires a bordering national state to a closed sea to coordinate how implementing artificial islands may impact the other state's rights and obligations to protect the marine environment. Beyond maintaining good neighbourly relations, the clear lack of an international framework and principles subsequently means that the properterising nature of artificial islands is understood and classified by the respective domestic jurisdictions where these structures are located.

Artificial islands, as engineered spaces in water, represent an asymmetry in how the law struggles to understand and treat these resources. Constructing artificial islands within national territorial waters for purposes which ranging from urban expansion to public infrastructure inevitably alters a coastline's natural function, access, and appearance of real property. The creation of new landmasses in and on water demonstrates how two resources located in the global commons do not perfectly accord to the traditional bundle of rights description of property. Beginning with land, it has been a natural resource that needs to be acquired through various means. Historically, land ownership vested in the state originated from violent conquests. The outcome of these often long-projected battles was the enclosure of land—previously an undisturbed resource of the commons.

Taking land through force manifested a sovereign's desire to control, possess, and exploit a specific natural resource. Through exploitation, a sovereign would remove land from the commons and turn it into private property. Subsequent political agendas and institutional arrangements upheld a system of private property over time. Developing a system of private property involving land presents a striking juxtaposition of simplicity and complexity. The simplicity of property law systems has been encapsulated in the right to exclude outsiders. The powerful right allows owners to control access to a clear physical boundary often determined by the state. But, the simplicity is tested in various factual scenarios involving 'lumpy packages of legal relations' [24]. Measuring the *in rem* right by the quality of the exclusive possession afforded to an individual landowner or occupier (the individual) against the rest of the world always seems to be an endless David and Goliath battle.

The exclusion strategy employed by the individual in the battle against the world and, more specifically, against certain persons is often coordinated by a formal and interrelated contract and tort paradigm. From the Legal Realist perspective, knowing how to lean into the *in rem* construct o of the exclusion strategy is helpful in coupling the relationship of land with the individual. In some instances, while the pattern of coupling is linear, as seen when confirmation by transfer or certificate issuance confirming the *in rem* rights by the State, the pattern can also be blurry. Novel and new forms of things can challenge the linear coupling pattern to be indistinct, yet the pattern will endure to deliver upon a private end. Artificial islands provide a simple linear coupling exercise of affording good quality exclusive possession *in rem*. Manufacturing and replicating a natural resource of the commons into a definable physical parcel of land called an island embeds the *in rem* and possessory rights for personal interests enforceable against the world. Indeed, the world at large is very much kept at a distance.

Artificial islands are an unsubtle case study of the exclusion strategy in property theory. The individual's entire private relationship with the island land space (of course, subject to any respective covenants touching and concerning the land and community titling living arrangements found in contract, deed or torts within the highly secured physical bounds of the island) makes the historical David and Goliath battle largely redundant. Its loss of appeal is reflected in how easily the individual's private rights in the land are of such good quality it eliminates (or at the very least dampens) the interface with the rest of the world. Yet, it is a worthy reminder that cocooned private property rights in artificial land can only supported by the critical natural resource of water.

Bodies of water dominated by artificial islands can distort the boundaries of the global commons. Water, like land, can also be open to limited access as a natural resource belonging to the commons, but owing to its non-static nature, it is a resource that is more difficult to treat as a thing [25]. Smith's assertion corresponds with the thinking that if the exclusion strategy is the definitive characteristic of private property, he challenges how a tangible resource, such as water, can become an ownable source. In contrast, as Larissa Katz points out, land has a 'legal and physical imprint' [13] that easily permits exclusion, classifying water as a thing to properties relies heavily on other rights.

Water rights, unlike land rights, are typically predominately concerned with access. Access to water and taking the resource itself or from it manifests the stick of exploitation in the bundle of rights. Obsession with the exclusion strategy from the outside is less prevalent concerning water. The force of water can often have unpredictable flow, and a force makes it a challenge and, at times, an impossible source to control and capture. Indeed, the high labour and costs required to control and capture water in the open natural environment are to be created, and access rights to water must be given. The access to the resource gives water its properties and nature, which is very of the exclusive agenda. Affording accessibility rights to water is at the expense of excluding others. Indeed, the duplicitous strategies of access and exclusion are demonstrated in an artificial islandscape.

Without water, artificial islands are engineered parcels of land that are not a special property case. To elevate artificial islands as unique, water access is needed to construct the excludable land and access rights that facilitate crossing and recrossing and allow residents to use and enjoy water. Further, access (easement) rights to transgress across water from land reflect the exclusion strategy for owners and occupiers. The access rights of residents for personal use and enjoyment of water, coupled with access rights across the water to the artificial island, reimagines water no longer a resource of the commons. From a simple standpoint, water access rights are as important as the exclusion strategy in propertising artificial islands. In the absence of access water rights that touch and concern artificial islands, more than the exclusion strategy, as the hallmark of propertising a thing, would be needed to sustain or preserve good quality ownership rights.

Excludability has consistently prevailed in the hierarchy of rights within the conceptual framework of property as a 'bundle of sticks.' Nevertheless, the predominant right of exclusion within the paradigm of private property should not be excessively emphasised when addressing the issue of property rights applicable to an artificially constructed island landscape. The reliability of exclusion as a strategy will remain pertinent in the legal discourse surrounding artificial islands. Individually delineated parcels, defined by physical boundaries established on the engineered land, will grant the owner and occupant *in rem* exclusionary rights against the world. However, in the context of artificial islands, the access to water and the exclusion strategy should not be viewed as two rights directly competing with or against each other. Instead, these two rights working together in and around the engineered land reflect an additive understanding of property in the artificial landscape.

Critically, access rights and the exclusion strategy also work together regarding land. However, the predominant focus on the exclusion strategy often downplays the other rights in the bundle. Land being unlocked has always been accorded to the agenda not to 'disturb rights of private owners' (*Amodu Tijani v Secetary, Southern Nigeria*, [1921]); in doing so, the remaining bundle of rights is secondary and often subordinate. The usefulness of those secondary rights from the collection sticks becomes apparent in supporting the exclusion strategy, along with the unilateral actions of the owner, which gives functionalism to a private property system. Operating to preserve simplicity and refine complexity through duties, privileges, and, in general, *in rem* rights, the sticks in the bundle need to be paired.

Individualising and pairing the sticks in the bundle allows for the artificial island scape to sit outside any novel category of property and, at a basic level, attempts to sidestep any issues of complexity (perceived or otherwise) within a private property system. Yet the dynamism of the artificial landscape provides a different context and reality whereby the pairing of sticks will need to be linked. Plucking the water access stick from its bundle linked with the excludability stick from the general property bundle is one way of reflecting the context and reality of artificial landscapes, whereby without the water stick of access, the economisation of the *in rem* stick by the law will invariably be less (see Chap. 19 for value). Furthermore, in grasping the convenience of a paired or linked stick concerning the artificial landscape, the holder of the new stick has a unitary relationship with the engineered land and the natural water resource.

In presenting the idea of linking sticks, it would require the institutional interest of the state to have formal legitimacy. The punitive rights that linked sticks would afford an individual would be a matter postulated by political and moral theory in the face of the collective good. Indeed, there are drawbacks to having linked sticks. One obvious limitation is that the notion of collective good cannot allow private property ownership to be a higher right that will provide one group (typically the 'uber-rich') power and control over the global commons. The jural relationship between the uber-rich and the other groups that make up the collective (in the context of artificial landscapes) would be one of a power imbalance. Whatever challenges there may be to an individual's ownership power, the dispute will need to be adjudicated by a judge. Although the Legal Realist asserts that such disputes are likely to be determined by intuition, the more fact-specific grounds of decisions, which Realism promotes, are dominant in property law [5]. Adding to the facts are the social and cultural aspects that led judges to produce certain results. Indeed, the following two case studies, when read through a Realism lens, show how property rights are valued across two different jurisdictions and how these differences underline and direct the substantive law.

4 The Courts and Floating Structures

Courts have been known to use the bundle of rights picture as a pragmatic and functional approach to property law, emphasising the role of courts in shaping and/ or confirming property rights through judicial decisions. The critical legal question at a domestic level is whether or not a floating structure *realty*. The courts in the Netherlands and New Zealand jurisdictions specifically have been required to address this legal issue. The minimal judicial attention given to this query has produced two different outcomes, reflective of differences found in common law and civil property law systems. Indeed, both systems invite a separate and distinct analysis of substantive property rules and doctrines contained within a common law and civil tradition. While the respective domestic systems may not perfectly align, both systems were called upon to determine which category of property these structures belong to, *moveable* or *immovable*. The focal point of difference is that *civil law* holds that floating structures are personal and thus *moveable. In contrast,* the *common law* holds these as part and parcel of land, therefore *immovable*.

In 2012, in an extraordinary appeal (appeal in cassation), the Dutch Supreme Court of Justice was required to determine if a 'water house' is *movable* or *immovable* property within Article 3:4 of the Dutch Civil Code. The court's finding that a 'water house' is *moveable* (that is, a chattel, not realty) examined whether the 'water house' is permanently united with the ground and whether the intended nature and layout of the dwelling is to remain permanently in place (ECLI:NL:HR:2012:BV8198 (Supreme Court, 9 March 2012).

Interestingly, this decision turns on the court's application of Article 3:4 concerning the relational dimension the floating structure had with the ground and a seabed. The 'water house' was in a recreation park attached to two poles approximately 15 m long and driven seven meters into the subsoil. Two double brackets bolted onto the 'water house' allowed it to move freely up and down these poles with the changing water level. The 'water house' was connected to onshore utilities, such as electricity, telephone, water, gas, and sewage, by flexible connections and quick couplings. The Inspector of Taxation argued that the 'water house' was realty because the 'water house' and the recreation park were closely related and attuned to each other. In rejecting this argument, the court made it clear that the appeal would fail as an item intended to float must be held to be a ship within the meaning of Article 8:1 of the Civil Code.

The emphasis placed by the court on an item's ability to float meant no guarantees could be given that a 'water house' would ever be permanently united with the ground. Even though the 'water house' was in a recreation park and connected to utilities on the land, the court took the firm view that the recreation park (the land) should not be considered when determining whether a 'water house' is a *moveable* or *immovable* property under the Civil Code. Thus, drawing a bright line between what is *realty* and what is not. The court's determination has significant implications for settling urban life on the water, as it downplays the relationship of connection and purpose that a floating structure has with the land.

The reach of this decision was seen only two years later (ECLI:NL:GHSHE:2014:799.2014 (Court of Appeal-Hertogenbosh, 14 March 2014). In this latter case, the interested parties operated port terminals and storage areas in the municipality of Vlissingen. The port terminal comprised large (80 m, 85 m in width) and smaller (80 m, 39 m wide on the seaside and 13 m on the shoreside) pontoons that floated on the water. Both pontoons were designed to connect to the shore by two hooks. These hooks were part of the pontoon that fell over a steel tune (bank seat), bolted at both ends to a concrete abutment made from basalt blocks that separated land and water. Access to the pontoons was by means of a gangway.

The Court of Appeal determined that the pontoons cannot form part of the land (*immovable* property) and, again, emphasised that a structure needs a permanent connection with the ground and the nature and intent of the structure to remain permanently in place. The parties' factual argument concerning the purposeful relationship the floating platoons had with the road on the bank and that this was essential for their 'core business', in support of why the pontoons were part of the land, was rejected by the Court. Both of these decisions illustrate a reluctance to examine holistically the circumstances that could displace the characterisation of a 'water house' and pontoon as chattel. These decisions were informed by the Dutch Civil Code and the Court's application of it, which ignores one glaring problem concerning the right of superficies or an easement.

The law's broad qualification of a floating structure and construction as moveable property equated with vessels creates serious pragmatic and theoretical challenges for a system of private property. Classifying these structures as chattels, while seemingly straightforward, overlooks the complex interplay of rights and interests that arise in these unique maritime environments. This classification, rooted in the traditional dichotomy of movable and immovable property, fails to adequately address the *sui generis* nature of floating structures, which often possess characteristics of both. A vessel, by its very nature, is designed for mobility and is readily subject to relocation. Floating structures, however, while capable of some movement, are often intended for long-term or even permanent placement, functionally akin to land-based property. This distinction is crucial because it directly impacts the bundle of rights associated with these structures, particularly the interplay between excludability and access.

The Dutch approach, by focusing solely on the physical ability to float, effectively severs the floating structure from its surrounding environment and the network of dependencies that give it practical value. This narrow interpretation overlooks the functional integration of the structure with the land, the utilities it connects to, and the access it requires. It diminishes the importance of the *purpose* and *use* of the structure, focusing instead on a limited physical characteristic. This creates a legal fiction where a structure, while functionally tied to the land and essential infrastructure, is treated as a mere chattel, adrift in a private property law system.

This classification as movable property significantly weakens the "excludability" stick in the bundle of rights. While the owner may possess the structure itself, their ability to control access to it, and more importantly, the *water space* it occupies, becomes significantly compromised. Unlike land-based property, where the boundaries are clearly delineated and enforced by the state, the owner of a floating structure faces a much more ambiguous situation. The water surrounding the structure, while essential for its function and use, remains part of the public domain, accessible to others. This raises complex questions about the extent to which the owner can exclude others from approaching, mooring alongside, or otherwise interfering with the structure.

Furthermore, the movable property classification creates difficulties in establishing and enforcing rights related to access. Access to a floating structure often requires traversing the surrounding water, which raises questions about easements, rights of way, and the regulation of navigation. If the structure is deemed movable, it becomes difficult to establish any long-term or stable access rights across the water, as these rights are typically tied to land. This can create significant uncertainty for the owner and hinder the development and use of floating structures. Imagine a floating home connected to the mainland by a dock. If the home is movable property, can the homeowner claim a right of access across the water to their dock? Can they prevent others from anchoring nearby, obstructing their access, or interfering with their use of the surrounding water?

This is where the concept of "linked sticks," as previously discussed, becomes particularly relevant. The traditional "bundle of sticks" metaphor, with its emphasis on excludability, needs to be adapted to account for the unique characteristics of floating structures. In this context, the "excludability" stick must be inextricably linked to the "access" stick. The right to exclude must be complemented by a clearly defined right of access across the water, secured through easements, licenses, or other legal mechanisms. This "linked stick" would recognise the functional interdependence of the structure and the water space it occupies, providing a more robust and practical framework for property rights in these environments.

The absence of such a "linked stick" creates a legal vacuum that can lead to disputes, uncertainty, and ultimately, hinder the development of innovative maritime solutions. The Dutch cases, by prioritising the movable/immovable dichotomy over the functional realities of floating structures, illustrate this problem. By focusing on the ability to float, the court effectively ignored the crucial connection between the structure and the water space it occupies, leaving the owner with a weakened bundle of rights and uncertain access. This approach fails to recognise that for a floating structure to function as intended, the owner needs more than just possession of the structure itself; they also need a degree of control over the surrounding water space, including secure access rights. The legal framework must evolve to acknowledge this reality and provide a more nuanced approach to property rights in the context of floating structures.

Indeed, The Dutch court's decision, viewed through a Legal Realist lens, can be interpreted as reflecting a deep-seated cultural connection to water and a historical acceptance of floating dwellings. The Netherlands has a long tradition of living on and alongside water, with houseboats and other floating structures a common sight in its canals and waterways. The court's emphasis on the "movable" nature of these structures, even when connected to land-based utilities, could be seen as an affirmation of this cultural heritage. Legal Realism suggests that judicial decisions are not solely based on abstract legal principles but are also influenced by social context and prevailing cultural norms. In this case, the court's decision, while seemingly technical in its application of legal definitions, can be understood as implicitly acknowledging and reinforcing the Dutch cultural preference for a fluid relationship with water, a relationship that transcends the rigid categories of movable and immovable property. By upholding the movable status of floating structures, the court may be subtly preserving a traditional way of life and reflecting a societal value placed on the freedom and adaptability associated with living on water. This interpretation aligns with the Realist emphasis on examining the "actual workings" of the law and its impact on society, rather than focusing solely on formal legal rules. The court's decision, therefore, can be seen not just as a technical legal determination but also as a reflection of Dutch cultural identity and a pragmatic accommodation of a long-standing practice. However, the approach taken in New Zealand, which was factually analogous to the second case from the Netherlands, had a different focus and approach towards floating structures.

In 2000, New Zealand's Court of Appeal (Court of Appeal/Court) discussed how certain floating facilities can be *realty* [30]. The significance of the Court of Appeal's decision in property law is two-fold. First, the common law entertains how a floating structure may be part and parcel of land capable of affording rights and interests in and on the floating structure. Secondly, the alternative discussion of the exclusive right of occupation in floating facilities is not only confined to the seabed but also airspace when floating pontoons move with the tide.

The case of *Auckland Ports* involved two main terminals: multi-cargo wharves used to load and unload cargo and embark and disembark passengers in the Port at Auckland within Waitemata Harbour. The Port's berthage provided berths, pile moorings and swing moorings. Some berths have a finger at right angles to the central pier, and the pier, as a floating pontoon, is attached by rings to locate piles driven into the seabed. The pontoons moved vertically with the tide and can be accessed by foot over bridges hinged to the ground at the landward end and rest on the unattached pontoon. Utilities such as power, water supply and lighting are connected to the piers.

One of the grounds of appeal was whether certain facilities in the water space were *rateable property* as fixtures annexed to the land. The greater the *rateable property,* the greater the rates required to be paid. Although the Court of Appeal found that the floating jetties and pontoons were not structures or land owing to their location outside the city's boundaries, and thus not *rateable property*, the Court went some way in considering whether, if the structures were instead located within the City of Auckland's boundary, these would have been chattels (*moveable* property) or fixtures (*immovable* property).

Noting the absence of a precedent with comparable facts, the Court relied upon the traditional two-prong test examination of the *degree of annexation* and the *object of annexation* [18]. In identifying that, prior to their attachment by rings to the locating piles, pontoons were *moveable* property, the Court went on to opine that once pontoons are attached to the locating piles, these become integral parts of a composite jetty structure and, thus, *immovable*. In considering the jetty's composition, the NZ

Court (in contrast to the Netherlands) noted the rings keep the pontoons permanently in place, which makes them *immovable* and unable to be separated from the locating piles. Embedding locating piles in the seabed reflects the jetty's degree of annexation, and the use of the jetty's structure depends on maintaining that *degree of annexation*. Incorporating the pontoons into the jetty structure as a whole evidenced the *object of annexation,* which was to effect and fully enjoy the rights of those who occupied the water space.

While this case primarily deals with rating legislation, its core revolves around defining the scope of pontoons and determining what constitutes land used for loading and unloading, which has implications for access rights and the "linked stick" concept.

The Auckland City Council argued that the exemption from rates should apply only to the immediate quayside strip where goods are physically handled. They wanted a narrower definition to increase the rateable land within the port. The court, however, adopted a broader interpretation, recognising that modern port operations require extensive backup areas for container marshalling. The court determined that the process of loading and unloading begins when cargo is moved to stacking areas *awaiting* the vessel and concludes when imported cargo reaches similar stacking areas. This effectively extended the exempt pontoon area beyond the immediate quayside.

This broader definition has implications for access because it recognises that the *functional* wharf includes more than just the point of contact between ship and shore. The stacking areas, while not directly adjacent to the water, are essential for the access rights to load and unloading process. This suggests an implicit recognition of a "linked stick" principle: the right to use the quayside (access) is inherently linked to the right to utilise the adjacent land for marshalling and storage (excludability), even though those areas might be physically separated from the water. The court's decision acknowledges the interconnectedness of these rights in the context of modern port operations.

The case also considered the status of floating structures in Westhaven Marina. The council argued these should be included in the city's rateable property. The judgment discussed whether these floating structures were within the city's district and whether they qualified for the wharf exemption. However, the provided text does not give the final decision on this. If the floating structures were deemed part of the city's rateable property, the question would then become whether they are "piers or jetties used as a wharf," which would bring the "linked stick" concept into play again. If they are considered wharves, the access rights to the water surrounding them would be inextricably linked to the right to exclude others from the floating structures themselves.

This New Zealand case does reveal a different approach to property compared to the Dutch cases previously discussed. Viewed through the Legal Realist lens the Dutch court seemed to prioritise a formalistic application of legal categories, potentially reflecting a cultural affinity for the fluidity of water, the New Zealand court's decision demonstrates a more pragmatic, functional approach. The *Auckland*

City Council case emphasises the *actual operations* of the port and the interconnectedness of land and water in modern shipping. Rather than adhering to a strict, physical definition of real property court looked at the practical realities of cargo handling, recognising that the pontoon extends beyond the immediate quayside to include essential backup areas. This focus on the "real world" practices and economic necessities of port operations aligns with Legal Realism's emphasis on the factual context of legal disputes. The court's decision can be seen as a reflection of New Zealand's reliance on maritime trade and the economic importance of efficient port operations. In contrast to the Dutch cases, where cultural values surrounding water seemed to influence the court's interpretation of property rights, the New Zealand case highlights the influence of economic realities and practical considerations in shaping legal outcomes. This contrast underscores a key tenet of Legal Realism: that law is not merely a set of abstract rules, and that property law is an important means to achieve social and economic objectives.

5 Conclusion and Future Direction

As the world grapples with the increasing impacts of climate change, innovative solutions like artificial islands and floating structures (FS) have emerged as potential responses to land scarcity and displacement. While these engineered spaces offer promise, they also present complex legal and theoretical challenges. This chapter has explored the nature of property rights in artificial landscapes and FS within a system of private property, particularly focusing on the right of exclusivity.

The traditional understanding of property rights, grounded in the individualised right of exclusivity, may not be entirely suitable for these novel spaces. Unlike land, which is a relatively stable and finite resource, artificial islands and FS are dependent on water, a dynamic and often unpredictable element. This dependence necessitates a nuanced approach to property rights that accounts for the unique characteristics of these engineered structures.

To ensure the successful development and utilisation of artificial islands and FS, a clear and coherent property framework is essential. This framework should balance individual property rights with broader societal interests, such as environmental protection and equitable resource access. By fusing specific sticks from the bundle of rights, it is possible to create a new property regime within the context of artificial landscapes.

Future research should delve deeper into the specific legal and regulatory implications of artificial islands and FS. Comparative analysis of different jurisdictions can provide valuable insights into best practices and potential pitfalls. Additionally, interdisciplinary approaches, involving law, engineering, environmental science, and economics, can offer a more comprehensive understanding of the challenges and opportunities associated with these innovative spaces.

Indeed, as new land technologies continue to evolve, so too must our understanding of property rights. By embracing new technologies and adapting traditional legal concepts, we can ensure that artificial islands and floating structures contribute to a sustainable and equitable future.

References

1. Anderson G (2022) Property law—concepts and doctrine. Chatswood (New South Wales), LexisNexis
2. Anderson G (2022) Property law: concepts and doctrine. Chatswood (New South Wales), LexisNexis Australia
3. Bix B (2008) Legal realism. In: The New Oxford companion to law. Oxford University Press, Oxford
4. Blackstone W (1765–69) Commentaries on the laws of Engalnd. Clarendon Press, Oxford
5. Dagan H (2015) Doctrinal categories, legal realism and the rule of law. Univ Pa Law Rev 163:1889–1917
6. Edgeworth B (2021) Sickville and Neave Australian property law, 11th edn. LexisNexis, Chatswood, New South Wales
7. Fraser D (1988–89) What a long, strange trip it's been: deconstructing law from legal realism to critical legal studies. Aust J Law Soc 5:35–40
8. Gilmore G (1961) Legal realism: its cause and cure. Yale Law J 70(7):1037–1048
9. Gray J (1998) False dawn: the delusions of global capitalism. The New Press, New York-London
10. Gray K (1991) Property in thin air. Cambridge Law J 50(2):252–307
11. Hohfeld WN (1913) Some fundamental legal conceptions as applied to judicial reasoning. Yale Law J 23:16–59
12. Honore A (1961) Ownership in Oxford essays in jurisprudence. Clarendon Press, Oxford
13. Katz L (2008) Exclusion and exclustivity in property law. Univer Toronto Law J 58:276–313
14. Locke J (1884) Two tresties of government. Routledge and Sons, New York
15. McFarlane SDAB (2014) Defining property rights. In: Philosophical foundations of property law. Oxford University Press, Oxford
16. Merrill TW (1998) Property and the right to exclude. Nebraska Law Rev 77(4):730–731
17. Merrill TW (1998) Property and the right to exclude. Nebraska Law Rev 77(4):736
18. National Provincial Bank Ltd v Ainsworth (1965)
19. Penner JE (2020) Property rights: a re-examiantion. Oxford Universtiy Press, Oxford
20. Raddin M (1938) A restatement of Hohfeld. Harv Law Rev 51:1141–1164
21. Simith A (1723) Lectures on jurisprudence. Glasgow: s.n
22. Smith HE (2011) Property is just not a bundle of rights. Econ J Watch 8(3):279–291
23. Smith HE (2012) On the economy of concepts in property. Univ Pa Law Rev 160:2097–2128
24. Smith HE (2012) Property as the law of things. Harv Law Rev 125:1691
25. Smith HE (2012) Property as the law of things. Harv Law Rev 125(7):1691–1726
26. Smith HE (2015) The presistence of system in property law. Univ Pa Law Rev 163(7):2055–2083
27. Smith Y-CCAHE (2012) An economic analysis of civil versus common law property. Notre Dame Law Rev 88(1):1
28. Wang CWAB (2008) Colonization of the ocean and VLFS technology. In: Very large floating structures. Taylor and Francis, Abingdon
29. Yusuf BVAMS (2021) The legality of Denmark's artificial Island (Lynetteholm) in iternational law of the sea. Padjadjaran J Int Law 5(2):188–205

Case Law

30. Auckland City Council v Ports of Auckland (2000) 3 NZLR 614

31. Dolan v City of Tigard 512 US 374 (1994)
32. Knapp v Knapp (1944) SASR 257

Kim D. Weinert is a legal academic and lawyer. Kim researches the intersection of law, society, and culture. She has a particular focus on charity law, not-for-profit governance, and property law. Kim's not-for-profit research has been cited by the Queensland Supreme Court and the Western Australian parliament. Kim was the recipient of the Australian Postgraduate Award (APA) Scholarship at Griffith University, where she completed her doctoral thesis on speech legality in Australian culture. Kim's research interest in law and culture in Australia and the South Pacific has been published in international journals. The exploration of cultural legalities in Kim's research provides a unique perspective on the impact doctrinal law has on minority groups. Through this lens, Kim can identify and address numerous shortcomings in the law which paves the way for legal and governance reform. Exploring the traditional boundaries of doctrinal law and its theoretical foundations in relation to emerging and new societal challenges, her research unites several strands of legal inquiry linked to social justice and equality. Her charity law research has been cited by the Queensland Supreme Court, the Western Australian Parliament, and eminent legal charity scholars. The impact of Kim's research has produced an impact on how charity law is discussed and applied in Australia.

Buoyant Futures: Our Global Duty to Recognise Floating Cities as 'Territory' in International Law

Brydon T. Wang

Abstract This chapter explores the international legal framework required to support floating cities as adaptive solutions to rising sea levels and as alternatives to the displacement of coastal populations. While climate change poses significant risks to low-lying coastal areas, Small Island Developing States (SIDS) face an additional existential threat as sea level rises challenges established principles of statehood, sovereignty and maritime rights under international law. Floating cities, as innovative urban environments, offer a potential pathway for preserving statehood and addressing global environmental challenges. However, the classification of floating structures under frameworks such as the United Nations Convention on the Law of the Sea (UNCLOS) remains ambiguous, complicating their governance and integration into international law. This chapter argues that recognising floating cities as sovereign extensions of submerged territories should be considered an *erga omnes* obligation, reflecting the collective responsibility of states to address climate-induced displacement and uphold fundamental human rights. By examining structural characteristics, jurisdictional complexities and emerging legal norms, this chapter highlights the need for international cooperation to develop a unified framework for floating cities. Such recognition would ensure the preservation of maritime zones, support the rights of vulnerable populations and foster global resilience to environmental change.

Keywords Erga omnes · International law · Floating cities · Sovereignty · International environmental law

B. T. Wang (✉)
Centre for Policy Futures, The University of Queensland, St Lucia, QLD, Australia
e-mail: Brydon.Wang@uq.edu.au

Deal Advisory and Infrastructure, KPMG Australia, Brisbane, QLD, Australia

B. T. Wang et al. (eds.), *Large Floating Solutions*, Lecture Notes in Civil Engineering 636, https://doi.org/10.1007/978-981-96-5435-2_22

1 Introduction

Floating cities fall within the broader concept of 'buoyant urbanism', defined here as a spectrum of construction technologies that utilise the buoyancy force of water to enable urban development to respond to the incursion of water (floods, tidal inundation, sea level rises etc) or to allow a city to stretch out and float over its adjacent water body. These buoyant urban solutions, such as amphibious housing or Very Large Floating Structures (VLFS), have emerged as a pressing and necessary part of our toolkit for addressing the impact of climate change and coastal erosion on our urban populations.

Climate change is fundamentally altering the relationship between humanity and the environment, bringing with it rising sea levels, a wetter climate that is coupled with changes to precipitation patterns and increasingly severe flooding [10]. Populations living within the low-elevation coastal zone (LECZ) are particularly vulnerable to climate change risks. This zone encompasses coastal areas of continental land (including land hydrologically connected to the sea) and islands situated less than 10 m above sea level. More than a tenth of the world's population resides in this zone. The risks are particularly acute for large cities as nearly 'two-thirds of urban settlements with populations greater than 5 million fall, at least partly, in the zone' [15].

At the same time, river delta environments are experiencing progressive subsidence due to human activities through groundwater or hydrocarbon extraction processes and land reclamation, or through natural geological processes like compaction and sediment loading [5]. With global urbanisation continuing to exert pressure on available coastal land, our coastal cities (particularly around river deltas) and low-lying urban areas face a dual crisis of urban land scarcity and the threat of territorial loss to the increasing incursion of water in our cities.

The 2023 IPCC Synthesis Report highlights the escalating frequency and intensity of severe weather events, with devastating consequences for vulnerable communities worldwide. In some regions, adaptive capacity is nearing its limits, underscoring the urgent need for innovative climate-resilient development. For example, Small Island Developing States (SIDS) face existential threats, with some countries projected to lose significant portions of their territory by the middle of this century [17, 38, 9].

These environmental challenges compound existing socio-economic pressures, deepening poverty and threatening the livelihoods of island populations [35]. As land-based adaptation measures are increasingly rendered ineffective, communities are faced with displacement [4]. While some advocate for these climate-induced community relocations, buoyant urbanism offers a pathway for adaptation that transcends traditional terrestrial urban development, presenting a compelling solution to these challenges. Of these solutions, the more ambitious proposals for floating cities are not merely engineering feats but holistic responses to the socio-environmental demands of the twenty-first century. These floating city proposals are often conceived as modular, with adaptive designs that allow for resilience in the face of environmental changes, while offering mobility to enable unprecedented flexibility in urban planning [36].

The focus of this chapter is to explore the international legal frameworks necessary to support floating cities, balancing innovation, sovereignty and environmental governance. It begins by exploring the potential for floating cities to serve as an urban solution for vulnerable populations living in the LECZ. The chapter then turns its focus to SIDS, highlighting the additional existential threat of loss of status under international law and how floating cities might serve to address the loss of territory and sovereignty of submerged states. The chapter progresses to address the legal ambiguities surrounding the legal concept of territory, and the classification of floating structures and floating cities in international law. It then argues that there is an urgent need for recognition of these floating solutions as a viable alternative to climate-induced displacement, one that stems from a global duty owed by all states to assist SIDS as they face the loss of territory and related maritime zones crucial for their socio-economic well-being. Finally, this chapter suggests how a unified framework to support floating cities could be achieved, before presenting a conclusion that argues for the need for global efforts to embrace these floating solutions.

2 Floating Cities as an Urban Solution for the LECZ

The preceding chapters in this edited collection have raised the various potential application of floating structures to demonstrate how cities can adopt buoyant urban solutions to address the challenge of climate change. These contrast with more conventional approaches to mitigating sea level rises and the substantial resources that have been invested in risk reduction and hazard prevention. For example, the Netherlands has implemented a comprehensive system of dams, sluices, locks, dikes and storm surge barriers designed to protect low-lying coastal areas from the sea [37, 2]. Singapore has similarly constructed water control structures, including the *Marina Barrage*, a dam that serves multiple purposes of flood control, water supply and as leisure infrastructure, converting two bays (the Marina Bay and Kallang Basin) into a freshwater reservoir. Singapore has also investigated hybrid mangrove-rock revetments [12] and will embark on the *Long Island* project, where an artificial island will be constructed along its southeastern coast to protect low-lying shorelines from sea-level rise [18]. The Maldives has constructed artificial islands like *Hulhumalé* and protective infrastructure such as the *Great Wall of Malé* [8].

However, many of these conventional measures may be insufficient to counteract climate-driven erosion, coastal flooding and water incursion in the long-run without further build-up of infrastructure. For example, artificial islands are man-made landforms constructed in water bodies by piling materials like sand to create fixed structures above the waterline. The static nature of these artificial islands limits their ability to adapt to rising sea levels. These land reclamation projects are also vulnerable to subsidence. Many of the coastal cities that built airports partly on land reclaimed from the sea to address the lack of flat ground have experienced problematic ground subsidence [5]. In particular, the Kansai Airport artificial islands continue to experience larger in-situ subseabed consolidation exceeding what was anticipated in the laboratory, leading to significant settlement. Mesri and Funk [16] suggest that by the turn of this century, the Kansai Airport islands will settle between 17.6 m and 24.4 m—a particularly striking figure given the islands were designed to be about 4 m above sea level.

In contrast to artificial islands that are anchored to the seafloor (bottom-founded), floating cities and their component structures are buoyant and capable of moving with tidal and sea-level changes. They are also earthquake-resistant and do not lead to the complete destruction of the ecological environment below their footprint (i.e. the environment within the subjacent water column) [36]. Floating cities can take a number of structural forms, such as pontoon or semi-submersible structures [36]. These urban developments are designed to support permanent human settlement, and as such encompass residential, commercial and recreational spaces as discussed in Chaps. 2, 15 and 17. By incorporating renewable energy systems, advanced waste management and sustainable food production techniques, these floating city proposals aim to minimise their environmental impact and engage with sustainable practices. These innovative approaches to making a city more climate resilient demonstrate how human habitation can adapt to environmental changes without exacerbating existing challenges.

Further, these floating city proposals align with several Sustainable Development Goals (SDGs), including SDG 11 (*Sustainable Cities and Communities*), SDG 13 (*Climate Action*), and SDG 14 (*Life Below Water*). Given that they do not lead to wholesale destruction of the underwater environment and can incorporate and expand ecosystem services (see Chaps. 14 and 16), they also align with the *UN Environment Programme's* emphasis on ecosystem-based adaptation measures, which promote biodiversity, health outcomes and economic resilience. Consequently, these buoyant urban solutions offer vulnerable coastal populations the opportunity to remain connected to their socio-economic and cultural environments, while adapting to changing realities through sustainable and resilient floating solutions [35].

However, the legal recognition and regulation of floating cities remain underdeveloped, presenting both challenges and opportunities. Traditional legal systems that are anchored in land-based conceptions of property and sovereignty, struggle to accommodate the hybrid nature of floating cities, particularly its mobile nature and ability to be reconfigured. These structures challenge existing definitions of territory, jurisdiction and ownership (as discussed in Chaps. 20 and 21), necessitating a fundamental shift in legal thought. At the heart of the legal challenge lies the ambiguity in classifying floating cities. Are they territory, vessels, floating installations, artificial islands or extensions of naturally formed land? Each classification carries distinct implications for sovereignty, jurisdiction and governance. For example, under the *United Nations Convention on the Law of the Sea* (UNCLOS) [30], the regime of global law and order for the sea, artificial islands do not generate territorial seas or exclusive economic zones, while vessels are governed by flag state jurisdiction [33, 34]. These classifications, however, fail to capture the hybrid nature of floating cities, which are both mobile and yet designed for permanence and human habitation and can be built as extensions of the city over the water.

Domestically, floating cities face challenges in property law. As Weinert discusses in Chap. 21, the distinction between real property (land) and personal property (objects, such as vessels) does not easily apply to structures that are stationary (that is, they might be fixed in place) and yet have the potential to be mobile, or structures that may be moored and rise up and down vertically with water level changes. Case law, such as New Zealand's *Auckland City Council v Ports of Auckland Ltd* and the Netherlands' decisions on floating pontoons, highlight the inconsistencies in judicial interpretations [33, 34].

The regulatory landscape for offshore wind development in Australia offers valuable insights into the evolving governance of maritime spaces and the legal ambiguities surrounding floating structures. Offshore wind energy projects, like floating cities, require a legal framework that accounts for their unique relationship with maritime zones, environmental impact and jurisdictional challenges. Recent legislative developments, including the Offshore Electricity Infrastructure Amendment Regulations 2024 (Cth), illustrate how governments are refining regulatory frameworks to support large-scale offshore infrastructure. The introduction of feasibility licences and regulatory amendments to address overlapping applications, demonstrate a growing recognition of the complexities associated with allocating rights in maritime environments. These regulatory mechanisms highlight the need for legal clarity in balancing environmental protection, stakeholder interests and economic feasibility, a challenge equally relevant to floating cities.

Legal uncertainties will be challenging to untangle. Take for example the uncertainty in respect of the classification and governance of offshore wind infrastructure. A recent Federal Court decision (*Seadragon Offshore Wind Pty Ltd v Minister for Climate Change and Energy* [2024] FCA 1290) underscored the challenges of development in marine spaces, as the court overturned a ministerial refusal to grant a feasibility licence due to ambiguities in the *Offshore Electricity Infrastructure Regulations 2022* (Cth). In response, legislative amendments were enacted to provide greater regulatory certainty, but these amendments effectively reversed the successful judicial review of the ministerial refusal. These legal developments suggest that similar ambiguities of the legal status of floating cities will need to be addressed, particularly regarding zoning, property rights and long-term governance structures. At present, these ambiguities hinder investment and complicate the integration of floating cities into existing legal frameworks [6]. Just as offshore wind projects require coordinated national and international legal frameworks to ensure their viability, floating cities will depend on the evolution of maritime governance to support their integration into legal and regulatory systems.

3 The Threat to the Legal Status of Small Island Developing States

Shifting the discussion to global legal frameworks and the threat to SIDS, 'statehood' is the fundamental legal status that defines an entity as a subject of international law. Such statehood enables full participation in the international community, including membership in international organisations, treaty-making, and the assertion of rights and obligations under international law. The 1933 *Montevideo Convention on the Rights and Duties of States* outlines four criteria for statehood: a permanent population, a defined territory, a form of government, and the capacity to engage in international relations with other states. The loss of land due to sea-level rise challenges these criteria, particularly the requirement for a 'defined territory', raising significant implications for the statehood and maritime zones of SIDS.

While a state retains its statehood as long as it has 'defined territory'—that is, even where its defined territory is substantially diminished by coastal inundation—the total loss of landmass through submergence has profound implications under international law. As such, the fundamental threat to SIDS is that total loss of territory from inundation will lead to extinguishment of statehood, sovereignty and international legal personality. For the global community, the disappearance of a state without a successor challenges the established norms in international law and raises concerns about the protection of sovereignty, self-determination and basic human rights [39]. Moreover, United Nations membership, the ability to participate in treaties, and the allocation of maritime zones under UNCLOS are all inherently linked to a nation's statehood.

Recognising these challenges, the *International Law Commission* (ILC) began examining the issue of sea level rise in 2018 as part of its long-term programme of work. This initiative was formalised in 2019 with the establishment of an open-ended Study Group, which has since produced several key issues papers [31]. These papers address the implications of sea-level rise for the law of the sea, statehood and the protection of individuals affected by such changes. The ILC's work illustrates an emergent effort within the international community to tackle the unique challenges faced by SIDS as a result of climate change.

Legal scholars and international bodies have explored whether a state can retain its international legal personality despite the loss of its physical territory. The principle of state continuity, traditionally applied to states experiencing changes in government or territorial configuration, has been proposed as a basis for recognising states that lose their territory due to environmental changes [21]. This concept of 'de-territorialised statehood' has emerged as a potential solution, allowing a state to maintain sovereignty and international legal rights even in the absence of physical landmass. This could involve the relocation of government functions to offshore territories or the use of floating structures to preserve a physical presence within the boundaries of the submerged landmass.

However, the practical and legal implications of de-territorialised statehood remain uncertain. International obligations, such as those under the *International*

Covenant on Civil and Political Rights (ICCPR), are tied to individuals 'within its territory and subject to its jurisdiction,' raising questions about how such obligations would be upheld in a de-territorialised context. Additionally, the affixation of maritime baselines to preserve maritime zones for submerged states, though increasingly discussed, poses significant challenges.

4 Legal Ambiguity in Defining Territory and Maritime Zone Entitlements that Flow from Such Territory

There is a level of uncertainty at international law as to what is the definition of territory at international law. This fundamental question of what is 'territory' encompasses a range of issues that bears impact on how and when human intervention—either through artificial islands or floating structures (and other maritime installations)—can be 'built, used, and claimed by states, as well as how international law doctrines interact with such [purported "territories"]' [22].

However, the legal concept of 'territory' is one that is based on both population requirements (human needs) and historical development [25]. Indeed, Shaw's observations that the concept of territory is not just an articulation of relationship between subjects of international law, but also a '[reflection of] the relationship between the people and the geographical space they inhabit' is echoed separately in Cradduck and Weinert et al.'s work in Chaps. 19 and 20. But it is this historical development in response to technological and environmental changes that bears examination. Where emerging iterations of technology enable access to new areas, such as the Antarctic, the continental shelf, or artificial islands, the law must then consider how to incorporate these areas with its existing jurisdiction; and its corresponding impact on the concept of territory [22]. Floating cities represent the next phase of this evolution, challenging the international community to redefine the criteria for territory. While artificial islands provide a precedent for territorial recognition without maritime entitlements, floating cities demand broader considerations of mobility, sustainability and resilience.

Judge Alvarez observed in the *Corfu Channel* case that floating spaces could fall within the ambit of the definition of territory based on the advancement of technologies. His Honour posited:

> By sovereignty, we understand the whole body of rights and attributes which a State possesses in its territory, to the exclusion of all other States, and also in its relationship with other States. Sovereignty confers rights upon States and imposes obligations on them.
>
> These rights are not the same and are not exercised in the same way in every sphere of international law. I have in mind the four traditional spheres—terrestrial, maritime, fluvial [rivers] and lacustrine [lakes]—to which must be added new ones—aerial, polar and *floating* (floating islands). [emphasis added]

However, it is this intervention of technology and its effect on the concept of territory that states are cautious about. In the negotiations prior to the adoption of

UNCLOS, states intentionally drafted in language separating artificial islands from naturally formed land. This was to categorise these landforms with other human interventions and installations such as 'floating works, anchored buoys, etc.' [11], and to ensure that no ability to generate territorial sea would follow. Stoutenburg [27] notes, however, that while artificial islands might not generate territorial sea, they are still treated as part of the sovereign territory of a state, unlike artificial installations, such as floating structures, that would not typically even qualify as 'territory', and so lie one step removed from being able to generate maritime zones.

Under UNCLOS, maritime zones—including territorial seas, contiguous zones, exclusive economic zones (EEZs), and continental shelves—are measured from ambulatory baselines, which shift landward with rising sea levels. For SIDS, these zones are vital as their EEZs often constitute a significant proportion of their economic resources. For instance, the average EEZ of SIDS is 28 times the size of their land area, making access to ocean-based resources critical for the economic and environmental sustainability of SIDS [31].

In response to these challenges, four options for addressing maritime baselines in international law have been proposed [1]. First, is a hands-off approach where the current legal approach is preserved. This is the other side of the coin of the international legal concept of 'accretion', where territory is shaped and formed by natural processes, allowing states to acquire new territory [24]. In this first option, the loss of natural land through environmental forces is seen as part of the ambulatory nature of baselines. However, under this option, where sea level rises lead to the full submergence of states, this would result in the total loss of the entire territory of the low-lying state and the loss of its baselines and maritime zones.

Second, adopting straight baselines could help states maintain some maritime zones despite partial submersion. In this approach, baselines are not set up to cling to the granular contours of the coastline. This means that where low-lying areas are inundated, minor shifts of the coastline landwards may not cause the overall straightened baselines to retreat. This approach would be valuable to delta regions such as the Mekong and the Ganges–Brahmaputra–Meghna [14], and assist states with large coastlines, but will not assist states with entirely submerged territories.

Third, affixing current maritime baselines has been advocated by many Pacific island states through domestic legislation and international declarations, such as the 2015 *Taputapuatea Declaration on Climate Change* [20] and the *2018 Delap Commitment on Security of Commonwealth of Oceans* [7]. This position assumes that the original territory generating the maritime baseline is still present, albeit diminished and not fully submerged. However, without the international community supporting this approach, such public-facing announcements may not provide these Pacific island states with the legal certainty they desire.

The fourth option looks to fully affix baselines to preserve maritime zones for de-territorialised states. This approach would allow coastal states to retain their maritime zones even if their physical landmass is fully submerged. This last option offers a potential solution to inundation for these low-lying states, but it lacks consensus on its legitimacy and feasibility. While some nations and legal commentators support this approach to protect the rights of vulnerable states, the affixation of baselines has yet to be codified in international law, leaving the future of maritime rights uncertain. Critics argue that such measures conflict with the ambulatory nature of baselines under UNCLOS and could set a precedent that destabilises the broader legal framework governing maritime zones [35]. Fundamentally, this approach still leaves the state without a physical territory.

Given the unprecedented challenges posed by climate change, there is a growing recognition for innovative legal frameworks to address the specific needs of island states and the evolving concept of territory. This chapter argues that there is a fifth option: floating structures located within fixed maritime baselines, positioned over inundated land, could serve as 'defined territory' under the *Montevideo Convention* as a means of preserving statehood and sovereignty for nations facing the loss of their land. Recognising floating cities as territory is proposed as an *erga omnes* obligation—an obligation owed by all states to the international community [29]. Such recognition would enable SIDS to preserve their EEZs and maritime zones, ensuring their sovereignty and access to critical resources despite the loss of terrestrial landmass.

If we consider artificial islands present within the context of this question of statehood, there are parallels between floating cities and artificial islands that stem from both solutions emerging out of the unique intersection of technological advancement and evolving sovereignty norms. Saunders [22] observes that artificial islands can constitute territory under international law if they meet certain criteria, including being located within the state's EEZ, and exhibiting stability and state occupation. This interpretation could complement the proposed recognition of floating cities as sovereign surrogates of submerged territories, particularly for SIDS—although the full loss of natural land through inundation may mean that a floating city is not located within the extinguished state's EEZ (because neither the state nor EEZ would exist). However, while artificial islands are fixed structures built through reclamation, floating cities will struggle to meet the criteria of 'stability' given their buoyant structural characteristics challenge traditional conceptions of stability and permanence in territorial law. Accordingly, this legal solution necessitates an examination of how the law considers the geographic location and structural characteristics of these floating structures and cities.

5 Legal Ambiguity in Defining Floating Structures in International Law

The classification of floating structures under international law is fraught with ambiguity as it straddles multiple legal categories and frameworks. International law currently lacks a specific legal definition for floating cities, leaving their classification to be inferred from broader categories such as artificial islands, installations or vessels. While Article 11 of UNCLOS excludes offshore artificial islands, installations, and structures from affecting the calculation of baselines for maritime zones, it does not provide a definition for 'artificial islands' or 'installations' [22]. As such, the language of UNCLOS provides no clarity on the legal status of floating cities that are not bottom-founded (that is, what are they, and whether they can be 'territory'?). There are two primary questions at the heart of the definitional challenge for floating cities. The first is a matter of their geographic location; the second, their structural characteristics.

5.1 Geographic Location

Historically, marine utopian projects have aimed to establish sovereign colonies on the high seas, dating back to the 1950s [32]. More recently, *The Seasteading Institute* (TSI) has sought to assemble floating communities around atolls and archipelagos in international waters, creating self-governing micro-nations (see Chap. 15). While these projects explore alternative governance models, they remain subject to the structural and legal constraints of existing international frameworks.

As such, an important dimension in defining the legal status of floating structures is their geographic location, as the legal interpretation of these structures becomes increasingly complex when they are sited further from the coastline. For instance, under Dutch law, property law does not extend to the seabed within the state's EEZ [13]. Similarly, in the case of *re Duchy of Sealand* (Case No. 9 2565/77) involving an artificial structure constructed on concrete pylons that had been declared sovereign, the Administrative Court of Cologne rejected the expert opinion that suggested the construction of this artificial structure could be territory. Instead, the court suggested that there needed to be a 'creation of [a] new piece of the earth's surface'.

Saunders [22] notes that this continues the argument advanced by Schwarzenberger [23] and adopted by Shaw [24] that the acquisition of 'territory' is dependent on its adjacency to natural land; that is, human intervention to create new space can only be territory if it expands naturally formed land. Again, the clarification here is that this naturally created land must not be a low-tide elevation or uninhabitable rock and that the ambit of 'land' is defined strictly according to the law. Consequently, floating structures situated offshore (that is, not near the coastline) are not typically regarded as real property (land) (see Chap. 21 for more detailed discussion).

When floating city developments are intended to be located offshore, they are treated differently from those sited near the coast or within harbour facilities. However, the term 'offshore', in turn, lacks a universal legal definition, leaving it open to varying interpretations across jurisdictions. This sets further hurdles in integrating floating structures into existing legal frameworks, particularly regarding their structural characteristics and their role within harbour operations.

Floating cities deployed within a coastal state's EEZ will place them under the host state's jurisdiction. This challenges the business model of developers seeking autonomy from national laws. These developers, as proponents of floating city projects, often advocate for their classification as self-governing entities, akin to micro-nations. However, under international law, such claims are likely to conflict with the sovereignty rights of coastal states.

In areas beyond national jurisdiction (ABNJ), where no single state exercises sovereignty, floating cities face an entirely different set of challenges. Without clear legal recognition, these developments risk being treated as unregulated installations, subject to ad hoc governance mechanisms. This creates a potential vacuum in which issues such as liability, environmental protection and resource rights are inadequately addressed.

The ambiguity surrounding their legal status also complicates efforts to integrate floating cities into broader frameworks for sustainable development and maritime governance. For example, if a floating city straddles multiple jurisdictions or migrates due to environmental conditions, questions arise regarding which state has the authority to regulate its operations or impose legal obligations on its residents and operators [6]. This ambiguity creates the potential for disputes between states and complicates the enforcement of environmental and security laws. The lack of clarity regarding the status of floating cities also raises concerns about their use as platforms for activities such as tax evasion, unregulated resource extraction or human trafficking.

5.2 *Structural Characteristics*

As international law has yet to establish a comprehensive framework for defining floating structures, legal interpretation of the status of these structures often relies on their physical characteristics to determine the classification of the floating structure. This potential overlap between legal concepts such as ‘artificial islands’, ‘vessels’, and ‘platforms’ exacerbates definitional challenges.

UNCLOS, for instance, categorises artificial islands and other installations under Article 60 but does not extend these classifications to floating structures in a coherent manner. Under UNCLOS, artificial islands and installations are typically defined by their interaction with the seabed. Article 11 explicitly excludes offshore floating structures from the calculation of baselines, thereby precluding their influence on the delimitation of territorial seas or EEZs [34]. Likewise, the *1958 Convention on the Continental Shelf* limits installations that interfere with navigation, fishing or conservation activities.

However, ambiguities arise when floating structures are attached to, or installed above islands, low-tide elevations (LTEs), or other artificial islands. These could be amphibious in nature, resting on dry ground during low tides, but moving vertically to float above the sea-level during high tides. In these cases, could they be treated as artificial islands without the structural characteristic of being bottom-founded? Soons suggested (prior to the adoption of UNCLOS) a taxonomy of four categories of offshore structures: structures created through the aggregation of natural substances like sand, rocks and gravel (that is, artificial islands), floating structures anchored or moored, bottom-founded fixed structures resting on columns, piles or tubes, and a separate category of concrete structures [26]. However, the category of concrete structure and the lack of a category for amphibious structures leaves us with a taxonomy that is unclear and incomplete.

There is also an overlap between the question of location and the second question of the structural characteristics and functionality of a floating structure. A pontoon-based floating structure, moored with cables and designed for temporary purposes, can only be deployed in near-shore shallow waters, and as such would be located adjacent to naturally formed land. On the other hand, a semi-submersible structure (such as those typically used for oil rigs) allows for deployment of the floating structure in deeper, open waters. Such a structure that is intended for permanent habitation could be interpreted as an installation under certain UNCLOS provisions. These distinctions that are based on the structural characteristics of the floating installation can have profound implications for jurisdictional rights, liability frameworks and environmental regulations in ABNJs [34].

Given the complexity that exists in overlapping legal frameworks, small-scale floating structures designed for temporary use may be easier to categorise. However, this definitional challenge is compounded when we consider the agglomeration of these floating structures into larger urban formations to create the floating city. These assemblages as floating cities offer themselves as larger-scale developments designed for permanent or semi-permanent habitation, and present even greater challenges for international legal classification. Unlike standalone floating structures, assembled floating cities could integrate residential, commercial, and recreational spaces, further blurring the lines between artificial islands, vessels, and installations as each of these spaces could have different structural characteristics, mobility functions and connection types with adjacent naturally formed land or the seabed.

Likewise, their ability to remain buoyant and adaptable also adds a layer of complexity absent in fixed installations. Floating cities could potentially meet the criteria of vessels under international maritime conventions like the *International Convention for the Prevention of Pollution from Ships* (MARPOL). Indeed, there have been attempts by cruise operators to characterise their vessels as 'floating cities' [34]. Article 2(4) of MARPOL adopts an expansive definition of 'ship', encompassing floating platforms and installations. This broad definition ensures the inclusion of floating structures within its regulatory scope for environmental protection but creates conflicts in other legal contexts. If floating structures are treated as vessels, they would be subject to the jurisdiction of their flag state under Article 91 of UNCLOS.

However, such a classification is problematic for developments intended to remain stationary or semi-stationary. Lin et al. [13] propose that floating cities should be classified as vessels while in transit but redefined as artificial islands or installations when moored. This approach underscores the importance of structural characteristics in determining legal status, and courts and regulatory bodies will likely continue to rely on these structural characteristics to navigate the complex legal terrain of floating cities. Similarly, the modular and mobile nature of floating cities further challenges traditional classifications, as these cities often lack the fixed characteristics associated with artificial islands or installations. If individual components are designed for navigation but the overall assemblage is not, the floating city might fall within certain legal definitions while simultaneously evading others.

As mentioned, the structural characteristic of a floating structure is also tied to the inherent programming of its spaces and what functions such a structure may accommodate. Accordingly, the multifunctional nature and wide spectrum of proposed designs of floating cities further complicates their definition. For example, a floating city that includes elements such as harbours or industrial facilities may fall under the category of 'harbour works' if interpreted as permanent structures. This is important as when floating structures are interpreted as part of permanent harbour works, they may extend the low-water line of a coastal state, resulting in a 'seaward shift' of the baseline used to define maritime zones [13, 34]. While Article 11 of the UNCLOS also explicitly excludes offshore installations and artificial islands from the calculation of the low-water line, floating cities might not fall under the definition of an 'offshore installation' or an 'artificial island'. Similarly, terms such as 'platforms and other man-made structures at sea,' as used in Article 1 of UNCLOS concerning the definition of 'dumping', further complicate the classification of floating structures. Structures that restrict or alter tidal flows could also fall under different legal categories depending on their functional and structural characteristics, such as eligibility for registration under carbon credit or abatement schemes [33].

In the absence of a consistent legal framework, ambiguities surrounding floating structures are likely to persist, particularly as technological advancements lead to the development of hybrid structures that defy existing classifications. As assemblages of floating structures increasingly serve as platforms for residential, commercial and industrial activities, the need for a unified legal definition under international law becomes more urgent. The absence of a clear definition for floating cities creates uncertainty regarding their governance and the rights of states to exercise jurisdiction over them.

For low-lying island nations at risk of inundation, the legal definition of what is a floating structure and what is a floating city goes to the existential question of statehood. While these island nations may seek to deploy floating structures over their traditional waters or submerged landmasses, these structures, or rather their agglomeration into a floating city, do not enable these nations to retain claims over their maritime zones. Addressing this issue would require adjustments in international law to consider the interests of states affected by sea-level rise.

6 Recognising SIDS Floating Cities as an *Erga Omnes* Duty

The diverse interpretations and legal approaches to floating structures and cities underscore the need for a harmonised international framework. Without clear definitions and consistent legal treatment, floating cities risk becoming entangled in jurisdictional disputes and regulatory gaps. For example, a floating city that transitions from a mobile vessel to a stationary installation could face conflicting obligations under maritime and terrestrial laws.

Harmonising international and national legal frameworks would provide greater certainty for developers, operators and residents of floating cities [6]. It would also help address broader issues such as environmental protection, resource rights and the equitable distribution of maritime benefits. As floating cities evolve from conceptual projects to tangible realities, the international legal community must prioritise clarity and consistency to support their sustainable development.

The principle of *erga omnes* obligations provides a framework for addressing these gaps, enabling the international community to collectively safeguard the sovereignty and maritime rights of submerged states through innovative solutions like floating cities. As the author has previously shared in the 4th World Conference on Floating Structures (held in Hong Kong, 2–4 December 2024) where this proposal was initially socialised with scholars of floating solutions:

> The concept of *erga omnes* obligations in international law refers to duties that states owe to the international community as a whole. *Erga omnes* obligations encompass norms such as the prohibition of genocide, slavery, and racial discrimination. These norms have a higher status in international law, often overlapping with *jus cogens* norms (fundamental principles accepted by the international community that no state is permitted to derogate from). The enforcement of *erga omnes* obligations allows any state to invoke the responsibility of another state for breaches, even if it is not directly affected by the violation. [35]

The recognition of *erga omnes* obligations is crucial in the context of rising sea levels, where entire populations face the prospect of losing their land, nationality and sovereignty. These obligations transcend the traditional state-centric framework, enabling any state to invoke responsibility for violations that undermine collective interests, such as the protection of vulnerable populations and the prevention of statelessness. At present, the concept has already been considered for populations that may lose their statehood due to inundation [40]. This chapter extends the legal concept of *erga omnes* to the recognition of floating cities used by submerged states as 'territory' given the pressing need for legal recognition and governance of floating cities in response to climate change.

Central to this extension is the evolving recognition of *erga omnes* obligations in international environmental law, which increasingly acknowledges the universal impact of climate change, framing obligations to mitigate and adapt to global environmental challenges as duties owed by each state to the global community at large. These obligations encompass collective norms essential for preserving the environment and addressing climate-induced displacement. This principle of collective responsibility is particularly critical to addressing the disproportionate impacts of climate change on vulnerable nations, where SIDS contribute minimally to global greenhouse gas emissions yet bear the brunt of its consequences.

Floating cities, which embody innovative solutions to rising sea levels and urban displacement, highlight the intersection of environmental, human rights, and maritime legal frameworks and seeks to address some of the most pressing challenges in contemporary international law. Floating cities also represent a forward-thinking adaptation strategy with global significance. They demonstrate how urban environments can adapt to rising sea levels and extreme weather events while contributing to biodiversity conservation and carbon sequestration. As models for climate resilience, floating cities align with ecosystem-based adaptation strategies emphasised by the IPCC. These strategies offer co-benefits, such as improved health outcomes and economic resilience, further underscoring the need for collective interest in their success.

Recognising floating cities as sovereign extensions of submerged states reflects the moral and legal duty of all states to mitigate these impacts and support adaptive solutions that may help create floating habitats and community space. Such recognition would ensure that no single nation bears the burden of global climate change. The *International Court of Justice* (ICJ), in its 1996 *Advisory Opinion on the Legality of the Threat or Use of Nuclear Weapons*, underscored the shared responsibility of states to prevent environmental harm with global repercussions, affirming that the protection of the environment is a matter of universal concern [3].

The preservation of statehood and the right to nationality in the face of rising sea levels is a critical dimension of this collective responsibility. The ICJ's dictum in the 1970 *Barcelona Traction* case (ICJ GL No 50) highlighted that the protection of basic human rights, including the right to nationality, engages the legal interest of all states. Statelessness, often resulting from the loss of territorial sovereignty, leads to severe human rights violations, including restricted access to healthcare, education and freedom of movement. The prevention of statelessness aligns with *erga omnes* principles by addressing the discriminatory treatment and socio-economic exclusion of stateless individuals.

By recognising these cities as extensions of partially-submerged territories within affixed maritime baselines, vulnerable populations are given the opportunity to remain connected to their traditional lands and waters. For example, communities in SIDS such as Tuvalu have historically rejected climate-induced relocation [40], underscoring the importance of maintaining ties to their cultural and geographic heritage. However, the popularity of the recently introduced first-of-its-kind climate change visa by Australia in June 2025 for citizens of Tuvalu under the *Australia-Tuvalu Falepili Union* indicates that rising sea levels may eventually require such relocations. By supporting the development of floating cities, the international community offers an alternative to such climate-induced displacements and fulfills its collective duty to uphold fundamental human rights and prevent statelessness.

The recognition of floating cities as territory supporting statehood under *erga omnes* principles addresses the multifaceted challenges posed by climate change. First, it preserves the sovereign rights and maritime entitlements of submerged states, ensuring their access to vital resources within affixed maritime baselines. Second, it upholds the rights of displaced populations, including the right to nationality and self-determination, which are integral to the *International Bill of Rights* and the *UN Charter*. Third, it advances global climate adaptation efforts by establishing floating cities as a scalable and replicable solution for mitigating the impacts of rising seas on the environment and communities.

The universality of the need for the successful design and deployment of floating cities transcends national boundaries, making their development a matter of collective interest. By pioneering innovative solutions to global environmental challenges, these cities contribute to the shared benefit of humanity, fulfilling the criteria for *erga omnes* obligations as elucidated by legal scholars like Tams [28]. This collective benefit reinforces the necessity of integrating floating cities into international legal frameworks that support their development, governance and sustainability.

To solidify the role of floating cities within international law, the international community must take concerted action to address existing legal gaps and ambiguities. This includes: first, developing a unified legal framework. International bodies such as the United Nations and the International Maritime Organisation should collaborate to create a cohesive framework for the governance of floating cities, recognising their role as sovereign entities linked to submerged states. Second, there should be a corresponding codifying of affixed maritime baselines to ensure that maritime rights are preserved despite rising sea levels. UNCLOS should be amended or interpreted to allow for the recognition of affixed baselines for states that adopt floating cities as extensions of their territory. Third, the international community must strengthen global cooperation and prioritise financial and technical assistance to support the development of floating cities in vulnerable regions, fostering equitable adaptation strategies and promoting sustainable development.

Consequently, this chapter argues that floating cities are *de lege ferenda*—that is, a law whose time has come—echoing a similar observation made by Papadakis in 1977 in relation to artificial islands [19]. It is critical for this need to recognise floating structures as territory to be dealt with by a new international convention.

7 Conclusion

Floating cities present a transformative opportunity to address the dual challenges of climate-induced displacement and the loss of sovereignty faced by Small Island Developing States. These buoyant urban environments challenge traditional notions of territory and statehood, necessitating innovative legal approaches to their recognition and governance. Current international frameworks, such as UNCLOS, lack the clarity needed to classify and regulate floating cities, resulting in gaps that undermine their potential to support vulnerable states and populations.

This chapter has argued for the recognition of floating cities as sovereign extensions of submerged territories under *erga omnes* obligations. Such recognition aligns with the collective responsibility of the international community to mitigate the impacts of climate change, protect fundamental human rights and preserve maritime zones for affected states. Floating cities not only address immediate concerns of displacement but also serve as global exemplars of sustainable urban development and climate resilience, particularly for cities located in the low-elevation coastal zone.

To address these challenges, the international community must prioritise the development of a unified legal framework that explicitly defines floating cities within the context of UNCLOS and related international instruments. Such a framework should integrate the concept of affixed baselines to preserve the maritime rights of submerged states and establish governance mechanisms for floating cities in areas beyond national jurisdiction. By aligning these efforts as an *erga omnes* obligation, the international community can ensure that floating cities fulfill their promise as innovative, inclusive and globally significant responses to climate change.

References

1. Armstrong C, Corbett J (2021) Climate change, sea level rise, and maritime baselines: responding to the plight of low-lying atoll states. Glob Environ Politics 21(1):89–107
2. Bijker WE (2002) The Oosterschelde storm surge barrier: a test case for Dutch water technology, management, and politics. Technol Cult 43(3):569–584
3. Black-Branch JL (2021) The treaty prohibiting nuclear weapons: legal challenges for military doctrines and deterrence policies. Cambridge University Press
4. Bronen R (2011) Climate-induced community relocations: creating an adaptive governance framework based in human rights doctrine. NYU Rev L Soc Change 35:357
5. Cavalié O, Sladen A, Kelner M (2015) Detailed quantification of delta subsidence, compaction and interaction with man-made structures: the case of the NCA airport, France. Nat Hazard Earth Syst Sci 15(9):1973–1984
6. Cradduck LM, Wang BT, Weinert KD (2024) Underpinning the value of buoyant urbanism: returning Torrens to the seas. In: International conference on offshore mechanics and arctic engineering, vol 87875. American Society of Mechanical Engineers, p V009T13A006
7. Delap Commitment (2018) Securing our common wealth of oceans. Majuro, Marshall Islands, 2 March
8. Gagain M (2012) Climate change, sea level rise, and artificial islands: saving the Maldives statehood and maritime claims through the constitution of the oceans. Colo J Int'l Envtl L Pol'y 23:77
9. IPCC (2007) IPCC fourth assessment report (AR4): climate change 2007—synthesis report
10. IPCC (2023) Summary for policy makers. In: Lee H, Romero J (eds) Climate change 2023: synthesis report. IPCC, Geneva, Switzerland, pp 1–34
11. Johnson DHN (1951) Artificial Islands. Int Law Q 4:203
12. Lee S, Xu S, Kua I (2023) How the Tiny Island city-state of Singapore fights rising sea levels. Bloomberg
13. Lin FY, Spijkers O, Van der Plank P (2021) Legal framework for sustainable floating city development: a case study of the Netherlands. In: WCFS2020: proceedings of the second world conference on floating solutions, Rotterdam. Springer Singapore, pp 433–460
14. Magnan AK, Oppenheimer M, Garschagen M, Buchanan MK, Duvat VK, Forbes DL, Ford JD, Lambert E, Petzold J, Renaud FG, Sebesvari Z (2022) Sea level rise risks and societal adaptation benefits in low-lying coastal areas. Sci Rep 12(1):10677
15. McGranahan G, Balk D, Anderson B (2007) The rising tide: assessing the risks of climate change and human settlements in low elevation coastal zones. Environ Urban 19(1):17–37
16. Mesri G, Funk JR (2015) Settlement of the Kansai international airport islands. J Geotech Geoenviron Eng 141(2):04014102
17. Mimura N (1999) Vulnerability of island countries in the South Pacific to sea level rise and climate change. Climate Res 12(2–3):137–143
18. Ng KG, Begum S (2023) Long Island to be reclaimed off East Coast could add 800ha of land, create Singapore's 18th reservoir. The Straits Times
19. Papadakis N (1977) The international legal regime of artificial Islands, vol 2. Brill Archive
20. Polynesian Leaders Group (2015) Taputapuatea declaration on climate change. Papeete, French Polynesia, 16 July
21. Rayfuse R (2011) International law and disappearing states-maritime zones and the criteria for statehood. Envtl Pol'y L 41:281

22. Saunders I (2019) Artificial islands and territory in international law. Vand J Transnat'l L 52:643
23. Schwarzenberger G (1957) Title to territory: response to a challenge, 51 AM. J INT'L L 308:313
24. Shaw MN (1982) Territory in international law. Neth Yearb Int Law 13:61–91
25. Shaw MN (1986) Title to Territory in Africa. International legal issues. Oxford, Clarendon Press
26. Soons AH (1974) Artificial Islands and installations in international law. University of Rhode Islands Law of the Sea Institute
27. Stoutenburg JG (2015) Disappearing island states in international law. Brill
28. Tams CJ (2005) Enforcing obligations Erga omnes in international law, vol 44. Cambridge University Press
29. Tanaka Y (2021) The legal consequences of obligations Erga omnes in international law. Neth Int Law Rev 68(1):1–33
30. UN Convention of the Laws of the Sea (UNCLOS) (1982) Montego Bay, 10 December, 1833 UNTS 3
31. UN International Law Commission (2023) Sea-level rise in relation to international law
32. Wang B (2019) Floating cities: the future or a washed-up idea?. The Conversation
33. Wang B (2023a) Regulatory reform and the potential use of floating artificial reefs under blue carbon and ocean carbon methodologies. In: International conference on ocean, offshore and arctic engineering No. 42nd
34. Wang B (2023b) Development of the legal definition of the floating city: judicial interpretation of structural characteristics of floating homes and developments. In: World conference on floating solutions. Springer Nature Singapore, Singapore, pp 139–157
35. Wang B (2024) Floating cities and buoyant urbanism in international law: inundation, territory and Erga omnes obligations. In: World conference on floating solutions No. 4
36. Wang CM, Wang BT (2006) Colonization of the ocean and VLFS technology. In: Very large floating structures. CRC Press, pp 15–34
37. Watson I, Finkl Jr CW (1990) State of the art in storm-surge protection: the Netherlands Delta project. J Coastal Res 739–764
38. Watson RT, Zinyowera MC, Moss RH, Dokker DJ (1996) Climate change 1995: impacts, adaptations and mitigation of climate change: scientific-technical analyses. Contribution of working group II to the second assessment report of the intergovernmental panel on climate change
39. Willcox S (2012) A rising tide: the implications of climate change inundation for human rights and state sovereignty. Essex Hum Rights Rev 9(1):1–19
40. Wong D (2013) Sovereignty sunk-the position of sinking states at international law. Melbourne J Int Law 14:346

Brydon T Wang is an expert on the trustworthy development and regulation of AI and Construction technology. He researches on trustworthy decision-making in infrastructure delivery and data governance, with a focus on how technology can help increase climate-resilience in cities. He is dually qualified in law and architecture and advises on multi-billion dollar infrastructure projects with KPMG. Brydon practised as a construction and technology lawyer with global top-tier law firm Allens Linklaters and has a previous career in architecture and contract administration on award-winning construction projects. He has over 15 years in the construction industry in both client-side (government) and contractor-side roles. Brydon was recently lead editor of the book, *Automating Cities: Design, Construction, Operation and Future Impact* (Springer, 2021) and co-edited *Large Floating Structures* (Springer, 2015), a book exploring how we can embrace floating structures in our cities to respond to the increased incidences of floods and sea-level rises. He was featured on ABC Radio National's *Future Tense* discussing offshore architecture and marine urban sprawl, and on Seeker's popular documentary, 'How close are we to Living in the Ocean?' Brydon's expertise in technology and cities also extends to data-focused technologies such as digital twins, BIMs, smart construction contracts, and the blockchain. He developed and taught the contract and technology law subjects at the Queensland University of Technology (data privacy, regulating the internet, EU AI Act). His PhD thesis on the 'Role of Trustworthiness in Automated Decision-making Systems and the Law' received the *QUT Faculty of Business and Law Executive Dean's Commendation for Outstanding Doctoral Thesis Award*.

MIX
Papier aus verantwortungsvollen Quellen
Paper from responsible sources
FSC® C105338

If you have any concerns about our products,
you can contact us on
ProductSafety@springernature.com

In case Publisher is established outside the EU,
the EU authorized representative is:
Springer Nature Customer Service Center GmbH
Europaplatz 3, 69115 Heidelberg, Germany

Printed by Libri Plureos GmbH
in Hamburg, Germany